PERCEPTS, CONCEPTS AND CATEGORIES

The Representation and Processing of Information

ADVANCES IN PSYCHOLOGY

93

Editors:

G. E. STELMACH

P. A. VROON

NORTH-HOLLAND
AMSTERDAM • LONDON • NEW YORK • TOKYO

PERCEPTS, CONCEPTS AND CATEGORIES

The Representation and Processing of Information

Edited by

Barbara BURNS
Department of Psychology
University of Louisville
Louisville, KY , U.S.A.

1992

NORTH-HOLLAND
AMSTERDAM • LONDON • NEW YORK • TOKYO

NORTH-HOLLAND
ELSEVIER SCIENCE PUBLISHERS B.V.
Sara Burgerhartstraat 25
P.O. Box 211, 1000 AE Amsterdam, The Netherlands

ISBN: 0 444 88734 2

This book is printed on acid-free paper.

Printed in The Netherlands

Table of Contents

Contributors

Numbers in Parentheses Indicate Chapter Number in Volume.

HERVE ABDI (8), School of Human Development, University of Texas at Dallas, Richardson, TX, 75083-0688 USA

SUSAN E. BARRETT (8), Department of Psychology, Lehigh Univeristy, Bethlehem, PA 18015 USA

ANGELA H. BECKER (12), Department of Psychology, Texas A & M University, College Station, TX 77843 USA

DORRIT BILLMAN (11), School of Psychology, Georgia Institute of Technology, Atlanta, GA 30332 USA

RANDY D. BRIGHT (15), Department of Psychology, University of Louisville, Louisville, KY 40292 USA

BARBARA BURNS (6), Department of Psychology, University of Louisville, Louisville, KY 40292 USA

TARA C. CALLAGHAN (5), Psychology Department, Box 74, St. Francis Xavier University, Antigonish, Nova Scotia, B2G 1CO, CANADA

C. MELODY CARSWELL (16), Department of Psychology, University of Kentucky, Lexington KY 40506 USA

FRANCINE M. DEUTSCH (13), Department of Psychology and Education, Mount Holyoke College, South Hadley, MA 01075 USA

STEPHEN E. EDGELL (15), Department of Psychology, University of Louisville, Louisville, KY 40292 USA

JAMES T. ENNS (2), Department of Psychology, University of British Columbia, 2136 West Mall, Vancouver, British Columbia, V6T 147 CANADA

EDWARD A. ESSOCK (1), Department of Psychology, University of Louisville, Louisville, KY 40292

LAURA A. FORD (15), Department of Psychology, University of Louisville, Louisville, KY 40292 USA

MORRE GOLDSMITH (3), The Institute of Information Processing and Decision Making, University of Haifa, Haifa 31999, ISRAEL

DIANA HEISE (7), Department of Psychology, Indiana University, Bloomington, IN 47405 USA

DENNIS H. HOLDING (17), Department of Psychology, University of Louisville, Louisville, KY 40292 USA

RUTH KIMCHI (3), Department of Psychology, University of Haifa, Haifa 31999 ISRAEL

JUDITH F. KROLL (13), Department of Psychology and Education, Mount Holyoke College, South Hadley, MA 01075 USA

MARY E. LASSALINE (9), Department of Psychology, University of Michigan, 330 Packard Rd, Ann Arbor, MI 48104 USA

GREGORY R. LOCKHEAD (4), Department of Psychology, Duke University, Durham NC 27706 USA

DOUGLAS L. MEDIN (9), Department of Psychology, University of Michigan, 330 Packard Rd, Ann Arbor, MI 48104 USA

IRWIN D. NAHINSKY (10), Department of Psychology, University of Louisville, Louisville, KY 40292 USA

PAK CHUN NG (15), Department of Psychology, University of Louisville, Louisville, KY 40292 USA

THOMAS K. NOONAN (15), Department of Psychology, University of Louisville, Louisville, KY 40292 USA

ALBERT F. SMITH (14), Department of Psychology, State University of New York at Binghamton, Binghamton, NY 13901 USA

LINDA B. SMITH (7), Department of Psychology, Indiana University, Bloomington, IN 47405 USA

JILL M. SNIFFEN (8), Department of Psychology, Lehigh University, Bethlehem, PA 18015 USA

THOMAS B. WARD (12), Department of Psychology, Texas A & M University, College Station, TX 77843 USA

EDWARD J. WISNIEWSKI (9), Department of Psychology, University of Michigan, 330 Packard Rd, Ann Arbor, MI 48104 USA

Preface

Ten years ago Farah and Kosslyn (1982) argued that the "most important distinction derived from the computational view of thought is between structures and processes". They continued their argument stating that structures and processes cannot be examined in isolation and concluded that "converging operations" are required to isolate the structure-process pair that can explain a particular finding. "A structure can be observed in different experiments in which it is operated on by different processs. Characteristics that are observed in all processing contexts can then most parsimoniously be assumed to belong to the structure, not the various processes" (1982, p. 129). The continued interest and vitality of this distinction was supported by a recent edited volume by Shepp and Ballesteros (1989) entitled *Object Perception: Structure and Process.*

The current volume focuses on this distinction between structure and process within the study of percepts, concepts and categories. The volume is further organized into four major areas: Early Visual Representation and Processing; Percepts, Concepts, Categories and Development; Categories, Concepts, and Learning; and Higher-Order Representation and Processing. Contributors were selected to illustrate various aspects of this distinction such that there would be broad coverage of the implications of the structure-process distinction for our understanding of percepts, concepts and categories. Research programs of many of the contributors have been closely identified within this framework of characterizing and evaluating structure and process in the study of percepts, concepts and categories. Other researchers have not explicitly dealt with this distinction previously but have contributed knowledge to various aspects related to this distinction. A diversity of positions as to the description and utility of distinguishing structures and processes is evident in this volume. At the same time it is clear that researchers from diverse areas of study, from *simple* structure and process involved in perceptual organization and texture to *complex* structure and process associated with reading graphs and chess expertise, utilize such a distinction in similar ways. To make more salient such dissimilarities and similarities among the authors, commentaries on each chapter written by fellow contributors were solicited and follow each chapter in the volume.

I am most grateful to all of the contributors for their willingness to consider their own work in reference to this structure-process framework and for their commentaries, as well as for their excellent chapters which are significant contributions to the literature. Support for the preparation of this book was received from the President's Research Initiative, University of Louisville and the Department of Psychology, University of Louisville. I am indebted to Mrs.

Margaret Biegert who worked with great enthusiasm and professionalism on the preparation of this volume. I am also grateful to Lora Schlewitt who helped in preparation of the author index. Finally, I wish to recognize the important contributions of my spouse, Ed Essock, to this volume. I am most appreciative of his thoughtful and critical comments throughout the structure and process of preparation of this book and I am deeply grateful for his support.

Farah, M.J. & Kosslyn, S.M. (1982). Concept Development. *Advances in Child Development and behavior, 16*, 125-167.

Shepp, B.E. & Ballesteros, S. (Eds.) (1989). *Object perception: structure and process.* Hillsdale, NJ: Lawrence Erlbaum Associates.

PART A. Early Visual Representation and Processing

Percepts, Concepts and Categories
B. Burns (Editor)

1

An Essay on Texture: The Extraction of Stimulus Structure from the Visual Image

EDWARD A. ESSOCK
University of Louisville

I. OVERVIEW

Ultimately, the processing that is possible to perform on the visual stimulus is limited by the information that is present in the stimulus image. In this essay I first consider what information actually exists in the stimulus image for a visual system to potentially extract, and then consider how this stimulus structure is initially processed to provide meaningful regions that are the basis of the perceptual organization of forms and objects. I present an overview of the ideas that have appeared in the computer vision literature as well as those in the human vision literature, as, at this initial level, the task confronting either type of vision system is really the same. Researchers in artificial vision have the inherent luxury of being able to sit back and ask, "of all the types of structure in the stimulus image apparent to me, which ones will I choose to utilize?" Researchers investigating human vision, of course, are forced to play the hand that they are dealt and ask, "which types of stimulus structure, presently apparent to me or not, does the human visual system actually use?" For those of us who work in the realm of human vision, examining what approaches have been explored by those with seemingly unlimited degrees of freedom may serve to generate new ideas, or at least a new perspective, on the types of structure that are present in an image and what associated processing is required of the vision perceptual system.

At the first step in the visual perceptual process the situation for an artificial system or a biological system is equivalent. The human visual system is presented with an image on the retina. This retinal image simply functions as a mapping of light intensity across retinal distance. In a typical human perception experiment "the stimulus", for example a red triangle, is the interesting part of an otherwise rather homogeneous intensity map representing the full visual stimulus image. This stimulus image at the level of raw input is thus represented as a surface when intensity is plotted across distance in two dimensions. The intensity surface, however, is not sampled continuously across its x and y extent, but is sampled discretely due to the nature of the neural hardware. Receptors (i.e., their receptive fields) have finite extent and thus the image is, with some particular spatial resolution, necessarily sampled across space. In the case of computer vision, the visual image falls on an artificial surface and is scanned, measuring intensity over an array of discrete positions. Thus, in either biological or computer vision systems, the starting point for the processing task is the same: a set of spatially distinct locations where intensity is measured over a discrete area. The measurement may differ and the grid of measurements may differ (e.g., lattice shape, spacing, spatial resolution, and so on), but in this basic sense, the neural image produced by the transduction by the retinal receptors is equivalent to the intensity array of pixels of the artificially scanned image in computer vision. The task facing any vision system is to tease out meaningful information from the structure of this intensity surface to build organized perceptions.[1]

[1]For the reasons stated above, there is no real distinction to be made in the present context between the visual input and its discretization in the case of human vision and in the case of computer vision. For convenience, the terms "intensity image" and "pixel" will be used in both the computer and human contexts.

The remainder of this chapter is in three parts. The first section considers the structure that exists in the intensity image and reviews the types of image structure that have been termed "texture". The next section considers the general nature of the segmentation task and the way in which texture can provide the basis for image segmentation. The third section reviews models of texture-based segmentation in human vision and similar methods of segmentation employed in various computer vision systems. The segmentation task considered in the third section is broken into two parts: the *stimulus features* used in human and computer systems; and the *representation and processing* of this information to provide texture-defined regions.

II. IMAGE STRUCTURE AND TEXTURE

The extensive investigations into texture analysis in both computer and human vision suggest the importance of texture analysis in both disciplines, but the literature also serves to underscore what a difficult problem texture analysis poses. Indeed, it is even difficult to define the term "texture" as it can easily refer to such diverse patterns as blades of grass in a lawn, rectangles and lines formed by the bricks and mortar of a brick wall, or even the pattern of "noise" on a black and white television screen. Not surprisingly therefore, there is no single comprehensive definition of texture that seems to apply well in all situations. Many conceptualizations of texture utilize the idea of local elements and/or their properties that occur repeatedly in the image structure. In various formulations these elements are called "texture primitives", "texture elements", "texture features" or "textons".

In computer vision, use of the term texture most often excludes the properties of color and sometimes even intensity. In these cases texture refers only to multi-pixel spatial relations irrespective of intensity (or of color, if multiple filtered intensity maps, for example RGB, are available). Consistent with the human vision literature this distinction between texture and color/intensity will not be made here. Furthermore, in computer vision "texture analysis", usually implies that the analysis is performed on single, static images. If motion (e.g., flow fields) or depth (e.g., stereo, radar or other range information) are to be analyzed, other single-image measures typically are not. Human vision texture research has not excluded color and more recently stereoscopic depth and motion have also been considered (Nakayama and Silverman, 1986). In this chapter "texture" is considered to included all of these sources of information.

Where is texture in the intensity image? Clusters of pixels (or even single pixels) whose intensity deviates relative to that of their neighbors convey structure in the stimulus image. Such a region in the intensity image may look something like a mountain peak, ridge, trough, or even isolated dots; it may have smooth, abrupt, or jagged intensity transitions to its surrounding neighborhood; and the region may have a small or large intensity difference relative to its surrounding neighborhood. A given example of such a local event in the intensity surface may function as a constituent of a significant edge that makes up part of an object's form *directly*. Alternatively, an identical local event may serve instead as a constituent of a particular texture element (or be a single-pixel texture element itself). The texture element may, in association with many other such elements, serve to form a region of relatively homogeneous texture that differs from the texture of an adjacent region. This boundary is a texture-based edge and hence the local event that is a constituent of a texture element *indirectly* form an object's edge. The fundamental difference is that in the image events that make up an image edge directly, their exact position is important; they form an edge located at a specific position directly. In the present context, these images events are called "structure elements". In the other case, that of the "texture elements", the *exact* position is irrelevant. In other words, the local phase, or position relative to neighboring image events, is irrelevant. This notion of position-insensitivity is a fundamental premise to currently prominent models of human visual search (e.g., Treisman and Gormican, 1988) and preattentive texture discrimination (e.g., Julesz, 1984a).

Stating that the local position of a texture element, for example a dark line, is not relevant is not to say that global location can be completely ignored. A good example of this distinction is found in use commercially to cover screw heads in budget furniture. The visual impression of the "wood" veneer can be described as a nonuniform pattern of dark, non-random lines superimposed upon a background of generally homogeneous color. Over areas of moderate size, these lines (and the inter-line spaces) tend to have predominant orientation, width and spacing. Some manufacturers of plastic-veneer particle board furniture provide the customers with gummed-back stickers to place on the veneer to cover screw heads. These stickers approximately match the simulated wood in color, and the grain is approximated by dark striations of a width and spacing that is *typical* of that found across the wood. By orienting these stickers to approximately match the orientation of the grain, they indeed blend into the wood texture exceedingly well. The point is that an extremely good match is provided by approximately matching color and general orientation, even though no provision is made for positioning the particular grain lines accurately relative to one another (i.e., in terms of line spacing and line continuation, or in terms of general differences in curvature, orientation, number, and width). That is, local phase is disregarded by the visual system as well as by the manufacturer. If an observer focuses attention on these areas and examines them element by element, the two areas are distinguishable, but in terms of texture analysis, the two areas are indistinguishable preattentively by the human visual system.

This distinction based upon positional insensitivity initially seems similar to a common distinction in the computer vision literature between "macrotexture" and "microtexture". Van Gool, Dewaele, and Oosterlink (1985) define macrotexture as the "pattern generated by the larger primitives" and microtexture as "texture within the primitives themselves". Similarly, macro- and microtextures are defined by Davis and Mitiche (1980) in terms of relative size of the texture elements, with macrotextures having elements "several pixels in diameter". A similar point of view is that texture can be comprehensively defined as bi-leveled, with a high "macroscopic" level that is viewed as "structural" which specifies the arrangement of the primitives and a low "microscopic" texture level that is concerned with the "probabilistic" or "anarchical" aspects of texture (Gagalowicz and Ma, 1985). In these and similar dichotomies prevalent in the computer vision literature, the distinction is really based on levels of analysis (e.g., larger, versus smaller, elements), with positional importance ascribed more to the larger elements than to the smaller elements. Thus, although a general position-sensitivity versus position-insensitivity dichotomy is implied, it is based on size and therefore quite distinct from the *functional* structural/textural dichotomy used here. However, the comparison to the macroscopic/microscopic dichotomy does serve to point out the hierarchy that is also implicit in the present view of an intensity surface and in the present texture/structure distinction. At the lowest level, pixel intensities form the structure of elements of all sizes. Position is important, but trivially, as the elements are formed by intensity relationships between neighboring pixels. It certainly seems reasonable to consider this level as "*microstructure*" although this is *not* texture and therefore completely distinct from the notion of "microtexture". Texture, as defined here earlier, occurs at the next level where, for these elements made from local events, either relative position is important and the element contributes to a structural element (e.g., an edge of some form), or relative position is irrelevant and the element is therefore a texture element. An additional consideration is that when events that often serve as texture elements are lined up in very regular ways (such as tiles or bricks of a wall), structural factors emerge that are otherwise not present. Since position clearly is crucial to this emergent structure (e.g., the location of the bricks cannot be random), such elements that could form texture in other instances form superordinate *structure* elements and *not* texture elements in these cases (cf. Beck et al., 1983).

The macrotexture/microtexture distinction of others often corresponds to another distinction made in the computer vision literature between "structural" and "statistical" approaches to specifying textures (Gagalowicz and Ma, 1985). Macrotextures (i.e., large elements) tend to be specified in terms of the specification of the texture primitives and of their "placement rules". Microtextures (i.e., small elements) tend to be specified in terms of statistical measures.[2] In practice, the term macrotexture is often used only for those highly repetitive patterns (e.g., a tile wall) that here are viewed as patterns with "emergent structure" rather than as textures at all. It is these and other very regular patterns for which the approach can truly be termed a pure structural approach (Haralick, 1979). For most textures (including the less-structured usages of "macrotexture"), approaches in computer vision are really "structural-statistical" (Haralick, 1979) in that the spatial interrelations of explicitly defined primitives are specified by probabilities (Derin and Cole, 1986; Van Gool et al., 1985; Haralick, 1979). This viewpoint leads to a distinction between "strong" and "weak" structural texture measures (Haralick, 1979; Davis, 1979). Haralick (1979) coined the terms to refer to the "spatial interaction" between primitives. Strong measures address the non-random spatial interactions that specify how the primitives spatially co-occur, whereas weak measures address the properties of the texture elements themselves.

Computer vision approaches utilizing weak texture measures actually tend to resemble region-formation methods in that these texture methods tend to directly group together nearby pixels with similar properties such as gray level or "edgeness" (Rosenfeld and Davis, 1979) or size of homogeneous area (Zucker, Rosenfeld and Davis, 1979). Since the present view of texture specifically ignores all positional information of one texture element relative to another, it is clearly not a "strong" texture measure approach. Although somewhat similar to weak measures, since the present view emphasizes texture primitives it is not truly typical of the "weak" texture measures of computer vision either.

Finally, it is worth noting a general similarity of the present approach to some of the pure statistical approaches of computer vision. The computation of primitive-density, as proposed by Julesz (1984a), is similar to measuring frequency of occurrence of a given primitive type within a given distance of an instance of

[2]For example, in his review Haralick (1979) identifies eight statistical approaches: "autocorrelation functions, optical transforms, digital transforms, textural edgeness, structural elements, spatial gray tone co-occurrence probabilities, gray tone run lengths, and autoregressive models".

the primitive type. Hence, density (i.e., local number over a neighborhood) is generally similar to co-occurrence probabilities, particularly generalized co-occurrence matrices (Davis, Clearman and Aggarwal, 1981) where the spatial co-occurrence of each of a particular type of primitive relative to the others is catalogued. Although some pure statistical approaches seem to contain information quite similar to the information contained in a texture primitive's density map, the information is typically much lower-level (usually pixel-level). It is typically a highly local measure computed across the entire image followed by some region-growing algorithm.

III. THE TASK OF SEGMENTING THE IMAGE

A fundamental processing step of visual perceptual organization is the formation of regions by grouping together smaller local areas which display similar characteristics. Segmentation of the raw visual image into these regions of relatively homogeneous characteristics is an important step to the organizing and understanding of the visual input for both the human visual system and for computer vision systems. Research in the areas of visual search and preattentive texture segmentation suggests that the human visual system can process the visual field in parallel, very rapidly, and without scrutiny of focused attention, to provide grouping of the image into regions of similar properties (e.g., Julesz, 1984b; Treisman, 1985). Meanwhile, research in computer vision has implemented a very wide variety of image segmentation methods. The task to be handled by processing the image is the same in either type of vision system.

William James (1890) described the visual image confronting an infant as a "great booming, buzzing, confusion." This classic description of the task of perceptual organization is really a description of the task of segmenting an image. The description applies equally well to the segmentation task facing either a human visual system or a computer visual system. A vision system receives an image digitized into a mass of juxtaposed individual intensity values which must be organized to escape from the overwhelming "booming, buzzing confusion." This array of pixel intensities is such a rudimentary abstraction of the vision sense that extracting areas with common properties is possibly the most important step in a vision system's processing. How effective a machine, or biological, vision system can ultimately be, may very well be fundamentally limited by how skillfully it can link the pixels together to form regions. However, producing unfragmented, uniform regions of common attributes that accurately delineate corresponding areas in the world remains a very difficult task in computer vision. Understanding how this is done in the human visual system also remains a very difficult task (see, e.g., Julesz, 1989). The human visual system serves as an "existence proof" to stimulate researchers in computer vision, whereas the difficulties of the computer vision systems underscore the importance and amazing nature of human texture-based preattentive segmentation.

In computer vision, the basic way in which texture-defined regions are formed is by linking together pixels that share a value of some property. The property may be pixel intensity (color), in which case regions of uniform intensity (color) are produced, or the property may reflect a *pattern* of pixel intensities, in which case regions of uniform texture are produced. The former properties directly reflect pixel attributes and the latter (except in artificial special cases) reflect multi-pixel relations measured over a pixel's neighborhood. Thus, a region segmentation process delineates regions that are relatively homogeneous with respect to one or more attributes of texture (again, with texture broadly-defined here to include the term "color"). When performed on the basis of low-level processing of image attributes (i.e., prior to the introduction of knowledge of objects or of structural constraints), such a segmentation is typically termed a "low-level segmentation."

It is widely held in the computer vision literature that, in principle, a low-level segmentation system that is based on pixel properties (i.e., intensity or color) should be improved by adding what that literature terms "texture features" and thereby capitalizing on the considerable untapped low-level information in the image. However, exactly which aspects of local structure should be used in a computer vision segmentation system to achieve this more-complete accessing of the image information is far from resolved. This is true in spite of the extensive literature on computer texture analysis and the striking diversity of the computer vision approaches.

A prominent view of how human preattentive texture segmentation processing works is that texture features, or texture elements, are extracted (e.g., the intensity or size of "blobs") and that variations in the distributions of the texture features across the image leads to the formation of perceptual regions of relatively homogeneous texture properties, much as just described for computer vision. This general framework which emphasizes extracted texture features is consistent with viewpoints expressed by Julesz, Treisman, Marr, and Beck, as well as many others. This emphasis on the first order statistics of texture elements differs somewhat from other views of human preattentive texture segmentation which presume the analysis of only the graded output of a set of linear filters (e.g., Bergen and Adelson, 1988; Nothdurft, 1991; Landy and Bergen, 1991; Malik and Perona, 1991). The next section considers these different views in more detail.

IV. METHODS AND MODELS OF TEXTURE-BASED SEGMENTATION

In this section the various approaches to extracting image structure and processing that information to obtain a texture-based image segmentation are surveyed. Various methods implemented in computer vision systems, and various models or viewpoints of the task performed in human vision, are presented. The aim is to step back from the details of human segmentation research controversies as much as possible to survey in broad terms what structure and processing is involved in the segmentation task.

The segmentation task is divided into two parts for the present consideration. The first part considers the texture features that have been utilized in both computer vision methods and human preattentive segmentation models. The second part considers the region formation process and is presented in three subsections: the utilization of feature maps; extraction of texture-defined boundaries and regions; and the methods of linear-filter models.

A. Features Used in Texture-Based Segmentation

1. Human vision

The literature describing attempts to specify the visual features used by the human visual system to segment images has its roots in what was termed "similarity grouping", "perceptual grouping", "texture segmentation", or "texture discrimination" (see, Wertheimer, 1923; Beck, 1967; Beck et al., 1983; Olson and Attneave, 1970; Julesz, 1981). In the cognitive psychology literature on visual search, this processing is akin to the preattentive aspect of the preattentive/focal processing dichotomy first defined by Neisser (1967). As is true for many others recently (e.g., Treisman and Gelade, 1980), I make no distinction in this chapter between the processing referred to within these areas. In this chapter they are considered to be examples of the process of preattentive image segmentation.

As was alluded to before, there are currently two main views as to how the human visual system performs texture-based segmentation. The most prominent view over the past thirty years is a feature-based view as presented by Beck, Attneave, and others (e.g., Beck, 1982; Beck, Prazdny and Rosenfeld, 1983; Olson and Attneave, 1970). For a period of time, Julesz favored a global, statistical view, but more recently has developed one of the most cited feature-based models, termed "texton theory" (Julesz, 1981). These models all hold that special texture elements, or features, are first extracted from the image and that one location of the image is compared to the next in terms of the prevalence of a particular texture feature or, in some models, combination of features. This approach is very common in the computer vision literature as well. Generally, many of these models of human texture segmentation analyze image events considered as "blobs", and the features of blobs such as length and color. The first step in specifying such a model then, is to provide a list of the features extracted, or, in other words, the list of particular image events that can provide the basis for human texture-based segmentation of image regions. Given the complexity of the task, it is particularly noteworthy that the current list of candidates is relatively short. The second step in specifying such a model is to specify how the distribution of these features across the image (their first-order statistics) is processed to extract discontinuities of their distribution. Such discontinuities or steep gradients represent texture-edges, boundaries between two areas that differ in the amount of some texture feature present.

Over the past few years, an alternative type of model has presented a strong challenge to the traditional texture segmentation models based on the first-order statistics of texture elements and texture features. These models propose that no special system is needed at all to extract texture information. These "linear filter" models propose that the basic image analysis hardware, a set of size-tuned filters (e.g., Wilson and Bergen, 1979), can itself provide for texture segmentation without the use of first-order statistics of special texture features. In theory these two approaches are diametrically opposed: one extracts specific features and counts up (in all-or-nothing, integer counts) where the features are in the image, whereas the other approach measures the graded amount of output from general filters measured at all points across the image. However, when late nonlinear stages such as rectification or threshold are added to the linear filter stage as is typical (see Malik and Perona, 1991), this distinction is blurred and the difference between the two types of model is more in name than real. Even the extraction of features itself, is not a way to distinguish first-order feature models from linear-filter models. This is because even though linear-filter models do not extract features explicitly as do the first-order feature models, the linear-filter models implicitly specify features by the shapes of the particular filters selected. In this sense the linear-filter models all use a size (frequency) feature, most employ oriented size-tuned filters and hence an orientation feature, and some bring in other properties such as color, as well (Gorea and Papathomas, 1991).

Of the first-order feature models, possibly the most explicit comprehensive model of human texture segmentation was formulated by Julesz (1981, 1984a, 1984b). Julesz proposed that texture consists of small, localized elements (or of single pixels) that he called textons. In the second heuristic to the theory, Julesz (1984a) explicitly listed the texture features extracted by the human visual system in his model for static two-dimensional images as: "(A) Elongated blobs--for example rectangles, ellipses, line segments, with specific colors [brightnesses], angular orientations, widths and lengths; (B) Terminators (i.e., ends of lines) of line segments; [and] (C) Crossings of line segments." Julesz (1984b) has noted that "binocular disparity, velocity and flicker" were also "powerful textons" (see Nakayama and Silverman, 1986; McLeod, Driver and Crisp, 1988). That the features of color, orientation and width provide texture discrimination was by no means unique to texton theory (e.g., Beck, 1967; Treisman and Gelade, 1980). The real distinction of texton theory was (and is) in the proposed method of extracting texture edges on the basis of texton density (see Section 4.2 below).

Julesz's other two textons, terminators and crossings, have come under considerable fire subsequently. Several researchers have concluded that crossings do not function as features for texture segmentation (Treisman, 1985; Enns, 1986; Gurnsey and Browse, 1987; Voorhees and Poggio, 1988; Nothdurft, 1990). Bergen as well (e.g., Bergen and Adelson, 1988), has concluded that a confounding of size is really the basis for texture discrimination in the prior studies that supported crossings as textons. In the case of terminators, while some researchers doubt that they serve as textons (Gurnsey and Browse, 1987; Voorhees and Poggio, 1988; Bergen and Adelson, 1988; Nothdurft, 1990), others contend that terminators can, at least to some extent, serve as textons (Treisman and Paterson, 1984; Treisman and Souther, 1985; Enns, 1986). In a more recent report Julesz (1989) stated his view that the main textons are the elongated blobs and the duals that they form (the inter-blob areas). He emphasizes as before color/brightness,[3] size (spatial frequency), length, width and, when relevant, binocular disparity, velocity and flicker rate.[4] He states (Julesz, 1989) that these are adequate to explain

[3]Although most all researchers list brightness and color separately as two features, it is redundant to do so as "brightness" is contained in the term "color". How many dimensions utilized preattentively are contained in the term "color" is an open empirical question.

[4]No distinction is made in this chapter between velocity and flicker rate, as from a local spatial perspective, there is no difference in terms of local temporal modulation. Whether these temporal modulation types provide distinct preattentive dimensions is another open empirical question.

segmentation of most textures "without having to introduce crossings or terminators," but he also states that in addition to properties of blobs "there appears also to be some hidden textonal properties, related somehow to 'closure' and 'corner' that cannot be described adequately by 'terminators' and 'crossings,' and neither can be modeled by Laplacian spatial filters followed by squaring."

Closure has often been considered by others as an additional stimulus feature that can serve as the basis for texture-based segmentation (e.g., Treisman, 1985; Enns, 1986). Treisman (1985) suggests that something similar to blob convexity or closure (specifically where "closure" can vary by degree) is a feature utilized by preattentive processing. She has also suggested curvature, number/proximity, and other properties related to closure as segmentation features as well (Treisman and Gormican, 1988; Treisman, 1988). Segmentation and visual search are typically thought of as a process of early vision, yet two other segmentation features have been reported that are particularly noteworthy due to their apparently sophisticated nature. Enns (Chapter 2, this volume) has reported convincing evidence that some types of 2-dimensional cues of 3-dimensional structure pop-out preattentively. Gurnsey, Humphrey and Kapitan (in press) report compelling evidence that subjective contours are processed preattentively (cf. Section 4.2.2.). Both features are possibly difficult to fit into the traditional conceptualization of preattentive vision.

Although typical first-order models (e.g., the models of Julesz, Treisman, and Beck) all contend that color is a dimension used in preattentive discrimination, it was not until recently that this was demonstrated to indeed be so. Prior studies undoubtedly confounded a luminance cue with the particular color features used, and hence the basis of "pop-out" or segmentation could well have been the luminance cue rather than chromatic contrast. More recently however, several studies have used equiluminant patterns and suggest that the preattentive system indeed can utilize color information distinct from luminance information (e.g., McIlhagga, Hine, Cole and Snyder, 1990; Smallman and Boynton, 1990; Nagy, Sanchez and Hughs, 1990). The strength of color as a preattentive feature appears to be based on the separation of the two regions in (psychological) color space. However, some of the initial studies suggest that the strength of the segmentation cue may be attenuated in the S-cone direction (McIlhagga et al., 1990; see also Nagy et al., 1990). Also noteworthy is that colors were found (Smallman and Boynton, 1990) to segregate well whether or not they were associated with color names ("basic colors"). Clearly, a full understanding of the preattentive effectiveness of color differences awaits the comprehensive assessment of color space in conditions of preattentive processing.

A difficult issue in the specification of this list of features used in human preattentive segmentation is the determination of the features that specify the *shape* of the texture elements. It is difficult to specify the shape of small closed elements in such simple terms, and, possibly for this reason, the problem is often implicitly ignored in the human literature by using only test stimuli (i.e., images) that are made of elements of identical shape (obviating a role of shape). To the extent that shape is addressed, a typical approach is to state that the features used for preattentive segmentation are features of "blobs". In models that specify only the length and width of blobs (Julesz, 1984b), or even area (Voorhees and Poggio, 1988), shape is left only vaguely specified. Treisman (1985) specifies length, while specifically ignoring width (cf. Section 5.0). She also proposes that shape as defined by aspect ratio ("height-to-width ratio") does *not* function as a segmentation feature. Clearly, additional research is needed to determine the correct formulation of shape as utilized by the human preattentive system.

The preattentive features thought to be used in human preattentive segmentation are similar to Marr's specification of the primitives in his theory of vision. Marr (1982) stated that the primitives of the "raw primal sketch" are "edges, bars, blobs, and terminations, and these have attributes of orientation, contrast, length, width and position." He also included the intervening space between items as features (cf. Beck et al., 1983; Julesz, 1984b; 1989). With the notable exception of position, this list is quite similar to those established for the human preattentive processing system. However, in addition to these features, Marr also states that spatial arrangements of these elements contribute to the primal sketch. Here his ideas depart in a fundamental way from the other first-order models of preattentive vision. In his list of texture features Marr included six "image properties" which are local measures of properties that are derived from a grouping process. That is, in Marr's theory, only the elements that are similar are grouped and then six local descriptors are calculated: average local intensity, average local size (including length and width), local density, local orientation, local distances separating the items, and local orientation of the virtual lines joining neighboring pairs of similar items.

However strong the similarities, the differences between Marr's model and those of the psychologists are even more striking due to Marr's consideration of precise spatial relations between the features and the explicit linking of spatially disparate features in this fashion. This, of course, is the antithesis of the first-order models of preattentive texture segmentation. Indeed, these differences are so strong that Marr's notions of the primal sketch possibly should not be compared to the preattentive processing system, but rather to the focal processing system. However accurate for certain aspects of human form vision, it might not have been most appropriate for Marr to assume that his model of vision is equally applicable to all aspects of human visual information processing, particularly human texture discrimination (cf. Schatz, 1977). As stated by Julesz (1984b), "Marr assumed that forms would be built up from the same building blocks as are used in early vision, whereas I am skeptical that the textons of preattentive vision are also the units of form perception."

In summary, among those researchers that utilize models other than pure linear filtering models, the list of texture features used by the human visual system in processing texture is fairly small. There is general agreement (among the first-order models) on five features: color/brightness, orientation, size and/or shape (sometimes including length, width, length-to-width ratio, or area), motion, and disparity. Other possibilities (but less widely supported) are closure, terminators, curvature, and the more recent suggestions of subjective contours and inferred 3-dimensional properties. This situation, then, is not too far removed from the situation that Julesz (1984a) described when he compared the analyzing of all textures in terms of a small set of primitives to the processing of color on the basis of three cone types. Furthermore, the image properties implicitly extracted by size-tuned oriented filters of the linear-filter models do not seem that different from the image properties on this list of features from first-order models.

2. *Computer vision*

This section considers the texture features employed in various computer vision approaches to texture analysis and contrasts them with those features employed in the models of human processing of texture described in the prior subsection. Only the features themselves are considered in this section, the utilization of these features is considered in the next section.

The types of stimulus structure extracted from the image in typical computer vision approaches is generally similar to that extracted in human vision models of segmentation. Some computer vision methods are based on linear filtering of the image by symmetric or elongated size-tuned filters (Marr and Hildreth, 1980; Voorhees and Poggio, 1988). Thus, at this lowest level, these computer vision methods are essentially identical to the linear-filter models of human texture-based segmentation. Many other computer vision methods extract features and are comparable to the first-order feature models of human vision. They are similar in that the extraction of features such as intensity/color, orientation, length and width are prevalent. They differ, however, in that computer vision methods have considered a wider, and possibly more comprehensive, range of features, and are applied to real-world images of greater complexity. Possibly the human-vision first-order models will have to incorporate some of these additional features when natural images are used routinely.

Some computer vision studies have been specifically designed to employ the same features as certain models of the human preattentive segmentation system. For example, terminators, crossings, orientation, and length have been used in some studies (Rearick, 1985; Caelli, 1985). In such feature-based studies the texture elements can be delineated syntactically or obtained by convolution masks used to extract preconceived features. The most common of these *a priori* primitives are lines and edges (Caelli, 1985; Derin and Cole, 1986; Gagalowicz and Ma, 1985) and spots (Zucker et al., 1975). The main alternative to the use of preconceived primitives is the extraction of homogeneous areas from the image by, for example, simple thresholding of pixel gray scale or similar local edge-based techniques. Typical "weak" features used to describe these pixel clusters are statistics of intensity, size, elongatedness, curvature, area, perimeter, compactness, eccentricity (shape), orientation, dispersedness and homogeneity (Wang, Hanson and Riseman, 1986; Tomita, Shirai and Tsuji, 1982; Hong, Dyer and Rosenfeld, 1980; Tsuji and Tomita, 1973).

Studies by Laws (1979, 1980) using convolution masks are very often cited and seem to have become an unofficial standard to which the performance of other computer vision texture analysis approaches are compared (e.g., Rearick, 1985; Wang et al., 1986; Lee and Price, 1982; Pietikainen and Rosenfeld, 1983; Vilnrotter, Nevatia and Price, 1986; VanGool et al., 1985). Laws produced a variety of two-dimensional masks, some of which resembled spot, edge, and local average masks. Thus, the masks varied in size and resembled detectors of spots, ripples, stripes, and so on. The four most effective masks resembled spot-, edge-, line- and V-shapes. The success of these particular masks was corroborated by Pietikainen et al., (1982). It is interesting to note that the most successful masks from various computer vision studies appear to extract features which, like the "blobs" of human vision segmentation models (e.g., Marr, 1982; Julesz, 1984a), tend to resemble the physiological "masks" observed in the receptive field profiles at early stages of biological vision.

The use of features related to the property of shape has been much more extensive in the computer vision literature than in the human vision studies. Possibly this is because human vision studies have typically used highly contrived textures made of elements of identical shape and have simply deferred the issue of shape. It is also possible that there is an implicit assumption that the essence of shape is directly based on striate cortex receptive field shape, and thus width, and to some extent length (end-stopped receptive fields), are the only features of shape likely to be utilized. However, this assumption may or may not be born out as more shape dimensions are considered in the human vision texture literature and, more importantly, in the physiological literature.

Another technique commonly used in computer vision is to use convolution masks to extract edges and then to use the edges to characterize texture. Occasionally edges are treated as primitives as noted above (Pietikainen and Rosenfeld, 1982), or linked together to form bounded primitives (Hong et al., 1980; Vilnrotter et al., 1986). However, edges are typically used in statistical approaches which seek to specify the positioning relationships (e.g., co-occurrence probabilities) between edges (as opposed to using the edges for the measurement of features per se). One of the most interesting edge-related features is "edgeness" (Rosenfeld and Thurston, 1971), measuring edge magnitude per unit area. This measure is low for coarse textures and high for fine textures. Coarseness has also been measured in terms of the extrema per unit area (Rosenfeld and Troy, 1970) and intensity variability along particular directions (Wang et al., 1978). Comparable measures are usually not even referred to in the human preattentive segmentation literature. The coarseness of an area, as defined by local standard deviation of intensity, gradient magnitude or extrema density, is an interesting local feature that may be the type of feature that feature-based human segmentation models may need to consider when more complicated stimulus images are employed.

In computer vision segmentation, "color" information is extracted in one of two ways: (1) the *position* in three dimensional color space is considered, or (2) distance along one or more specific *directions* of three-space is used, with each direction derived in isolation. Position in color space has been used with good success for both edge detection (Robinson, 1977; cf. Nevatia, 1977) and for pixel linking via clustering (Schacter et al., 1976; Davis and Rosenfeld, 1976; Sarabi and Aggarwal, 1981). More frequently however, segmentation studies use "color features" (again, a set of single *directions* in color space), rather than true color space. These latter studies typically derive linear and nonlinear transformations of the RGB tristimulus coordinates, with each transformation producing three new color features. Since in this context one is not concerned with color space, per se, any number of the dimensions can be utilized from any number of repeated transformations. Thus, depending on the particular color constituency of the prominent or "important" objects in the particular scene, one or another of the coordinates will be useful (Ohta, Kanade and Sakai, 1980; Ohlander, Price and Reddy, 1978). For typical images of outdoor scenes, those directions with dominant wavelengths in the yellow-green or green area of color space are most effective for segmentation. For example, in an examination of over 100 transformed features, Ohta et al. (1980) found [(2G-R-B)/2] and [R-B] to be among the most useful. This utilization of color in terms of features, rather than color space, provides the basis for two types of procedure. Rooted in the work of Ohlander (1975), one method performs histogram analysis on each feature separately to extract a region of a prominent color. In the second method, a pixel's values on all of the particular dimensions are used to specify a point in n-dimensional feature-space. Of course, this feature space is usually quite dissimilar to 3-dimensional true (human) color space because of the arbitrary number (and types) of dimensions selected. Often these features are used in conjunction with non-color features (e.g., a "texture" measure), which further distinguishes this multidimensional feature space from 3-dimensional color space.

The analysis of color in computer vision approaches seems to suggest other approaches that might be profitably considered in analyzing how human preattentive segmentation utilizes color information. It is not yet clear whether preattentive distinctiveness is solely determined by a single feature, distance in 3-dimensional (psychological) color space, or along a set of separate dimensions with each considered separately as they are in much of the computer vision research. For example, conceivably, color could be preattentively analyzed *separately* on dimensions such as hue, saturation, and brightness; S-, M-, and L-cone directions; or on ecologically pragmatic directions as it is in the computer vision literature (Ohta et al., 1980). However, from work in other contexts one might expect that humans do not necessarily preattentively analyze color into separate dimensions (e.g., Burns and Shepp, 1988). As mentioned in the last section, in the event that it is a single value that is used (i.e., difference in the location of two areas in 3-dimensional color space), whether all directions are equally effective preattentively remains to be determined.

Related to these issues concerning color is the question of how brightness is used in human preattentive segmentation. Is it simply subsumed by the specification of "color", or is it a separate feature? If a separate feature, how is it specified? Invariably edge contrast, edge brightness level, and brightness of the primitives are confounded in the simple patterns typically utilized. Yet to be clarified is whether preattentive segmentation is based on the edge contrast as assumed by Marr (see also, Treisman and Gormican, 1988; cf. Beck, Sutter and Ivry, 1987, for Rayleigh contrast), or primitive brightness as typically assumed in human vision models.

B. Texture Boundaries and Region Formation

The three most often cited first-order models of preattentive texture segmentation are those of Beck (e.g., Beck et al., 1983), Treisman (e.g., 1985) and Julesz (e.g., 1981). In the last section it was seen that the lists of texture primitives utilized by various first-order models are quite similar. The models differ considerably, however, in terms of how they propose that the extracted texture primitives are then processed to provide segmented regions. In this section, an overview of how segmented regions may be obtained within first-order models, linear filter models and computer vision models is presented.

1. Feature maps of first-order models

The various first-order models of human preattentive segmentation present different views of how the various features are analyzed. Treisman (1985) contends that her view of this differs from that of Neisser, Beck, Marr, and Julesz in that they each propose that the entire preattentive representation is a single map (i.e., a single spatial representation). She proposes that each value of a segmentation feature (e.g., the values red, vertical, or rightward moving) is represented separately in its own feature map. Furthermore, she contends that although organized spatiotopically, the preattentive system is unable to access even the coarsest information about position within these maps. Her view entails a set of distinct maps, one for each feature, and a master map of image locations (in three spatial dimensions). The preattentive system has access to only the pooled response (i.e., a single value representing the average activity, or strength, of that feature) summed over the entire spatial mapping of the feature's presence in the image (cf. Gurnsey and Browse, 1989). The information available preattentively from feature maps does not tell where a feature is located, nor how the instantiations of various features are linked; features are spatially "free-floating" (i.e., relative position, or phase, is ignored) and not conjoined. She likens her view of distinct features maps to Barrow and Tenenbaum's idea from the computer vision literature of separate intrinsic images (Treisman, 1985).

The master map is a set of spatial coordinates (3-dimensional) that can be scanned serially with focused attention (Treisman and Gormican, 1988). The map indicates where (and how many) features occur, but not *which* features exist at a given location. The master map of locations is available preattentively specifying "where". Attention selects a spot on the master map and conjoins the specific features that are represented at that point via links to the specific feature maps. These links provide the "glue" that brings relative position into play.

Like many others, Treisman does not draw a distinction between the bases of "pop-out" in visual search and texture discrimination (e.g., Treisman, 1986). Her model of preattentive vision presents an explicit account of how pop-out occurs in visual search, yet the model does not make explicit how regions of texture or texture-boundaries are formed preattentively. Treisman (Treisman and Gormican, 1988) has proposed only that "difference detectors" are activated by a difference of features in adjacent elements and that they code these "local boundaries" of texture in a relational code (e.g., "darker than"). In fact, the "pooled response" aspect of her model would seem to make the extraction of texture-based regions or boundaries from the feature maps impossible. If the model is tenable for segmentation as well as visual search, these boundaries must be extracted preattentively. It would seem that either some "where" information concerning texture edges must be represented in the distinct feature maps or some additional "what" information concerning the features' texture edges (or a composite measure of edges from all features) must be represented in the master map. Current versions of Treisman's model have yet to address this issue directly. Such location information is, however, specifically provided in certain variants of Treisman's model (e.g., Pashler, 1987; Wolfe et al., 1989). For example, in Wolfe et al.'s (1989) "guided search" version of Treisman's model, the feature maps provide some "where" information. The feature maps make available for parallel processing the set of locations at which a particular feature occurs (e.g., red). This modification of Treisman's feature maps makes the guided search model a plausible basis of preattentive segmentation, but segmentation has not yet been addressed by this model either.

Several variations of Treisman's basic model (including Treisman, 1988) have been proposed which question the independence of certain features. Although similar at a basic level, these models depart form Treisman's earlier model in terms of how search for multiple features (either conjoined or separate) is mediated (Treisman and Gormican, 1988; Quinlan and Humphreys, 1987; Wolfe et al., 1989; McLeod et al., 1988; Nakayama and Silverman, 1986; Enns, 1986). Unfortunately for the purposes of this chapter, these models focus on how feature differences lead to successful visual search, but do not specify how feature differences lead to texture-boarders and region segmentation.

Beck's view (e.g., 1982) of the integration of features is at the opposite extreme from that proposed in Treisman's model. In Beck's model (e.g., 1982, Beck et al., 1983), "difference signals" are generated if two areas differ in terms of any texture feature. The location of these difference signals is known preattentively and the segmentation decision is made on the basis of the magnitude and number of these signals at a given location. Hence this preattentive segmentation occurs most readily if the two areas differ on many features and have few features in common (see also, Enns, 1986; Nothdurft, 1991). In effect, the areas are compared in n-dimensional feature space. Although not phrased this way by Beck, it is as if there is a master map which plots how far apart a point is from its neighbor in feature space. In other words, whereas Treisman's model explicitly keeps information derived from each feature distinct preattentively, Beck's model explicitly combines all such information preattentively.

The views presented by Julesz (e.g., 1984a) and by Marr (1982) are intermediate to those presented by Treisman and Beck. They propose that some features like orientation, length, width and color are properties of a particular element (e.g., a red vertical line) rather than distinct themselves. Although Treisman and Julesz both emphasize that this is a significant distinction between their models, the distinction can be viewed as merely semantic. Treisman's model treats each feature in the same way (i.e., each has its own feature map), but, so too, does Julesz's model. His model computes the density of each feature and thus provides an implicit density map for each one (with each map apparently computed in the same way). Thus, in terms of how the features are utilized, it seems that there is no real difference in how the two models treat texture primitives: in both models all features (whether properties or parts) have their own map.

The theories of Julesz and Treisman have come to be more and more alike as they have evolved (see particularly, Julesz, 1986, but also Bergen and Julesz, 1983, as well as more recent papers; Treisman, 1985 and Treisman and Gormican, 1988). In these papers Julesz takes several positions like those of Treisman summarized earlier in this section: (1) preattentive vision provides "where" information without providing "what"[5]; (2) "blobs" as textons are redefined more along the lines of Treisman's "closure"; (3) it is emphasized that the presence of a texton is detected better than its absence; (4) conjunctions of features are considered similarly, stating that the features are not "glued" together outside of the aperture of focal attention; and (5) the size of the area over which density is computed scales with the size of the texton difference. This last issue has been taken up by several other authors recently who suggest that the size of the area that can be searched in parallel preattentively increases as the feature difference increases (e.g., Julesz, 1984a; Enns, 1986; Treisman and Gelade, 1988;

[5]That is, the resultant information within Treisman's model, that of the master map, is comparable. However, a distinction can be drawn on the basis of the representation within the feature maps (Treisman and Gormican, 1988).

Treisman, 1988; Nagy et al., 1990; Nothdurft, 1991). In fact, some of these models propose that the image is "chunked" in the sense that the area of parallel preattentive processing is narrowed and moved to successive locations across the image (see also Pashler, 1987). Thus, from the current literature, the preattentive/attentive "dichotomy" appears to be more of a continuum than a true dichotomy.

To determine texture boundaries, texton theory emphasizes differences between regions in terms of the number of a given texton type computed over a specific area (i.e., density), although the magnitude of a difference within a dimension, such as orientation, is also considered (cf. Nothdurft, 1991). Unfortunately, Julesz, like Treisman and Beck, never specified exactly *how* texture differences (texton density gradients in his model) lead to the extraction of texture-based edges and regions. For ideas about this next step, the extraction of texture regions or texture boundaries, we must look elsewhere.

2. Texture-boundary and texture-region formation

Although the first-order feature models described in the last subsection did not specify exactly how it is done, it is apparent that various methods *could* specify locations (i.e., boundaries) where adjacent areas differ in terms of density of texton, position in n-dimensional feature space, or in a master map of texture discontinuities. However, no matter which method is used, the texture-based boundaries that are initially extracted are highly unlikely to be complete boundaries, delineating the full perimeter of textured regions. Extraction of texture boundaries is such a difficult problem that the extracted boundaries would undoubtedly be fragmented at this initial stage and require a filling-in process guided by these fragments in order to form completed regions (e.g., Grossberg and Mingolla, 1985). The complexity of the image structure conveying the texture information is illustrated in Figure 1.1. Figure 1.1a shows the original image of a real-world scene. Figures 1.1b and 1.1c show "feature maps" for two orientation features, 45° and 100°, respectively. These two feature maps serve to illustrate how much work is yet to be done by the segmentation system. Complete, uniform regions of texture corresponding to grass, bush, tree, roof or windows, for example, do not yet pop-out form Figures 1.1b and 1.1c. Figure 1.1d shows a density measure applied to a feature map such as the one in Figure 1.1b. By counting the number of occurrences of the texture feature over a local neighborhood, a density map is formed from which texture-based edges (i.e., fragments) can be extracted. Both of these particular feature maps suggest important discontinuities between roof and sky and between tree and sky. The

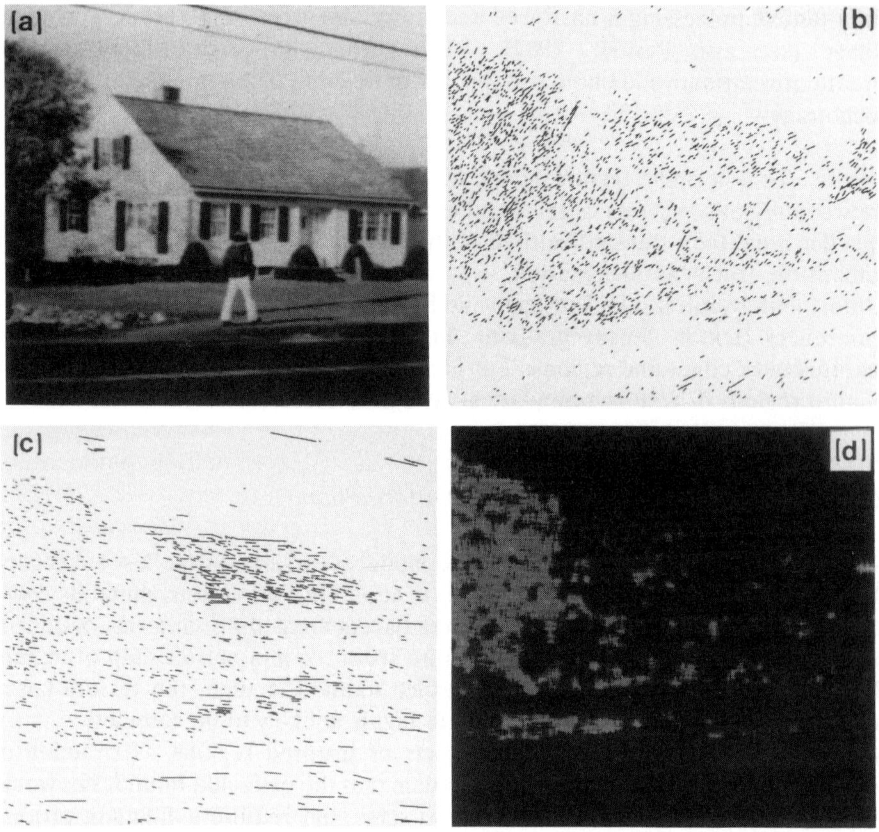

Figure 1.1. An image of a natural scene to be segmented (a); some of the structure in the image (b) and (c); and a density map resulting from the initial processing of a feature map such as in 'b'. See text.

feature of Figure 1.1c is particularly good at indicating roof, yet an accurate rendering yielding a filled-in parallelogram with complete, straight edges obviously is not available in this single map. Clearly, information needs to be extracted from multiple maps and then combined to extract the tree region, or even a well-demarcated roof region. If the segmentation system could extract fragments of texture-based boundaries in each of many feature maps, then their combination into a master map would provide considerably more information. Contiguous fragments from different maps could be linked to extend a given boundary fragment and overlapping fragments would strengthen the texture-based edge extracted at that location.[6]

An alternative conceptualization, more along the lines of Beck's approach, is to extract all boundary fragments simultaneously based on separation in n-dimensional space. In either case, it seems that two steps are necessary: (1) the extraction of as many fragments of the boundaries of textured regions as possible, and (2) a stage of filling-in to complete these regions, using the extracted boundary fragments to initiate and guide the formation of segregated regions (cf. Grossberg and Mingolla, 1985).

Methods to fill-in or "grow" regions are common in both the computer vision and human vision literature. In human vision, the fundamental example is the blind spot. The region of the retina where the nerve fibers leave the eye has no receptors, yet the corresponding "blind spot" in the visual world is perceptually filled-in to match the surroundings such that a discontinuity is undetectable. Pathological scotomas and artificial scotomas (from image stabilization) fill in as well (reviewed by Paradiso and Nakayama, 1991). In computer vision, approaches to this problem are often distinguished as edge-based or region-based. Edge-based techniques link edges to form explicit boundaries containing implicitly defined regions (e.g., Medioni, 1982; Hanson and Riseman, 1978). Region-based techniques link pixels to form explicit regions which form implicit boundaries (e.g., Ohlander, 1975). Either type of technique can be local or global in nature. In either technique, the goal is to provide a region that is both accurate and uniform (i.e., completely filled-in without holes or being "spotty"). The more prevalent region-growing techniques (Zucker, 1976; Haralick and Shapiro, 1985) form a region by linking together neighboring pixels that have sufficiently similar properties (Zucker, 1976). An alternative approach (Ohlander, 1975) is to link together pixels that cluster in feature space (i.e., the image locations whose pixels contribute to a pronounced peak in an n-dimensional histogram). Hybrid

[6]For example, the tree region is probably distinguished by having high values on many orientation features (i.e., high magnitude, yet no single dominant orientation) and on certain green colors. However, color is a surprisingly hard feature on the basis of which to segment an area in a natural scene. The green of the tree will be quite variable depending upon lighting (direct sunlight or shadows from other leaves), which leaves are twisted to reveal the color of the top side or the color of the bottom side, and whether particular leaves are old or new growth (areas of new growth are often much brighter and yellower).

techniques can combine the local and global aspects of these techniques (Kohler, 1984). In addition, "relaxation techniques" can be used in either edge-, or region-based, techniques to fill-in fragments in the regions or edges. Relaxation techniques allow the spatial propagation of information or constraints from other sources to be utilized in parallel, usually iteratively (reviewed by Rosenfeld, 1978). Particular limitations characterize these various computer vision segmentation techniques (e.g., see the review by Haralick and Shapiro, 1985), which surely apply as well to the human vision situation. Simple region-growing algorithms tend to over-merge areas, joining areas that are better left as separate regions. Conversely, clustering techniques tend to have overly jagged boundaries as well as false internal boundaries causing holes in the region. Balancing these complementary problems of over-merging and fragmenting, while providing accurate region borders, remains a difficult problem for computer vision.

The human visual system is faced with the analogous problem of joining areas of similar texture to form a perceptually uniform region. That powerful linking or filling-in processes indeed operate in human vision is evident by the blindspot demonstration (see Paradiso and Nakayama, 1991) and by many perceptual illusions (Arend, Buehler and Lockhead, 1971; and see Grossberg and Mingolla, 1985; Arend and Goldstein, 1987). For example, the Craik-O'Brian-Cornsweet illusion provides striking evidence of the tendency of the human visual system to give spatially bounded areas uniform perceptual attributes. A brightness or color difference just inside a border can lead to the perception of the entire region as one uniform illusory brightness or color. An analogous filling-in process has been reported for texture where a *texture edge* creates an illusory filling-in of an inner area bounded by the edge and thus causes physically identical texture regions to appear to be uniform areas, each of a different texture (Nothdurft, 1985; Sagi and Hochstein, 1985; Muller, 1986).

Typically, the region to be filled-in is not surrounded by a completed boundary, but instead the preattentive vision system has been able to extract only fragments of texture boundaries from the complex and noisy intensity image. Cohen and Grossberg (1984) have proposed a very comprehensive model of the way in which the human visual system forms contours and uses the contours to aid in filling-in regions (see also, Grossberg and Mingolla, 1985). They propose that "boundary contours" are formed from extracted luminance edges and subjective contours, and that the boundary contours are linked by a "boundary completion process" which encodes orientation and contrast but ignores the sign of the edge (see also Shapley and Gordon, 1985; Marr, 1980). They also propose a companion "feature system" in which feature properties are detected locally and are essentially tabulated in terms of number (i.e., much as in the models of preattentive segmentation) and which determine the visible percepts. Feature attributes spread via a relaxation process to establish feature boundaries that code magnitude (contrast) and polarity of the boundary but not its orientation (cf. Paradiso and Nakayama, 1991). The spread of the feature is stopped either by an insufficient amount of the locally-detected feature to maintain the spreading or *by encountering a boundary contour.* Rather direct evidence obtained by Kawabata (1984) suggests that preattentive segmentation boundaries are formed first and then guide the subsequent perceptual filling-in process (cf. Paradiso and Nakayama, 1991).

The general process of Cohen and Grossberg's model (1984) of human contour formation and filling-in is quite similar to several approaches employed in computer vision to solve corresponding problems (Boldt and Weiss, 1987; Medioni, 1982; Haralick and Shapiro, 1985). Filling-in based on an iterative averaging process (as proposed by Cohen and Grossberg) is quite comparable to the "hybrid linking region growing" techniques of computer vision reviewed by Haralick and Shapiro. The use of a boundary contour to restrict the region growing process has been employed successfully in computer vision segmentation (Latty et al., 1985; Strong and Rosenfeld, 1973).

In summary, the first-order models of human texture-based segmentation have not advanced beyond the step of region boundary extraction. However, in this subsection it was seen that human perceptual phenomena and perceptual models (particularly by Cohen and Grossberg) are quite integral to the segmentation task and provide a natural extension of the preattentive texture segmentation models. This extension suggests that a type of region-growing based on the local prominence of the various segmentation features is needed to fill-in the regions up to the location of whatever texture-based segmentation boundaries (or structural edges) have been extracted.

3. Linear-filter models

Proponents of the linear-filter models of human preattentive segmentation claim that these models are simpler and less contrived than the models which extract a list of texture features. Typical of this type of model is a set of filters, analogous to channels or mechanisms in models of human threshold spatial vision (e.g. Wilson and Bergen, 1979). The filters are of various sizes (i.e., size-tuned) and usually of various orientations. The weighting functions of the filters are convolved with the image, with each producing a different filtered image. This set of output maps from the set of linear filters must then be analyzed somehow to segment the image.

A prominent example of a linear-filter texture segmentation model is that of Landy and Bergen (1991). They used narrow-band filtered-noise stimulus patterns rather than the typical micropatterns, and developed a model that segments based on gradients of orientation structure in the image. Their model can be considered as consisting of seven processing steps: (1) The image is filtered with a set of four oriented filters (V,R,H,L) of the appropriate size-tuning for the stimulus patterns; (2) The output of each filter is squared; (3) Each filter's response is averaged over a small spatial region (and subsampled) to provide "local oriented energy maps"; (4) Two opponent orientation signals are created (H^2-V^2 and R^2-L^2); (5) To make these measures reflecting local image contrast into texture measures, the responses are normalized to the local oriented contrast; (6) An edge operator is applied to produce texture maps reflecting edge gradient magnitude; (7) A nonlinearity is used to suppress the weak responses. Thus, this is a model typical of the class of models that extracts only filter energy from the image (i.e., no texture primitives are explicitly extracted), and predicts human texture discrimination performance successfully. The texture maps that the model produces for the four orientations, and the extraction of locations of large gradients in the orientation maps, greatly resembles the areas of "discontinuity" or feature "difference" that the first-order models loosely define (as described earlier). Landy and Bergen state that their model "does not require more elaborate processing mechanisms such as the detection of textons, linking of edges, and region growing". In spite of this claim, however, the model does not seem qualitatively different from the models based on the first-order statistics of features described earlier: (1) This linear-filter model clearly extracts (albeit graded) what those models would call four texture features (horizontal, vertical, 45° and 135° orientations); (2) the exact local position of the energy output of the linear filters is ignored when the local spatial averaging is performed; and (3) the measure utilized, normalized filter energy per unit area, is essentially a "density" measurement. With respect to these three points, it appears that the segmentation produced by this linear-filter model is actually based on something quite similar to discontinuities in the densities of four texture features.

The purported *qualitative* difference between the models of human texture-based preattentive segmentation that are based on the first-order statistics of texture features extracted from the image and those that are based on a linear filtering of the image, do not hold up when the issue is scrutinized. At the level of extracting image structure, the linear filter models are indeed linear. The first-order models don't really address how the texture elements are extracted (one would presume that the image structure is initially extracted by similar linear filters and then quantized into features). The first-order models are concerned with the counts of certain features and thus are soon clearly nonlinear in nature. That linear-filter models are that different is questionable for at least two reasons. First, in some sense, the linear filters extract features (see above). The particular weighting functions convolved with the image are effectively templates ("masks") that are fit at each location in the image and the extent to which the template fits the image structure at that location is noted. In this sense, a set of specific features, one by each filter, is being extracted, the only difference being that the features are of graded magnitude rather than counted. Secondly, once these graded feature maps are obtained, nonlinearities are invariably introduced into the linear-filter model. That is, both types of model are nonlinear and both types of model consider specific features; the linear-filter model proponents simply promote the linear filtering stage of their models (a stage that seems implicit in biologically-plausible first-order models).

The need for the introduction of nonlinearities into these linear-filter models is presented quite convincingly by Malik and Perona (1991). Their model is extremely powerful, predicting the salience of texture boundaries on a wide range of classic micropatterns. It is a model using a large bank of filters which includes six orientations and two types of radially symmetric filters, each at twelve sizes (spatial frequencies) and all duplicated in "on" and "off" phase (a total of 192 filters, a value to be contrasted with the length of the list of features of first-order models). They argue convincingly that an early nonlinearity is essential in a biologically plausible model and that a halfwave rectification of the filters' output is best. They also argue that a second nonlinearity following rectification to suppress weaker responses within a local spatial neighborhood (which functions something like a threshold), is optimal. Other linear-filter models employ essential nonlinearities as well (e.g., Voorhees & Poggio, 1988; Bergen and Adelson, 1988; Gorea and Papathomas, 1991; Caelli, 1985). Thus, it is accurate to think of these models as nonlinear models that use linear filtering (i.e., in no way are they "linear filter-models", but rather "linear-filter models"). Using linear filters to extract image structure is not new to computer vision, computational models of human texture discrimination (Caelli, 1985), or even to first-order models which simply assume extraction of orientation and 'blobs' (e.g., Julesz, 1984a). What is new is the completeness of these models (i.e., they don't stop at the level of region-extraction as do the first-order models) and their biological plausibility.

V. CONCLUSIONS

From the last subsection we see that the controversy between the first-order texture-feature models and the linear-filter models boils down, not to whether linear or not, but really to what features are used and how the nonlinearities lead to formation of textured regions. Neither type is strictly linear. First-order models presume that extracted features are counted all-or-none (e.g., thresholded) and then calculate some graded measure such as density. In some fashion that is currently left ill-specified, regions which differ on this measure are formed. Linear-filter models extract a graded measure of the strength of some image structure (specifically, the mask profile), and then, at some point, use a nonlinearity to threshold or similarly "suppress" parts of the image and thereby form regions. Considered in this way, possibly the strongest difference between the two types of model as currently developed, is the list of masks (i.e., the "filters" of one type of model and the "features" of the other type) that the models employ. That is, for the features of orientation and spatial frequency (size), the models are of the same ilk and there is no conflict between them. The difference is just that some models are more complete and specified computationally whereas others are general. For features such as motion, disparity, intensity or color (cf. Gorea and Papthomas, 1991) the models are also not inconsistent (these computational models simply aren't yet extended to these obvious features). However, for features such as terminators or closure (or the more exotic features that are only occasionally considered in the first-order models; e.g., see Treisman and Gormican, 1988), there is a real difference between linear-filter models and those first-order models that do incorporate such features. Indeed, these are precisely the features to which advocates of linear-filter models point to emphasize the simpler, less contrived, nature of their models.

Based on the physiology, it would appear that other features are appropriate for a biologically-plausible linear-filters model as well. In particular, length, luminance, chromatic contrast, and disparity seem appropriate extensions due to their roles as fundamental properties of early cortical receptive fields.[7] Alternatively, just because in the context of a current biologically-plausible linear-filter model, a particular feature seems implausible, is simply to say that no currently-known receptive-field property corresponds to it, not that such a property could never be found in the future. The point here is not to say that crossings, for example, are likely to be a feature utilized in preattentive

[7]Subjective contours have been reported as a feature that is also detected at a relatively early stage of cortical processing, V2 (von der Heydt and Peterhans, 1989).

segmentation (they probably aren't), but that the list of biologically-plausible filters/features is not necessarily closed on the basis of currently known electrophysiology.

Thus, the human preattentive texture-based segmentation literature has come full circle. The distinction between models of texture discrimination, grouping, and visual search processes used to be made on the basis of which items were in each model's alphabet of texture features. From the perspective expressed in the last section of this chapter, there is no *fundamental* distinction to be made between the linear-filter and first-order models. Thus, the main difference between these models is that the list of features in typical first-order models is presently a little longer than the list of features incorporated in current biologically-plausible computational filter models. Other issues (e.g., the basis of search asymmetries, the specific nonlinearities and their computational details, and the extent to which the preattentive/attentive distinction is truly dichotomous) will guide the development of future models by adding constraints, but the most fundamental issue in texture-based segmentation remains the issue of which types of structure in the stimulus image are extracted by the early processing of the human visual system.

ACKNOWLEDGEMENTS

I thank Alan Hanson and Ed Riseman for their support of my work at the Laboratory for Computer Vision, University of Massachusetts, Amherst.

REFERENCES

Arend, L. E., Buehler, J. N., and Lockhead, G. R. (1971) Difference information in brightness perception. Perception and Psychophysics, 9(3B), 367-370.

Arend, L.E., and Goldstein, R. (1987) Lightness models, gradient illusions, and curl. Perception and psychophysics, 42, 65-80.

Beck, J. (1967) Perceptual grouping produced by line figures. Perception and Psychophysics, 2, 491-495.

Beck, J. (1982) Textual segmentation. In J. Beck (Ed.), Organization and representation. Hillsdale, NJ: Erlbaum.

Beck, J., Prazdny, K., and Rosenfeld, A. (1983) A theory of textural segmentation. In J. Beck, B. Hope, and A. Rosenfeld (Eds.), Human and machine vision. New York: Academic Press.

Beck, J., Sutter, A., and Ivry, R. (1987) Spatial frequency channels and perceptual grouping in texture segmentation. Computer Vision, Graphics, and Image Processing, 37, 299-325.

Bergen, J. R., and Adelson, E. H. (1988) Early vision and texture perception. Nature, 333, 363-364.

Bergen, J. R., and Julesz, B. (1983b) Rapid discrimination of visual patterns. IEEE Transactions on Systems, Man and Cybernetics, SMC-13, 857-863.

Boldt, M., and Weiss, R. (1987) Token-based extraction of straight lines. University of Massachusetts COINS Technical Report 87-104, Amherst, MA.

Burns, B., and Shepp, B. E. (1988) Dimensional interactions and the structure of psychological space: The representation of hue, saturation and brightness. Perception and Psychophysics, 43, 494-507.

Caelli, T. M. (1985) Three processing characteristics of visual texture segmentation. Spatial Vision, 1, 19-30.

Cohen, M. A., and Grossberg, S. (1984) Neural dynamics of brightness perception: Features, boundaries, diffusion and resonance. Perception and Psychophysics, 36, 428-456.

Davis, L. S. (1979) Computing the spatial structure of cellular textures. Computer Graphics and Image Processing, 11, 111-122.

Davis, L. S., and Mitiche, A. (1980) Edge Detection in Textures. Computer Graphics and Image Processing, 12, 25-39.

Davis, L. S., Clearman, M., and Aggarwal, J. K. (1981) An empirical evaluation of generalized co-occurrence matrices. IEEE Transactions on Pattern Analysis and Machine Intelligence, PAMI-3, 214-221.

Derin, H., and Cole, W. S. (1986) Segmentation of textured images using Gibbs Random Field. Computer Vision, Graphics and Image Processing, 35, 72-98.

Enns, J. (1986) Seeing textons in context. Perception and Psychophysics, 39, 143-147.

Gagalowicz, A., and Ma, S. D. (1985) Sequential synthesis of natural textures. Computer Vision, Graphics and Image Processing, 30, 289-315.

Gorea, A., and Papathomas, T. V. (1991) Texture segregation by chromatic and achromatic visual pathways: an analogy with motion processing. Journal of the Optical Society of America Series A, 8, 386-393.

Gurnsey, R., and Browse, R. A. (1987) Micropattern properties and presentation conditions influencing visual texture discrimination. Perception and Psychophysics, 41, 239-252.

Gurnsey, R. and Browse, R.A. (1989) Asymmetries in visual texture discrimination. Spatial Vision, 4, 31-44.

Gurnsey, R., Humphrey, G.K., and Kapitan, P. Parallel discrimination of subjective contours defined by offset gratings. Submitted.

Hanson, A. R., and Riseman, E. M. (1978) Segmentation of natural scenes. In A. Hanson, and E. Riseman (Eds.), Computer Vision Systems. New York: Academic Press.

Hanson, A.R., and Riseman, E.M. (1980) Processing cones: A computational structure for image analysis. In S. Tanimoto and A. Klinger (Eds.), Structured Computer Vision.

Haralick, R. M. (1979) Statistical and structural approaches to texture, Proceedings of the IEEE, 67, 786-804.

Haralick, R. M., and Shapiro, L. G. (1985) Survey: Image segmentation techniques. Computer Graphics and Image Processing, 29, 100-132.

Hong, T. H., Dyer, C. R., and Rosenfeld, A. (1980) Texture primitive extraction using an edge-based approach. IEEE Transactions Systems Man Cybernetics, SMC-10, 659-675.

James, W. (1890) The principles of psychology. New York: Holt, Rinehart and Winston.

Julesz, B. (1981) Textons, the elements of texture perception and their interactions. Nature, 290, 91-97.

Julesz, B. (1984a) A Brief outline of the texton theory of human vision. Trends in Neuroscience, 7, 41-45.

Julesz, B. (1984b) In G. M. Edelman, W. E. Gall, and W. M. Cowan (Eds.), Dynamic Aspects of Neocortical Function. New York: Wiley.

Julesz, B. (1986) Texton gradients: The texton theory revisited. Biological Cybernetics, 54, 245-251.

Julesz, B. (1989) AI and early vision-Part II. SPIE: Human Vision, Visual Processing, and Digital Display, 1077, 246-268.

Kawabata, N. (1984) Perception at the blind spot and similarity grouping. Perception and Psychophysics, 36, 151-158.

Kohler, R. R. (1984) Integrating non-semantic knowledge into image segmentation processes. University of Massachusetts COINS Technical Report 84-04, Amherst, MA.

Landy, M. S., and Bergen, J. (1991) Textural segregation and orientation gradient. Vision Research, 31, 679-691.

Latty, R. S., Nelson, R., Markham, B., Williams, D., Toll, D., and Irons, J. (1985) Performance comparisons between information extraction techniques using variable spatial resolution data. Photogrammetric Engineering and Remote Sensing, 51, 1459-1470.

Laws, K. I. (1979) Texture energy measures. Proceedings Image Understanding Workshop.

Laws, K. I. (1980) Textured image segmentation. Image Processing Institute, Report 940. Los Angeles: University of Southern California.

Lee, H. Y., and Price K. E. (1982) Using texture edge information in aerial image segmentation. In R. Nevatia (Ed.), Image Understanding Research, Final Technical Report. Los Angeles: University of Southern California.

McIlhagga, W., Hine, T., Cole, G. R. and Snyder, A. W. (1990) Texture segregation with luminance and chromatic contrast. Vision Research, 30, 489-495.

McLeod, P., Driver, J., and Crisp, J. (1988) Visual research for a conjunction of movement and form is parallel. Nature, 332, 154-155.

Malik, J., and Perona, P. (1990) Preattentive texture discrimination with early vision mechanisms. Journal of the Optical Society of America Series A, 7, 923-932.

Marr, D. (1980) Visual information processing: The structure and creation of visual representations. Philosophical Transactions of the Royal Society of London Series B, 290, 199-218.

Marr, D. (1982) Vision: A Computational Investigation Into the Human Representation and Processing of Visual Information. San Francisco: Freeman.

Marr, D., and Hildreth, E. (1980) Theory of edge detection. Proceedings of the Royal Society of London, Series B, 204, 301-328.

Medioni, G. G. (1982) Segmentation of images into regions using edge information. In R. Nevatia (Ed.), Image Understanding Research, Final Technical Report. Los Angeles: University of Southern California.

Muller, M. (1986) Texture boundaries: Important cues for human texture discrimination. In Proceedings of the Seventh International Conference on Pattern Recognition, IEEE, 464-468.

Nagy, A. L., Sanchez, R. R., and Hughes, T. C. (1990) Visual search for color differences with foveal and peripheral vision. Journal of the Optical Society of America Series A, 7, 1995-2001.

Nakayama, K., and Silverman, G. H. (1986) Serial and parallel processing of visual feature conjunctions. Nature, 320, 264-265.

Nevatia, R. (1977) A color edge detector and its use in scene segmentation. IEEE Transactions on Systems, Man, and Cybernetics, SMC-7, 820-826.

Neisser, U. (1967) Cognitive Psychology. New York: Appleton-Century-Crofts.

Nothdurft, H. C. (1985) Sensitivity for structure gradient in texture discrimination tasks. Vision Research, 25, 1957-1968.

Ohlander, R. B. (1975) Analysis of natural scenes, Ph.D. Dissertation, Computer Science Department, Carnegie-Mellon University, Pittsburgh, PA, June.

Ohlander, R., Price, K., and Reddy, D. R. (1978) Picture segmentation using a recursive region splitting method. Computer Graphics and Image Processing, 8, 313-333.

Ohta, Y., Kanade, T., and Sakai, T. (1980) Color information for region segmentation. Computer Graphics and Image Process, 13, 222-241.

Olson, R., and Attneave, F. (1970) What variables produce similarity grouping? American Journal of Psychology, 83, 1-21.

Pashler, H. (1987) Detecting conjunctions of color and form: Reassessing the serial search hypothesis. Perception and Psychophysics, 41, 1911-201.

Paradiso, M. A., and Nakayama, K. (1991) Brightness perception and filling-in. Vision Research, 31, 1221-1236.

Pietikainen, M., and Rosenfeld, A. (1982) Edge-based texture measures. In Proc of Sixth International Conference on Pattern Recognition, IEEE.

Pietikainen, M., and Rosenfeld, A. (1983) Experiments with texture classification using averages of local pattern matches. IEEE Transactions on Systems, Man, and Cybernetics, SMC-13.

Quinlan, P. T., and Humphreys, G. W. (1987) Visual search for targets defined by combinations of color, shape, and size: A examination of the constraints on feature and conjunction searches. Perception and Psychophysics, 41, 455-472.

Rearick, T. C. (1985) A texture analysis algorithm inspired by a theory of preattentive vision. In Proceedings of the Sixth International Conference on Pattern Recognition, IEEE, 312-317.

Robinson, G. S. (1977) Color edge detection. Optical Engineering, 16, 479-484.

Rosenfeld, A. (1978) Iterative methods in image analysis. Pattern Recognition, 10, 181-182.

Rosenfeld, A., and Davis, L. S. (1979) Image segmentation and image models. Proceedings of the IEEE, 67, 764-772.

Rosenfeld, A., and Thurston, M. (1971) Edge and curve detection for visual scene analysis. IEEE Transaction on Computers, C-20, 562-569.

Rosenfeld, A., and Troy, E. (1970) Visual texture analysis. Technical Report 70-116, University of Maryland, College Park, MD.

Sagi, D., and Hochstein, S. (1985) Lateral inhibition between spatially adjacent spatial-frequency channels? Perception and Psychophysics, 37, 315-322.

Sarabi, A., and Aggarwal, J. K. (1981) Segmentation of chromatic images. Pattern Recognition, 13, 417-427.

Schacter, B. J., Davis, L. S., and Rosenfeld, A. (1976) SIGART Newsletter, 58, 16-17.

Schatz, B. R. (1977) The computation of immediate texture discrimination. MIT Artificial Intelligence Laboratories Memorandum 426, Cambridge, MA.

Shapley, R., and Gordon, J. (1985) Nonlinearity in the perception of form. Perception and Psychophysics, 37, 84-88.

Smallman, H. S., and Boynton, R. M. (1990) Segregation of basic colors in an information display. Journal of the Optical Society of America Series A, 7, 1985-1994.

Strong, J. P., and Rosenfeld, A. (1973) A region coloring technique for scene analysis. Graphics and Image Processing, 16, 237-246.

Tomita, F., Shirai, Y., and Tsuji, S. (1982) Description of textures by a structural analysis. IEEE Transactions Pattern Analysis and Machine Intelligence, PAMI-4, 183-191.

Treisman, A. (1985) Preattentive processing in vision. Computer Vision, Graphics and Image Processing, 31, 156-177.

Treisman, A. (1986) Properties, parts and objects. In K. Boff, L. Kaufman, and J. Thomas (Eds.), Handbook of Perception and Human Performance, Volume 2: Cognitive Processes and Performance. New York: Wiley.

Treisman, A. (1988) Features and objects: The fourteenth Bartlett Memorial Lecture. Quarterly Journal of Experimental Psychology, 40A (2), 201-237.

Treisman, A., and Gelade, G. (1980) A feature integration theory of attention. Cognitive Psychology, 12, 97-136.

Treisman, A., and Gormican, S. (1988) Feature analysis in early vision: Evidence from search asymmetries. Psychological Review, 95, 15-48.

Treisman, A., and Paterson, R. (1984) Emergent features, attention and object perception. Journal of Experimental Psychology: Human Perception and Performance, 10, 12-31.

Treisman, A., and Souther, J. (1985) Search asymmetry: A diagnostic for preattentive processing of separable features. Journal of Experimental Psychology: General, 114, 285-310.

Tsuji, S., and Tomita, F. (1973) A structural analyzer for a class of textures. Computer Graphics and Image Processing, 2, 216-231.

VanGool, L., Dewaele, P., and Oosterlinck, A. (1983) Texture analysis anno. Computer Vision Graphics and Image Processing, 29, 336-357.

Vilnrotter, F. M., Nevatia, R., and Price, K. E. (1986) Structural analysis of natural textures. IEEE Transactions on Pattern Analysis and Machine Intelligence, PAMI-8, 76-89.

Voorhees, H., and Poggio, T. (1988) Computing texture boundaries from images. Nature, 333, 364-367.

von der Heydt, R. and Peterhans, E. (1989) Mechanisms of contour perception in monkey visual cortex. I. Lines of pattern discontinuity. Journal of Neuroscience, 9, 1731-1748.

Wang, R., Hanson, A. R., and Riseman, E. M. (1986) Texture analysis based on local standard deviation of intensity. In Proceedings of the Seventh International Conference on Pattern Recognition, IEEE, 482-488.

Wertheimer, M. (1923) Untersuchungen zur Lehre von der Gestalt II. Psychologische Forschung, 4, 301-350.

Wilson, H. R., and Bergen, J. R. (1979) A four mechanism model of threshold spatial vision. Vision Research, 19, 19-32.

Wolfe, J. M., Cave, K. R., and Franzel, S. C. (1989) Guided search: An alternative to the feature integration model for visual search. Journal of Experimental Psychology: Human Perception and Performance, 15, 419-433.

Zucker, S. (1976) Region growing: Childhood and adolescence. Computer Graphics Image Process., 5, 382-399.

Zucker, S. W., Rosenfeld, A., and Davis, L. S. (1979) Picture segmentation by texture discrimination. IEEE Transactions on Computers, C-24, 1228-1233.

Commentary

An Essay on Texture: The Extraction of Stimulus Structure from the Visual Image, E. A. Essock

RUTH KIMCHI & MORRIS GOLDSMITH
University of Haifa

In his chapter, Essock provides an enlightening discussion of the role of texture in visual image segmentation, focusing on the extraction of texture information from the visual image and its utilization in the segmentation task. Bringing together work from both the psychological and computer vision research literatures, he first clarifies the term "texture" and its relationship to other types/levels of stimulus structure, and then goes on to consider the nature of the segmentation task itself. He surveys the leading models of texture-based segmentation in human and computer vision, bringing out their similarities and differences. In doing so, he critically analyzes the distinction between "first-order feature" and "linear-filter" models of the segmentation process, arguing that despite superficial, implementational differences between the two types of models, both may be seen to involve the extraction of textural features from the stimulus image. The differences boil down to the exact lists of detected features. Thus, Essock concludes that the fundamental issue in texture-based segmentation remains which types of stimulus structure are extracted by early processing in the human visual system.

The problem of identifying textural features provides a good example of how the mutual constraints between structure and process may be utilized in psychological research. In our chapter (Kimchi & Goldsmith, this volume), we emphasized that although many types of stimulus structure may be physically or logically present in the visual stimulus, those actually used in the human perceptual system can be identified through their observed processing consequences across a variety of experimental tasks. This, of course, has been the approach taken by Beck (1982), Julesz (1981), Treisman (1985) and others working within a psychological research framework. Thus, as Essock points out, the relative lack of constraint in computer vision research may have its advantages, but as far as modeling human perception is concerned, it is also a weakness.

Note that a fundamental assumption of the entire body of research presented in the chapter is that the initial stimulus may be properly described in terms of a static retinal image, and that this "booming, buzzing, confusion" provides the basis for all further processing. It may be worth keeping in mind that other perceptual psychologists, especially those sympathetic to a more ecological, Gibsonian point of view, would disagree with this basic characterization. Perhaps the parallel between human (retinal) and computer (pixel) images, while enabling the extensive cross-fertilization between human and computer vision research, has at the same time unduely limited the types of stimulus descriptions and processes being considered. We have no immediate answer to this question, but feel it does warrant further thought.

Beck, J. (1982). Textural segmentation. In J. Beck (Ed.), Organization and representation in perception. (pp. 285-317). Hillsdale, NJ: Erlbaum.

Julesz, B. (1981). Textons, the elements of texture perception and their interactions. Nature, 290, 91-97.

Treisman, A. (1985). Preattentive processing in vision. Computer vision, Graphics, and Image Processing, 31, 156-177.

Percepts, Concepts and Categories
B. Burns (Editor)

2

The Nature of Selectivity in Early Human Vision

JAMES T. ENNS
University of British Columbia

I. OVERVIEW

The doctrine of attention is the nerve of the whole psychological system, and that as men judge of it, so shall they be judged before the general tribunal of psychology.

E. B. Titchener (1908)

The greatest obstacle to the study and understanding of vision may well be its immediacy--as long as our eyes are open a coherent and complete view of the world lies before us. All of this is accomplished without a hint of effort on our part. It is only when the researcher attempts to model the visual system with mathematical tools (e.g., Tsotsos, 1988; Grossberg, 1988) or to implement a functioning visual system with hardware (e.g., Horn, 1986; Marr, 1982; Nevatia, 1982) that the enormous complexity of vision becomes apparent. One of the hard-learned lessons of the past quarter-century of vision research is that vision does not involve simply "copying" the visual world in some form for inspection by the mind. Rather, our percepts are based on a highly selective set of operations that are applied to the images presented to our eyes by the physical world.

In this chapter, I will consider the selective nature of processing at the earliest stages of vision in humans. However, before doing so I would like to set the stage by discussing general issues surrounding selectivity. What is selectivity? Why is it necessary for every information processing system to be selective? These questions will be addressed in the first section of the chapter. The second section will use this framework to address the question of selectivity in early human vision. What is early vision? How is it selective? What role does early vision play in the larger scheme of a general-purpose biological visual system? The conventional answers to these questions will be reviewed, followed by answers given in recent work from my laboratory. This will result, in the third section, in a proposal for a revised view of early vision.

II. SELECTIVITY IN VISUAL INFORMATION PROCESSING

Selectivity refers to the processing of some information at the expense of other information. Broadly speaking, visual selectivity can be based on *spatial* considerations (information must be sampled over some spatial region), on *temporal* considerations (any sampling scheme has certain temporal parameters), and on *geometric properties* (the sampling of information implies some form of coding). But why must the human visual system be selective in these ways? Why is it not possible to simply consider all the information available to the eye? Within the field of psychology, this question has traditionally received one of two answers. The first begins by considering the stimulus that must be interpreted by the system. The second looks for an answer in the nature of the processing system. However, more recently a third answer has been gaining prominence. It concerns the nature of the task that the visual system must accomplish. In this section, I will consider these three answers in turn.

A. The Stimulus--Why Must Information Be Selected?

The logical need for selection becomes apparent when two observations about the visual stimulus are juxtaposed. First, the potential information available to a visual system is unlimited. Second, a visual organism uses this information as the basis for action. Therefore, to be useful to an organism behaving in real time, the visual system must choose to process information that is relevant for behavior and disregard other information. What is the nature of the information from which this choice is made? Is it already structured, or must the visual system impose its own structure? This question has divided perception psychologists for most of this century.

J. J. Gibson (1966) is eloquent in representing the view that the stimulus is inherently structured and that efficient perception depends entirely on registering the appropriate structure. He says

> *... the available stimulation surrounding an organism has structure, both simultaneous and successive, and this structure depends on sources in the outer environment. If the invariants of this structure can be registered by a perceptual system, the constants of the neural input will correspond to the constants of the stimulus energy... (p. 663)*

For Gibson, the environment is organized independently of any perceptual system. It is the task of a perceptual system to tune itself to the existing structure. This tuning occurs at a number of temporal levels ranging from the evolution of species, to the development of individuals, to the learning of a skill, and to the adaptation of specific mechanisms in response to changing environments.

The issue of selectivity in processing, in this view, can be reduced to the study of two questions. The first is the question of what structure is available to the perceptual system in a given environment. The second is the question of how an organism becomes attuned to the physical laws that are relevant in a given situation. Consider the consequences of this view for the study of the everyday problem of catching a thrown ball. We could begin by observing that the trajectory of the ball is determined by the gravitational and mechanical laws governing rigid objects in a three-dimensional space. These laws would have to be spelled out and examined in order to determine their interrelations. Our interest in how humans learn to "catch the ball" would then lead us to the question of how they learned these laws of motion. Importantly, for this approach, we would not be concerned with whether and how subjects formed internal representations of ball-throwing incidents, nor in whether they formed expectations from these incidents for future use. The latter questions are not of interest because the selectivity of perception is thought to be completely governed by the physical world.

An alternative position on the role of the stimulus in determining perceptual structure is represented by Garner (1983), whose main interest is in the relations between the structure of the stimulus and the nature of the processes underlying perception. This can be seen in his statement that

> *... when we are interested in process, we are also interested in the stimulus properties that lead to process differences, and when we are interested in stimulus structure, we are interested in processing consequences of the differences in structure. (p.2)*

The central task for Garner is to determine those stimulus dimensions for which selective processing is possible and those for which it is not. Out of this question has emerged his taxonomy of *integral* and *separable* dimensions. Separable dimensions are those which can be attended to without interference from orthogonal variation in other dimensions (e.g., shape and hue). Integral dimensions are those for which such selective attention is not possible (e.g., brightness and hue). A consequence of such a "failure of selection" is that integral stimuli can often be recognized on the basis of an emergent property, with the consequence that they can be responded to more efficiently than stimuli defined by a correlation of separable dimensions.

Garner clearly sees the structural laws governing various stimulus dimensions to be a consequence of the processing system rather than strictly reflective of the physical laws underlying the visual world. In fact, it is this distinction between logical separability and empirical separability that his taxonomy makes explicit. For example, although stimuli can be defined equally well by orthogonal variations in shape and hue or in brightness and hue, the finding that these variations are not treated in an equivalent manner by the visual system tells us something about its design. The structure of the processing system is apparent in the visual structure that is perceived.

B. The Perceptual System--Why Must the Perceiver Be Selective?

The second traditional answer to the question of selectivity points directly to the processing system. An inescapable fact of biology is that sensory and perceptual systems are resource-limited. The brain is an organ consisting of a large but finite number of neurons. Each perceptual system has both spatial and temporal limits on the kind of information it can process.

One way to begin to appreciate the complex factors influencing processing selectivity is to imagine the decisions that must be made in designing a new visual system. Consider first the boundary conditions under which the receptors should operate. Which portion of the electromagnetic spectrum should they be sensitive to? What will be their spatial and temporal resolution? Having made decisions on these questions, there are even more difficult decisions to be made downstream. What kind of geometrical codes will be used to reduce the information to a manageable and useful amount? How will these tokens be processed? What rules will apply when multiple sources and kinds of information are available?

Broadbent (1958) has inspired much of the modern interest in the selective nature of human information processing. His ideas have been strongly influenced by information theory, in particular, the notion that humans can be studied in the same way that one would study the characteristics of an electronic communication channel. Broadbent's notion of a *filter*, that is, a mechanism to switch the flow of information from one channel to the other, has guided much of the research on attention conducted in the past two decades.

Treisman's (Treisman & Gelade, 1980; Treisman, 1988) well-known feature integration theory is a direct descendant of Broadbent's simpler filter theory. The drawing of her theory shown in Figure 2.1 is proposed as an explanation of human performance in visual search and object identification tasks. The "filter" in this theory is now a "spotlight of attention" that can be directed by the stimulus in a bottom-up fashion, or by voluntary control in a top-down fashion. For example, bottom-up visual search is possible when search items are defined by simple feature differences in the image. These features can be detected in parallel over the visual field simply by checking for unique activity in the relevant spatiotopically organized "feature maps." Top-down guidance of visual search is required when the target item is defined by a conjunction of simple features in the display. However, top-down search is also limited by severe spatial and temporal constraints. It can only operate over a limited region of the visual field at a time, and therefore, must move serially from location-to-location in order to cover the entire field of view.

In this view, then, the issue of processing selectivity becomes the study

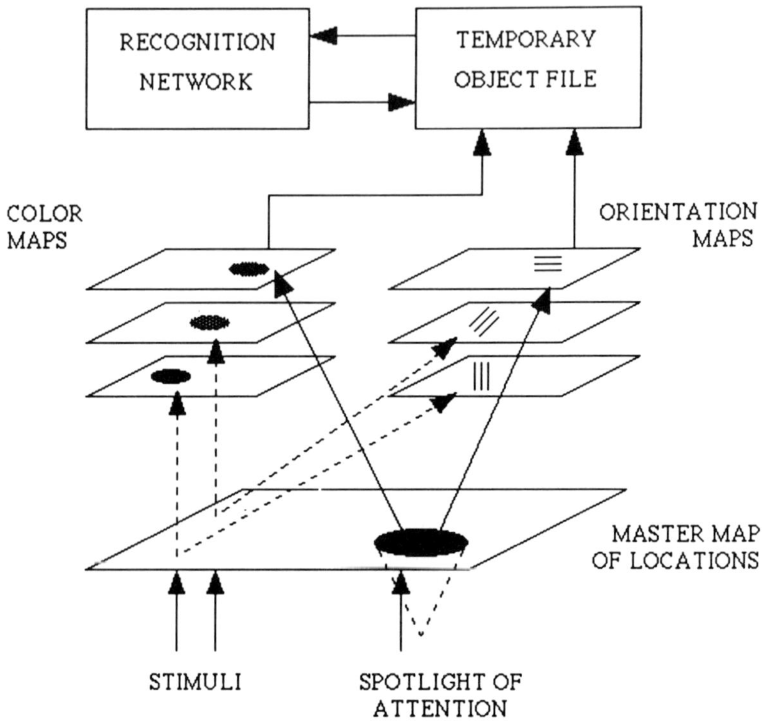

Figure 2.1. A sketch of Treisman's (1986, 1988) feature integration theory.

of the capacity limits of the processing system. Historically, these have been studied in three ways (Kinchla, 1980): by examining information tradeoffs in situations where multiple sources of information compete for limited processing resources (e.g., visual search), by examining situations in which sources of information are difficult to ignore (e.g., the Stroop task), and by studying the manner in which a processing system with limited resources switches from one source of information to another (e.g., dichotic listening).

C. The Task--How Does the Task Impose Selectivity?

The third reason for selectivity resides in the *function* of the visual system for the larger organism (Marr, 1982) or in the *task* being performed by that system (Ullman, 1984). Many visual tasks logically require that some information be processed before other information. Take for example, the simple problem of determining the spatial relation of OUTSIDE-INSIDE for the two objects in

Figure 2.2. This is a relation that is relevant to many everyday perceptual tasks. For problem A, the immediacy of our perception may deceive us into thinking that this is a trivial relation to determine. However, solving the task for problem B shows that more is involved. Correct responses for this type of problem are slow and effortful, suggesting a sequence of processing stages (Ullman, 1984).

Ullman (1984) emphasizes that the problem of selectivity is not peculiar to biological perceptual systems. He argues, for instance, that careful study of the complexity of the task faced by a perceptual system (be it biological or electronic) will give us a new appreciation for visual processing.

> *The apparent immediateness and ease of perceiving spatial relations is deceiving. As we shall see, it conceals in fact a complex array of processes that have evolved to establish certain spatial relations with considerable efficiency ... (p. 99)*

Ullman goes on to argue that, in many cases, a mechanism similar to a serial spotlight of attention is needed to solve spatial relations. It is important to note that this reason for proposing a spotlight differs fundamentally from that of Treisman (see above). The spotlight is proposed *not* because of human capacity limitations in visual processing, but because the task itself would be insoluble without it, even with unlimited processing resources. The important point is that

A B

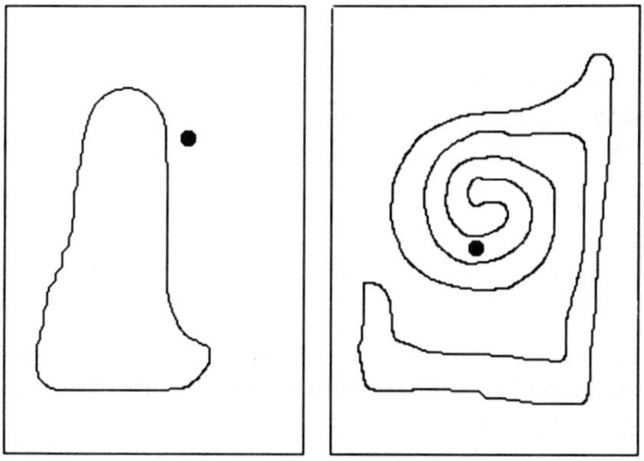

Figure 2.2. Is the dot INSIDE or OUTSIDE the closed figure? In A the answer is immediate and effortless, whereas in B the same answer is arrived at only after time-consuming and effortful processing.

certain steps in the task (e.g., object individuation) must be completed before others can be begun (e.g., determination of spatial relations between objects). The determination of spatial relations themselves require sequential steps such as *boundary-tracing* or *region-coloring* (Ullman, 1984).

Processing selectivity is studied, in this view, by first determining the function performed by a processing system. Next, the minimal steps and processes necessary for solving a task should be determined. Once these two steps have been undertaken, it is useful to study the particular way in which a task is solved by an actual perceptual system. However, in so doing, it is important to distinguish clearly between the formal analysis of the task, which must apply to all solutions to the task, and the particular algorithm for the solution that may be used by a given visual system (Marr, 1982).

To summarize, in this section I have reviewed three perspectives on why information processing is selective: there is unlimited information in the stimulus, there are capacity limitations in the processing system, and many visual tasks logically require it. I turn now to a consideration of one subsystem of human vision--that of early vision. It is my view that any processing system will be understood more readily and more completely if it is studied from the benefit of all three perspectives. I will support this view by drawing from work on early vision that is being conducted in my laboratory.

III. EARLY VISION

Since the days of Helmholtz (1867/1962), vision researchers have distinguished between early and later stages of visual processing. For researchers of physiology , "early" vision is usually used to refer to processes that are located in the retina and up to the first stages of cortical processing; later vision concerns the latter stages of cortical processing (Zucker, 1987). For psychophysicists,"early" refers to preattentive processes, and "later" to attentive processes (Treisman, 1986). Computational vision researchers use "early" to refer to bottom-up (data-driven) processes and "later" to refer to top-down (knowledge-influenced) processes (Tsotsos, 1988). The assumption that all of these views share is that "early vision" describes high-speed, specialized biological hardware (sometimes called "wetware") that operates automatically and in parallel across the visual field. "Later vision" refers to the slower, but vastly more varied, visual routines that can voluntarily be brought to bear on a visual problem. Although there have been some efforts to link the physiology and psychophysics of early vision (Treisman, Cavanagh, Fischer, Ramachandran, & von der Heydt, 1990; Zucker, 1987) and the psychophysics and computational aspects (Tsotsos, 1988), these efforts have been largely speculative to date. In this paper I will focus on the psychophysical data.

A. The Conventional View of Preattentive Vision

Conventional theories of preattentive vision tend to stress its non-selectivity rather than its selectivity (Kahneman, 1973; Neisser, 1967; Sperling, 1960). Preattentive vision is said to be of unlimited capacity, to operate in parallel over the entire visual field , and to register visual information in a non-categorical form. However, it is also clear that these claims are not meant to be taken literally--they are made in order to contrast the preattentive system with the attentive system, which is limited in capacity, limited in its spatial extent, and highly categorical. Preattentive vision itself is described as a set of operations that transform the visual input, rapidly and in parallel, into primitive visual features (or tokens) that can be used by later visual processes. Therefore, even in the conventional view, preattentive vision is selective in a temporal sense (only rapidly computable features will be represented) and in a categorical sense (light intensity is transformed into feature information). It is only in the spatial sense (processing occurs simultaneously over the visual field) that the conventional view sees preattentive vision as being truly non-selective.

Some of the features believed to be the primitives of form perception in this view are properties of elongated bars such as orientation, length, curvature, and motion direction (Beck, 1982; Julesz, 1984; Neisser, 1967; Treisman, 1986; Treisman et al. , 1990). The underlying assumption is that preattentive vision can operate rapidly and in parallel only because it has these very simple properties to compute and because each type of element is represented in a different spatiotopic map. The acknowledged price for this combination of speed, simplicity, and independence, however, is that only a very rudimentary analysis of the visual world can be accomplished preattentively. For example, in order for even simple spatial relations to be determined between features (e.g., relative location), slower and more costly attentive processes must be invoked.

One of the main psychophysical methods used to explore preattentive vision is the visual search task (Julesz, 1984; Neisser, 1967; Treisman, 1986). In this task, observers try to determine as rapidly as possible whether a target item is present or absent in a display. Target items that are detected rapidly and show little dependence on the total number of items in the display are assumed to contain a distinctive feature at the preattentive level. No attentive operations are required for their detection--the target simply "pops out" of the display. In contrast, other targets are more difficult to find, with search time depending strongly on the total number of items in the display. These targets are considered to be conjunctions of elementary features, requiring the serial operations of the attentive system for their detection.

B. Reassessment of the Conventional View

Recent findings are beginning to show that this picture of early vision is too simple and must be revised in several important ways. To begin with, the dichotomy between serial and parallel processes has been challenged by the observation that search rates range continuously in practice from very fast (i.e., less than 10 ms per item) to very slow (i.e., more than 100 ms per item). Although several attempts have been made to account for this finding while still holding to two separate subsystems (Treisman and Souther, 1985; Julesz, 1986), these efforts are now leading to proposals that search rates reflect processes that vary continuously in speed as a function of target and distractor similarity (Treisman & Gormican, 1988; Duncan & Humphreys, 1989; Humphreys, Quinlan & Riddoch, 1989; Wolfe, Cave, & Franzel, 1988).

The view that early representations are geometrically simple and fragmented has been challenged by reports that rapid search is possible under many circumstances for targets defined only by a conjunction of their features. These include conjunctions of binocular disparity and motion (Nakayama & Silverman, 1986), motion and form (McLeod, Driver, & Crisp, 1988), saturated colors, large forms, and distinctive orientations (Treisman, 1988; Wolfe, et al. , 1989), and spatial relations among line elements that are sufficiently long (Duncan & Humphreys, 1989; Humphreys, Quinlan & Riddoch, 1989).

Finally, there are numerous studies showing the context sensitivity of visual features. For instance, the detection of texture boundaries is influenced not only by the presence of distinctive feature differences, but also by the spatial relations among those features (Enns, 1986; Julesz, 1986; Nothdurft, 1985; Taylor & Badcock, 1988). Irrelevant variation of features within texture regions has also been shown to slow down texture segmentation and correlated variation among features has been shown to speed it up (Callaghan, 1989; Callaghan, Lasaga, & Garner, 1986). These findings are clearly not consistent with the independent spatial coding of visual primitives.

Given this discrepancy between the conventional view and research findings, where should one begin to look for order in early vision? How might one begin to narrow the potential list of features that might be registered there? What level of visual complexity can be processed, even in principle, by rapid parallel processes? In my laboratory, we confronted these questions by considering them from the perspective of each of the three different approaches to the study of selectivity outlined in the preceding section.

We began by considering the possible limitations placed on early vision by the stimulus and came away with very few insights. "Stimulus" seemed simply to be a term that included all the possible ways that one could parse the visual world into features. Even a cursory glance at the range of visual features that have been considered for various schemes in computer vision (e.g., Horn, 1986; Marr, 1982; Nevatia, 1982) suggested there were too many of these to expect a winnowing of alternatives from this perspective alone.

An examination of the processing system on its own also did not help to narrow the field. Neurophysiological studies of visual cortex had suggested some of the features that had already been proposed (e.g., oriented edges, gabor filters, moving edges). Psychophysical evidence had suggested others (e.g., lines, colors, volumetric primitives). However, the irksome problem here seemed to be that there was no end in principle to the kinds of features that might yet be proposed and tested. We had hoped, perhaps somewhat naively, that the question of primitive visual features could be placed on a firmer footing.

A new look at the possible primary function of early vision, however, gave us stronger direction. We reasoned that when humans and other mammals initially encounter a visual scene in a naturalistic setting, they are usually involved in some activity for which this scene provides biologically useful information (Enns, 1990). For instance, they may be navigating over an uneven terrain, they may be in search of food, or they may be on the lookout for predators. Therefore, we thought it was at least worth considering the idea that early vision had evolved as a high-speed early-warning system for the delineation of *objects* in a *three-dimensional world*. If early vision was able to recover even a limited number of properties of the three-dimensional scene, it would be able to guide immediate actions such as eye movements, as well as the more flexible processes further down the visual stream. If this were true, then differences in object attributes might be the most natural basis for visual search.

But which object attributes should be considered? To answer this question we found we had to retrace our steps and once again take up the problem of how the nature of the stimulus and the processing system might constrain the list of potential object attributes registered in early vision. We began by asking, to what extent are the rapid and parallel processes of human early vision (i.e., the processor) able to apprehend the shape and layout of objects in a three-dimensional world (i.e., the stimulus)? A deceptively simple-minded, but extremely important, clue to this question is that the visual system gains knowledge about the three-dimensional environment through the two-dimensional image of light projected onto its receptor surface. That is, the three-dimensional world of objects (which I will hereafter call the *scene*) can be processed by the early vision system only after light reflected from it has been projected onto the retinal receptors of the eye (hereafter the *image*) following the laws of optics and projective geometry.

C. Scene-Based Versus Image-Based Features

Our analysis of the possible function of early vision (i.e., object description), the nature of the early processing system (i.e., the rapid, parallel processes of early vision), and some important characteristics of the stimulus (i.e., light from the scene is projected onto an image), suggested to us that a plausible place to begin looking for features of early vision might be at the level of scene-based object properties. It is worth emphasizing how different this is from the conventional view of early vision. There the assumption is that, in order to be rapid, early vision must be based on the simplest possible geometric elements. Our reasoning suggested that this assumption might be too strong and unwarranted. Early vision might be able to process more complex properties, if these properties could be computed rapidly and if they were relevant to the task of extracting three-dimensional structure from an image.

What are the geometric constraints on the problem of recovering three-dimensional scene information from an image? In general, if a set of opaque objects is illuminated by a distant point source, the two-dimensional array of image intensities is completely determined by four factors: (i) the direction of lighting, (ii) the orientation, shape, and location of surfaces (iii) the reflectance properties of the surface, and (iv) the direction from which the scene is viewed. However, the complete recovery of all these quantities from a single image is impossible in principle, since there exists a large set of possible scenes that would give rise to any particular image. In order to recover a unique set of scene-properties it is necessary to simplify the problem by imposing constraints. The basic idea is that if the constraints are well-chosen, they will permit one scene candidate to be selected from the large set of possible scenes, thereby establishing a one-to-one correspondence between scene and image.

The scene-recovery problem can be simplified in essentially three ways. First, the number of possible scenes can be limited so that there is only a small equivalence class of scenes for each possible image. Second, several of the scene properties can be fixed at a constant value, leaving a smaller number of scene properties to be recovered. Finally, the correspondence between image and scene properties can be limited--in the extreme, every possible image property would correspond to a unique scene property. Thus, any analysis of the scene-recovery process, including that accomplished by early vision, will have to address three questions: (i) What are the constraints on the scene domain? (ii) What are the constraints on the image domain? and (iii) What are the possible correspondences between image and scene properties?

D. The Rapid Recovery of Scene-Based Properties in Early Vision

In this section I will summarize a series of studies that have begun to map out the scene-recovery capabilities of early human vision. The reader should be cautioned that this report will be incomplete, as the work is still underway. Furthermore, many of the most interesting results lead to other questions that remain unanswered. However, I take some comfort in the observation that these questions would not have been asked within the conventional view of early vision.

Overview of general method and data analysis. The method used in each experiment was the well-known visual search task (e.g., Treisman & Gormican, 1988; Treisman, 1988; Wolfe, Cave, & Franzel, 1989). The observers' task on each trial was to search for a single target item among a total of 1, 6 or 12 items, or 2, 8, or 14 items, depending on the experiment. Upon detecting the presence of the target, the observer made a manual response by pressing a key. If no target was detected, an alternative key was pressed. The target was present on a random one-half of the trials, distributed randomly on an imaginary 4 x 6 grid subtending approximately 10° x 15°. Each item subtended less than 1.5° in any direction and was randomly jittered in its grid locations by +/- 0.5° to prevent search being based on item collinearity. In the experiments to be reported, target and distractor items were composed of black/white images (see Figures 2.3 to 2.11).

A Macintosh computer was used to generate the displays, control the experiments, and collect the data (Enns, Ochs, & Rensink, 1990). Each trial began with a fixation symbol lit for 500-750 ms, followed by the display, which remained visible until the observer responded. The display was followed by a feedback symbol (plus or minus sign), which served as the fixation point for the next trial. Observers were instructed to maintain fixation and to keep errors below 10%. Five to ten observers with normal or corrected-to-normal vision completed 4-6 sets of 60 test trials in each condition of each experiment.

Although observers tended to be quite accurate overall (each observer made fewer than 10% errors on average), there were systematic differences in accuracy. Consistent with other reports, target present trials led to more errors than target absent trials (Klein & Farrell, 1989; Humphreys et al. , 1989). Most important for present purposes, however, was the observation that errors tended to increase with response time (RT), indicating that observers were not simply trading accuracy for speed.

RT data were analyzed the same way in each experiment. First, simple regression lines were fit to the target present and target absent data for each observer. Second, the estimated slope parameters were submitted to analyses of variance in which Condition (A, B, etc.) and Trial Type (present, absent) were the effects of interest. Finally, Fisher's *LSD* tests determined the reliability of pairwise slope differences in the context of significant main effects and interactions. The reported *t*-tests, therefore, are tests of differences in RT slope based on the pooled error variance and degrees of freedom from the main analysis.

As pointed out earlier, there is no sharp boundary between fast and slow rates of search. In this paper, I will follow the convention of using "rapid search" to refer to target-present search rates (RT slopes) of less than 10-15 ms per item. This speed is well below accepted estimates of attentional movement across the visual field (Jolicoeur, Ullman, & MacKay, 1986; Julesz, 1984; Treisman & Gormican, 1988).

1. Direction of lighting

The experiments in this section show that early vision is sensitive to several important properties associated with scene lighting. Subjects searched for target items defined by the spatial relations between shapes differing in luminance (Enns & Rensink, 1990a). As can be seen in Figures 2.3 and 2.4, some target and distractor items corresponded to projections of simple three-dimensional blocks, while others were not as easily interpreted as three-dimensional objects. These experiments were designed to ask first, whether early vision would distinguish among these items, and if so, which scene properties it was sensitive to.

Experiment 1 demonstrated that early vision was indeed sensitive to some spatial relations among image features. Consider Condition A in Figure 2.3, where items correspond to blocks differing in both orientation and lighting. Observers' search rates were relatively rapid, regardless of the number of items. In contrast, observers were much slower to find the target when items did not correspond to three-dimensional objects --search for similar relations of luminance and location among the polygons in two-dimensional items was much slower (see conditions B and C).

Experiment 2 tested for a similar sensitivity to luminance relations among the polygons, as shown in the upper panel of Figure 2.4. (Note that in this and in all future figures containing data, only the average search rate will be presented.) Search was rapid this time for items corresponding to three-dimensional blocks differing only in lighting direction (Condition A). Similar items that could not be given a three-dimensional interpretation resulted in much slower search (Conditions B and C). Therefore, these results show that rapid

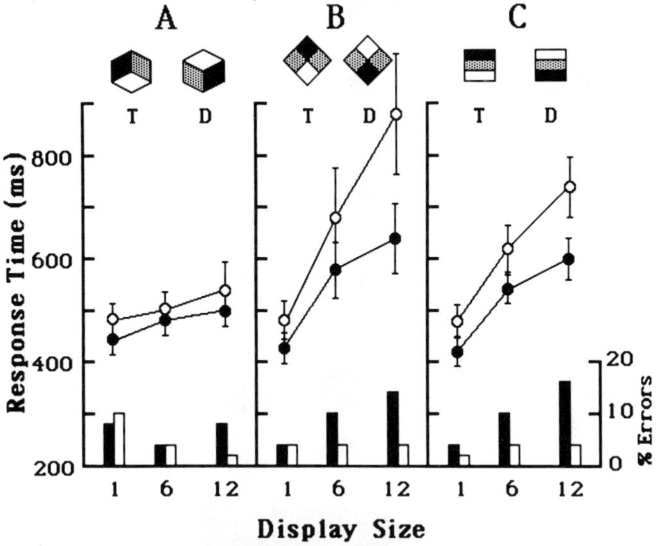

Figure 2.3. Experiment 1: Spatial relations among shapes of different luminance lead to rapid search when they can be easily interpreted as three-dimensional objects (Enns & Rensink, 1990a). The target (T) and distractor (D) items in the three conditions (A-C). Filled circles and bars represent target-present trials; open circles and bars represent target-absent trials. Response time values are mean +/- SEM. Display size indicates the number of items present in a trial. Average search rates were 6 ms per item for both target-present and target-absent in Condition A, 19 and 35 ms per item in Condition B, and 15 and 23 ms per item in Condition C.

search is possible when items can be interpreted as three-dimensional objects and are distinguished only by differences in lighting.

Experiment 3 used the diagnostic of search asymmetry (Treisman & Gormican, 1988) to examine early vision's sensitivity to lighting direction more closely. A search asymmetry is obtained when a simple exchange between the target and the distractor item causes a large change in search rates. The usual interpretation of such a finding is that the easily found item contains a feature not

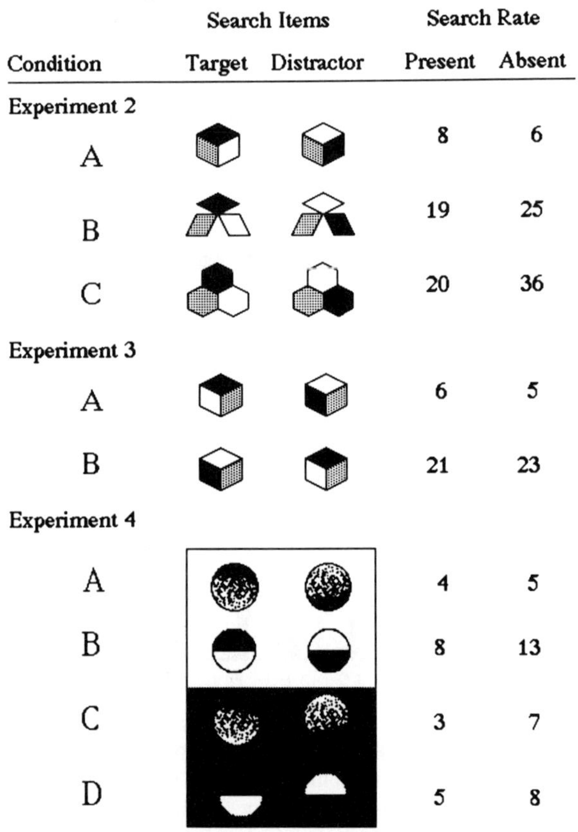

Figure 2.4. Experiment 2: Differences in the direction of lighting result in rapid search when items can be interpreted as objects (Enns & Rensink, 1990a). Experiment 3: A search asymmetry indicates that early vision registers deviations from top-lighting (Enns & Rensink, 1990a). Experiment 4: Shading gradients can be used by early vision to achieve object-background segregation and to detect surface curvature (Aks & Enns, in press). Display sizes were 1, 6, and 12 in Experiments 2 and 3; 2, 8, and 14 in Experiment 4.

present in the item that is slower to find. In this experiment, the conditions differed only in the roles the two items played as targets and distractors (see middle panel in Figure 2.4). This resulted in the primary direction of lighting being from above in condition A (all distractors were top-lit), and from below in Condition B (all distractors were bottom-lit).

The results showed that when the distractors were lit from above, search for a bottom-lit target was quite easy (Condition A). With the opposite arrangement of lighting direction, search was much more difficult (Condition B). It can therefore be concluded that early vision processes differences in lighting direction most efficiently when the primary source of light is overhead in the scene. This is consistent with the naturalistic environments that human visual systems have evolved in.

Experiment 4 examined the sensitivity of early vision to luminance gradients in the image (Aks & Enns, in press). Luminance gradients are of interest because they convey information about a number of intrinsic properties associated with scene lighting (see lower panel in Figure 2.4). One such property is surface convexity--a curved surface will give rise to an image gradient because light strikes different regions of the surface at different angles. A second property is object-background segregation--a curved object on a background of similar reflectance will give rise to a local change in contrast polarity (the top of the object will be lighter than the background, the bottom will be darker). The experiment tested whether each of these properties was sufficient on their own to allow for rapid search, or whether both features were required.

The influence of apparent curvature was tested by comparing search for a luminance gradient (Condition A and C) with search for an abrupt change in contrast (Condition B and D). The influence of contrast polarity was tested by comparing search under two background conditions (Conditions A and B were a white background, and therefore contained no change in contrast polarity; Conditions C and D were on a gray background, producing a reversal of contrast polarity within each item). The results showed that each factor alone was sufficient to yield rapid search. The luminance gradient resulted in rapid search even in the absence of a reversal in contrast polarity (Condition A), while the contrast polarity reversal caused rapid search even in the absence of a luminance gradient (Condition D).

Taken together, these experiments show that early vision is sensitive to scene lighting. Large differences in the direction of lighting are readily detected (Experiments 1 and 2), there is a strong preference to assume that the light source is overhead (Experiment 3), and the shading gradients that result from scene lighting are used by the early vision system to achieve object-background segregation and to detect surface curvature (Experiment 4).

These findings are relevant to computational models for the recovery of lighting direction in a scene and object shape from image shading. For instance, one algorithm for finding the direction of light is able to do so by using only the orientations of the lines and the intensities of the three regions at each vertex in the image (Horn, 1977). Our results indicate that early human vision can solve the same problem, but only when the vertex appears in the context of a three-dimensional object. Existing algorithms for shape-from-shading range from those that examine only local regions of luminance gradation in the image (Horn, 1977; Pentland, 1984), to those that also make use of distributed information such as the outlining contour of an item (Grossberg, 1983; Koenderink & van Doorn, 1980). Our data suggest that a local analysis of contrast polarity is sufficient, as is an analysis of the luminance gradient in the absence of a reversal in contrast polarity.

There are also physiological implications of these findings. If complex scene properties can be computed rapidly when they are environmentally relevant, then specialized biological hardware (i.e., "wetware") must exist for their computation. Simulated neural networks designed to perform shape-from-shading analysis suggest that single neurons with properties similar to those in primary visual cortex are sufficient for this task (Lehky & Sejnowski, 1988). A simulated neural network, trained on luminance gradients as "input," was able to learn the correspondence between simple luminance gradients in the image and surface curvature in the scene. Inspection of the behavior of the "hidden" units (those lying between the input and output layers), revealed a striking resemblance to the simple edge-detecting neurons in the primary visual cortex of cat and monkey. This suggests that units at the earliest stages of cortical processing may already be tuned to surface curvature.

2. *Direction of viewing*

The experiment reported in this section was designed as an initial investigation into the sensitivity of early vision to the property of viewing direction. It was inspired by the often-reported finding that objects are easier to apprehend, at least under attentive viewing conditions, when they are below the line of sight than when they are above it (e.g., Kaufman, 1974). Experiment 5 asked whether these effects could already be seen in early vision (Enns & Rensink, 1990a). Items in

	Search Items		Search Rate	
Condition	Target	Distractor	Present	Absent
A			10	11
B			18	23

Figure 2.5. Experiment 5: Early vision is indifferent to viewing direction (Enns & Rensink, 1990b). Display sizes were 1, 6, and 12.

conditions A and B were identical to the same conditions in Experiment 3, except that they were now viewed from below the line of sight (see Figure 2.5). Although such viewing conditions resulted in somewhat slower baseline responses than viewing from above, viewpoint had no significant effect on search slopes. Viewed from below, the items replicated the pattern of results found for the items viewed from above in Experiment 3.

Other results that we have obtained show that these effects generalize to blocks rotated 60° and 90° from those used in this experiment. We can conclude then, at least for the limited set of objects that we have tested, that early vision is indifferent to the direction from which an object is viewed. A strong statement concerning this scene property, however, will require tests for a larger range of objects and under a wider variety of viewing conditions. Nonetheless, it is instructive to consider the possibility that object representations may be viewpoint independent, even at these early stages of processing (Biederman, 1985; Marr, 1982).

3. Object orientation

In these experiments we systematically examined the property of three-dimensional orientation. Many of the target and distractor items we used were projections of rectangular objects differing in the three-dimensional orientation of their principal axes. These items always contained the same set of image-based lines and polygons. Therefore, if rapid search was possible, it would depend necessarily on the spatial relations among item elements that capture the three-dimensional orientation of the corresponding objects.

Condition	Search Items: Target / Distractor	Search Rate: Present	Search Rate: Absent
A		6	9
B		9	12
C		12	14
D	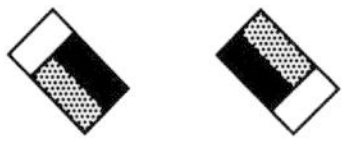	26	39

Figure 2.6. Experiment 6: Differences in three-dimensional orientation are detected quite rapidly even when no lighting information is given (Enns & Rensink, 1990b).

Experiment 6 demonstrated that rapid search was possible for line relations consistent with differences in three-dimensional orientation (Enns & Rensink, 1990b). We began by examining the relative contributions of shading and line relations to visual search. The shaded polygons in Condition A (Figure 2.6) could be detected very rapidly, regardless of how many distractor items were in the display. This replicated the finding in Experiment 1 that blocks differing in lighting and orientation could be rapidly detected. Systematically omitting the shading from these polygons in Conditions B and C eliminated lighting direction as a diagnostic feature--yet the search rate increased only slightly. This small increase lay in sharp contrast to the items in Condition D. Here luminance relations were preserved, but three dimensional orientation differences were eliminated. The consequence was that search was much slower.

To test the status of the three-dimensional orientation feature in early vision more rigorously, we compared the influence of image-based and scene-based orientation directly. In Experiment 7, the target item in each condition corresponded to a block oriented upward and to the left of the line of sight (Enns & Rensink, 1990b). Rotating and reflecting the target image generated seven different sets of distractors, with each condition having different diagnostic orientations (see Conditions A-G in Figure 2.7). For example, none of the lines in the target in Condition A differed from the lines used in the distractor. Only a difference in three-dimensional orientation distinguished the target item--a rotation about the x-axis (vertical) and a rotation about the y-axis (horizontal). In contrast, the target in Condition B differed from the distractor in two ways. First, differences in the image orientations of the component lines distinguished the target from the distractor. Secondly, the target differed in its

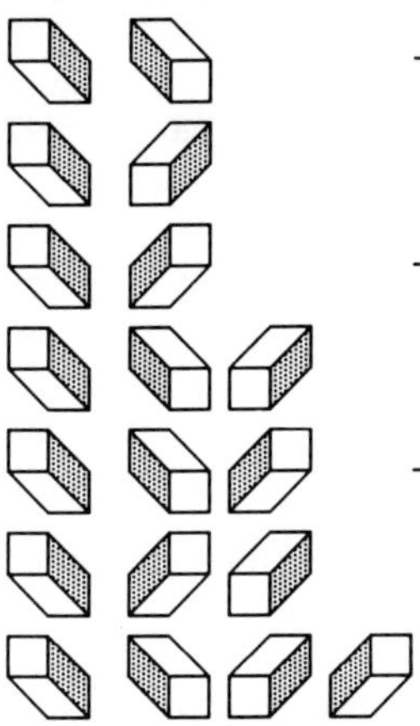

	Target Distractors	Features H	V	I	Present	Absent
A		+	+		6	7
B			+	+	-1	0
C		+		+	-1	0
D			+		7	17
E		+			15	47
F				+	1	17
G					22	87

Figure 2.7. Experiment 7: Three-dimensional orientation and two-dimensional orientation have comparable influences on visual search (Enns & Rensink, 1990b). Multiple regression models based on the diagnostic features of horizontal scene orientation (H), vertical scene orientation (V), and image orientation (I) were used to predict search rates. Display sizes were 1, 6, and 12 in both experiments.

three-dimensional orientations--by a rotation around the x-axis of the image plane (vertical). Conditions C to G can be analyzed in a similar fashion.

The search data were analyzed with multiple regression models in which the three orientation features (rotation about the y-axis, rotation about the x-axis, and image orientation) were used to predict mean search rates in each condition. This analysis showed that the two scene-based features (x-axis and y-axis rotation) were at least as useful as the image-based one (orientation) for improving search. A comparison of Conditions D and G in Figure 2.6 helps to illustrate these findings at a more intuitive level. Condition D is the same as condition G except that a third distractor is present in Condition G. If search is determined solely by image orientation, this third distractor should *increase* the distinctiveness of the target, and thereby increase search speed. In fact, as shown by the larger search slopes, search was slowed down considerably. It thus appears that the increased heterogeneity of the scene-based orientation contributed by the third distractor has actually interfered with rapid search.

	Search Items		Search Rate	
Condition	Target	Distractor	Present	Absent
A			28	42
B		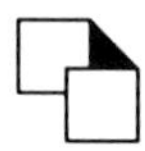	52	78
C	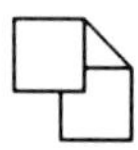	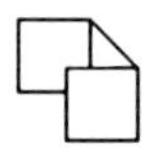	52	86

Figure 2.8. Experiment 8: Not all objects that differ in three-dimensional orientation can be detected rapidly (Enns & Rensink, 1990b). Display sizes were 1, 6, and 12.

In contrast to the previous experiments, the Experiment 8 showed that not all spatial relations that capture three-dimensional orientation, are sufficient for early vision to complete the scene-recovery process. This experiment repeated Conditions A-C in the Experiment 6 with bracket-like objects instead of blocks. As can be seen in Figure 2.8 , search was much slower than it was for the blocks in each case. Furthermore, removal of lighting direction information now caused search to slow down considerably (compare Condition A with Conditions B and C).

We were somewhat surprised to learn that early vision was sensitive to some line drawings but not to others. Why should search be rapid for drawings of simple convex blocks, but not for drawings of U-shaped brackets that had equivalent differences in three-dimensional orientation. Was this a reflection of limits on the kinds of geometric operations that could be performed by early vision? To answer this question we conducted a series of experiments to explore the spatial relations that early vision could use to recover the three-dimensional orientation of objects from line drawings.

Experiment 9 asked whether visual search was sensitive only to the most general kind of spatial relation--that of topology (Enns & Rensink, 1991). It has been suggested that topological relations between features can influence processing at the preattentive stage, but that they are the only kinds of relations to do so (Chen, 1982; 1990). If this is the case, the rapid detection of line relations in the previous experiments should be explicable purely on the basis of topological considerations.

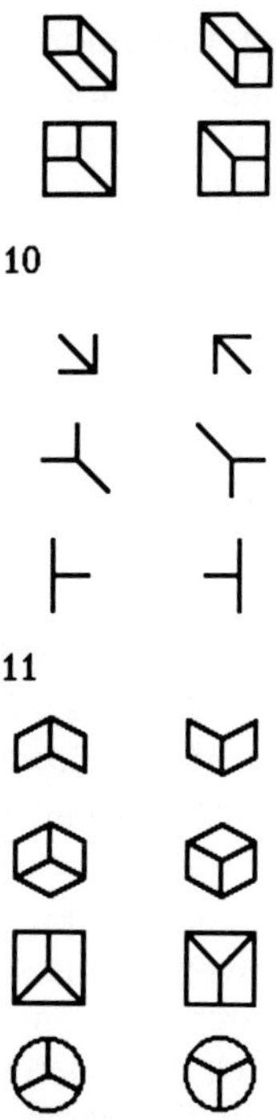

Condition	Search Items: Target	Search Items: Distractor	Search Rate: Present	Search Rate: Absent
Experiment 9				
A			7	12
B			51	96
Experiment 10				
A			10	13
B			18	24
C			37	66
Experiment 11				
A			16	19
B			16	21
C			42	67
D			48	77

Figure 2.9. Experiment 9: Early vision is sensitive to quantitative differences in line relations--not only topology. Experiment 10: Not all trihedral junctions receive equal treatment in early vision. Experiment 11: Early vision is sensitive to the entire system of relations in a search item. Display sizes were 1, 6, and 12 in each experiment. All experiments are from Enns and Rensink (1991).

The items in Condition A corresponded to simple blocks of different three-dimensional orientation, while the items in Condition B corresponded to truncated pyramids in which the line of sight was accidentally aligned with two of the surfaces (see upper panel in Figure 2.9). Quantitatively, the items in the two conditions differed considerably: the lines forming the L-junctions in Condition B were twice as long as those in Condition A, and two of the arrow-junctions in Condition A were replaced by T-junctions. Topologically, however, items in both conditions were the same. The results showed that search was much faster in Condition A than in Condition B, demonstrating that early vision is sensitive to more than just topological relations. Quantitative factors, such as the angles between connected lines, are also important.

Are the differences between Conditions A and B in Experiment 9 attributable to the different kinds of line junctions in the items? To find out, we measured search rates for each of the trilinear junctions contained in these items in Experiment 10. As shown in the middle panel of Figure 2.9, each pair of items consisted of the same line segments, so that the target and distractor differed only in the relations between segments. Arrow-junctions yielded the fastest search rates, Y-junctions were significantly slower, and T-junctions resulted in the slowest search of all. These results suggest that the difficulty of search in previous experiments was related directly to the kinds of junctions present in the items. The slow search for the items in Experiments 8 and 9A was possibly attributable to the presence of T-junctions.

To explore this hypothesis more thoroughly, Experiment 11 measured search rates for the line drawings in the lower panel of Figure 2.9. Here, targets always differed from distractors by a 180° rotation of the central Y-junction. These junctions were embedded in different contexts in order to generate items that varied in the number and type of other junctions present. For instance, in Condition A, one arrow-junction was added to the central Y, resulting in rapid search. In Condition B, two more arrow-junctions were added, but these did not influence the already rapid rate of search. In contrast, the presence of a single T-junction in Condition C slowed search by a factor of three. Finally, when all three arrow-junctions were replaced by T-junctions in Condition D, search was even slower.

Taken together, these results confirm that early vision has a preferential sensitivity for some trilinear junctions, and that it is sensitive to the entire system of line relations in an item. But why should it be selective in these particular ways? An answer can be found in an analysis of the correspondence between junctions in a line drawing and three-dimensional corners in a scene.

This scene-to-image correspondence problem has received a great deal of attention in the field of computational vision, where much work has been done on the *blocks world* (Huffman, 1971; Clowes, 1971; Mackworth, 1973; Waltz, 1972). This is a scene domain of polyhedral objects consisting only of trihedral corners (i.e., corners formed from three polygonal faces). The corresponding image domain is the orthographic projection of objects onto the image plane. By using line drawings, objects are assumed to have uniform reflectances on all visible surfaces. Furthermore, viewing direction and the direction of lighting are held constant, with the two directions being made coincident in order to avoid shadows. This leaves surface orientations and locations as the only variable scene properties.

The *blocks world* approach begins with the observation that each line in the image corresponds to one of three different kinds of edge in the scene: convex, concave, or object boundary. To interpret a line drawing correctly, each line must be labeled as corresponding to a particular kind of edge, with the labeling being consistent for all lines in the image. Algorithms to carry out the line-labeling process (e.g., Horn, 1986; Mulder & Dawson, 1990; Waltz, 1972) all rely on the fact that there are three kinds of trilinear junctions possible in an image: arrow-junctions, in which the largest angle between lines is greater than 180°; Y-junctions, in which at least two angles are greater than 90°, and T-junctions, in which one angle is exactly 180°. Examples of the three trilinear junctions are shown in Figure 2.10. As is evident from the figure, each trilinear junction may correspond to more than one kind of corner in the scene. The interpretation process therefore must eliminate junction interpretations that are inconsistent with the other junctions in the image.

An important observation from this analysis is that the three trilinear junctions differ markedly in the kind of information they carry about the three-dimensional structure. T-junctions most often correspond to occlusion, and as such, will signal only that two surfaces differ in their relative depth--they carry no information about surface orientation. However, arrow- and Y-junctions can be used to recover the orientations of the surfaces at the corresponding corner, provided that the surfaces are mutually orthogonal to one another. Perkins' (1968) law states that for an arrow-junction corresponding to an orthogonal corner, the sum of the two smallest angles must be at least 90°; for Y-junctions each of the angles must be at least 90°. Perkins (1968) also showed that if corners were assumed to be orthogonal, their three-dimensional orientations could be calculated directly from the angles about the arrow- and Y-junctions.

Interestingly, the items that subjects found easiest to detect in the previous experiments corresponded to objects with mutually orthogonal surfaces (see Figures 2.6-2.9). Furthermore, the one set of items in which the Y-junction did not correspond to such a corner resulted in slow search (i.e., the truncated pyramid in Experiment 9B, see Figure 2.9). This suggests that the orthogonality of corners may have an important influence on search rates.

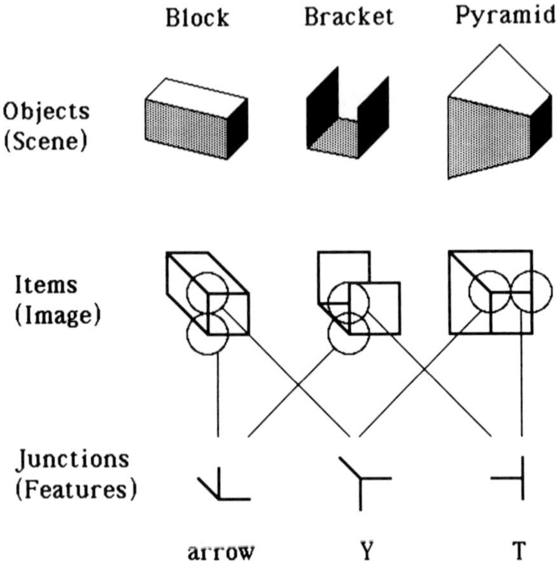

Figure 2.10. In line drawings of polyhedral objects, lines represent edges and bounded regions represent planar surfaces. Junctions involving three lines fall into three classes: arrow-, Y-, and T-junctions. However, there is no unique correspondence between a given junction and its correct three-dimensional interpretation. This can only be determined by considering the system of relations in an item.

This hypothesis was put to the test in Experiment 12 (see Figure 2.11). Here subjects searched for items containing corners that violated the orthogonality constraint (Enns & Rensink, 1991). In Condition A, items had the same outline as those in Experiment 9A, but the smallest angle of the internal Y-junction was made less than 90°. To control for the possible effects of the non-parallel orientations of the resulting lines, Condition B used drawings with similar Y-junctions, but in which parallel line orientations were maintained. These items had the same internal structure as those in Experiment 11B (Condition B), but the small angle of the Y-junction and the two wings of the arrow-junctions were both made less than 90°. Search was slow in both cases. Therefore, it appears that the parallel processes that extract orientation from line drawings can also detect when arrow- and Y-junctions violate the orthogonality constraint. In such a case, there is a slowdown in search similar to that caused by T-junctions that correspond to accidental alignments.

	Search Items		Search Rate	
Condition	Target	Distractor	Present	Absent
A			35	65
B			37	66

Figure 2.11. Experiment 12: Early vision is sensitive to the orthogonality constraint on trihedral junctions (Enns & Rensink, 1991). Display sizes were 1, 6, and 12.

But why should early vision be sensitive to the orthogonality of corners? Corners in the natural world are rarely formed from perfectly orthogonal surfaces. However, if there is no other way to determine three-dimensional orientation, the visual system may well assume mutual orthogonality in order to get a "quick and dirty" first approximation. There is a great deal of psychophysical evidence that humans assume orthogonality in line drawings of both familiar and unfamiliar objects (Butler & Kring, 1987; Perkins, 1972; Shepard, 1981). They even "see" rectangular corners when they know orthogonality has been violated (Kubovy, 1986). In addition to these reasons, orthogonal angles may also be natural defaults simply because they lie midway on the range of all possible angles between two surfaces.

4. *Inter-object relations*

The experiment reported in this final section is a preliminary investigation of the sensitivity of early vision to the relations between objects in a scene. The experiments on sensitivity to line relations have shown that some information concerning occlusion is registered preattentively--the presence of T-junctions in line drawings prevents the depicted objects from being rapidly interpreted. In Experiment 13 we asked whether the relation of occlusion between objects was registered explicitly in early vision.

The baseline task involved searching for a gap in a line that was left or right of center amongst distractor lines that had a centered gap (see Condition A in Figure 2.12). Search was very rapid in this task, as might be expected, given that subjects could simply base their search on the presence of a longer line in the display. However, filling the gap in these items with the image of a shaded cube made the search task very difficult (Condition B). Condition C showed that the

Condition	Search Items (Target, Distractor)	Search Rate Present	Search Rate Absent
A		3	8
B	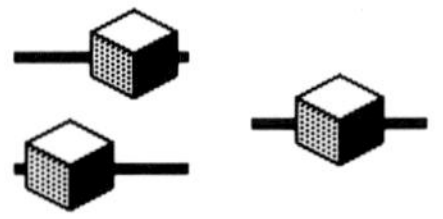	36	83
C		5	8
D	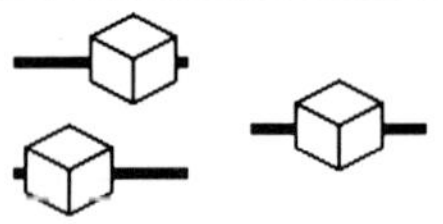	15	36
E	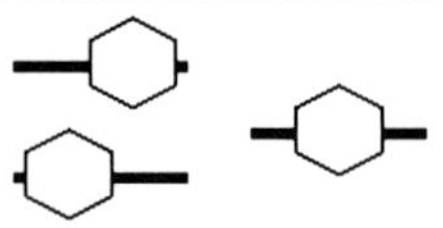	10	13

Figure 2.12. Experiment 13: Early vision is sensitive to the inter-object relation of occlusion (unpublished data). Display sizes were 2, 8, and 14.

difficulty was not caused simply by the presence of the cubes. In this condition, the same number of cubes were present in the display, although they were left unattached from the black lines--search was again very rapid. The final two conditions showed that line relations alone could make the search task difficult. Filling the gap with a line drawing of a cube (Condition C) resulted in slower search than filling the gap with the hexagonal outline of the cube (Condition D).

These results suggest that early vision is interpreting the inter-object relation of occlusion. When image features are arranged to be consistent with an interpretation of occlusion, early vision is unable to overcome this interpretation. Consequently, search cannot be governed only by the simple feature differences that distinguish the items.

IV. A REVISED VIEW OF EARLY VISION

These visual search experiments have demonstrated that early vision is capable of more sophisticated processing than has generally been assumed. Contrary to the conventional view that early vision is non-selective, I have shown that it is extremely selective in the properties it derives from the image. In addition to being sensitive to specific properties of simple geometric elements, it is sensitive to several complex image properties that correspond to intrinsic properties of the scene. How should early vision be characterized in light of these new findings?

It is helpful to recall the three-sided analysis of selectivity that was undertaken earlier in this chapter. I began by surmising that the primary function of early vision, within the larger framework of a biological vision system, was to delineate attributes of objects in the three-dimensional world. If this were possible, early vision would give a "headstart" to the more flexible processes further along the visual stream that could render complete descriptions of objects. An examination of the nature of the early vision processing system, however, put serious constraints on the kinds of computations that could be undertaken at this stage. Early vision was believed to involve specialized wetware for high-speed parallel processing. Therefore, computations were limited to those that were environmentally relevant and yet could still be conducted rapidly on the basis of image features in small local regions. Finally, an examination of the stimulus provided to this visual system indicated that object properties in the scene would have to be extracted from the projection of light onto the retinal image. This meant that the scene-to-image correspondence problem would have to be faced by any computations designed to extract object properties in parallel.

At a general level, the experiments reviewed in the previous section have already provided considerable evidence that early vision is able to extract scene-based properties from the image. But these findings do not in themselves point directly to how this is accomplished. As has been discussed at several points in this paper, physical constraints on the scene-to-image correspondence problem are often sufficient in principle to provide interpretations of scene-properties such as lighting direction, surface curvature, and three dimensional orientation from image elements. However, the computational complexity of the algorithms that have been developed in computer vision to recover these properties makes it very unlikely that they are actually used by early vision. For example, line-labeling of a polyhedral scene is a so-called *NP-complete* problem (Kirousis & Papadimitriou, 1988). This means that the time required for consistent labeling grows exponentially with the number of lines and junctions in the image (Garey &

Johnson, 1979). Any such algorithm is impractical for a real-time vision system (Tsotsos, 1988). If line relations are being used to determine three-dimensional structure in early human vision, or if luminance gradients are used to determine surface curvature, early vision must do so by way of "quick and dirty" estimates. In other words, the requirement of a complete and accurate representation must be given up in exchange for the requirement of speed of processing.

What would be required of such a system? To begin with, it should make extensive use of local measurements, since these can be computed in parallel across the image. Second, since NP-completeness comes about from the need to consider all possible combinations of all possible local interpretations, a rapid parallel system would only be able to examine relatively few candidates. Such a system could provide a rapid "first pass" at scene interpretation and pass the rest of the two-dimensional description on to higher-level processes. The interpretations formed at this stage would not likely form a complete reconstruction of the scene. Rather, they would be expected to recover scene-based properties at a relatively sparse set of locations in the visual field. This information would, nevertheless, still be useful for processes further along the visual stream. To date, Ron Rensink and I have put forward a proposal in this spirit for how early vision may be able to determine three-dimensional orientation from lines in the image (Enns & Rensink, 1991; in press). We hope that similar models will be developed in the near future for the other domains of early vision, including shape-from-shading and scene organization.

In conclusion, there are four important implications for a revised view of early vision that have emerged from the work I have summarized. First, it appears to be unnecessarily restrictive to assume that the parallel processes of early vision operate only on simple geometric elements. Although there must indeed be an initial stage which analyzes the retinal input in this way, our findings show that there must also be subsequent stages based on more complex properties. These properties are obtained neither by taking purely local measurements at each point in the image, nor by the operation of global processes that operate on the entire image (Horn, 1986). Rather, they are calculated by processes operating locally in neighborhood regions spread over the image. In this context, it is interesting to note that Walters (1987) found that junction type could effect the perceived brightness of a line, with apparent brightness increasing with line length to a maximum of 1.5°. It may well be that a similar spatial limit exists for recovery processes at preattentive levels.

Second, the elements of early vision may be characterized by environmental relevance. The experiments show that the elements of early vision describe at least some properties of the three-dimensional scene, including lighting direction, three-dimensional orientation, and object occlusion. As several researchers have pointed out (e.g., Weisstein & Maquire, 1978, Walters, 1987), the early determination of scene properties, even if incomplete, would facilitate processes further along the visual stream. It will be interesting to see which other properties can be recovered. Preliminary reports suggest that length may be registered only after size constancy mechanisms have operated on the image (Ramachandran, 1989). Color constancy may also operate at these levels (Land, 1977).

Third, the elements of early vision must be rapidly computable if they are going to be of use to the larger organism. As I have argued, preattentive processes cannot afford the time required for complete interpretations such as those given by line-labelling. But how should time be managed for "quick and dirty" processes? Should they simply be allowed to run to completion on a given input, or should they given some fixed span of time in which to "do their best." Visual search data suggest the latter. In all the experiments I reported, the intercepts of RT functions remained essentially the same, no matter how complex the items used or how steep the RT slope. If recovery processes are carried out in parallel, this implies that a fixed amount of time is allotted for their operation. Since information across the visual field is transmitted at a finite speed, this time constraint also provides an upper limit on the size of the neighborhood over which information is integrated. Future studies will need to examine this issue by varying systematically the distance between relevant sources of information within the search items.

Finally, the criteria of feature complexity, environmental relevance, and processing speed should be used to test other modules of early vision. The existence of features at preattentive levels for the interpretation of depth from single images suggests that other modules of early vision be studied from this new perspective. It would be interesting to determine, for example, whether motion perception or stereopsis is able to use the spatial relations we have studied here (Cavanagh, 1987). A comparison of the features used by various modules may help shed light on how they operate, and how they are related to one another.

V. EPILOGUE: STRUCTURE AND PROCESS IN EARLY VISION?

This chapter has focused on questions of stimulus structure and information processing in early human vision. The question of stimulus structure in this area is quite simple: What are the visual features that are registered in an initial glance at a scene? The question of processing can also be reduced to a simple one: What processes are available for the rapid registration of these features? However, one of the main themes of this chapter is that it has been very difficult, if not impossible, to answer these questions without considering a third question, namely, What is the *function* served by these features and processes? Once this question is answered, it becomes easier to constrain the set of possible answers to the first two questions.

I began this chapter with a consideration of the way in which each of these issues--function, processing capability, and stimulus structure--force early vision to be selective. To begin, the function that early vision is designed to accomplish for the organism probably imposes the greatest constraint on its selectivity--early vision is designed to recover "quick and dirty" descriptions of objects and scenes in order to guide immediate actions and the more flexible processes of attentive vision. Therefore, at a minimum, early vision must be fast acting and wide-ranging over the visual field. This consideration, in turn, provides strong motivation for a particular kind of processes that are contained in early vision-- in order to be fast, computations must be based on spatially parallel measurements and have little top-down control. Finally, the question of which features are registered must be addressed. The potential number of "features" in a visual image is unlimited--some choice must be made if the organism is going to be able to act rapidly on any of the information before it. One way to reduce the large potential set to a more manageable one is to consider only those features that are (i) environmentally relevant to the organism and (ii) can be recovered by rapid and parallel processes.

The work I have summarized has shown that early vision, working under these tight constraints, is sensitive to information regarding the direction of light in the scene, the three-dimensional orientation of surfaces, and the inter-object relation of occlusion. These are each features of the visual world that are potentially very useful to the organism. Note, however, that they do not necessarily correspond to geometrically simple features in the image. This suggests that the conventional assumption of geometrical simplicity for the features of early vision should be replaced by the more plausible assumptions of computational speed and parallelism.

ACKNOWLEGEMENTS

The research summarized in this chapter was funded by grants from the Natural Sciences and Engineering Research Council of Canada and the University of British Columbia. I am grateful to Lana Trick, Debbie Aks, and Diana Ellis for providing helpful suggestions on earlier drafts of this chapter.

REFERENCES

Aks, D. J., & Enns, J. T. (in press). Apparent depth influences visual search for the direction of shading. Perception & Psychophysics.

Beck, J. (1982). Textural segmentation. In J. Beck (Ed.), Organization and representation in perception (pp. 285-317). Hillsdale, NJ: Erlbaum.

Biederman, I. (1985). Human image understanding: Recent research and a theory. Computer Vision, Graphics, and Image Processing, 32, 29-73.

Broadbent, D. E. (1958). Perception and communication. London: Pergamon.

Butler, D. L., & Kring, A. M. (1987). Integration of features in depictions as a function of size. Perception & Psychophysics, 41, 159-164.

Callaghan, T. C. (1989). Interference and dominance in texture segregation: Hue, geometric form, and line orientation. Perception & Psychophysics, 46, 299-311.

Callaghan, T. C., Lasaga, M. I., & Garner, W. R. (1986). Visual texture segregation based on orientation and hue. Perception & Psychophysics, 39, 32-38.

Cavanagh, P. (1987). Reconstructing the third dimension: Interactions between color, texture, motion, binocular disparity, and shape. Computer Vision, Graphics, and Image Processing, 37, 171-195.

Chen, L. (1982). Topological structure in visual perception. Science, 218, 699-700.

Chen, L. (1990). Holes and wholes: A reply to Rubin and Kanwisher. Perception & Psychophysics, 47, 47-53.

Clowes, M. B. (1971). On seeing things. Artificial Intelligence, 2, 79-116.

Duncan, J., & Humphreys, G. W. (1989). Visual search and stimulus similarity. Psychological Review, 96, 433-458.

Enns, J. T. (1986). Seeing textons in context. Perception & Psychophysics, 39, 143-147.

Enns, J. T. (1990). Three dimensional features that pop out in visual search. In D. Brogan (Ed.), Visual search (pp. 37-45). London: Taylor & Francis.

Enns, J. T., & Rensink, R. A. (1990a). Influence of scene-based properties on visual search. Science, 247, 721-723.

Enns, J. T., & Rensink, R. A. (1990b). Sensitivity to three-dimensional orientation in visual search. Psychological Science, 1, 323-326.

Enns, J. T., & Rensink, R. A. (1991). Preattentive recovery of three-dimensional orientation from line drawings. Psychological Review, 98, 101-118.

Enns, J. T., & Rensink, R. A. (in press). A model for the rapid interpretation of line drawings in early vision. To appear in D. Brogan (Ed.), Visual search II. London: Taylor & Francis.

Enns, J. T., Ochs, E. P., & Rensink, R. A. (1990). VSearch: Macintosh software for experiments in visual search. Behavior Research Methods, Instruments, & Computers, 22, 118-122.

Garey, M. R., & Johnson, D. S. (1979). Computers and intractability: A guide to the theory of NP-completeness. New York: W. H. Freeman.

Garner, W. R. (1983). Asymmetric interactions of stimulus dimensions in perceptual information processing. In T. J. Tighe & B. E. Shepp (Eds.), Perception, cognition, and development. Hillsdale, NJ: Erlbaum.

Gibson, J. J. (1966). The senses considered as perceptual systems. Boston: Houghton Mifflin.

Grossberg, S. (1983). The quantized geometry of visual space: The coherent computation of depth, form, and lightness. Behavioral and Brain Sciences, 6, 625-692.

Grossberg, S. (Ed.) (1988). Neural networks and natural intelligence. Cambridge, MA: MIT Press.

Helmholtz, H. von. (1867/1967). Treatise on physiological optics (Vol. 3). In J. P. C. Southall (Ed. and Trans.). NY: Dover.

Horn, B. K. P. (1977). Image intensity understanding. Artificial Intelligence, 8, 201-231.

Horn, B. K. P. (1986). Robot vision. Cambridge: MIT Press.

Huffman, D. A. (1971). Impossible objects as nonsense sentences. In R. Meltzer and D. Michie (Eds.) Machine Intelligence 6. New York: Elsevier, 295-323.

Humphreys, G. W., & Quinlan, P. T., & Riddoch, M. J. (1989). Grouping processes in visual search: Effects with single- and combined-feature targets. Journal of Experimental Psychology: General, 118, 258-279.

Jolicoeur, P., Ullman, S., & MacKay, M. (1986). Curve tracing: A possible basic operation in the perception of spatial relations. Memory & Cognition, 14, 129-140.

Julesz, B. (1984). A brief outline of the texton theory of human vision. Trends in Neuroscience, 7, 41-45.

Julesz, B. (1986). Texton gradients: The texton theory revisited. Biological Cybernetics, 54, 245-261.

Kahneman, D. (1973). Attention and effort. Englewood Cliffs, NJ: Prentice-Hall.

Kaufman, L. (1974). Sight and mind. NY: Oxford.

Kirousis, L., & Papadimitriou, C. (1988). The complexity of recognizing polyhedral scenes. Journal of Computer and System Sciences, 37, 14-38.

Klein, R., & Farrell, M. (1989). Search performance without eye movements. Perception & Psychophysics, 46, 476-482.

Koenderink, J. J., & Doorn, A. J. van. (1980). Photometric invariants related to solid shape. Optica Acta, 27, 981-996.

Kubovy, M. (1986). The psychology of perspective and renaissance art. Cambridge, UK: Cambridge University Press.

Land, E. H. (1977). The retinex theory of color vision. Scientific American, 237, 108-128.

Lehky, S. R., & Sejnowski, T. J. (1988). Network model of shape from shading: Neural function arises from both receptive and projective fields. Nature, 333, 452.

Mackworth, A. K. (1973). Interpreting pictures of polyhedral scenes. Artificial Intelligence, 4, 121-137.

Marr, D. (1982). Vision. San Francisco: W. H. Freeman.

McLeod, P., Driver, J., & Crisp, J. (1988). Visual search for a conjunction of movement and form is parallel. Nature, 332, 154-155.

Mulder, J. A., & Dawson, R. J. M. (May 1990). Reconstructing polyhedral scenes from single two-dimensional images: The orthogonality hypothesis. In P. K. Patel-Schneider (Ed.), Proceedings of the 8th Biennial Conference of the CSCSI (pp. 238-244), Palo Alto, CA: Morgan-Kaufmann.

Nakayama, K., & Silverman, G. H. (1986). Serial and parallel processing of visual feature conjunctions. Nature, 320, 264-265.

Neisser, U. (1967). Cognitive psychology. Englewood Cliffs, NJ: Prentice Hall.

Nevatia, R. (1982). Machine perception. Englewood Cliffs, NJ: Prentice-Hall.

Nothdurft, H. C. (1985). Sensitivity for structure gradient in texture discrimination tasks. Vision Research, 25, 1957-1968.

Pentland, A. P. (1984). Local shading analysis. IEEE Transactions: PAMI, 2, 170-186.

Perkins, D. N. (1968). Cubic corners. M.I.T. Research Laboratory of Electronics Quarterly Progress Report, 89, 207-214.

Perkins, D. N. (1972). Visual discrimination between rectangular and nonrectangular parallelopipeds. Perception & Psychophysics, 12, 396-400.

Ramachandran, V. S. (1989). Is perceived size computed before or after visual search? Paper presented at the Psychonomic Society, Atlanta, Georgia.

Shepard, R. N. (1981). Psychophysical complementarity. In M. Kubovy & J. R. Pomerantz (Eds.), Perceptual organization (pp. 279-342). Hillsdale, NJ: Erlbaum.

Sperling, G. (1960). The information available in brief visual presentations. Psychological Monographs, 74 (11, Whole No. 498).

Taylor, S., & Badcock, D. (1988). Processing feature density in preattentive perception. Perception & Psychophysics, 44, 551-562.

Titchener, E. B. (1908). Lectures on the elementary psychology of feeling and attention. NY: Macmillan.

Treisman, A. (1986). Features and objects in visual processing. Scientific American, 255, 106-115.

Treisman, A. (1988). Features and objects: The fourteenth Bartlett memorial lecture. Quarterly Journal of Experimental Psychology, 40A, 201-237.

Treisman, A., & Gelade, G. (1980). A feature integration theory of attention. Cognitive Psychology, 12, 97-136.

Treisman, A., & Gormican, S. (1988). Feature analysis in early vision: Evidence from search asymmetries. Psychological Review, 95, 15-48.

Treisman, A., & Souther, J. (1985). Search asymmetry: A diagnostic for preattentive processing of separable features. Journal of Experimental Psychology: General, 114, 285-310.

Treisman, A., Cavanagh, P., Fischer, B., Ramachandran, V. S., & von der Heydt, R. (1990). Form perception and attention: Striate cortex and beyond. In L. Spillman & J. S. Werner (Eds.), Visual perception (pp. 273-316). New York: Academic.

Tsotsos, J. K. (1988). A 'complexity level' analysis of immediate vision. International Journal of Computer Vision, 1, 303-320.

Ullman, S. (1984). Visual routines. Cognition, 18, 97-159.

Walters, D. (1987). Selection of image primitives for general purpose visual processing. Computer Vision, Graphics, and Image Processing, 37, 261-298.

Waltz, D. L. (1972). Generating semantic descriptions from drawings of scenes with shadows. AI-TR-271, Project MAC, M.I.T. (Reprinted in P. H. Winston (Ed.), 1975, The psychology of computer vision (pp. 19-92), New York: McGraw-Hill.

Weisstein, N., & Maquire, W. (1978). Computing the next step: Psychophysical measures of representation and interpretation. In A. R. Hansen & E. M. Riseman (Eds.), Computer vision systems (pp. 243-260). New York: Academic.

Wolfe, J. M., Cave, K. R., & Franzel, S. L. (1988). Guided search: An alternative to the feature integration model for visual search. Journal of Experimental Psychology: Human Perception & Performance, 15, 419-433.

Zucker, S. W. (1987). Early vision. In S. C. Shapiro (Ed.), The encyclopedia of artificial intelligence (pp. 1131-1152). NY: John Wiley.

Commentary

The Nature of Selectivity in Early Human Vision, J. T. Enns

EDWARD A. ESSOCK
University of Louisville

Enns' chapter addresses the distinction between structure and processing within the area of visual search. In visual search, certain types of stimulus structure are processed rapidly, and in parallel, across the image. Thus, structure drives processing. The only issue, then, is the specification of the types of stimulus structure that are included in the list of image features processed preattentively.

Enns takes an ecological approach to this issue and speculates that the stimulus structure that conveys information about 3-dimensional object shape should be so useful to an organism that this information should be processed early and in parallel. Enns reviews a careful program of research in which he has begun to assess this conjecture. He summarizes data from thirteen experiments using blocks-world stimuli in visual search experiments. Together the experiments make a very convincing package which suggests that visual search is indeed faster when the stimuli can be interpreted as 3-dimensional objects as opposed to when they cannot.

Demonstrating that a particular feature is the basis for preattentive performance in a given situation has proven to be a torturous process. This is clearly evident in the continuing controversy concerning the features of crossings, terminators, and closure (see Essock, this volume). Because of this difficulty, the biggest issue concerning acceptance of Enns' exciting research is bound to be whether his stimuli are truly distinguished on the basis of a new feature, that of 3-dimensional object properties, rather than on the basis of established features such as size, orientation, or intensity. Clearly Enns has been very careful in the design of his stimuli, however, providing an iron-clad defense of a given preattentive feature is insidiously difficult. Enns' conclusion that 3-dimensional structure is processed preattentively is truly an important position and is bound to provide the impetus for considerable research. As Enns points out, this finding presents considerable embarrassment to the conceptualization of preattentive processing as "early vision", embracing only simple image features. On the other hand, certain recent models of visual search (e.g., Wolfe, Cave and Franzel, 1989) do suggest that top-down processing may indeed play a role in preattentive processing and might therefore more readily accommodate Enns' findings. It will be particularly interesting to see what future research offers with respect to linear-filter models of visual search, and whether their linear analysis of the image can suggest an alternative account of Enns' data.

Enns has made a very strong case for expanding the current conceptualization of visual search and preattentive processing. He argues convincingly that the types of image structure that are processed preattentively are more complex than typically embraced by the term "early vision." His ideas are certain to add a new dimension to the research in preattentive processing for years to come.

Wolfe, J.M., Cave, K.R., and Franzel, S.L. (1989). Guided search: An alternative to the feature integration model for visual search. Journal of Experimental Psychology: Human Perception and Performance, 15, 419-433.

Percepts, Concepts and Categories
B. Burns (Editor)

Structure and Process in Perceptual Organization

RUTH KIMCHI
MORRIS GOLDSMITH
University of Haifa, Israel

I. INTRODUCTION

A particularly stubborn and enduring issue in the psychology of perception concerns the way in which perception might be organized -- the primacy of "wholes" versus "parts." Two basic positions on this topic can be traced back to the controversy between two schools of perceptual thought: Structuralism and Gestalt.

The Structuralists (e.g., Wundt, 1874; Titchener, 1909), were rooted firmly in British empiricism with its emphasis on atomism and associative mechanisms, and were also influenced by nineteenth-century physiology. They held that the basic units of perception are independent local sensations and their physiological counterparts -- specific nerve energies. In their view, every sensory whole must be built up from a conglomerate of elementary sensations, and the perception of segregated, organized units corresponding to objects in the physical world is achieved only by associations learned through experience.

The Gestaltists (e.g., Kohler, 1929, 1930/1971; Koffka, 1935/1963; Wertheimer, 1923/1955), on the other hand, argued against both atomism and learning in perception, asserting the primacy of extended units and organization in the percept. A basic tenet of the Gestalt view is that a specific sensory whole is qualitatively different from the complex that one might predict by considering only its parts. The whole quality is not just one more added element or factor as was proposed by Ehrenfels' (1890) Gestaltqualitat, nor does it arise "as a secondary process from the sum of the pieces as such. Instead, what takes place in each single part already depends upon what the whole is" (Wertheimer, 1923/1955, p.5). That is, the quality of a part is determined by the whole in which that part is integrated. According to the Gestalt theory, the perception of distinct organized units is not the product of sensory elements tied together by associative learning, but is, instead, a direct result of electrical field processes in the brain responding to the entire pattern of stimulation.

While the Gestalt view eventually lost favor, perhaps largely due to its implausible physiological aspects (Pomerantz & Kubovy, 1986), the modern psychology of perception has continued to grapple with the problem of perceptual organization. On the one hand, the basic flavor of the structuralist approach has been retained in most current models of perception, especially models of pattern and object recognition (see Treisman, 1986, for an extensive review). Such "analytic" models assume that objects are identified, recognized, and classified by detecting combinations of elementary features, parts, or components. At the same time, in the last 15 years or so, perceptual organization and the Gestalt view of perception (excluding their physiological theory) have recaptured the interest of cognitive psychologists (e.g., Kubovy & Pomerantz, 1981; Beck, 1982; Boff, Kaufman, & Thomas, 1986, Vol. 2; Gopher & Kimchi, 1989; Shepp & Ballesteros, 1989). This revival includes work on such issues as perceptual grouping, part-whole relationships, processing of global and local aspects of visual patterns, object-superiority effects, configural-superiority effects, texture discrimination, and event perception. It is also expressed in the growing usage of the term "wholistic" rather than "analytic" to describe perception (e.g., Uttal, 1988).

Unfortunately, a clear understanding of the current work on "wholistic" perception may be hampered by the looseness with which the term is used in the literature, often without a clear theoretical or operational definition. There are, in fact, at least two different usages of the term "wholistic" with regard to perception. The first, and more common usage, is considered to be in the spirit of Gestalt theory, and refers to the primacy of *wholistic properties* in perception. In this usage, the terms "wholistic" and "global" are often used interchangeably to express the hypothesis that the initial information-processing step in the identification, discrimination, or classification of objects involves processing of global properties, rather than local or component properties (e.g., Navon, 1977, 1981; Uttal, 1988).

The other usage of "wholistic" perception is quite distinct. It refers to the notion that the *unitary whole*, rather than its properties (whether wholistic or component), is the primary unit for processing. In its strong version, such a notion seems to entail that at some level of processing, "properties" as such have no immediate psychological reality. This usage is most common among investigators working on dimensional interaction (e.g., Shepp & Ballesteros, 1989). Note that from this point of view, the primacy of wholistic properties suggested by the other usage would be equated with "analytic" processing, because properties (though wholistic ones) would have a definite psychological reality. A very thoughtful discussion of this notion of wholistic processing can be found in Kemler Nelson (1989).

We do not intend to propose that one conceptualization of wholistic processing is in any sense "better" than the other. Both, when clearly presented, may be entertained as viable hypotheses regarding characteristics of the human information-processing system, although for the sake of clarity it would be helpful if the two were termed differently. However, even within these two usages there remains a lack of conceptual clarity which we suggest may be remedied using the concepts of psychological stimulus structure and mode of processing. It seems that the former notion of wholistic processing, which refers to the primacy of wholistic properties, often fails to take into account the psychological structure of the stimulus. The "unitary whole" conceptualization, on the other hand, being developed within a framework which emphasized stimulus structure, does take it into account, but then it is sometimes unclear what remains to be accounted for by wholistic processing which is not already accounted for by the stimulus structure itself. Some attempts to explicate this latter usage in terms of structure and process can be found in a number of chapters in a recent book edited by Shepp and Ballesteros (1989), and the interested reader is referred to that book for more information.

In this chapter we will focus on the former usage of wholistic perception which refers to the primacy of wholistic or global properties. We will attempt to show how research within the dominant paradigm used to investigate this notion (the "global/local" paradigm) could be bolstered by a more careful analysis in terms of structure and process. In order to do so, we will begin with a metatheoretical discussion on the utility of a structure/process distinction, and to anticipate, support a distinction drawn largely from Garner's (1974) work on dimensional interaction. We will then go on to use this distinction in order to elucidate some basic conceptual problems having to do with stimulus structure within the "global/local" paradigm, and thereby hope to further clarify the utility of the structure/process distinction itself.

II. STRUCTURE AND PROCESS

A. Basic Concepts

A prominent conceptual distinction between structure and process in the cognitive psychology literature stems from the work of Garner, presented in his seminal book *The Processing of Information and Structure* (1974). Broadly speaking, this distinction refers to *stimulus structure* on the one hand, and to various *modes of processing* available to the perceiving organism on the other. Structure is tied to the stimulus, while process is tied to the perceiver. However, considering structure to be a stimulus concept does not necessarily mean that it is the physical structure which is being referred to. Rather, stimulus structure refers to those stimulus aspects which have human information processing consequences, and as such, it is *psychological* stimulus structure, defined *psychophysically*.

Consider, for example, Garner's identification of "integral" and "separable" stimulus structure on the one hand, and a possible processing mechanism such as "selective" processing on the other. Phenomenologically, certain stimuli which vary along what Garner termed integral dimensions (e.g., hue, brightness, and saturation) are perceived unitarily (e.g., as a single color), and a change in any one dimension appears to produce a qualitatively different stimulus (e.g., a different color). In contrast, stimuli varying along separable dimensions (e.g., size of circle and angle of a radial line inside it) are perceived as having distinct dimensions or attributes.

Garner (1974) used several converging operations to distinguish these two types of stimulus structure operationally. Integral dimensions: (a) show facilitation when the dimensions are correlated and interference when combined orthogonally in speeded classification, (b) produce a Euclidean metric in direct similarity scaling, and (c) are classified on the basis of overall similarity relations. Separable dimensions: (a) show neither facilitation nor interference in speeded classification, (b) produce a city-block metric in direct similarity scaling, and (c) are classified on the basis of separate dimensions.

Having identified integrality and separability as aspects of stimulus structure, Garner could then address how such structure might impose constraints on available processing mechanisms. He pointed out that selective processing is impossible for stimuli composed of truly integral dimensions, but it is optional, or may even be primary, for stimuli composed of separable dimensions.

While this example may serve to illustrate Garner's approach, we run the risk of overlooking its complexity. Thus several points should be emphasized: First, as we mentioned earlier, although Garner asserted that there is structure in the stimulus which can be described independently of the perceiver, it is still a psychophysical structure that he referred to. Second, although he emphasized the role of stimulus structure in constraining possible modes of processing, he at the same time recognized the flexibility of human information processing, as expressed in his notions of primary and secondary processes (Garner, 1974), and mandatory and optional processing (Garner, 1976). Third, he also recognized the mutual constraints that stimulus properties impose on processing and vice versa (e.g., Garner, 1978).

B. Arguments For and Against a Structure/Process Distinction

From the preceding discussion, we see that the structure/process distinction can be used to capture the relatively invariant constraints imposed upon human information processing by the relevant psychological properties of the stimulus on the one hand, and the relatively flexible modes of processing given structural constraints on the other. Many investigators who share the basic spirit of Garner's approach have followed his lead and emphasized the importance of psychological stimulus structure and modes of processing in understanding adult human information processing, as well as developmental trends in perception and cognition (see, e.g., Shepp & Ballesteros, 1989).

Nonetheless, in metatheoretical discussions regarding issues of representation, the very meaningfulness of a structure/process distinction has been called into question (e.g., Anderson, 1978). Several theorists have pointed out the necessary interdependence of cognitive representations (or data structures) and the processes which operate upon them, emphasizing the fact that any number of equivalent processing models may be derived for any psychological task. Different processing models may simulate human cognition equally well, because differences in representational structure may be compensated for by complementary differences in processing operations and vice versa (see, e.g., Palmer 1978; Rumelhart & Norman, 1988). This has led some theorists to the conclusion that the "true" allocation of structure and process in otherwise equivalent models is in principle undecidable, and therefore such efforts would best be directed elsewhere (Anderson, 1978).

Why have so many researchers in perception and cognition, then, maintained their faith in the utility of a structure/process distinction, despite such a fundamental objection? We suggest that the basic reason stems from the fact that "structure" and "process" as used by Garner and others have somewhat different meanings than when the same terms are used to refer to aspects of perceptual/cognitive *representational systems* (although the different senses are sometimes confused).

In one sense, the structure/process distinction following Garner is a conceptual distinction which provides a framework for studying the perceptual/cognitive system in its relation to the real world. It serves to highlight the mutual constraints between the structure of the stimulus to be processed on the one hand, and the different modes of processing available to the perceiving organism on the other.

In a different sense, however, representations (or data structures) and processes (or operations), as aspects of cognitive representational systems, are *functional entities* in information processing theories or "process models." Information processing theories attempt to describe what is actually going on inside the head when the organism is engaged in various perceptual and cognitive activities. The very general claim is that what is taking place is the processing of information, which is characterized in terms of cognitive representations and the processes which operate upon them (e.g., see Palmer & Kimchi, 1986). Thus, a fundamental issue for the information processing theorist is the nature of the internal representations and processes, and how to represent them in psychological theory.

Historically, the basic objection to the structure/process distinction was raised within the context of the "mental imagery debate" (see Block, 1981; Kosslyn, 1986), between proponents of "analog" versus "propositional" representations for modeling mental images. It is probably no coincidence that the debate involved researchers whose work was clearly tied to computer simulation modeling. At the detailed algorithmic level at which a "sufficient" computer simulation model is developed, decisions about the nature of data structures and the operations which access and modify them cannot be ignored (though they may be designated as "irrelevant" to the essence of the model).

The central issue with regard to structure and process in cognitive representational systems is whether the nature of internal representations and processes can be resolved by behavioral data (e.g., Anderson, 1978; Pylyshyn, 1979). Most cognitive psychologists do have faith in such an enterprise, at least at some level of description. We propose that it is precisely in undertaking this endeavor, at the appropriate level of description, that the conceptual distinction between structure and process presented earlier can be useful. Here psychological stimulus structure and mode of processing are theoretical *concepts* or explanatory constructs. As such, they are not embodied in the system itself, but rather in the theorist's understanding of the system. This conceptual distinction is useful in delineating those stimulus and organismic aspects which must be taken into account by any complete theory of human information processing, providing a working *framework* in which specific process models can be developed.

The relationship between the "conceptual" structure/process distinction and process models can be described as follows: Both stimulus structure and processing aspects, as well as their mutual interaction, need to be represented in any complete information processing model of the perceptual/cognitive system. That is, analysis guided by the conceptual distinction between structure and process provides useful constraints on *what* aspects of human information processing must be embodied in the process model. However, it need not predetermine *how* those aspects are to be represented in the model. How stimulus and processing aspects are actually represented in cognitive representational systems, in terms of data structures and operations, is a separate issue which pertains to specific process models.[1]

[1]We have confined our discussion of representational systems to fit those of "traditional" processing models, since historically the objection to the structure/process distinction was raised within this framework. More recently, however, the connectionist approach to modeling information processing has also been pointed to as having "dissolved"

There remains an issue, however, as to whether even at the conceptual level the distinction between structure and process might not be so much excess baggage. It could be argued that structure and process are simply two overlapping views of the performance characteristics of the perceptual/cognitive system. For instance, if stimulus structure is described in terms of all the information relevant to the perceiver, then such a description may be considered to be an adequate description of the perceptual system (e.g., Gibson, 1966, 1979). Alternatively, a description of the perceptual system might be given solely in terms of all the possible modes of processing by which an organism can process stimuli.

Most cognitive psychologists are not comfortable with either approach, for several good reasons: A description solely in terms of psychological stimulus structure, while it may constitute a higher-level mapping theory of perception, is concerned entirely with what the mapping is to the exclusion of how this mapping might be achieved (see Palmer & Kimchi, 1986). Also, many cognitive psychologists are unhappy about having to include in the stimulus structure (albeit perceived structure) properties that seem to require some organismic knowledge system (e.g., Rosch, 1978). On the other hand, a description completely in terms of processing fails to enhance our understanding of the structural constraints relevant to the human information processing system. Therefore, including both stimulus structure and process in psychological theory would seem to be desirable.

Perhaps the most important reason to maintain the structure/process distinction, however, is its relation to process models, as discussed above. We emphasized that while the analysis of human performance in terms of structure and process at a conceptual level need not constrain the allocation between representations and operations in possible process models, it does provide a different and important constraint -- what must be included in the model (regardless of how). Thus, the distinction between structure and process can provide a practical framework for the actual work of the cognitive psychologist, both in directing research, and in the interpretation of experimental results.

the distinction between structure and process, for different reasons. It is enough to say here that the issue of whether or not the structure/process distinction is meaningful in connectionist models, as in traditional process models, is an issue regarding representational systems. As such, it too has no bearing on the "conceptual" structure/process distinction to which we subscribe in this chapter.

C. A Methodological Criterion for Identifying Structure and Process

To avoid vagueness, there is still a need for a criterion on which to base the structure/process distinction. We suggest that such a criterion is available, at least as a first approximation. By tying structure to the stimulus and process to organismic mode of processing, converging operations can then be used to support the distinction. On the one hand, stimulus structure can be identified by the convergence of performance characteristics across information processing tasks with given stimuli. The set of converging operations for identifying integral and separable dimensions provided by Garner (1974) is a good example.

Conversely, mode of processing can be identified by the convergence of performance characteristics across stimuli for given information processing tasks. For example, a "wholistic" mode of processing in high speed classification tasks has been inferred from classification performance based on overall similarity across both integral and separable stimuli (J.D. Smith & Kemler Nelson, 1984). Further, as a means of dealing with the inherent mutual interaction between structure and process, it is possible to use a "boot-strapping" strategy to differentiate modes of processing once there exist well-defined operational criteria for distinguishing between different stimulus structures. A good example can be found in the work of L.B. Smith and Kemler (1978), Foard and Kemler Nelson (1984), and Kemler Nelson (1989). These workers initially abstracted a criterion of "privileged axes" from the well-defined converging operations for dimensional integrality and separability. They could then use this criterion to infer modes of processing (analytic or wholistic) across stimuli (separable and integral) in several different kinds of tasks.

III. STRUCTURE AND PROCESS IN WHOLISTIC PERCEPTION

A. The Primacy of Wholistic Properties

A visual object, viewed as a whole, has both wholistic properties and component parts.[2] Wholistic properties are properties that depend on the interrelation between the component parts. The Gestaltists' claim that the whole

[2]A thorough discussion of the various ways of analyzing a stimulus into properties, parts, features, dimensions, etc. is beyond the scope of this chapter. Comprehensive treatments may be found in Garner (1978) and Treisman (1986).

is more or at least different from the sum of its parts can perhaps be captured by the notion of wholistic properties such as closure, symmetry, and certain other spatial relations between the component parts. As we have mentioned earlier, one sense of "wholistic" perception is embodied in the claim that processing of wholistic properties precedes processing of component properties. Within this conceptualization, the *global precedence hypothesis*, put forward by Navon (1977), is considered by many cognitive psychologists to be a modern version of the Gestaltists' claim about the primacy of wholistic or global processing in perception (e,g., Pomerantz, 1981; Kimchi, 1982; Treisman, 1986; Robertson, 1986; Uttal, 1988). This hypothesis has generated a wealth of empirical research which has nonetheless left the issue still unsettled and somewhat confused.

We will first present the hypothesis along with the framework in which it was formulated, and the experimental paradigm used to test it. We will then critically analyze some basic assumptions underlying much of the research within this paradigm in terms of structure and process, and support this analysis with experimental evidence. Our analysis will then allow us to examine the extent to which this line of research has been able to shed light on the primacy of wholistic vs. component properties.

B. The Global Precedence Hypothesis

Posing the question "Is the perceptual whole literally constructed out of the percepts of its elements?" Navon (1977) proposed that "perceptual processes are temporally organized so that they proceed from global structuring towards more and more fine-grained analysis. In other words, a scene is decomposed rather than built up" (Navon, 1977, p. 354).

The global precedence hypothesis was formulated within a framework in which a visual scene is viewed as a hierarchical network of subscenes interrelated by spatial relationships (e.g., Winston, 1975; Palmer, 1977). The globality of a visual feature corresponds to the place it occupies in the hierarchy: features at the top of the hierarchy are more global than those at the bottom, which in turn are more local. Consider for example the structure of a human face. The face as a whole has global features (e.g., shape, expression, etc.) as well as a set of local features or component parts (e.g., eyes, nose, dimples, etc.). In turn, the component parts when considered as wholes also have global features and a further set of local features or component parts. The global precedence hypothesis claims that the processing of a scene is global-to-local. That is, global

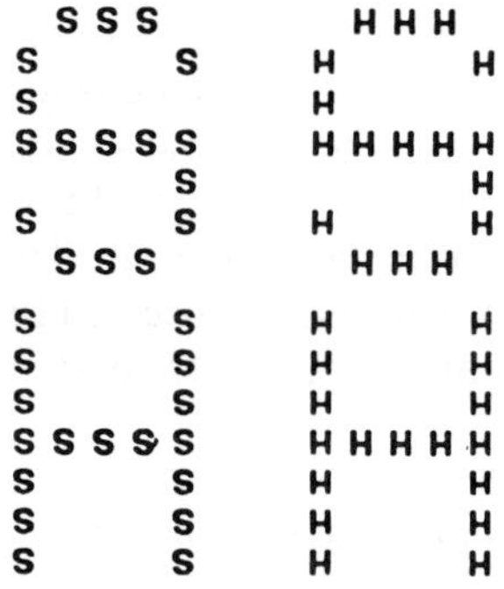

Figure 3.1. Example of the compound letters used in the global/local paradigm. (Adapted from Pomerantz, Pristach, & Carson, 1990.)

features of a visual object are processed first, followed by analysis of local features.

This hypothesis has been tested experimentally by studying the perception of hierarchically constructed patterns, in which larger figures are constructed by suitable arrangement of smaller figures. An example is a set of large letters constructed from the same set of smaller letters (see Figure 3.1). Within the framework discussed above, the large letter is considered a higher-level unit and the small letters are lower-level units. Both levels have a set of global and local features. However, the features of a higher level unit (either global or local) are considered to be more global than features of lower level units by virtue of their position in the hierarchy. Thus, the modifier "more" should precede local and global (see also Ward, 1982).

The choice of hierarchical patterns for testing the global-to-local hypothesis is seemingly well motivated, and the rationale is as follows: The global configuration and the local elements can be equally complex, recognizable, and codable, and one level cannot be predicted from the identity of the other (Navon, 1977, 1981). Once the global configuration and the local elements are equated, except for their level of globality, performance measures such as relative speed of identification, and/or asymmetric interference can be used in order to infer the precedence of one level or the other.[3]

[3]Whether the paradigm properly allows the inference of temporal *precedence* from findings of perceptual *dominance* is an interesting issue in itself. We will hereafter use "precedence" when referring to the original formulation of the theoretical hypothesis, but

Note however, that the local elements of hierarchical patterns are not the local features of the global figure in the same way as eyes, nose, and mouth are the local features of a face, or in the way that vertical and horizontal lines are the local features of the letter H. Thus, the global-to-local hypothesis that is in fact tested by hierarchical patterns is the following: The features of a high-level unit (e.g., the shape of the larger figure) are processed first, followed by analysis of the features of the lower-level units (e.g., the shape of the small figure) (Navon, 1981; Kimchi, 1982).

By a set of converging operations, Navon (1977) demonstrated the perceptual dominance of global configurations. In one experiment (Navon, 1977, Experiment 2), he asked subjects to respond to an auditorily presented name of a letter while looking at a hierarchical letter. The subject's auditory discrimination responses were affected (interfered or facilitated) by the global level of the visual stimuli but not by the local one. In another experiment (Navon, 1977, Experiment 3), Navon employed a Stroop-like interference paradigm and found that conflicting information between the local and the global levels (e.g., a large H made up of small S's) had an inhibitory influence on identification of the local letter but not on the identification of the global letter. Navon and others (e.g., Broadbent, 1977) interpreted these findings as evidence for the inevitability of global precedence in visual perception.

Other researchers have used similar stimuli (i.e., hierarchical letters composed of many small letters) and employed identical or similar experimental tasks (e.g., Stroop-like interference, target search, speeded classification) to explore the generality of global precedence in what we refer to here as the "global/local paradigm." These studies demonstrated important boundary conditions of the phenomenon and pointed out certain variables that can affect global versus local dominance. Such variables include overall stimulus size (e.g., Kinchla & Wolfe, 1979), sparsity of the local letters (e.g., Martin, 1979), "clarity" or "goodness" of form (e.g., Hoffman, 1980; Sebrechts & Fragala, 1985), retinal location (e.g., Pomerantz, 1983; Grice, Canham, & Boroughs, 1983; Kimchi, 1988), spatial uncertainty (e.g., Lamb & Robertson, 1988; Kimchi, Gopher, Rubin, & Raij, in press), and exposure duration (e.g., Paquet & Merikle, 1984). Several studies also have examined the locus of the phenomenon -- perceptual, attentional, or preattentive (e.g., Miller, 1981; Boer & Keuss, 1982; Paquet & Merikle, 1988).

will use the term "dominance" to refer to empirical findings.

C. The Structure of Hierarchical Patterns

Our present goal is not to evaluate or interpret the experimental results obtained within the global/local paradigm, but rather to question some basic assumptions underlying much of the research with hierarchical patterns. A basic presupposition seems to be that there are two distinct perceptual levels corresponding directly to the global figure and local elements, and that the critical question is which level gets processed first. We suggest that the supposed correspondence between perceptual and experimenter-defined levels of pattern structure may hold in some cases and not in others; therefore a clear notion of how hierarchical patterns are structured perceptually is an important prerequisite for asking meaningful questions about how such structure may be processed.

Experimenters have characterized hierarchical patterns as having two distinct levels of pattern structure: global configuration and local elements. (We use the term "pattern" to refer to the entire stimulus, i.e., to both levels at once.) In the perceptual domain, however, three phenomenal aspects can be identified: overall form, figural parts, and texture. Whenever small figures are positioned near each other in such a way that their positions form the pattern of a larger figure, the two experimenter-defined levels of pattern structure (i.e., the global configuration and the local elements) are present regardless of the number and/or the relative size of the elements. However, Kimchi (1982; Kimchi & Palmer, 1982) has claimed that the mapping from the two levels of pattern structure in the *stimulus* domain into meaningful levels in the *perceptual* domain depends critically on the number and the relative size of the elements.

Phenomenologically, patterns composed of many relatively small elements (many-element patterns) are perceived as overall form associated with texture. Patterns composed of few relatively large elements (few-element patterns) are perceived as overall form and figural parts. The local elements of many-element patterns lose their function as individual parts of the form and are relegated to the role of "material" (Goldmeier, 1936/1972) or "texture" (Kimchi, 1982; Kimchi & Palmer, 1982), and do not interact with the form of the pattern. That is, the global form and the local elements of many-element patterns are phenomenally independent: Replacing the elements of the patterns by other elements does not affect the perception of its overall form. On the other hand, the local elements

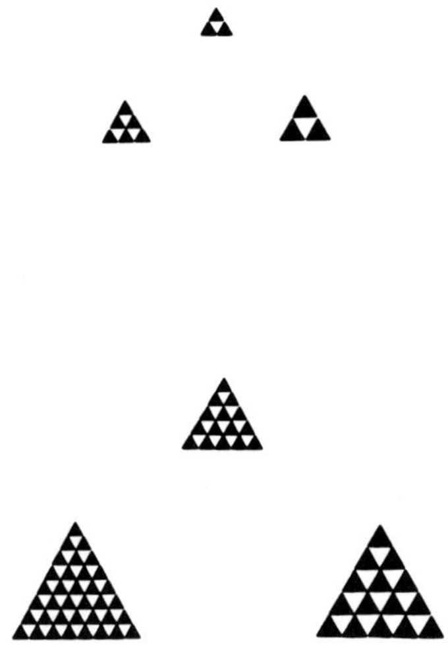

Figure 3.2. Example of the stimulus triads used in Kimchi & Palmer's (1982) study (Experiment 1) with adults, and in Kimchi's (1990) study (Experiment 1) with children. (Adapted from Kimchi, 1982.)

of few-element patterns are perceived as figural parts of the overall form.[4]

Kimchi (1982; Kimchi & Palmer, 1982, 1985; Kimchi, 1988, 1990) used several converging operations to support this distinction operationally. In a forced-choice similarity judgment task subjects were presented with stimulus triads composed of a standard pattern and two comparison patterns. One comparison pattern was a proportional enlargement of the standard pattern (i.e., an enlargement in which the size of both the global configuration and the local elements are increased by uniform dilation). The other comparison pattern was a particular sort of unproportional enlargement in which the global configuration is enlarged but the measurements of the elements are unchanged (see Figure 3.2). Few-element patterns were judged to be more similar to their proportional enlargements which preserved both the global and the local structures as well as the relationships between them. Many-element patterns, on the other hand, were

[4]A similar phenomenal distinction between two types of patterns was proposed independently by Pomerantz (1981, 1983). In what Pomerantz termed "Type P" patterns, only the position of the local elements matters for the overall form. In "Type N" patterns both the position and the nature of the local elements matter.

judged to be more similar to their unproportional enlargements which preserved the global form as well as the texture of the pattern (Goldmeier, 1936/1972; Kimchi & Palmer, 1982, Experiment 1; Kimchi, 1990, Experiment 1).

The relative salience of the local element in few- and many-element patterns was examined using a similarity judgment task involving stimulus triads in which the global configuration was pitted against the local elements (see Figure 3.3). Few-element patterns were judged to be more similar to a same-element pattern in which the same elements are arranged to form a different configuration than to a same-configuration pattern (i.e., a pattern in which different elements are arranged to form the same configuration), while many-element patterns were judged to be more similar to a same-configuration pattern than to a same-element pattern (Kimchi & Palmer, 1982, Experiment 2; Kimchi, 1990, Experiment 2). Subjects' preferences for verbal descriptions of the patterns were also consistent with the similarity judgments. When presented with descriptions in which the global configuration was the grammatical subject and the local elements were the grammatical object (e.g., "A triangle made of triangles") and descriptions in which the global configuration and the local elements had a reversed role (e.g., "Triangles arranged to form a triangle"), subjects preferred the former kind of descriptions for many-element patterns, and the latter kind of description for few-element patterns (Kimchi & Palmer, 1982, Experiment 4).

In a parametric study using the two similarity judgment tasks described above, the number of elements and their relative size was varied systematically. The results showed that the critical number of elements for which the switch in the similarity judgments occurred was 7 ± 2, both for adults (Kimchi, 1982; Kimchi & Palmer, 1982), and for children as young as three years of age (Kimchi, 1990).

Converging evidence was then obtained for the perceptual separation/nonseparation of global and local levels in hierarchical patterns as a function of the number of elements in the pattern. In a speeded classification task involving a set of four patterns created by orthogonally combining two types of global configuration and two types of local elements (see Figure 3.4), subjects were required to classify the patterns according to either global form or texture. Few-element patterns showed a pattern of results which is typical of integral dimensions: Facilitation was obtained when the global configuration and the local elements were combined redundantly, and interference was obtained when they were combined orthogonally. Many-element patterns, on the other hand, showed a pattern of results typical of separable dimensions: No facilitation was obtained when the global configuration and the local elements were combined redundantly,

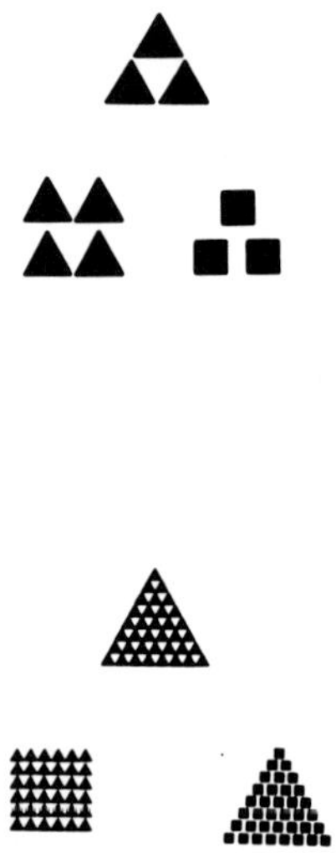

Figure 3.3. Example of the stimulus triads used in Kimchi & Palmer's (1982) study (Experiment 2) with adults, and in Kimchi's (1990) study (Experiment 2) with children. (Adapted from Kimchi, 1982.)

and no interference was obtained when they were combined orthogonally (Kimchi & Palmer, 1985, Experiments 1 and 3).

Few-element and many-element patterns also produced reliably different patterns of results in a simultaneous comparison task in which subjects were required to determine whether two simultaneously presented patterns were the same or different at the global or at the local level (Kimchi, 1988), and in an identification task employing a Stroop-like interference paradigm (Kimchi & Merhav, in press), using the same stimuli as in the speeded classification task. It is interesting that the requirement to classify the same patterns according to global and local forms (rather then in terms of global form and texture) did not affect the pattern of results obtained with the few-element patterns, both with the speeded classification task (Kimchi & Palmer, 1985, Experiments 2 and 4), and with the simultaneous comparison task (Kimchi, 1988). This could be expected from the relation between number of pattern elements and the "appearance" of texture. Inasmuch as a critical number of elements (around 7 + 2) is required for texture perception, there is simply no perceived texture in few-element patterns. For many-element patterns, on the other hand, there is a difference between texture and local form. While global form and texture of many-element patterns

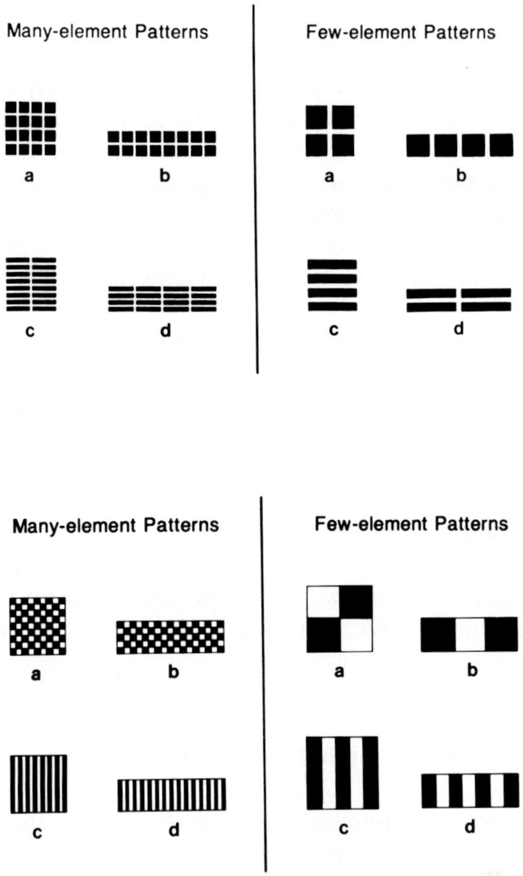

Figure 3.4. The four sets of stimuli used in Kimchi & Palmer's (1985) study, and in Kimchi & Merhav's (1990) study. The stimulus pairs used in Kimchi's (1988) study were created from the two upper sets. (Adapted from Kimchi & Palmer, 1985.)

were found to be perceptually separable, the requirement to classify such patterns in terms of global and local forms did result in interference between the levels (Kimchi & Palmer, 1985, Experiments 2 and 4; Kimchi, 1988; Kimchi & Merhav, in press).

Further evidence for perceptual separation/nonseparation between the global and local levels of many- and few-element patterns respectively, has been obtained with similarity judgments and speeded classification using different stimuli (Klein & Barresi, 1985).

In sum, this body of evidence suggests that there is good reason to distinguish the perceptual structure of patterns composed of many, relatively small elements, from that of patterns composed of few, relatively large elements. The two types of stimuli show clearly distinguishable performance characteristics across different tasks, and across subjects (adults and children). In particular, the local elements of few-element patterns are perceived as figural parts of the overall form, and the global and local levels are perceptually integral. On the other hand, the local elements of many-element patterns are perceived as textural molecules, and the overall form and the texture of such patterns are perceptually separable.

The specific perceptual mechanisms underlying these two types of perceptual structure have yet to be determined (but see Kimchi & Palmer, 1985, for some suggestions), and the research findings reported here do not dictate how such mechanisms should be modeled. Nonetheless, we see here a good example of how the empirical identification of perceptual stimulus structure provides constraints for any cognitive/perceptual model which must then allow for such structure. We propose that the converging evidence for the difference in the perceptual stimulus structure of few- and many-element patterns has several important implications for the global/local paradigm, and for the issue of structure and process in general. We first discuss the implications for the global/local paradigm.

D. Implications for the Global/Local Paradigm

The choice of hierarchical patterns for testing the global precedence hypothesis was guided by the assumption that the global configuration and the local elements comprise two distinct structural levels which are statistically independent, and that each constitutes a stimulus in its own right. The global configuration and the local elements could then be equated, except for their level of globality, and the perceptual precedence of one level or the other inferred from performance measures such as relative speed of identification and/or asymmetric interference. However, the finding that the perceptual separation of configural and elemental levels of hierarchical patterns depends on the number and relative size of the elements -- that local elements are sometimes perceived as distinctive parts of the overall form, and at other times, as textural molecules -- challenges the validity of this assumption, and has implications for interpretation of the experimental findings obtained in the global/local paradigm.

First, the asymmetric interference effects used to infer global or local precedence may depend on the relative perceptual separation of the global and the local levels. To the extent that the local elements are perceptually separable from the global configuration, local-to-global interference is much less likely than when the two levels are perceptually integral. As we have seen, the relative perceptual separation depends at least in part upon the number and relative size of the local elements. It follows, then, that positing precedence of the global level of structure (as operationalized in this paradigm) as a rigid perceptual law is hardly tenable.

Second, even if we were to find that global dominance is uniformly observed with *many-element* patterns, what could we infer from this with regard to the global precedence hypothesis? If the local elements of many-element patterns serve to define texture, then they may not be represented as individual figural units at all. Therefore it is not clear whether a faster identification of the global configuration should be accounted for by its level of globality, or rather, by a qualitative difference between figural unit and textural molecule.

There is also a further implication which stems from the characterization of the local level in many-element patterns as texture. It is frequently claimed that texture segregation occurs early in perceptual processing; it organizes the visual field and defines the units for further processing. Experimental evidence suggests that texture segregation depends on local properties of the texture molecules (e.g., Julesz, 1981; Beck, 1982; Treisman, 1985). This, in turn, would suggest that properties of the local elements in many-element hierarchical patterns are extracted prior to those of the global configuration, even though they do not seem to affect the speed of response to identify the global configuration.

Also, in view of the confounding between relative size and number of elements in many-element patterns, it may be that relative size alone could account for observed global dominance. That is, a local element may be processed more slowly simply because it is relatively small, rather than because of its level of globality (Navon & Norman, 1983; Kimchi et al., 1990).

Finally, in a more fundamental sense it can be argued that if the global precedence hypothesis is interpreted as positing the perceptual primacy of wholistic properties versus component properties, then patterns composed of few, relatively large elements should be better suited to test the hypothesis. That is because the local elements of such patterns seem to have psychological reality as component parts of the overall form, while the local elements of many-element patterns do not.

E. Implications Regarding Structure and Process

We now turn to the structure/process distinction. We feel that the study of few- and many-element patterns has several important implications for this issue.

First, it demonstrates the importance of defining properties *psychophysically* rather than just physically. The findings presented above clearly show that the same stimulus properties as defined by the experimenter (i.e., global configuration and local elements), are not treated equivalently by subjects, for few- and many-element patterns. The local elements are physically present in hierarchical patterns, and they have the same logical status, regardless of the number of elements. However, the *perceptual* property is either texture or figural part, depending on the number and relative size of the local elements.

Navon pointed out, and rightly so, that "strictly speaking, global precedence cannot be tested unless it is known what thc pcrccptual units are" (1981, p. 27), and in the absence of such precise knowledge, Navon suggested that "we have to rely on our common sense reinforced by our knowledge of Gestalt laws of organization" (p. 27). Yet, it has been demonstrated above that such informal criteria may still allow for a discrepancy between the effective perceptual units as defined by the experimenter, and those having psychological reality in the perceptual system. Clearly, such a discrepancy can cause the experimenter to commit an inferential error regarding the proper characterization of perceptual processing. For this reason, the stimulus structure used to test a processing hypothesis has to be carefully analyzed and supported in psychophysical terms.

A second point concerns the fact that logical relations in the stimulus structure should not predetermine processing hypotheses, because the logical structure of the stimulus does not necessarily predict processing consequences (see Garner, 1983; Kimchi & Palmer, 1985). Hierarchical patterns provide a clear case of asymmetry in the logical structure of the stimuli: Local elements can exist without a global configuration, but a global configuration cannot exist without local elements (e.g., Pomerantz & Sager, 1975), and this asymmetry holds both for few- and many-element patterns. However, the findings presented above serve to emphasize the danger of predicting performance or making processing assumptions on the basis of logical relations in the stimulus domain alone.

The danger in tying processing assumptions to logical stimulus structure may be further illustrated with regard to the notion of "emergent properties". Wholistic properties such as closure and symmetry are often termed "emergent

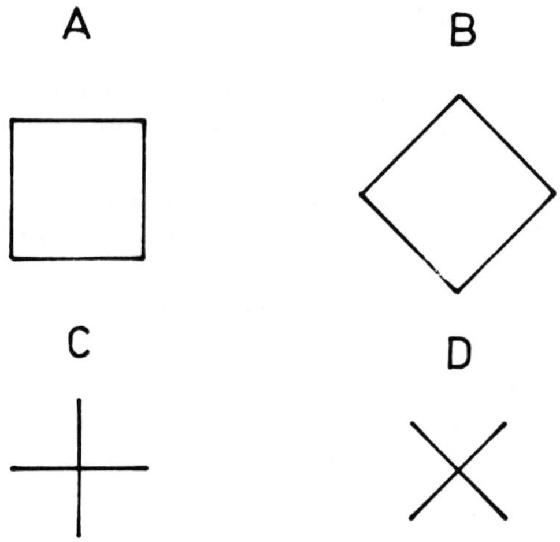

Figure 3.5. Example of a set of stimuli that share either wholistic or component properties: The pair A,B and the pair C,D share wholistic properties (closure and intersection, respectively). The pair A,C and the pair B,D share component properties (horizontal/vertical lines, and oblique lines, respectively). (Adapted from Lasaga, 1990.)

properties"(e.g., Garner, 1978; Pomerantz, 1981) because they do not inhere in component properties but rather they "emerge" from the interrelations between the components themselves (see Figure 3.5). Despite the semantic connotation of the term "emergent," however, there is no actual necessity for emergent properties to be perceptually *derived*. Emergent properties might be computed from relevant component properties, but it is also possible that they are directly detected by the perceptual system (i.e., without the component properties having a psychological reality of their own). The description of wholistic or configural properties as "emergent" is only supported as a description in the *stimulus* domain. In other words, both component and wholistic properties (whether "emergent" or not) must be treated as stimulus aspects. Whether wholistic properties then dominate component properties at a certain level of processing, or whether they are extracted earlier than component properties, are empirical questions in the *processing* domain. At present, there is some evidence that such "emergent properties" do indeed dominate component properties in discrimination and classification tasks (e.g., Pomerantz, Sager, & Stoever, 1977; Lasaga, 1989), and that they may be extracted at early stages of perceptual processing (e.g., Treisman & Paterson, 1984; Kolinsky & Morais, 1986).

F. More about Global Precedence

In the foregoing discussion of the global precedence hypothesis and the global/local paradigm used to test it, we focused on issues of stimulus structure. There are also *processing* issues related to this paradigm which include, for example, the distinction between processing dominance and temporal order of processing (e.g., Ward, 1983), serial processing versus relative speed of processing (e.g., Navon, 1981, Lasaga, 1989), and the possible difference in the underlying modes of processing reflected in speed of processing versus interference (e.g., Navon & Norman, 1983; Lamb & Robertson, 1988). We chose to focus on the structure issues because they bear heavily on the assumptions underlying the paradigm, and yet have been neglected by much of the research. Further, although structure and processing issues are to some extent independent of each other -- and this is a strength of the structure/process distinction -- we feel that structure issues are in some sense primary, since questions regarding processing can only empirically be decided once we have a clear idea of the stimulus structure which is being processed.

Another aspect to be considered when testing a processing hypothesis is the nature of the task. Any information processing activity involves processes at different stages of processing, and different experimental tasks can tap different stages of processing. For example, Kimchi & Palmer (1985) found, using a speeded classification task, that the dimensions of form and texture of many-element patterns were separable. However, in a simultaneous comparison task, asymmetric interference was observed when a potential conflicting output between these dimensions was present (Kimchi, 1988). Other similar findings have been reported demonstrating that with separable dimensions selective attention can be possible in one task, but not in another, depending on the likelihood of dimensional output conflict (e.g., Santee & Egeth, 1980). Stated more generally, dimensional analysis is a necessary but not sufficient condition for successful selective attention to a stimulus dimension. In a similar vein, local properties may be extracted prior to the stage of complete identification of the global configuration, depending on task demands and stimulus structure. However, it does not rule out the possibility that in early stages of perceptual processing, wholistic properties are available prior to the component properties.

At this point we would like to emphasize what may have been overlooked. Our analysis does not have any implication for the tenability of the global precedence hypothesis. As we understand it, the global precedence hypothesis postulates that wholistic properties such as closure, symmetry, and other spatial relations between component parts are available in the percept prior to component properties. This is an interesting and viable hypothesis that should continue to be examined. The arguments we presented in this chapter are relevant to the assumptions underlying the global/local paradigm and to the operationalization of globality. Essentially, we pointed out the importance of making a clear and careful analysis of stimulus structure before going on to make inferences about processing.

In view of the evidence that local elements of many-element patterns do not have a psychological reality as local or component properties, the choice of such patterns for testing the global precedence hypothesis seems to be unfortunate, despite its elegance in controlling for many intervening variables. Furthermore, as mentioned earlier, relative size alone may provide a reasonable account for obtained global dominance with many-element patterns (Navon & Norman, 1983; Kimchi et al., 1990). To the extent that globality is inherently confounded with relative size, the finding that "larger" properties are available earlier than relatively "smaller" properties (provided that eccentricity is held constant) would be informative. But certainly more than this is claimed by the global precedence hypothesis. The interesting and essential difference between wholistic and component properties is not their relative size.

Consider for example the stimuli in Figure 3.5. To distinguish the wholistic property of closure from the component vertical and horizontal lines on the basis of their relative sizes would seem to miss the point. Rather, the essential characteristic of wholistic properties, as we have mentioned before, is that they do not inhere in the components, but depend instead on the interrelations between them. We suggest that the ultimate precedence of wholistic versus component properties should be tested with stimuli whose structure does not allow for the most essential aspect of globality to be confounded with other, less central aspects. This is not an easy challenge.

IV. CONCLUDING REMARKS

In this chapter we have tried to show how the theoretical concepts of structure and process can help to further our understanding of the perceptual/cognitive system. We began with a discussion of the structure/process distinction in general, and then continued with an analysis of the global/local paradigm, which we hope served to illustrate more concretely the benefits of carefully considering both structure and process together.

The global precedence hypothesis is a processing hypothesis, and the major import of our analysis was to demonstrate that a clear notion of the perceptual structure of the stimulus is an important prerequisite for asking meaningful questions about processing. A basic assumption underlying the global/local paradigm is that the two levels of hierarchical patterns, the global configuration and the local elements, map directly into distinct perceptual units that differ only in their level of globality. However, we presented evidence that the local elements map into different perceptual units, and that the configural and elemental levels bear different perceptual relations to each other, in few- and many-element patterns. Such evidence severely weakens the plausibility of the forementioned assumption, and was shown to have important implications for the interpretation of obtained experimental findings. As a general rule, it is important to consider the perceptual structure of the stimuli used to test any processing hypothesis.

At the same time, the studies of few- and many-element patterns reported here serve to demonstrate that although a careful analysis of stimulus aspects per se may be necessary for understanding performance on a psychological task, it is not sufficient. Consider performance on a task involving hierarchical patterns. As discussed earlier, from strictly a stimulus point of view, the logical structure of such patterns predicts processing asymmetry. Yet it was found that the perceptual relations between configural and elemental levels of such patterns depend on the number of elements embedded in the patterns. The logical structure cannot account for this fact without redefining the "logically-given" relations. On the other hand, from strictly a processing point of view, it might be argued that the very fact that local elements of many-element patterns are perceived as "textural molecules" is precisely due to global precedence itself (e.g., Navon, 1981). This argument, however, has difficulty explaining why, in contrast to many-element patterns, the local elements of few-element patterns are perceived as figural parts. Here too, it seems that the processing hypothesis of global precedence alone cannot account for the different perceptual status of the local elements in few- and many-element patterns without redefining its notion of global precedence.

Clearly, then, stimulus structure and organismic mode of processing must be considered together if we are to understand performance on a psychological task. Further, recall that we refer to stimulus structure defined psychophysically, and as such it depends not only on the properties of the stimulus per se, but on characteristics of the perceptual system as well. For this reason, identification of psychologically relevant properties in the stimulus domain may also further our understanding of the processing mechanisms underlying stimulus structure. For instance, differences between few- and many-element patterns in terms of stimulus properties such as number and relative size of local elements can give us some clues as to the operation of possible mechanisms that may account for the differences in the perceptual structure of these two types of hierarchical patterns.

This leads us to one final important point. If perceptual stimulus structure constrains processing, but at the same time depends on processing, where does that leave us with regard to the distinction between structure and process? We might say that a more complete understanding of human information processing would require a distinction between three aspects: (a) the stimulus structure, (b) the processing of that structure, and (c) the processing mechanism accounting for that structure. The inclusion of stimulus structure in our psychological explanations allows for the mechanism accounting for such structure to be dealt with as a problem separate from how that structure may be processed in different psychological tasks. Thus, any complete process model for hierarchical patterns will have to account for the relative perceptual separation of configural and elemental levels when the number and the relative size of the elements vary. But as long as such effects are taken into account as stimulus structure, global precedence and other similar processing hypotheses can be tested using appropriate stimuli, notwithstanding temporary ignorance regarding the specific mechanisms underlying such structure.

In the long run, differentiating relatively invariant stimulus structure from relatively flexible modes of processing offers the far-reaching promise of guiding us towards the built-in structural constraints of the perceptual/cognitive system itself. Admittedly, it is no easy task to determine what in human performance should be accounted for by stimulus structure, and what by mode of processing. However, such an enterprise, difficult though it may be, is not without its rewards.

ACKNOWLEDGMENTS

The preparation of this chapter was partially supported by a grant from The Israeli Academy of Sciences and Humanities to the first author, and by a National Science Foundation Graduate Fellowship to the second author. It also benefitted from the facilities of the Institute of Information Processing and Decision Making, University of Haifa.

REFERENCES

Anderson, J. R. (1978). Arguments concerning representations for mental imagery. Psychological Review, 85, 249-277.

Beck, J. (1982). Textural segmentation. In J. Beck (Ed.), Organization and representation in perception (pp 285-317). Hillsdale, NJ: Erlbaum.

Block, N. (Ed.) (1981). Imagery. Cambridge, MA: The MIT Press.

Boer, L. C., & Keuss, P. J. G. (1982). Global precedence as a postperceptual effect: An analysis of speed-accuracy tradeoff functions. Perception & Psychophysics, 31, 358-366.

Boff, K. R., Kaufman, L., & Thomas, J. P. (1986). Handbook of perception and performance, Vol. 2. New York: Wiley.

Broadbent, D. E. (1977). The hidden preattentive processes. American Psychologist, 32, 109-118.

Ehrenfels, C. von. (1890). Uber Gestaltqualitaten. Vierteljahrschrift fuer Wissenchaftliche Philosophie, 14, 249-292.

Foard, C. F. & Kemler Nelson, D. G. (1984). Holistic and analytic modes of processing: The multiple determinants of perceptual analysis. Journal of Experimental Psychology: General, 113, 94-111.

Garner, W. R. (1974). The Processing of Information and Structure. Hillsdale, N.J.: Erlbaum.

Garner, W. R. (1976). Interaction of stimulus dimensions in concept and choice processes. Cognitive Psychology, 8, 98-123.

Garner, W. R. (1978). Aspects of a stimulus: Features, dimensions, and configurations. In E. Rosch & B. B. Lloyd (Eds.), Cognition and categorization. Hillsdale, NJ: Erlbaum.

Garner, W. R. (1983). Asymmetric interactions of stimulus dimensions in perceptual information processing. In T. J. Tighe, & B. E. Shepp (Eds.), Perception, cognition, and development: Interactional analysis (pp. 1-38). Hillsdale, NJ: Erlbaum.

Gibson, J. J. (1966). The senses considered as perceptual systems. Boston: Houghton-Mifflin.

Gibson, J. J. (1979). The ecological approach to visual perception. Boston: Houghton-Mifflin.

Goldmeier, E. (1972). Similarity in visually perceived forms. Psychological issues, 8 (1, Whole No. 29). (Originally published, 1936)

Gopher, D., & Kimchi, R. (1989). Engineering psychology. Annual Review of Psychology, 40, 431-455.

Grice, G. R., Canham, L., & Boroughs, J. M. (1983). Forest before trees? It depends where you look. Perception & Psychophysics, 33, 121-128.

Hoffman, J. E. (1980). Interaction between global and local levels of a form. Journal of Experimental Psychology: Human Perception and Performance, 6, 222-234.

Julesz, B. (1981). Textons, the elements of texture perception and their interactions. Nature, 290, 91-97.

Kemler Nelson, D. G. (1989). The nature and occurrence of holistic processing. In B. Shepp & S. Ballesteros (Eds.), Object perception: Structure & Process. Hillsdale, NJ: Erlbaum. pp. 357-386.

Kimchi, R. (1982). Perceptual organization of visual patterns. Doctoral Dissertation, University of California, Berkeley.

Kimchi, R. (1988). Selective attention to global and local levels in the comparison of hierarchical patterns. Perception & Psychophysics, 43, 189-198.

Kimchi, R. (1990). Children's perceptual organization of hierarchical patterns. European Journal of Cognitive Psychology.

Kimchi, R., & Palmer, S. E. (1982). Form and texture in hierarchically constructed patterns. Journal of Experimental Psychology: Human Perception and Performance, 8, 521-535.

Kimchi, R., Palmer, S. E. (1985). Separability and integrality of global and local levels of hierarchical patterns. Journal of Experimental Psychology: Human Perception and Performance, 11, 673-688.

Kimchi, R., & Merhav, I. (1991). Hemispheric processing of global form, local form, and texture. Acta Psychologica, 76, 133-147.

Kimchi, R., Gopher, D., Rubin, Y., & Raij, D. (in press). Dichoptic and binocular viewing: Effects of attention and task requirements. Human Factors.

Kinchla, R. A., Wolfe, J. M. (1979). The order of visual processing: "Top down," "bottom up" or "middle-out". Perception & Psychophysics, 25, 225-31.

Klein, R. M., & Barresi, J. (1985). Perceptual salience of form versus material as a function of variations in spacing and number of elements. Perception & Psychophysics, 37, 440-446.

Koffka, K. (1963). Principles of Gestalt Psychology. New York: Harcourt, Brace & World. (Originally published, 1935.)

Kohler, W. (1929). Gestalt Psychology. New York: Liveright.

Kohler, W. (1971). Human perception. In M. Henle (Ed. and trans.) The Selected Papers of Wolfgang Kohler. New York: Liveright. (Originally published in French, 1930.)

Kolinsky, R., & Morais, J. (1986). Evidence for early extraction of emergent properties in visual perception: A replication. Perceptual and Motor Skills, 63, 171-174.

Kosslyn, S. M. (1986). Toward a computational neurophysiology of high-level vision. In T. J. Knapp & L. C. Robertson (Eds.), Approaches to Cognition: Contrasts and Controversies (pp. 223-242). Hillsdale, N.J.: Erlbaum.

Kubovy, M., Pomerantz, J. R. (Eds.) (1981). Perceptual Organization. Hillsdale, NJ: Erlbaum.

Lamb, M. R., & Robertson, L. C. (1988). The processing of hierarchical stimuli: Effects of retinal locus, locational uncertainty, and stimulus identity. Perception & Psychophysics, 44, 172-181.

Lasaga, M. I. (1989). Gestalt and their components: Nature of information-precedence. In B. Shepp & S. Ballesteros (Eds.), Object perception: Structure & Process (pp. 165-202). Hillsdale, NJ: Erlbaum.

Martin, M. (1979). Local and global processing: the role of sparsity. Memory and Cognition, 7, 476-484.

Miller, J. 1981. Global precedence in attention and decision. Journal of Experimental Psychology: Human Perception and Performance, 7, 1161-74.

Navon, D. (1977). Forest before trees: The precedence of global features in visual perception. Cognitive Psychology, 9, 353-83.

Navon, D. (1981). The forest revisited: More on global precedence. Psychological Research, 43, 1-32.

Navon, D. & Norman, J. (1983). Does global precedence really depend on visual angle? Journal of Experimental Psychology: Human Perception and Performance, 9, 955-965.

Palmer, S. E. (1977). Hierarchical structure in perceptual representation. Cognitive Psychology, 9, 441-474.

Palmer, S. E. (1978). Fundamental aspects of cognitive representation. In E. Rosch & B. B. Lloyd (Eds.), Cognition and Categorization (pp. 259-303). Hillsdale, NJ: Erlbaum.

Palmer, S. E. & Kimchi, R. (1986). The information processing approach to cognition. In T. J. Knapp & L. C. Robertson (Eds.), Approaches to Cognition: Contrasts and Controversies (pp. 37-77). Hillsdale, NJ: Erlbaum.

Paquet, L. & Merikle, P. M. (1988). Global precedence in attended and nonattended objects. Journal of Experimental Psychology: Human Perception and Performance, 14, 89-100.

Paquet, L. & Merikle, P. M. (1984). Global precedence: The effect of exposure duration. Canadian Journal of Psychology, 38, 45-53.

Pomerantz, J. R. (1981). Perceptual organization in information processing. In M. Kubovy & J. R. Pomerantz (Eds.), Perceptual Organization (pp. 141-179). Hillsdale, NJ: Erlbaum.

Pomerantz, J. R. (1983). Global and local precedence: Selective attention in form and motion perception. Journal of Experimental Psychology: General, 112, 511-535.

Pomerantz, J. R., Sager, L. C., & Stoever, R. J. (1977). Perception of wholes and their component parts: Some configural superiority effects. Journal of Experimental Psychology: Human Perception & Performance, 3, 422-435.

Pomerantz, J. R., & Kubovy, M. (1986). Theoretical approaches to perceptual organization. In K. R. Boff, L. Kaufman, & . P. Thomas (1986). Handbook of perception and performance, Vol. 2 (pp 36:1-46). New York: Wiley.

Pylyshyn, Z. W. (1979). Validating computational models: A critique of Anderson's indeterminacy of representation claim. Psychological Review, 86, 383-394.

Robertson, L. C. (1986). From Gestalt to Neo-Gestalt. In T. J. Knapp & L. C. Robertson (Eds.), Approaches to Cognition: Contrasts and Controversies (pp. 159-188). Hillsdale, NJ: Erlbaum.

Rosch, E. (1978). Principles of categorization. In E. Rosch & B. B. Lloyd (Eds.), Cognition and categorization (pp. 27-48). Hillsdale, NJ: Erlbaum.

Rumelhart, D. E. & Norman, D. A. (1988). Representation in memory. In R. C. Atkinson, R. J. Herrnstein, G. Lindzey, R. D. Luce, (Eds.), Handbook of Experimental Psychology. New York: Wiley.

Santee, J. L. & Egeth, H. E. (1980). Selective attention in the speeded classification and comparison of multidimensional stimuli. Perception & Psychophysics, 28, 191-204.

Sebrechts, M. M. & Fragala, J. J. (1985). Variation on parts and wholes: Information precedence vs. global precedence. Proceedings of the Seventh Annual Conference of the Cognitive Science Society, 11-18.

Shepp, B. E., & Ballesteros, S. (Eds.) (1989). Object perception: Structure & Process. Hillsdale, NJ: Erlbaum.

Smith, J.D., & Kemler Nelson, D.G. (1984). Overall similarity in adults classification: The child in all of us. Journal of Experimental Psychology: General, 113, 137-159.

Smith, L. B., & Kemler, D. G. (1978). Levels of experienced dimensionality in children and adults. Cognitive Psychology, 10, 502-532.

Titchener, E. (1909). Experimental psychology of the thought process. New York: MacMillan.

Treisman, A. (1985). Preattentive processing in vision. Computer Vision, Graphics, and Image Processing, 31, 156-177.

Treisman, A. (1986). Properties, parts, and objects. In K. R. Boff, L. Kaufman, & . P. Thomas (1986). Handbook of perception and performance, Vol. 2 (pp 35:1-70). New York: Wiley.

Treisman, A., & Paterson, R. (1984). Emergent features, attention, and object perception. Journal of Experimental Psychology: Human Perception and Performance, 10, 12-31.

Uttal, W. R. (1988). On seeing forms. Hillsdale: N.J.: Erlbaum.

Ward, L. M. (1982). Determinants of attention to local and global features of visual forms. Journal of Experimental Psychology: Human Perception and Performance, 8, 562-581.

Ward, L. M. (1983). On processing dominance: Comment on Pomerantz. Journal of Experimental Psychology: General, 112, 541-546.

Wertheimer, M. (1955). Gestalt theory. In W. D. Ellis (Ed.), A source book of Gestalt psychology (pp. 1-16). London: Routhedge & Kegan Paul. (Originally published in German, 1923.)

Winston, P. H. (1975). Learning structural descriptions from examples. In P. H. Winston, (Ed.), The psychology of computer vision. New York: McGraw-Hill.

Wundt, W. (1874). Grundzuge der physiologischen psychologie. Leipzig: Engelmann.

Commentary

Structure and Process in Perceptual Organization, R. Kimchi & M. Goldsmith

JAMES T. ENNS
University of British Columbia

A Martian vision researcher, on a first visit to earth, would not be able to miss the fascination that Earthling researchers hold for a peculiar sort of stimulus. It consists of a large shape--often a letter of the English alphabet--made up of many smaller shapes--often also letters of the alphabet. Although the present popularity of this stimulus might at first suggest to the Martian that it was a recent innovation, careful reading of older Earthling articles would soon reveal its distinguished history as a behavioral probe into the hidden workings of the human visual system.

Since its recent re-introduction by Navon (1977), this stimulus has played an important role in several areas of visual perception, including work on attention, the time course of perception, multiple spatial frequency channels, the relative importance of parts and wholes in determining a final percept, interactions among processing streams, hemispheric differences in pattern perception, and the development of perception in children. Its wide popularity has sprung largely from the assumption that this stimulus permits the separation of perceptual processing into "local" and "global" streams. Although the theoretical concepts of *local* and *global* have varied from area to area, the operational definition has been clear--*local* refers to the small letters and *global* to the large one.

Kimchi and Goldsmith force all researchers to reconsider this assumption. Does this stimulus, in fact, separate the local from the global information? The empirical and logical evidence amassed argues for a strong "No." The authors show that it all depends on the number and size of the little elements (letters or shapes) that make up the big form (letter or shape). With a small number of elements, the local level is seen as an integral part of the global level. This subjective impression is supported by performance measures of similarity judgment, speeded classification, and speeded comparison. With a large number of elements, the local level is seen as a separable, or interchangeable, aspect of the global form. As such, the small elements do not have a perceptual status equal to the large form. The unmistakable conclusion is that the structure of these popular stimuli cannot be predicted on purely logical or physical grounds, as has been assumed for so long. The *psychological* structure depends on the visual system's interpretation--the elements can be either *form* (integral parts of the larger whole) or *material* (interchangeable placeholders with a characteristic texture).

What are the consequences of this result for future research and theory? Kimchi and Goldsmith spell out several, including the importance of defining stimulus structure psychophysically, the danger of predicting how a stimulus will be processed from its logical structure, and the inherent ambiguity of the conventional test for so-called global precedence. I would like to focus briefly on what I see to be an even more subversive implication--subversive even to the view espoused by Kimchi and Goldsmith. It concerns an assumption that is operative throughout this chapter, namely, that there is *one* stimulus structure to be uncovered by careful psychophysical testing.

Kimchi et al.'s results show that the interpretation of the identical local elements will change as a function of the surrounding context (i.e., the number and size of other local elements). Surely the representation of the same element has not changed at the level of the retina. However, and this seems equally clear, *some* representation has changed in order to produce the obtained behavioral data. To me, this argues for a view in which the phrase *stimulus structure* refers to nothing more (or less) than the representation available to the processing mechanism which governs the response. In other words, its meaning is inherently relative. In theory, the physical structure of the stimulus can be contrasted with its structure at the ganglion level, this structure can be compared with that at the striate cortex, an even more complex representation may be available to conscious awareness. Ultimately, it will be important to know *which* structure is responsible for a particular psychophysical outcome or for the percept that is experienced phenomenologically.

At this stage in the history of vision science, it is asking too much of any psychophysicist to know the correct mapping between behavioral tasks and level of processing (see Spillman & Werner, 1990, for chapters addressing this issue). However, some first approximations already exist and are hinted at by Kimchi and Goldsmith. For example, measures of preattentive vision such as "pop out" visual search and "immediate" texture segmentation allow stimulus structure to be examined prior to the onset of attentive processing mechanisms. The psychophysical technique of selective adaptation can also be used to study representations that may not be available for conscious inspection. The challenge, of course, will be to develop a series of such tools for probing vision at a large number of different levels.

The important contribution of Kimchi and Goldsmith toward this endeavor has been to show the importance of defining stimulus structure with converging psychophysical tests. What remains for future research is the design of tests that isolate the *multiple* stimulus structures in the visual system.

Navon, D. (1977). Forest before trees: The precedence of global features in visual perception. Cognitive Psychology, 9, 353-383.

Spillman, L., & Werner, J.S. (1990). Visual perception: The neurophysiological foundations. Academic Press.

Percepts, Concepts and Categories
B. Burns (Editor)

4

On Identifying Things: A Case For Context

GREGORY R. LOCKHEAD
Duke University

I. OVERVIEW

Two assumptions organize the work summarized in this article. One concerns process and the other concerns structure. Concerning structure, it is assumed that differences between objects in memory can be approximated by mapping similarity judgments into a geometric space, where the distance between points in the space is a monotonically decreasing function of the similarity between the objects. Concerning process, it is assumed that attention can be directed or moved about the memory space. A reviewed model proposes how judgments depend jointly on the location of attention when a stimulus occurs and on the distribution of potential stimuli in the space.

Data from many psychophysical and cognitive studies are consistent with the model. These include response times to judge similarities of birds and fruits, category judgments of animal names, response times and accuracy to identify tones, colors, rectangles, and line pairs, priming effects when people make same-different judgments of colored patches, and sequence effects. In every instance considered, performance is related to similarities among items that belong to the same category (a Within groups effect) and to similarities between items that belong to different categories (a Between groups effect). Objects and features, which are important for many other reasons, do not predict judgments. Rather, the context for the stimuli predicts both response choices and response times.

Perhaps you have sometimes thought you recognized a friend on the street, only to discover you were wrong when you greeted the stranger. This embarrassment has happened most often to me when I was in a novel place, one in which the context was very different from that in which I knew the friend. Perhaps the opposite has also happened to you. Perhaps you have readily and correctly recognized a friend, even though he was different in that he had shaved his beard and was sporting new glasses. And, most glaringly, you knew something was different about him but had no idea what it was. Many people report such experiences.

Such confusions demonstrate there is a global aspect to recognizing people and things. We often correctly identify a familiar object even though parts have been removed or added or changed, and we sometimes incorrectly identify an object or person seen in a novel context. This means that identification is not always a precise match between a stimulus and some template-like memory. Probably it is never an exact match. Identification is not a simple "yes" or "no" response that reflects the presence or absence of some precise physical event (cf. Labov, 1973; Lakoff, 1972; Lockhead, 1972; Rosch, 1975; Wittgenstein, 1958). Instead, just as for most categorization tasks, it is an approximation procedure. The selected response is chosen because the stimulus in its context is sufficiently like the memory for this response, and sufficiently unlike other available responses, that this response is the best choice available. That is, responses are largely determined by relations among the stimulus object, the context in which it appears, and memories of possibilities that might appear. This paper examines how classifications of simple stimuli and natural objects depend on such relations.

Things that are similar are likely to be classified together. This is a platitude; what makes things similar is not understood. To study this similarity problem, some researchers have examined what particulars of an object make it appear like some other object and what particulars make it appear different from other objects. A common such approach has been to ask what elementary qualities compose a thing and which of those elements make one thing appear different from, or similar to, other things.

This search for aspects has a long history. Perhaps it began with Plato's inquiry concerning elements. One reason it continues is it is supported by the introspectively obvious fact that aspects of things can be described and, thus, parts or elements of objects are psychologically real. Hence, many researchers ask what are the features or the elements or the configurations that make some objects distinctive from other objects, and they ask what is it of these particulars that results in objects being confused. This question is: what features make some objects similar and others different?

The focus of this article is a slightly different but again a frequently discussed approach to understanding how we identify things. Rather than ask what physical features of objects determine performance, this approach asks how performance in identifying stimulus X is determined by *perceived relations* between X and other things. The above search for elements assumes that, ultimately, it is the stimulus or aspect(s) of it that makes it easy or difficult to classify. This relational approach considers, instead, that *any* stimulus can be easy or difficult to classify. This is because it is not the stimulus or features of it that determines performance. Rather, it is relations between it and other, or potential other, things that are determining. This aspect of the arguments in this paper is consistent with Garner's (1962) concept of *assumed subsets*, the idea that choices are made in terms of what alternatives the subject considers likely.

A second and equally important aspect of the position reviewed here, and which is again different from analytic views, is people do not ordinarily decompose each stimulus into independent elements and then identify the stimulus in terms of those components or some combination of them. This is consistent with Miller and Johnson-Laird (1976, p. 46) who "would like to think that the form of an object percept occurs first" and that attributes are then available to be abstracted. The hypothesis is that a stimulus is initially processed as a separate thing, as what has been inelegantly called a *blob* to emphasize that the precision of locating a perceived stimulus in some representative mental space is often poor and that "Subsequent processing of the blob into its components...occurs in a subsequent stage if the task requires. For object identification tasks no later stages of processing are required" (Lockhead, 1972, abstract) as long as the set of potential stimuli has been learned by the observer.

The notion is that it is objects that are ordinarily perceived. This does not mean elements of perceived objects cannot be abstracted or are not important. These must be available and valuable since we use them in descriptions of perceived objects and we use them in discussing relations among objects. The notion is simply that many simple stimuli and natural objects or, more precisely in terms of the theoretical view to be pursued, locations in a representative mental space are perceived first. How precisely those locations must be identified depends on task demands and what other possibilities might occur, on the available context.

There is also a further restriction. It is not suggested that this notion generalizes to all objects. The ideas here are restricted to two classes of stimuli. This is to simple stimuli like those commonly used in psychophysical tasks and to more complex stimuli when they are naturally occurring objects or names of naturally occurring objects. It is likely the ideas expressed here do not generalize to objects made by people to satisfy some function. Features are essential to most functional, manmade objects; a chair without a place to sit is not a chair. Features may even be defining for artifacts (Barton & Komatsu, 1989). This is not the case for most naturally occurring objects; a man without legs is still a man (cf. Putman, 1975; Barton & Komatsu, 1989). Said differently, the concepts pursued in this paper concern objects that are judged in terms of their essences and may not predict performance with stimuli that are judged in terms of their functions.

To support the view that perceived similarity among natural objects determines performance, it is necessary to have measures of similarity and it is necessary to determine if and how other performance measures, such as errors and response times when stimuli are classified, are correlated with similarity measures. Also, documentation is needed, to the extent possible since the null hypothesis cannot be proven, that object identification is never an absolute event in regard to any aspect of an object. Rather, identification it is always a function of the perceived relations among objects.

The following sections review and integrate some previous work on this identification problem and present some new observations. The similarity problem is considered first, followed by demonstrations that similarity or some other measure of relations is needed in order to predict classification and identification. This is because, it will be concluded, physical measures of stimuli are not adequate.

II. SIMILARITY RELATES TO PERFORMANCE: SOME DEMONSTRATIONS AND MEASURES

The background to the issue of similarity is too extensive to review here. The intent of this section is only to document that similarity predicts performance with simple stimuli and natural objects in ways that are important to our understanding of how such objects are identified. For this purpose, the first study reviewed is one in which response times were recorded while people judged which of two words was more similar to a third word. People were asked to decide, as quickly as they could, which is more like a chicken, a duck or an eagle? Or, which is more like a plum, an apple or a goose? There were many such questions. The time people took to make the various decisions correlates with their judgments of the similarities between the objects. It took longer to decide which is more like a chicken, an eagle or a sparrow, than to decide which is more like a chicken, a goose or a sparrow, and it took very little time to decide that a plum is more like an apple than a parrot (Hutchinson & Lockhead, 1977).

These decision times correlated highly with measures of judged similarity. One general result of the study is that response times were long when the two comparison words were similar to each other. Thus, concerning Figure 4.1 which displays two dimensions of the similarity relations among a subset of the words studied, it is harder for people to report if sparrow or robin is more like parrot than to report if sparrow or eagle is more like parrot. Another general result is the decision times were long when the comparison words were each different from

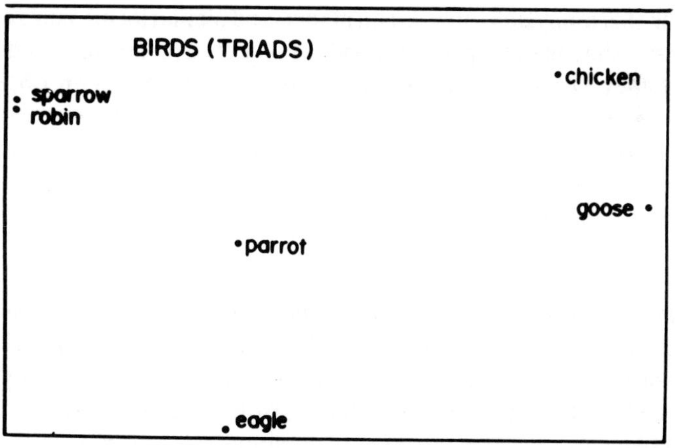

Figure 4.1. The similarity relations among a subset of words (from Hutchinson & Lockhead, 1977, with permission).

the referent word. It was easier to report if a parrot or eagle is more like robin, than if a chicken or eagle is more like robin.

Effects like these are reminiscent of the Weber-Fechner effect which is fundamental to classical psychophysics. When simple domains like loudness and brightness are studied, relative judgments become more variable when the difference in discriminability between stimuli is increased, and response times to discriminate the difference between stimuli increase when the physical difference between the stimuli is decreased. The results when people judged relations among words suggest a similar psychophysical function relating response times and similarity for these abstract items as the Weber-Fechner effect shows for tones and lights and other simple stimuli. If similarity is treated as discriminability, then psychophysical judgments of simple stimuli are consistent with these semantic judgments of words.

If this should be the case in general, then all that is needed for a psychophysical approach to be meaningful when stimuli are mapped according to similarity rather than according to a physical continuum is for there to be monotonicity between similarity and each dependent variable, such as response times and errors. The necessary assumption is that "The relations between objects in semantic memory are described by mapping those objects into a metric space according to the assumption that distance between points in that space is a monotonically decreasing function of the similarity distance between objects" (Hutchinson & Lockhead, 1977, p. 644).

For this approach, only performance and similarity measures are needed. The elements that make up the individual objects are not involved beyond the extent that they contribute to the similarity relations between objects. There would thus be no need to assume that elements, in and of themselves, contribute to performance. Indeed, because complex objects have many features, the very many interrelations between them can be so overdetermining of the object that removal of any one feature has little effect. This may be why removing or adding a beard contributes little to face identification.

An obvious counter example extends this argument from the context of the features in the face to the context of the face in the environment. This is that adding a beard to one face in a set of beardless faces greatly affects performance when the task is to detect the bearded person. But this feature change also changes similarities. What is needed experimentally is to separate similarity from features before any definitive answer to the identification question can be given. Some such deconfounding is described later in this article.

The study with words reviewed above shows that measures of similarity among semantic items can predict relative judgments concerning those items. It is appropriate next to ask if such similarity relations also predict classification. To examine if the similarity structure among items determines how readily people can classify learned groups of items, forty animal names were chosen as stimuli. People rated these words for similarity. On a scale of 1 - 7, they rated how similar, for example, is bat to wolf, and is koala to rabbit. The best fitting multidimensional scaling solution was made to all 780 possible ratings. A three-dimensional solution fit the data well. The relations among the 40 words on two of those dimensions are shown in Figure 4.2.

It is seen in Figure 4.2 that wolf is closer to eagle than to guinea pig or porpoise. This reflects the fact that people judged wolf to be more like eagle than like guinea pig or porpoise. The other distances in Figure 4.2 can be interpreted in this same manner. Words that are close to one another in the space were judged as, relatively, similar to one another. While it is not appropriate to assign meaningful labels to the axes of similarity solutions, one might consider that the solution in Figure 4.2 reflects the underlying dimensions of size and predacity suggested by Henley (1969). The third dimension of the 3-dimensional scaling solution, which is not shown in the figure, seems to reflect skin characteristics since it separates the animals in terms of skin, scales, feathers, or fur. Details of the analysis are in King, Gruenewald, and Lockhead (1978).

Before these similarity measures were taken, each of these 40 animal names had been assigned to one of four categories on the basis of intuition. (The people who made the similarity judgments had no knowledge of this.) These assignments are noted by open and filled circles and squares in Figure 4.2. Ten animals were classified as big and nice (filled circles), ten as big and nasty (filled squares), ten as little and nice (open circles), and ten as little and nasty (filled circles). For scientific terminology we call this the Big-Little-Nasty-Nice classification task, or BLNN.

The clusters in Figure 4.2 show some relation between similarity judgments and these independent classifications; items assigned to the same category tend to be near one another in similarity space. There are only two marked deviations from this generality. I had classified turkey as big but it was judged as similar to little animals, and I had classified bluejay as nasty but the subjects rated it as similar to nice animals. Otherwise, the fit is generally reasonable. This is convenient only for reasons of face validity; the fit does not matter for the purpose of the experiment which was conducted to learn if the

Figure 4.2. The similarity relations among 40 nouns (from King et al., 1978, with permission). Circles and squares denote words classified as nice and nasty, respectively. Filled and open labels denote words classified as big and little, respectively.

separation among words in similarity space predicts how readily people classify those words into categories.

To examine this question, twelve people were asked to sort various subsets of the words into two bins. For each of 12 different tasks, the subjects first learned the category assignment for each item by memorizing ten items listed on each of two cards. For example, in one task only names of "nice" animals ever occurred, i.e., filled versus open circles in Figure 4.2. One of the 20 words was selected randomly on a trial and displayed. The subject pressed the left of two buttons if the word belonged with one group and pressed the right button if the word belonged with the other group. Subjects were urged to respond quickly but also to try to not make errors.

In like manner, every cell was sorted against every other cell in the 2 x 2 matrix. There are six such tasks (four univariate tasks, e.g., nice-big vs nice-little words were sorted; and two redundant tasks, e.g., nice-big vs nasty-little). Too, every pair of cells was sorted against every other pair of cells. There are three such tasks (two orthogonal tasks, e.g., nice independent of size vs nasty independent of size; and one condensation task in which members of one diagonal were sorted against members of the other diagonal). Many researchers have reported data from such 2 x 2 tasks (e.g., Garner, 1974; Pomerantz, 1990) when there was only one stimulus per cell; whereas there were 10 different stimuli per cell in these tasks.

The median sorting times for each condition are shown in Figure 4.3. When only nice animals occurred and were classified as big or little, the median sorting time was 295 msec. This is one of the four univariate conditions; times for the other univariate conditions were 333, 297, and 299 msec. Sorting was reliably faster in the correlated cases than in these univariate conditions. The average

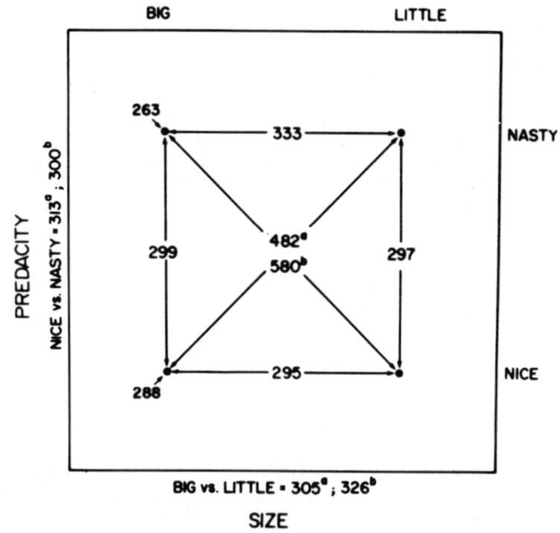

Figure 4.3. The times required for subjects to classify various subsets of the words in Figure 4.2 (from King et al., 1978, with permission).

time to sort big-nasty vs. little-nice was 263 msec, and the time to sort big-nice vs. little nasty was 288 msec. This improved sorting of correlated as compared to univariate stimuli has been called a redundancy gain in the literature (Lockhead, 1966; Garner, 1974). However, this result is reinterpreted, instead, as due to distances between stimuli in similarity and not due to redundancy (Lockhead, 1972).

Performance in each of the four orthogonal tasks (300, 305, 313, and 326 msec) was poorer than that in all but one of the four univariate tasks. This replicates the usual finding for integral stimuli using one stimulus per cell in these classification tasks. This result has often been called a filtering or interference effect in the literature. The reason is the theoretical guess that trial to trial variation in an irrelevant attribute of a stimulus somehow interferes with the processing of the relevant attribute. Here, this would be interpreted to mean, for example, that random variation in size interferes with the ability to judge predacity. This result is also interpreted as, instead, due to similarity differences (King et. al., 1978).

Performance was worst of all in the condensation task. Here, both little-nasty and big-nice animals were assigned one response, while little-nice and big-nasty animals had the other response. Those response times were 580 and 482 msec for the two different sets of 20 words available for this condition.

Similarity measures describe all of these performance differences. In general, performance was easy (fast and accurate) when words that belonged to separate bins were different in similarity, and performance was difficult when words that belonged in the same bin were separated in similarity. To examine this more formally, for each of the nine sorting tasks the similarity distance of each item in one category to that of every item in the other category was calculated. This is a measure of how the two categories are separated in similarity. This measure of the difference in similarity *Between* the categories is called *B*. *B* correlates -0.64 with response times in these data. The task is easier when *B* is large.

The similarity between each item and every other item *Within* each subgroup was also calculated. This measure of the spread or separation in similarity among items assigned to the same category is called *W*. *W* correlates 0.63 with response times. The further items of the same category were separated, the more difficult it was to classify them into the same response bin.

These measures of *B* and *W* were described by King et al. (1978) as follows. To define *B*, M is the number of items in category X, N is the number of items in category Y, and d_{mn} is the euclidean distance between the mth item in X and the nth item in Y, and

$$B = \frac{1}{MN} \sum_{m=1}^{M} \sum_{n=1}^{N} d_{mn}$$

To define W, m is an item in category M, and j is some other item in category M, the value of W in any single category is given by

$$W = \frac{\sum_{m=1}^{M-1} \sum_{j=m+1}^{M}}{\binom{M}{2}} d_{mj}$$

Joint effects of *B* and *W* were also examined. There is no theoretical reason *B* and *W* should be strongly related and they were not in these data ($r = 0.18$, n.s.). The multiple correlation among *B*, *W*, and response times for the six tasks that involve only two cells was calculated. These are the four univariate conditions (for example, big, nice names sorted against little, nice names) and the two correlated conditions (big-nice vs. little-nasty; big-nasty vs. little-nice). That correlation is 0.99.

Using this regression equation calculated for the two-cell conditions, performance in the four-cell conditions (orthogonal and condensation tasks) was predicted rather well (King, et al., 1978, Table 3). *B* and *W*, captured most of the variability in sorting times and in errors when people classified subsets of these words.

These results are encouraging support for a model of performance based on similarity. Similarity or discriminability measures provide a geometric interpretation of a psychophysical transfer function between objects and performance. Such transfer functions might eventually form a basis for a model of psychophysical relations among objects. Should this be so then, just as for psychophysical models of univariate data, physical differences between objects

```
A          . .           EASY

B          ..   .        HARDER

C          ..      .     HARDEST

D    .     ..            HARDEST


E    ..   ........   ..  EASY

F  ..     ........     ..HARD
```

Figure 4.4. Relative stimulus spacings used in various identification tasks.

would not be used to predict performance. Rather, perceived differences made apparent by some transfer function would relate stimuli to performance.

A. Categorizing Univariate Stimuli

Psychophysicists relate physical measures of stimuli to discriminability by means of psychophysical transfer functions. As examples, loudness and brightness are linear in psychological differences (when stimuli are equally spaced on a logarithmic scale) but not directly in linear physical differences.

Psychophysical transfer functions can provide uniform descriptions of judgments when people label sets of univariate stimuli. For example, Parducci and Perritt (1971) demonstrated that mean judgments of tones depend on the *distribution* of stimuli along the psychological continuum rather than directly in terms of sound pressure.

Stimulus distribution affects response variability as well as response means. Identifications of a stimulus are more variable when the stimulus range is larger (Gravetter & Lockhead, 1973). Some examples are shown in Figure 4.4. For each stimulus set depicted, a dot represents a loudness on an equidiscriminability (psychological) scale. The experimental task involved a series of trials using one of the stimulus sets, A through F, where subjects identified randomly presented tones.

Consider sets A and B. The quietest two tones in set B are harder to identify (more errors, slower response times, greater response variability) than are

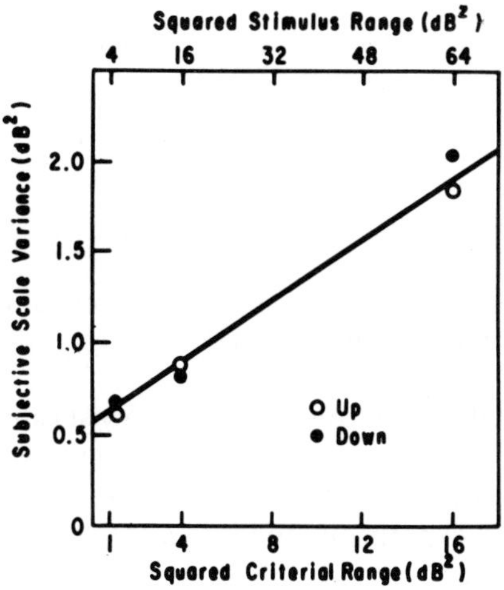

Figure 4.5. Variability of responses to the identical two stimuli in sets like A - C in Figure 4.4 when the spread stimulus was made louder [open circles] or quieter [filled circles] (from Gravetter & Lockhead, 1973, with permission).

the identical tones in set A. These same tones are even more difficult to identify in sets C and D. Thus, the larger the range over which stimuli vary, the more difficult it is for people to identify members of the set. Figure 4.5 displays some of these results showing that *response variability increases monotonically with stimulus range.*

As an aside, note how this observation may relate to theoretical concerns with channel capacity (Miller, 1956). Channel capacity is ordinarily described as a limited number of alternatives, e.g., 7 +/- 2, one can categorize reliably along a continuum. However, the results associated with Figure 4.4 suggest that stimulus range, rather than number of stimuli, is what is important. Range and number are confounded in those studies of channel capacity in which more and more stimuli are added to the set until subjects make identification errors. Such a procedure increases both range and number of stimuli. When these variables are deconfounded and pairwise discriminability is also controlled, it is seen that performance depends on range and not the number alternatives (Holland, 1968). Hence, the central stimuli in set E of Figure 4.4 are easier to identify than the identical stimuli in set F. Apparently, channel capacity is associated with stimulus range and not stimulus number.

The above demonstrations concern identification tasks. These are one-to-one mappings of stimuli onto responses; each stimulus is assigned a unique

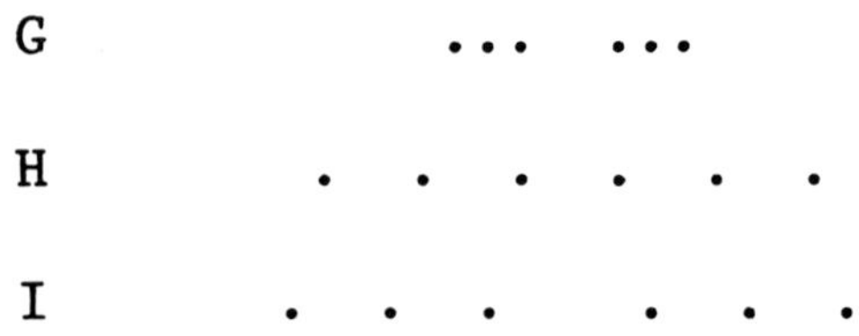

Figure 4.6. Stimulus spacings for which classifying the central two tones into different groups is difficult (set H) or easy (sets G and I).

category. This is a special case of categorization. Most categorization studies use many-to-one mappings; stimuli are categorized into groups and many stimuli are assigned the same response. Consider Figure 4.6 which depicts sets of tones that vary in loudness. For the experiment, tones in a set are randomly presented one at a time. The subject's task is to classify the quietest three stimuli into one bin and the loudest three into another bin. Although the central two stimuli in sets G and H are identical, these tones are easier to classify in G than in H. This is another demonstration the range effect (which, in turn, seems actually to be a sequence effect; cf. Lockhead, 1989).

The data reported in Figure 4.5 allow suggesting that G is easier than H because performance is proportional to range. If this is so, then in order to not change performance when classifying the central stimuli when stimulus range is increased, it is necessary to increase the group differences in proportion to the range change. This is done in set H. The result is the central stimuli in sets G and I are sorted with about equal facility (Gruenewald, 1978).

B. Categorizing Bivariate Stimuli

The above discussion shows that the classification of stimulus elements depends on the relations among all members of univariate stimulus sets. One conclusion from this work is that performance, discriminability, and relative similarity are closely related. The following summarized studies expand on this and examine if similarity also predicts the classification of more complex stimuli.

Some studies conducted for this purpose used rectangles that varied in height and width as stimuli. One study was an identification experiment like that with univariate tones reported in Figure 4.5 but now using subsets of rectangles as the stimuli. One result is rectangles are identified faster and with fewer errors when members of the set are more different from one another in similarity, i.e., when B is large.

Importantly, physical measures of the rectangles did not predict performance. Neither response times nor errors correlated with height, width, area, or diagonal extent. Just as when tones and animal names were classified as reported above, similarity predicts response times when rectangles are classified and physical features do not predict performance. Specifically, when there was one rectangle per category (identification), performance was better when B was large (Monahan & Lockhead, 1977, Exp. 1), and when there were several rectangles per category (classification), performance was better when B was large *and* when W was small (Gruenewald 1978).

In summary, B and W predict anchor effects in absolute identification tasks, interference effects in same-different tasks, and range effects in both relative and absolute judgments, and these are the case whether the stimuli are loudnesses, rectangles, or nouns. Such a consistent organization of results makes appealing the idea that similarity structure is important to performance.

However, similarity is not a satisfying concept. It is a response measure and thus only provides response-response correlations. Essentially, it is a dependent variable that can provide converging arguments when response times and errors and other dependent variables are also measured, but it does not allow interpretations of causation. Another reason to be dissatisfied is that similarity is a relative measure rather than an absolute measure. This is necessarily the case because ratings are done on an arbitrary scale (usually 1-7) and values assigned depend on what alternative stimuli are used in the set.

We often want absolute measures and we generally want to know how and why things occur. For these goals, predicting performance on the basis of similarity is not enough. This is one reason many theorists have proposed stimulus attributes such as features, symmetry, configurations, prototypes, and templates as explanations of how things are processed. Another reason to search a different solution is that similarity judgments do not always conform to mathematical assumptions regarding them. Tversky and Gati (1982) note several such cases, e.g., people judge Cuba is more like Russia than Russia is like Cuba, and argue against using similarity models. Their suggestion may or may not be appropriate. The processes involved in producing similarity judgments are not known: It might be that a different mode of data collection is needed, or perhaps the structure of the stimuli set examined is important to averaged judgments (Krumhansl, 1978), or perhaps countries are processed as manmade objects and differently than natural objects, or many other possibilities. Whatever turns out to be the eventual case, it may be premature to reject the best measure yet available for predicting performance.

Furthermore, although feature and other such approaches are indeed attractive, it is argued ahead that an answer to what determines classification does not lie in the study of attributes of natural objects. Nor is it logically necessary it should. For example, in order to understand gravitational constant, *g*, we now know it is not necessary to know the contributions of physical features of the falling objects. Introspections notwithstanding, independent physical features of objects also may not determine similarity or classification. To examine this, the approach here to learning what determines similarity and classification is to build on what psychophysics has learned. Rather than attempt to directly infer features, mechanisms, and processes from introspections of physical stimuli, the approach here examines psychological structures. Rather than relate physical measures to responses in order to discover the postulated determining features, it is suggested that measures of discriminability or similarity structures might be a more useful referent for understanding the classification of natural objects.

At least as a starting point, the studies above show we generally cannot predict classification or identification from features but we can predict these from similarity measures. Continued such efforts might help us better understand what lawfulness exists between behavior and similarity. If so, we might then work backwards, from similarity to the stimuli, to learn what it is of stimuli or receptor systems or prior experience or their combinations that produces the perceptual or memorial structures of interest. While it is of course too early to know if such an approach will be valuable [see Shepard's theory (1981; 1991) and Medin's (1989) cautions], I take the results summarized in this paper as encouraging.

III. PHYSICAL MEASURES DO NOT PREDICT PERFORMANCE; PSYCHOPHYSICAL MEASURES DO

There are several reports in the literature that elements or features determine classification. Features and elements are confounded with similarity in at least some of those studies. Further, in all cases I know in which these factors have been separated experimentally, perceptual measures have predicted performance and physical measures have not. Some examples are reviewed ahead; other examples may be found in Crist (1981), Gruenewald (1978), Lockhead & King (1977) and Shepard (1981).

A. Simultaneous Context Effects

In all demonstrations of context effects discussed previously in this paper, stimuli were presented successively and so were separated in time. Thus, each stimulus was compared with some memory. Temporal separation is not necessary for context to determine judgments. Simultaneously available identical stimuli are sometimes also seen as different. Too, simultaneous, different objects are sometimes seen as identical. What is needed for such results is for the contexts of the spatially separated objects to be appropriately different.

This is well known in the area of simultaneous brightness contrast. If one gray patch is placed on a white surround and an identical gray patch is placed on an adjacent black surround, then these identical grays will appear different. That on the white ground appears darker than that on the black ground. This demonstrates that the intensity of the patch does not directly determine its appearance. Rather, relations (context) among luminances determine the appearance of any part of the field (Lockhead, 1988; Gilchrist, 1990). Depending on the context, coal can appear white and writing paper can appear black.

Brightness contrast demonstrates that objects available for simultaneous comparison are nonetheless perceived in terms of their contexts. This fact is commonly attributed to inhibitory sensory mechanisms. Whether these mechanisms are different from those involved when objects are compared in memory is not known since a mechanism for memory comparisons is not known. One thing that is known is assimilation between judgments occurs more often than contrast when successive stimuli are judged (Lockhead, 1983). This does not mean temporal and spatial comparison effects are different, however. Simultaneous brightness comparisons can also result in assimilation. Whether there is contrast or assimilation depends on the structure of the stimulus array. Assimilation often occurs in simultaneous brightness comparisons when the context contains gradual, rather than abrupt, luminance gradients (Arend, Buehler, & Lockhead, 1971). Possibly, the mechanisms involved in comparing items that are simultaneously presented and items that are successively presented are not deeply different.

B. Redundancy

It has frequently been suggested that redundancy is important to identification. Of all of the arguments given in support of an elemental approach, this one is probably the most frequent. The idea is that the more physically independent sources of the same information there are, the easier it is to identify the object. This sounds reasonable and is certainly correct when the redundant materials are perceptually independent and the source of errors is related to the observer's inability to obtain certain stimulus information. For example, if auditory information is redundant with visual information, and if either the view of the stimulus is obscured or its sound is masked, then redundancy clearly can be useful. This class of redundancy has been called a state limitation (Garner, 1974). This class of redundancy is not considered here.

The class of redundancy that is considered here was described by a study reported in 1955 by Eriksen and Hake. They measured how redundantly combined information affects identification. In one condition of their study, Eriksen and Hake had people identify ten lightnesses. These were 10 achromatic patches that occurred randomly one at a time on a fixed background and were identical in all regards except lightness. The grays differed from one another in equally discriminable lightness steps (Munsell notation). In a second condition, people similarly identified ten patches that differed only in hue. These are called univariate sets because stimuli differ from each other along only one dimension. People made many identification errors in both of these univariate tasks. In a third condition, Eriksen and Hake used ten patches that were redundant. These patches covaried in lightness and in hue. Knowledge of the hue perfectly predicted the brightness, and vice versa. These bivariate, redundant stimuli were identified more accurately than was either set of univariate stimuli, the lightnesses or the hues.

This result has been taken as evidence that redundant information improves performance. The common, essential idea is people can identify the stimulus on the basis of its lightness, or its hue, or both. Over the past 35 years, this idea has been demonstrated to have considerable appeal. Some variation of it has been used in many process models by many people. Perhaps the most common variant is the horse-race model in which it is assumed that a stimulus is analyzed by the subject into its components (its hue and lightness for the above example), that the components are then each evaluated independently (and in parallel in most models), and that the subject's response is based on the best processed of the two attributes. A variety of related models propose some modification of this view (Biederman & Checkosky, 1970; Pomerantz, 1986; Treisman & Gelade, 1980; others).

A different interpretation of Eriksen and Hake's finding is also possible. Perhaps it is not redundancy of independent features that was important in that study. Instead, perhaps discriminability between stimuli in the set determined

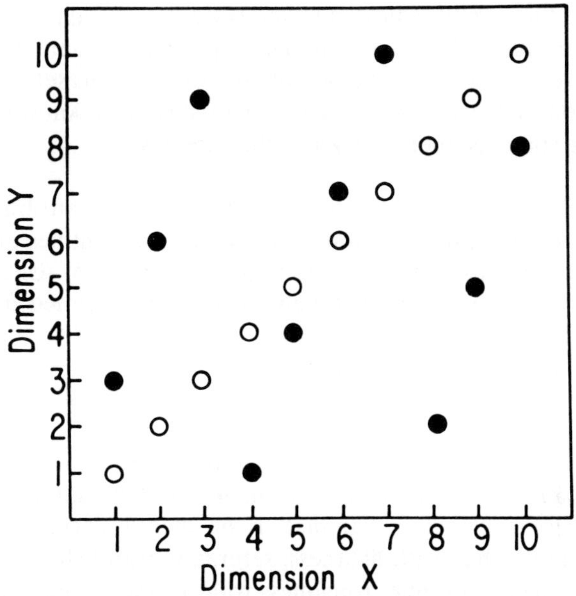

Figure 4.7. Stimulus combinations used for linearly paired (open circles) and scattered (filled circles) redundant stimuli (from Lockhead, 1970, with permission).

performance. Consider Figure 4.7. If lightness and hue are dimensions X and Y, any row and any column represent Eriksen and Hake's univariate hues and brightnesses. The open circles in Figure 4.7 represent their redundant stimuli.

Now consider Figure 4.7 geometrically. The spatial distance between adjacent, redundant stimuli is greater than that between adjacent, univariate stimuli. According to Pythagoras, the redundant stimuli (the open circles) are separated from one another 1.414 times more than are the univariate stimuli (any row or column). If identification performance is determined by discriminability (which might be reflected as the physical distance between stimuli), then these redundant stimuli should be easier to identify than any univariate set and the amount of this improved performance should be predictable. This is indeed the case for at least some stimulus dimensions (Lockhead, 1966).

It happens that predictions of response times by this distance interpretation and by the attribute or horse race model are identical. This means these two classes of models cannot be discriminated here (Lockhead, 1970). A different procedure is needed for this. In order to determine if distances or redundant features better account for the improved performance with these correlated sets, the stimuli indicated by the filled dots in Figure 4.7 were used in an identification study. Just as for the linearly correlated set (the open circles), these stimuli are also perfectly redundant. In both stimulus sets, each value of X is paired with only one value of Y. Accordingly, if subjects know the value of one dimension of a stimulus, then they know the stimulus.

According to the horse-race or any other available attribute model, these stimulus sets are identical. For either set, the subject identifies one or the other attribute of the stimulus. According to a discriminability-by-distance model, the sets are different. The scattered stimuli are further separated from each other than the linearly correlated stimuli. Hence, so the filled-O stimuli should be better identified than the open-O stimuli.

To examine this, these sets were compared using various stimulus sets. In every case, the filled dot stimuli were easier to identify (faster and fewer errors) than the open dot stimuli. Indeed, performance on the scattered, correlated set was regularly very much better than that on the linearly, correlated set. This was found for pairings of hue with lightness, saturation with lightness, loudness with roughness, roughness with hue, roughness with lightness, and lightness with dot position. It was also found for three-dimensional patterns with eye lightness, pupil position, and eye hue correlated in a schematic face, and was found for four dimensional patterns where orientations of lines representing eyes, nose and mouth were correlated in schematic faces (Lockhead, 1970, 1979). To my knowledge, no existing model of physical redundancy or elements can describe these results.

C. The Forms of the Physical and Similarity Spaces

Consider when hue and lightness are the dimensions in Figure 4.7. Whether these dimensions are plotted in the figure in terms of physical measures (wavelength and luminance) or discriminability measures (JND steps) the figures would look pretty much the same, particularly if lightness is graphed in log energy. This fact makes it difficult to determine if physical measures or psychological measures determine the above results. The studies summarized above demonstrate that something about relations or differences between stimuli determines performance, but those studies do not demonstrate the nature of those relations.

These possibilities can be separated in cases where physical and perceptual measures are not monotonically related when dimensions are combined. This often occurs. Then, differences between stimuli are qualitatively different when they are mapped in physical coordinates than in discriminability (or psychological or similarity) differences.

To demonstrate this, consider when long wavelength lights are used for one dimension and short wavelength lights are used for the other dimension (examples using more ordinary dimensions are given after this). Suppose the attributes chosen from these dimensions are a red and a yellow light (dimension X) and a green and a blue light (dimension Y). Each bivariate stimulus is then generated by shining the two relevant lights on the same spot. Figure 4.8 describes the resulting 2 x 2 matrix. It is seen that the two compound, redundant stimuli on the major diagonal are both white (red light plus green light makes white; yellow light plus blue light also makes white). These redundant stimuli are indistinguishable. It is also seen that the two compound, redundant stimuli on the minor diagonal are a purple and a lime. These two stimuli are perceptually very different.

Hence, here is a situation in which physical and psychological spaces are very different and in which no feature theory or redundancy argument can account for the results. The purple and lime stimuli are redundant and are easy to identify. However, the two white stimuli are equally redundant but these are hopelessly confused. Also, members of each of the four univariate sets are easy to classify. Simply, neither features nor redundancy predict performance.

The point of this exercise is to note that when stimulus dimensions interact physiologically or psychologically, then performance in judging complex stimuli cannot be predicted from knowledge of only the separate elements. We must know their interactions. This becomes immediately obvious when colored lights are combined and then scaled for similarity. The members of one diagonal (the whites) are very similar and those of the other diagonal (the purple and the lime) are not. The physical stimulus structure does not reflect either the psychological structure or performance.

One might argue that the demonstration in Figure 4.8 is unfair, perhaps because the stimuli were arbitrarily selected. This would emphasize the fact that "psychological dimension" is not a well defined term (Garner, 1978; Nickerson, 1978). If this objection means the only acceptable dimensions are ones that provide data in agreement with some theory, then that would not be acceptable. Fortunately, this is not a concern because there are many available examples for

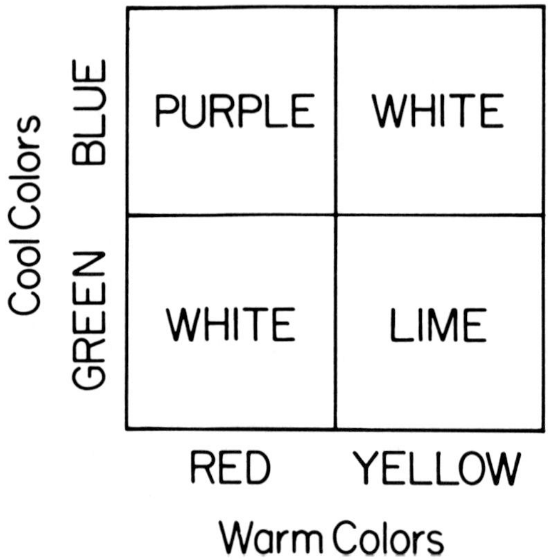

Figure 4.8. Combining long (warm) and short (cool) wavelength colors produces multidimensional lights of various appearances.

the point made by Figure 4.8 that are not liable to the objection that the dimensions are somehow inappropriate.

One such example is shown in Figure 4.9. Here, each bivariate stimulus was generated by placing two independent, vertical lines side by side. Dimension X is the length of the left line and dimension Y is the length of the right line. Extent is an historically frequent dimension. Using all possible combinations of these left and right extents provides a square physical matrix of 49 stimuli.

The central 25 members of this set of line pairs were judged for similarity. People rated the similarity of each stimulus to every other stimulus on a scale of 1 to 7. Those averaged judgments were scaled using a Guttman-Lingoes nonmetric scaling method for finding the smallest coordinate space for a configuration of points. The resulting coefficient of alienation for a euclidean space of two dimensions was 0.14 (the one for three dimensions was 0.08; Monahan & Lockhead, 1977). The locations of these 25 stimuli in the resulting two dimensional similarity space are shown in Figure 4.10.

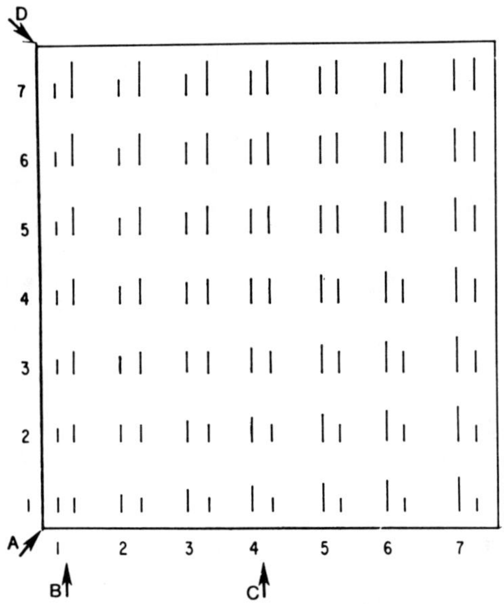

Figure 4.9. A two-dimensional set of 49 stimuli generated by pairing 7 lines of length x, the left line, with 7 lines of length y, the right line. Four subsets used in identification experiments are indicated by arrows; these are redundant sets composing the positive (A) and negative diagonal (D), and univariate sets forming the left (B) and central (C) columns. A scattered, redundant set according to the concept of the filled circles in Figure 4.7 was also tested. (From Monahan & Lockhead, 1977, with permission).

Note that the physical space for these line pair stimuli is square but the similarity (or psychological or psychophysical) space is U-shaped. As for the colors in Figure 4.8, one space does not directly map onto the other. These perceptual differences between stimuli cannot be directly predicted by combinations of the physical features that produced them. The physical and psychological structures are simply different. For example, stimuli 2-2 and 5-5 are widely separated in physical space (Figure 4.9) but are near one another in similarity (Figure 4.10), while stimuli 4-4 and 3-5 are close in physical space but far apart in similarity. There are many such examples.

This stimulus set can be used to examine if classification performance is better predicted by physical structure or by psychological structure. This is because the nonmonotonic relations between the physical and psychological distance measures allow asking if stimulus attributes or if similarity distances better relate to performance.

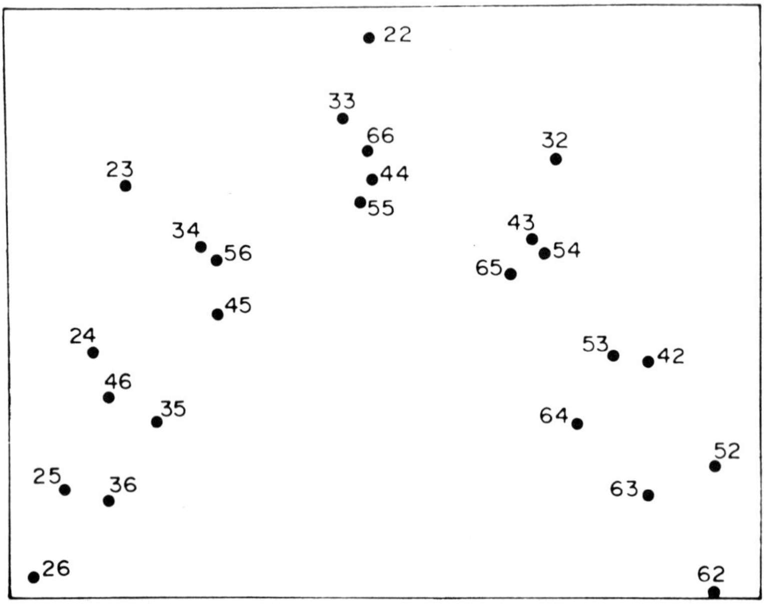

Figure 4.10. Similarity relations among central 25 stimuli of Figure 4.9. Numbers identify stimuli in Figure 4.9. For example, stimuli 22, 33, 44, 55, and 66, from the positive diagonal, were judged as relatively similar and thus are near one another in the similarity space (From Monahan & Lockhead, 1977, with permission).

For this purpose, people were asked to identify members of various subsets of the line-pair stimuli in Figure 4.9. These subsets were: (A) the 7 redundant stimuli composing the positive diagonal, (B) the 7 univariate stimuli composing the left column, (C) the univariate stimuli making up the middle column, (D) the 7 redundant stimuli on the negative diagonal, and (E) 7 redundant stimuli that were scattered about the physically denoted matrix in the manner of the filled dots in Figure 4.7.

The detailed results are published elsewhere (Monahan & Lockhead, 1977) and only selectively summarized here. Members of one redundant set, the major diagonal, were the most difficult of all sets to identify, while members of the other two equally redundant sets, the minor diagonal and the scattered set, were the easiest of all to identify. Too, one non-redundant set, the middle column, was nearly as easy as the scattered and minor diagonal sets, while the other non-redundant set, the left column, was nearly as difficult as the positive diagonal set. *Redundancy is not related to performance in these data.*

The redundant sets (A, D, and E) were made up of identical elements. In each set, one and only one stimulus had each stimulus feature and every feature was used once in each subset. The difference between the sets is simply which features were combined with which others. Because performance on the three sets was very different although there are no feature differences between them, no theory about features or about independent elements can account for these data. *Elements are not related to performance in these data.*

What does predict performance is *relations* between stimuli as measured or as reflected by the results of the similarity scaling study. The distance in similarity between items within each set predicts more than 90% of the variance in the averaged data. In addition to predicting these average performances across sets, similarity also predicts performance on individual stimuli within sets. This is the case for both response times and errors (Monahan & Lockhead, 1977). *Similarity relations predict performance in these data.*

Models of configuration have also been based on this interpretation that perceived relations are important. Those models provide a different conclusion than the one here. The relative similarity view here is that perceived differences between objects determine performance. A configuration view is that particular characteristics of the stimulus determine performance.

To contrast these views, note in Figures 4.8 and 4.9 that different combinations of physical features often result in different configurations or appearances. Some complex stimuli appear very different than others. Such observations have led to the suggestion that the particular *configuration* is what is important. That, for whatever reason, some configurations are intrinsically easier to identify than some others (Pomerantz & Garner, 1973, Pomerantz, 1990).

Whether or not this is sometimes so, such a *configuration theory cannot explain these line-pair data.* To demonstrate this, note that the identical configuration was used in four conditions. The middle stimulus in the physical matrix, 4-4, was used in the positive diagonal (P), negative diagonal (N), middle column (M), and scattered (S) stimulus sets. When 4-4 was a member of set N, it was the easiest stimulus or configuration in that set to identify. However, when 4-4 was a member of set P, it was the most difficult stimulus in that set to identify. More specifically, error rates on P, N, M, and S were, in order, 52%, 2%, 33%, and 38%, and median response times were 1580, 890, 905, and 1190 msec.

Such large differences should not occur if performance is determined by the configuration. In that case, 4-4 should be identified about equally well or poorly in all conditions. It is not. Instead, 4-4 is easy or difficult to identify depending on its perceptual similarity to other stimuli the set. It is context, not configurations or elements or redundancy, that determines performance in these data (Pomerantz & Lockhead, 1991).

These results and conclusions are for identification data. To assure they are not restricted to a particular task, performance on a variety of classification tasks was tested with subsets of these line-pair stimuli (cf. Lockhead & King, 1977). Again, similarity predicted performance and, also again, no available measures of redundancy, configurations, or elements correlated with either accuracy or response times, except when both measures made the same prediction. It can thus be concluded that *the similarity structure of the stimulus set determines classification and identification performance.*

This conclusion does not consider the possibility that both configuration and similarity might determine performance. It might be that some stimuli are inherently easy or difficult to identify and that similarity is additionally important. There is support for this in the literature. Prototypic colors are easier to respond to than are colors rated as poor (Rosch, 1975) and dot patterns or configurations rated as good (Garner, 1974; Palmer, 1991; Pomerantz, 1991) are easier to respond to than are other patterns in the set. Probably many factors are involved in all judgments and perhaps the above observations should be limited to stating that similarity relations are one of these and to emphasizing that the magnitudes of their effects (e.g., 2% vs. 52% errors for the same configuration in different sets) are large.

D. Prototypes

The idea of the prototype is that a representative item stands as the best exemplar of a set of items that belong to one category. Robin is a good bird, as determined by subjects' ratings, and penguin is not; chair is a prototypic item of furniture and ashtray is not; one red patch is a good color but many other colors that are also classified as red are not good.

Based on such findings and a set of priming experiments, Rosch (1975, 1978) importantly decided membership in a category is graded. Some items are better members of a category than are others. She also concluded that decisions about objects are made in reference to prototypes. Her initial and perhaps most compelling arguments were made with colors as stimuli. That work is considered next.

The purpose of these following arguments is not to contest prototypes (or configurations or goodness or symmetry) as important in various situations. I believe they are all important although perhaps not all for the same reason. Rather, the purposes here are to show that structural measures can be useful in a wide variety of studies, including studies of prototypes, and to show that evaluations of some theories might be improved by considering the spatial metaphor pursued next and described further in the discussion.

Eleanor Rosch and others who have followed her lead have nicely shown that there are good or prototypic colors and these are essentially identical across cultures and subjects' ages. Adults in New Guinea who have an impoverished color vocabulary, children living in a poor section of Massachusetts, and students at Harvard University all make essentially the same color choices when they are asked to select "good" colors. Apparently for all people, excluding the color-blind, there are some good reds, a few good blues, etc., and there are very many poor colors.

To demonstrate one importance of these prototypes (the good colors), Rosch did this study: Present (or prime) a person with the word "red" or the word "blue" or the word "blank." Following this, present a pair of colored patches. The observer is instructed to press one button if the two colors are identical and to press the other button if the two colors are different. For my current purposes, the important aspect of Rosch's results is that it takes less time for subjects to report that two prototypic red patches are identical following the word "red" than following the word "blank". Rosch interpreted this as evidence that the prime, the word "red" in the above example, serves as a category reference. The idea is that "red" provides a color to be "called up" and that priming this category reference point facilitates performance.

Consider a slightly different possibility. Consider that the discriminability space might be what is important and prototypes are irrelevant to performance in this same-different judgment task, except to the extent they provide a locus for attention in the discriminability space. That is, the possibility explored is Rosch's results demonstrate that performance is faster when the attended locus in memory and the stimulus are similar. Possibly, a prototype is not otherwise involved in these studies.

To test this possibility, Susan Gaylord, Nancy Evans and I replicated Rosch's experiment except for three differences. First, poor colors as well as good colors were used as primes. Interestingly, Rosch never conducted this control condition. She only primed with a prototype word or the work "BLANK". Second,

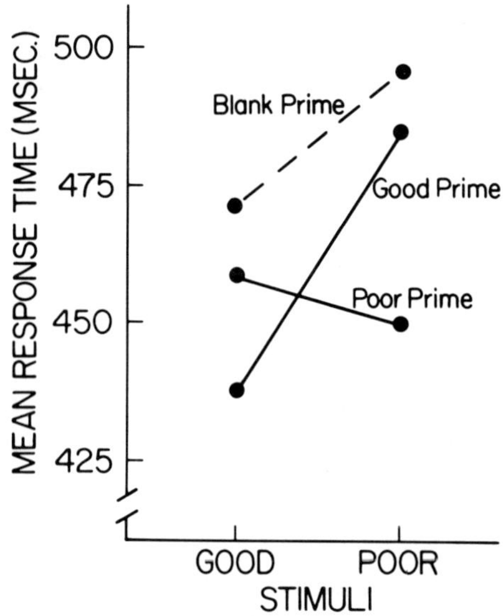

Figure 4.11. The time required for people to report that two identical good or identical poor colors (the X axis) were in fact identical when they were preceded by good or poor primes or no prime.

actual color patches were used as primes. Rosch only used words because of her interest in category priming (we also did this as reported ahead). Third, when the subject was primed with a color, only colors belonging to that category could occur, either two prototypes, two poor colors, or one of each. When Rosch primed with "red", the observer might sometimes be shown a red patch and a blue patch.

The results are shown in Figure 4.11 (Lockhead, Gaylord, & Evans, 1977). The important result is that the prime performs its cuing function whether it is a prototype or a poor color. The correct baseline condition here is when the blank preceded the stimuli. This is because, due to familiarity, or ease of naming, or the effect of a prototype, or any of many other unknowns, some stimuli may be simply easier to respond to than others. In general, good colors were indeed responded to faster than poor colors in this study.

Compared to the baseline condition, when the prime was a good color it took less time to say that two good colors are identical than to say that two poor colors are identical. This is just as Rosch reported. But when the prime was a poor color the reverse happened. It then took less time to say that two poor colors are identical than to report that two good colors are identical.

These results are consistent with one of Rosch's important arguments. Namely, categories are structured. Performance depends on something other than sheer category membership. It depends on perceived relations among items in the category. These results also replicate Rosch's demonstration that performance on good colors is better following a good prime than following no prime. However, the results are not consistent with the inference that it is the prototypicality or goodness of the prime that is important to "same" judgments. That theory cannot explain why poor primes facilitate responses to poor colors.

What again seems to be important is similarity. The distance in similarity between the prime and the stimulus predicts the performance differences. Responses are fast if the stimulus occurs in the perceptual region of the prime, and responses are relatively slow if these two locations are different. Again, measures of independent objects, of the prototype in this case, are not sufficient to predict performance and might not even be involved.

In a following experiment, Evans (1979) showed that this result is not restricted to the use of real objects, i.e., of colored patches rather than words as primes or as stimuli. Using nouns as primes, Evans (1979) showed that *neither prototypicality nor typicality determine performance in categorization tasks*. She used words like "FRUIT", "ORANGE", "PRUNE", and "BANANA" to prime stimulus pairs like "APRICOT - TANGERINE" and "RAISIN - RAISIN". The results were that judged similarity between the prime and the stimulus predicted performance, while neither prototypicality nor typicality were predictive when effects of similarity were partialed out.

Such studies show that responses are facilitated when stimuli are primed by a cue that is similar to them. This conclusion is consistent with Rosch's finding that priming with a prototype facilitates performance on good members of the category. But the reason for this result is apparently not associated with prototypes. Rather, it is because performance is facilitated when the prime is similar to the stimulus, and prototypes are similar to good category members.

IV. DISCUSSION

One essential assumption for this work is that differences between objects in memory can be approximated by mapping similarity judgments into a geometric space. A second essential assumption is that attention can be directed within the space. To combine these assumptions, consider a simile. Suppose attention is like a spaceship flying in discriminability or stellar space, and suppose the task of the pilot of the spaceship is to categorize stars or objects in the space. Further suppose this is accomplished by the spaceship "Attention" flying toward each anticipated stimulus as if to be prepared for it, and flying toward each actual stimulus as if to get a better look at it.

According to these outlandish assumptions, classification would be easy if stars that belong to different galaxies (categories) are well separated in the space. Then B is large. Classification would also be easy if stars that belong to the same galaxy are tightly clustered. Then W is small. In either case, classification would be easier when the spaceship is near the star to be judged than when Attention and this stimulus are widely separated.

Compare the stimulus sets described by Figures 4.4C and 4.4A. For either condition, consider that the extreme stimulus was just presented and so the spaceship Attention moved toward it, and thus away from the other two stimuli. The average magnitude of this movement is greater in 4.4C than 4.4A. When one of the other stimuli next occurs, performance will be poorer in condition 4.4C than 4.4A because Attention is further from the locus of the stimulus. This describes the range effects reported in Figure 4.5.

This interpretation would mean the range effect seen in Figure 4.5 is actually a sequence effect. It occurs because successive stimuli are more different, and more often different, when the range is larger. Since the spaceship Attention moves toward each stimulus in order to classify it, attention will be relatively near the locus of the just previous stimulus when the current stimulus is presented. Thus, in large range conditions, attention will be relatively far from the next stimulus, on average. Because performance is generally poorer when attention is located very differently than the stimulus, classification should be imprecise when successive stimuli are very different, as is the case when the range is large (cf. Lockhead, 1983).

This process describes both averaged results and sequential results. Performance is poorer on trials that successive stimuli are more different (Lockhead & Hinson, 1986), and because successive stimuli are more different on average in conditions in which range is larger, performance is poorer in conditions of large stimulus range (Figure 4.5). This may be a general result. These same effects are also seen in absolute judgments and magnitude estimations (Lockhead, 1983). Probably they occur in all psychophysical tasks. According to the metaphor, this would be because the spaceship is more often located advantageously on trials that successive stimuli are similar, and successive stimuli are more similar, on average, in conditions for which the stimulus range is small.

These effects are also seen in priming studies if enough time is provided after the prime for Attention to move toward that locus in space before the stimulus is presented. When the stimulus then occurs, Attention does not have to move much if the stimulus is near the prime, but must move a lot if the stimulus is very different from the prime. Because movement requires time, this predicts performance is slower when the prime is a prototype and the stimulus is very different from it, than when both are prototypes. This is the result and it is consistent with Rosch's suggestion that category membership is decided by initially accessing the prototype in her studies (Rosch, 1978).

However, the model here also predicts performance is faster when the prime is distant from the prototype and the stimulus is similar to the prime, than when the prime is a prototype and the stimulus is different from it. This is the result (Figure 4.11 for colors as stimuli; Evans, 1979 for words as stimuli). This is not consistent with the idea that access to the category occurs via the prototype. Rather, these results are consistent with the interpretation that all of these priming effects are due to perceived differences between successive events.

Even though the spaceship model has some success in predicting performance, such metaphors can be misleading and should be interpreted cautiously, if at all. The idea of attention being moved about in discriminability space describes several observations and can serve as a shorthand for predicting the results considered, but it cannot be considered an explanation. We do not know what produces similarity. Some unknown correlate might be the cause of all of the data summarized here. Indeed, this lack of understanding, along with only a few inconsistencies in similarity measures, has led some authors to reject similarity as a theoretical tool. While rejection might well be correct, some parallels indicate it may be premature.

Again consider the gravitational constant, g. It was believed for a long time that features of objects determined the rate at which they fall. It was thought that large rocks fall faster than small rocks which fall faster than feathers. Too, physicists obtained slightly different measures of gravitational acceleration when they measured objects falling in different atmospheres and at different locations on earth. But eventually, as the apocryphal Leaning Tower of Pisa story emphasizes, it was learned that perceptually obvious features do not determine g.

Perhaps analogously, what makes objects perceptually similar is also not understood. Perceptual features are not the reason. Too, we get different measures of similarity when an object is judged a different contexts. Possibly, it is premature to reject similarity unless it is replaced with a more successful model. Any such replacement must account for such facts as these: Similarity measures describe performance when people judge relations among nouns (Figure 4.1) and when they identify tones (Figures 4.4, 4.5, 4.8, 4.10 and associated text). Similarity between, *B*, and within, *W*, categories predicts performance when people classify items into bins (Figure 4.6) and when they classify words into learned groups (Figures 4.2 and 4.3). Similarity between successive items, as revealed by trial-by-trial analyses, predicts performance when people identify objects (Lockhead, 1983; Lockhead & Hinson, 1986), make magnitude estimations of attributes (Lockhead, 1983), and make same-difference judgments following a prime (Figure 4.11).

It is important to stress that the metaphor used here relates to similarity space, not physical space. Unless they are correlated with similarity measures, physical measures do not predict performance very well (cf. Figures 4.7 - 4.10 and the associated text). Measures of features, dimensions, templates, configurations, prototypes, and combinations of these do not correlate with performance. Instead, all of the results reviewed here and all similar studies I know in the literature are consistent with the idea of Attention being moved in *perceptual* space to discriminate among regions in the space.

This conclusion that physical measures do not predict classification performance does not mean physical measures have no psychological value. While they may not be important for classifying natural objects, physical measures are important in other ways. For example, features are central for describing objects and relations between them. Dogs are four legged mammals with a tail and fur; fish are cold-blooded vertebrates with gills and who live in water; and so on. Possibly, perceived relations among known possibilities (similarity) are important for classification, while physical features are important for taxonomy and communication. Further, when the stimulus is not known features of natural objects might then be primarily important as people search their knowledge to decide what some novel object might be.

Independent of any merit of the above conjectures, perceived context largely determines response times and errors when people classify natural objects or the names of natural objects. This means it is necessary to estimate the transfer function relating objects to perceptions in order to predict classification. This conclusion has long been accepted as true when simple stimuli are judged.

When tones and lights are the stimuli in psychophysical studies, it is regularly agreed that psychological measures are better related to performance than are physical measures. For instance, rather than reporting the energy of a stimulus light we always report its luminance, which is energy adjusted by the responsiveness of the eye to different wavelengths, and we always report its context. Similarly, the extended arguments here are that psychophysical measures also predict performance when complex stimuli are judged and that all of our dependent measures depend partly or entirely on context.

ACKNOWLEDGEMENTS

Supported by AFOSR-87-0353 to Duke University.

REFERENCES

Arend, L.E., Buehler, J.N. and Lockhead, G.R. (1971). Difference information in brightness perception. Perception & Psychophysics, 9, 367-370.

Barton, M.E. & Komatsu, L.K. (1989). Defining Features of Natural Kinds and Artifacts. Journal of Psycholinguistic Research, 18, 433-447.

Biederman, I. & Checkosky, S.F. (1970). Processing redundant information. Journal of Experimental Psychology, 83, 486-490.

Crist, W.B. (1981). Matching performance and the similarity structure of the stimulus set. Journal of Experimental Psychology: General, 269-296.

Eriksen, C.W., and Hake, J.W. (1955). Multidimensional stimulus differences and accuracy of discrimination. Journal of Experimental Psychology, 50, 153-160.

Evans, N.J. (1979). Priming in semantic categories: an evaluation of the effects of similarity and prototypicality. Ph.D. dissertation, Duke University.

Garner, W. R. (1962). Uncertainty and Structure as Psychological Concepts, NY: Wiley.

Garner, W.R. (1974). The processing of information and structure, Hillsdale, NJ: Lawrence Erlbaum.

Garner, W.R. (1978). Selective attention to attributes and to stimuli. Journal of Experimental Psychology: General, 107, 287-308.

Gilchrist, A. (1990). The perception of surface blacks and whites. In I. Rock (Ed.), The Perceptual World, W.H. Freeman, NY.

Gravetter, F. & Lockhead, G.R. (1973). Criterial range as a frame of reference for stimulus judgments. Psychological Review, 80, 203-216.

Gruenewald, P. (1978). Similarity and the classification of multidimensional stimuli. Ph.D. dissertation, Duke University.

Henley, N.M. (1969). A psychological study of the semantics of animal terms. Journal of Verbal Learning and Verbal Behavior, 8, 176-184.

Holland, M. (1968). Channel Capacity and Sequential Effects: The influence of the immediate stimulus history in recognition performance. Ph.D. Dissertation, Duke University.

Hutchinson, J.W., and Lockhead, G.R. (1977). Similarity as distance: a structural principle for semantic memory. Journal of Experimental Psychology: Human Learning and Memory, 3, 660-678.

King, M.C., Gruenewald, P., and Lockhead, G.R. (1978). Classifying related stimuli. Journal of Experimental Psychology: Human Learning and Memory, 4, 417-427.

Krumhansl, C. (1978). Concerning the applicability of geometric models to similarity data: The interrelationship between similarity and spatial density. Psychological Review, 85, 415-563.

Labov, W. (1973). The boundaries of words and their meanings. In: C.J. Bailey and R. Shuy (Eds.), New ways of analyzing variations in English. Washington, D.C.: Georgetown University Press.

Lakoff, G. (1972). Hedges: a study of meaning criteria and the logic of fuzzy concepts. Papers from the Eighth Regional Meeting, Chicago Linguistic Society. Chicago: Chicago Linguistic Society.

Lockhead, G.R. (1966). Effects of dimensional redundancy on visual discrimination space. Journal of Experimental Psychology, 72, 95-104.

Lockhead, G.R. (1970). Identification and the form of multidimensional discrimination space. Journal of Experimental Psychology, 85, 1-10.

Lockhead, G.R. (1972). Processing dimensional stimuli: a note. Psychological Review, 79, 410-419.

Lockhead, G.R. (1979). Holistic versus analytic process models: a reply. Journal of Experimental Psychology: Human Perception and Performance, 5, 746-755.

Lockhead, G.R. (1983). Sequential Predictors of Choice in Psychophysical Tasks. In S. Kornblum & J. Reguin, Preparatory States and Processes. NY: Erlbaum.

Lockhead, G.R. (1988). Modeling temporal and spatial differences. The Behavioral and Brain Sciences, 11, 302-303.

Lockhead, G. R. (1989). Category bounds and stimulus variability. In B. Shepp and S. Ballesteros (Eds.), Object Structure and Process, 267-296. Norwood, NJ: Erlbaum.

Lockhead, G. R., & Hinson, J. (1986). Range and sequence effects in judgment. Perception & Psychophysics, 40, 53-61.

Lockhead, G.R., Gaylord, S., and Evans, N.J. (1977). Priming with nonprototypical colors. Paper presented at the Eighteenth Annual Meeting of the Psychonomic Society, Washington, D.C., November.

Lockhead, G.R. and King, M.C. (1977). Classifying integral stimuli. Journal of Experimental Psychology: Human Perception and Performance, 3, 436-443.

Medin, D.L. (1989). Concepts and Conceptual Structure, American Psychologist, 44, 1469-1481.

Miller, G.A. (1956). The magical number seven, plus or minus two. Psychological Review, 63, 81-97.

Miller, G.A., and Johnson-Laird, P.N. (1976). Language and Perception, Belknap, Harvard University Press.

Monahan, J.S. and Lockhead, G.R. (1977). Identification of integral stimuli. Journal of Experimental Psychology: General, 106, 94-110.

Neisser, U. (1976). Cognition and Reality. San Francisco: W. H. Freeman and Company.

Nickerson, R.S. (1978). Comment on W.R. Garner's "Selective Attention to Attributes and to Stimuli." Journal of Experimental Psychology: General, 107, 452-456.

Palmer, S (1991). Goodness, Gestalts, Groups, and Garner: Symmetry subgroups as a theory of figural goodness. In G. Lockhead & J. Pomerantz, (Eds.), The Perception of Structure (pp. 23-39), Hillsdale, NJ, Erlbaum.

Parducci, A. (1965). Category judgment: a range-frequency model. Psychological Review, 72, 407-418.

Parducci, A. and Perritt, L.F. (1971). Category rating scales: effects of relative spacing and frequency of stimulus values. Journal of Experimental Psychology, 89, 427-452.

Pomerantz, J.R. (1986). Visual form perception: An overview. In E.C. Schwab, & H.C. Nusbaum (Eds.), Pattern Recognition by Humans and Machines, Vol 2: Visual Perception (pp.1-30). New York: Academic Press.

Pomerantz, J. (1991). The structure of visual configurations: Stimulus versus subject contributions. In G. Lockhead & J. Pomerantz, (Eds.), The Perception of Structure (pp. 195-210), Hillsdale, NJ, Erlbaum.

Pomerantz, J.R. and Garner, W.R. (1973). The role of configuration and target discriminability an a visual search task. Memory & Cognition, 1, 64-68.

Pomerantz, J.R. and Lockhead, G.R. (1991) Perception of structure: An overview. In G.R. Lockhead G. & Pomerantz, J. (Eds.),The Perception of Structure,(pp. 1-20), Hillsdale, NJ, Erlbaum.

Putman, H. (1975). Is semantics possible? In H. Putnam (Ed.), Mind, language and reality: Philosophical papers, 2, NY, Cambridge University Press, 139-152.

Rosch, E. (1975). Cognitive representation of semantic categories. Journal of Experimental Psychology: General, 104, 192-233.

Rosch, E. (1978). Principles of categorization, in E. Rosch and B.B. Lloyd (Eds.), Cognition and Categorization. Hillsdale, N.J., Erlbaum.

Shepard, R.N. (1981). Psychophysical complementarity, In Kubovy, M. and Pomerantz, J. (Eds.), Perceptual Organization, Hillsdale, NJ, Erlbaum.

Shepard, R. (1991). Integrality Versus Separability of Stimulus Dimensions: Evolution of the Distinction and a Proposed Theoretical Basis. In G. Lockhead & J. Pomerantz, (Eds.), The Perception of Structure (pp.53-72), Hillsdale, NJ, Erlbaum.

Tversky, A. & Gati, I. (1982). Similarity, Separability, and the triangle inequality. Psychological Review, 89, 123-154.

Treisman, A.M. & Gelade, G. (1980). A feature integration theory of attention. Cognitive Psychology, 12, 97-136.

Wittgenstein, L. The Blue and Brown Books. New York, Harper, 1958.

Commentary

On Identifying Things: A Case for Context, G. R. Lockhead

C. MELODY CARSWELL
University of Kentucky

Any researcher who attempts to understand and predict identification performance is faced with the fundamental problem of describing how to-be-identified objects differ from one another. The present chapter compares two possible representations of such differences: a physical, attribute-based representation and a perceptual, similarity-based representation. While the two representations may sometimes show great overlap, they frequently do not, and when they do not, the perceptual similarity-based representation proves to be the better predictor of behavior. This generalization, that performance is better predicted by psychological difference measures, is the basis for the cautionary message of the present chapter.

Lockhead's cautions echo the lessons of classical psychophysics in which psychological measures of discriminability are better than most physical measurement scales for predicting performance with simple stimuli. Likewise, exploring psychological representations of similarity, and discovering the relation of such representations to performance, may prove to be an important step in understanding how we identify complex natural objects. Lockhead argues that we should first establish the relationship between psychological similarity measures and performance, and only then should we look for the determinants of the similarity representation.

Thus, the present paper clearly emphasizes structural aspects of identification performance, forcing us to reconsider our frequent attraction to the physical, attribute-based description of our stimuli as explanations for such performance. One of the attractions of the use of attributes as an explanatory device has been the difficulty associated with obtaining measures of psychological differences among complex objects. The approach described in the present article assumes that subjects' similarity judgments can be represented using spatial extent as an analog for the psychological differences (dissimilarity) between objects. This representation of object relationships is called a similarity space, and it is proposed that attention moves about in this space seeking out anticipated or probable objects. The dynamics of attention, within the context of any particular similarity space, may thus determine the time and accuracy associated with identification responses, as well as performance in categorization tasks. Such a model, combining psychological similarity structure and assumptions regarding attentional processing, can be successfully used to explain such specific performance phenomena as priming effects, range effects, "redundancy" effects, and sequence effects.

PART B. Percepts, Concepts, Categories and Development

Percepts, Concepts and Categories
B. Burns (Editor)

5

Structure in the Process of Seeing

TARA C. CALLAGHAN
St. Francis Xavier University

I. INTRODUCTION

In her book *Wisdom and the Senses* Joan Erikson has written, "It is important to realize that all knowledge begins with sensory experience. The role of the senses, then, is to inform the mind" (1988, p. 25). While many psychologists who study perception would agree with this emphasis on the fundamental role of sensation in knowledge acquisition, there are differences of opinion as to exactly how it is that this feat of informing the mind is accomplished. One way to establish how theories view the essence of perception is to ask how each defines the concept of perceptual structure. I begin this chapter with an examination of some diverse views on this issue. In the sections that follow I have three aims: 1) to emerge with an accommodating view of perceptual structure that encompasses strong trends from both adult and developmental research, 2) to present some developmental data that are consistent with this emergent view, and 3) to outline findings from a new research area that asks how perceptual structure may influence the process of drawing. This research area promises to provide insights into one way that basic sensory processes inform the more complex cognitive processes of mind.

II. WHAT IS, WHERE IS, STRUCTURE?

A. Structure is Directly Picked up from the Stimulus

One answer to the question of how mind is informed by the senses has been given by J. J. Gibson (1966) who stresses that the organism directly picks up information about the stimulus (i.e., structure) from the stimulus itself. All information, be it lower order (e.g., colour) or higher order (e.g., edibility), is said to be present in the stimulus flux. The organism need only select the information. In order to account for important developmental differences in perception, E. J. Gibson (1969, 1987) has pointed out that the structure picked up by the organism becomes more differentiated over development. Thus, as Gibson and Gibson (1955) pointed out long ago, we become better at the process of selecting structure, not at the process of creating structure. Much of the work in the Gibsonian tradition has focused on a detailed description of the structure inherent in the stimulus flux, and more recently there has been an account of how much of the potentially rich structure is available innately (see E. J. Gibson, 1987 for a summary of this research). Thus, from the Gibsonian viewpoint, structure is a stimulus concept, and mind is informed directly from the stimulus.

B. Structure is Derived from the Stimulus Through Processing

A second answer to this query of perception comes from the information processing tradition. Many diverse viewpoints comprise this tradition, however, all see the process of informing mind as being indirect in the sense that the organism's processing capacities intervene between the stimulus and the experience of seeing. There is a strong focus on how stimulus structure changes during processing, even though it is acknowledged that the stimulus provides necessary initial input. One of the most widely accepted models of this process is Treisman's Feature Integration Theory. In her original formulation of the theory (Treisman & Gelade, 1980), object perception was assumed to involve at least two stages; preattentive and attentive. It was suggested that in the preattentive stage simple stimulus features are registered by the visual system, automatically and in parallel across the field. These features are then conjoined, during the attentive stage, to form the more global object percept through the process of focused attention on a spatial location. In a recent modification of the theory (Treisman & Gormican, 1988), it was argued that the construct of attention was potentially operative at all points of the perceptual process. Although the stage metaphor has been dropped in the recent account, the focus in Treisman's model, and in most others of the information processing tradition, is still on how stimulus structure changes along with processing by the organism.

Although the precise source for structural change is rarely made explicit by information processing theorists, a couple of themes emerge from the vast literature. First, stimulus structure changes as processing proceeds. This change is not necessarily unidirectional. Stimulus structure may go from parts to wholes, from wholes to parts, or even from parts to wholes and back to parts again as processing continues over time. Second, the mind is the processor. Thus, changes in stimulus structure are assumed to be caused by the organism. The mind is not so much informed by anything in this view as it is continuously updated by its own processing efforts, which take their germinal seed for structure from the stimulus itself. Third, since the focus in most information processing theories is on processing, constraints on the particular structure that is produced are assumed to be caused by limitations of the processor. Thus, as far as one can ascertain from the writings of many of the information processing theorists, structure is an organismic concept, and the mind is indirectly (in)formed from the stimulus during processing.

C. Structure is Selected from the Stimulus in the Limiting Process of Seeing

There is an important exception to this position found in the work of W. R. Garner (1974). Unlike most information processing theorists, Garner emphasizes both stimulus and processing factors. On the stimulus side, Garner and his colleagues (c.f, Garner, 1974; Garner and Felfoldy, 1970) have shown that particular combinations of physical attributes appear to tightly constrain the structure that can be perceived by the organism. Thus, in a typical perceptual processing task attributes like hue and brightness invoke a process of comparison that is based on overall, or wholistic, similarities between stimuli in the stimulus space. Garner calls stimuli that trigger this process integral. On the other hand, attributes like circle size and angle of radial line invoke a comparison process that is based on the dimensional, or attribute, similarities between stimuli. These stimuli are called separable. In their early work Garner and his colleagues employed a variety of converging processing tasks to ask what processing options were available to the observer with these two types of stimuli. It was found that a distinct pattern of results, indicative of either integral or separable structure, was found for particular combinations of physical attributes regardless of the particular task. To explain these findings, Garner (1974, p. 120) argues that logically both structures are potentially available to the observer because they are inherent in the stimulus. However, since the primary process invoked for integral stimuli is based on overall similarity, and for separable stimuli on dimensional similarity, the usual structure is wholistic, unitary for integral and componential, dimensional for separable stimuli. Thus, according to Garner, structure is a stimulus concept. Nonetheless, the organism is seen to be actively engaged during perceptual processing in the work of selecting structure from the stimulus.

Although not always recognized by researchers in this tradition, Garner did not throw out the processing baby when he strongly emphasized the importance of the stimulus. On the processing side, Garner (1983) reported that some combinations of physical attributes (e.g., hue and form) appear to result in different perceived structures, depending on the task demands. Speeded tasks, or any others that place high processing demands on the observer, typically promote integral structures for these stimuli. Nonspeeded, or low demand, tasks typically promote separable structures. This finding clarified Garner's position on perceptual structure. Structure is in the stimulus, and processing allows the observer to select the structure. The particular structure available for selection depends not only on the unique physical attributes that make up the stimulus, but also on the demands placed on the organism in the task situation.

D. Structure is Selected from the Stimulus and Limited by Developmental Level

In 1976 Shepp and Swartz discovered that developmental level of the observer also limits the potential for selecting particular structures from the stimulus. They suggested the Separability Hypothesis to account for their finding that young children perceive an integral structure for stimuli that older children and adults perceive as comprised of separable dimensions. It was argued that with development, the child becomes better able to access the dimensional structure inherent in the stimulus.

Kemler (1983) reviewed the literature relevant to the Separability Hypothesis and concluded that under typical task demands young children prefer to access an integral structure, whereas older children and adults prefer a dimensional structure. However, observers at all ages can access any structure given special conditions (e.g., training). What develops in perception, according to Kemler, is the ability to analyze stimuli according to component, meaningful dimensions, a process she labels dimensionalization of the stimulus. Interestingly, Ward and Vela (1986) suggest that over development observers also become better at integrating separate dimensions. This conclusion follows a finding that although both young children and adults perceive a wholistic structure for colour stimuli, 5-year-olds give evidence of having a *less* wholistic structure than adults. Thus, we have the suggestion that both dimensionalization and integration abilities may improve with development.

Smith (1989) presents a comprehensive model of perceptual classification that makes specific claims about exactly what is developing in perception, and clarifies the concept of structure in the context of the developing organism. In the model, it is proposed that not only does the observer's ability to selectively attend, and maintain attention to component dimensions improve over development, but so too does the observer's concept of similarity. Very roughly speaking, young children's concept of same is less precise than the older child's. So when you ask children to classify together objects that are the same, the young child will group together two objects that are roughly the same over all component dimensions, while the older child will group together only those objects that are identical on all component dimensions. Smith (1989, p. 127) makes an important assumption about representation that I believe helps to clarify how structure and process are related in her model. She states that representations are built up from features and exist as a cohesive unit once they are formed. Further, she argues that our conscious impression of objects is that they are cohesive wholes. Thus, in Smith's model, there is a wholistic structure both at the level of representation *and* at the level of consciousness (see also Treisman, 1986). The parts, or features, of an object can only be consciously accessed from the representation with some effort. Thus, once a representation is formed from features it is that representation that undergoes processing. Furthermore, the representation stays roughly the same across development. What changes are the operations mentioned above - attention to dimensions and concept of similarity - which are applied to the representation. Thus, in this developmental model, it is considered that structure is in the stimulus and processing the representation of that stimulus allows for access to various levels of that structure.

E. An Emergent View of Perceptual Structure

If we consider all of the viewpoints on stimulus structure together - direct perception, information processing, and developmental - an interesting view of perceptual structure and processing emerges. First, it is clear that as Gibson (1966) and Garner (1974) have argued the stimulus world (i.e., the physical world) is structured. Once an observer beholds this world, information is registered in what is typically called a representation. The representation contains all of the information necessary for accessing the structure(s) inherent in the physical stimulus. Second, the processing task for the observer is to select structure from the stimulus representation. Here, the emergent view parts with the direct perception view since selection is from the representation and not the stimulus itself. If the process of selection is frozen in time, as we assume it is when the observer makes a response in a typical perception task, then the pattern of responses will suggest that a particular structure has been accessed. These frozen frames of processing are equivalent to the representation of the stimulus at that point in time. Thus, the structure accessed at a given point in processing is the representation of the stimulus at that point. Presumably, if we sample a number of different points over the time course of processing we can keep track of changes in the structure available for selection across levels of processing. Third, the process of selection is one that is influenced by a number of factors in an nontrivial way. These factors include the particular dimensions that comprise the stimulus itself, the demands of the task, and the age of the observer. As is apparent in this description, the emergent view is not offering new ideas about structure and process so much as it it offering a new way of combining old ideas. In this view then, structure is inherent in the stimulus and in the multiple representations of the stimulus, and processing allows the observer to access one of the available structures.

In the section that follows, I present evidence to support this emergent view of structure selection. I will draw primarily on results from my lab to do this, including published data as well as preliminary data from ongoing research, but will also incorporate relevant findings from the literature.

III. STRUCTURE IN THE PROCESS OF SEEING

There are two major questions that underlie my research program and that can help to focus this review of the evidence. The first asks whether the structure that is accessed by the observer is influenced by the level of processing. By level of processing I mean extensiveness of processing, with low level implying less extensive processing. I assume, along with others (Kahneman, 1973; Smith, 1989; Treisman, 1986), that performing a task where one must access the components of the stimulus requires more extensive processing than a task that can be performed without access to components. Whether one calls the less extensive processing preattentive, early, or unconscious (Treisman & Gelade, 1980), and the more extensive processing attentive, late, or conscious is not so relevant as the realization that processing implies change over time for each perceptual event. What I propose changes over time is the nature of the structure that is available for access by the observer.

The second question that motivates this research is the developmental question that was discussed in the previous section. That is, whether the ability to access structure in the stimulus is limited by the developmental level of the observer. As mentioned, we have plenty of evidence that perceived structure is limited by development (e.g., Kemler, 1983; Shepp & Swartz, 1976; Smith, 1989). What is unique about the present research program is that it asks the question of whether development limits selection of structure within the context of levels of processing. Thus, the question posed asks whether the structure accessed by the observer changes across levels of processing, and if so, does it change in the same way at all stages of development? If the available structure does change across levels of processing then we may see a change from integral to separable structure, or vice versa, as we go from less to more extensive processing. If there are developmental differences in the nature of this change, then we may see that a given pattern (i.e., integral to separable with more extensive processing) holds for one developmental level but not another.

A. Outline of Methodologies and Experimental Logic

1. Texture Segregation Task

There are two types of procedures used in my lab to address this question. The first asks observers to indicate where a boundary lies in a textured array and traces whether this boundary perception is interfered with by the variation of irrelevant information in the array. This task combines the logic of Garner and Felfoldy's (1970) speeded sorting task with Treisman and Gelade's (1980) texture segregation task. Typically, observers are presented with 36-element arrays that contain one quadrant that differs from the rest on the basis of differences between the elements on one of the target dimensions. Observers are asked to indicate the different quadrant by pressing one of four computer keys (adults) or by touching the discrepant quadrant (children). Figure 5.1 illustrates the types of arrays used in these experiments. The arrays in the figure contain fewer elements than actual stimuli. The presentation size of the entire array is approximately 5.7 degrees of visual angle in adults studies and 8.5 degrees for developmental studies.

There are two major types of arrays - control and low similarity - illustrated in Figure 5.1. Control arrays contain discrepant quadrants with elements that differ from background elements in a level on one dimension and share a level with background elements on a second dimension. Low similarity arrays also have discrepant quadrant elements that differ on one dimension, but now all the elements vary randomly across the array on the second, irrelevant dimension. Following the logic of Garner and Felfoldy (1970), evidence for interference (i.e., longer reaction time, RT) for boundary judgements in low similarity as compared to control arrays is taken to suggest that the perceived structure is integral. Equivalent performance (i.e., equivalent RT) for boundary judgements in both array types suggests separable structure.

2. *Speeded Sorting*

The second task was first used by Garner and Felfoldy (1970) with adults and subsequently by Shepp and Swartz (1976) with children and involves speeded classification. In this task observers are asked to sort stimuli according to value on component dimensions. There are three major types of stimulus manipulations. Decks to be sorted may contain stimuli that vary on one dimension only and are identical on the second (control), vary on two dimensions in a redundant fashion (correlated), or vary on both the target dimension and an irrelevant dimension (orthogonal, equivalent to low similarity in texture segregation task). As outlined by Garner and Felfoldy (1970) interference (i.e., longer sorting times) for orthogonal as compared to control decks, and redundancy gains (i.e., shorter sorting times) in correlated as compared to control arrays indicates integral structure. Equivalent sorting times across all types of decks indicates separable structure. Much of the data presented below comes

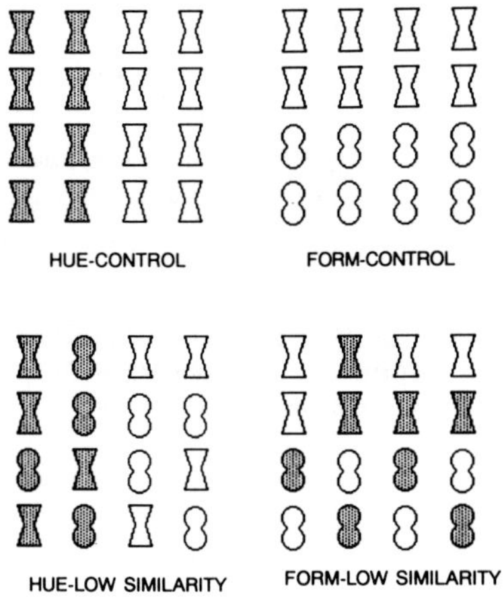

Figure 5.1. Examples of stimulus arrays used in interference experiments.

from existing speeded classification studies with children and adults, though there are some preliminary data from my lab that will be included where relevant.

3. *Stimuli*

The results from experiments employing two types of stimuli will be summarized below. One type combines hue and geometric form dimensions in the stimulus, and the other combines hue and line orientation. In all cases where it is possible I will present data for stimuli where the discriminability of each of the component dimensions is equivalent (i.e., hue/form, or hue/orientation). This is to ensure that interference effects are genuine and not due simply to the overpowering salience of one of the dimensions. (Equivalence of discriminability is evaluated by comparing judgement RTs for each of the two control arrays.) The particular hue values were obtained either with the variation of Munsell colour swatches (10R 4/12, 2.5R 4/12, or 7.5RP 4/12), or by the variation of Panatone inks that were matched to approximate Munsell values. The forms were either circle/square or the straight/curved novel shapes from Figure 5.1. The line orientations that were varied were horizontal, vertical, left and right diagonal.

Since objects must first be segregated from the background before they can be compared with other objects on the basis of component dimensions, it is assumed that the texture segregation task requires less extensive processing than the speeded sorting task. Thus, by comparing perceived structure for two types

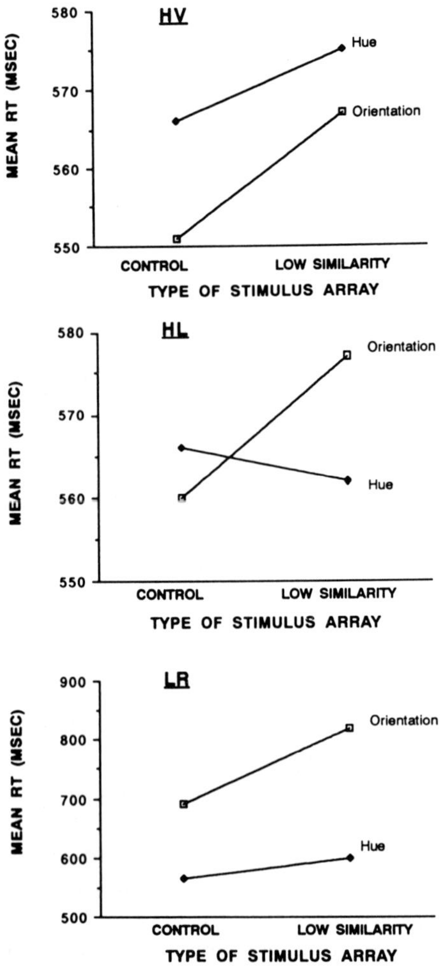

Figure 5.2. Mean Reaction Time (msec) for Correct Segregation Responses to Hue/Orientation Arrays (Callaghan et al., 1986). Note scale difference for vertical axis for LR graph.

of stimuli across these tasks, and across levels of development, we have the means of addressing the questions of whether structure changes across levels of processing and development.

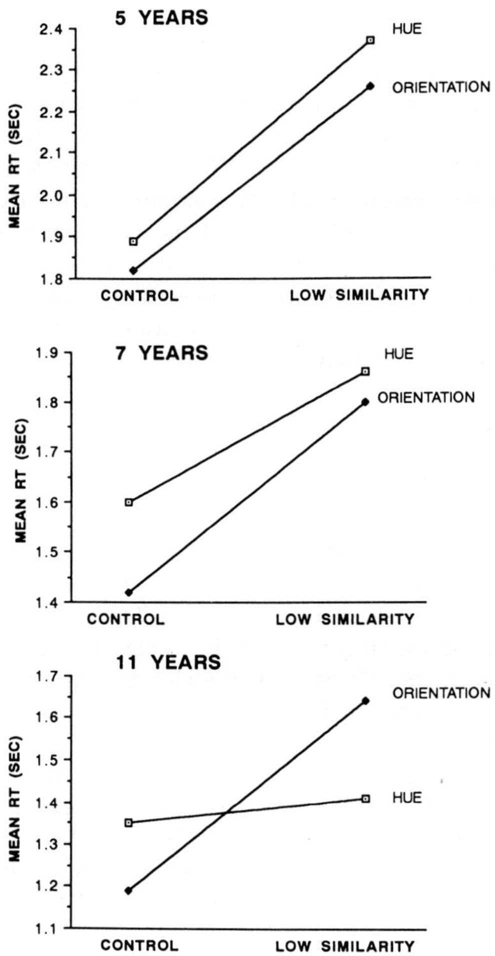

Figure 5.3. Mean Reaction Time (sec) for Children's Correct Segregation Responses to Hue/Orientation Arrays from Callaghan (in press).

B. Results

1. Hue and Orientation

First consider adult data from texture segregation experiments conducted by Callaghan, Lasaga and Garner (1986). Figure 5.2 shows mean RTs for correct segregation responds to hue/line orientation arrays, when hue and line orientation are of equal discriminability. When horizontal/vertical and left/right diagonal lines are varied in low similarity arrays (top and bottom graph of Figure 5.2) there is overwhelming evidence for interference of boundary judgements, both when hue judgements are being made and orientation varies, as well as when orientation judgements are being made and hue is varied. When horizontal/left diagonal lines are varied (middle graph of Figure 5.2) there is asymmetric interference. What this means is that processing based on one of the component dimensions is interfered with by irrelevant variation on the other dimension, but processing of the second dimension is not impaired by irrelevant variation. In these results, hue variation interferes with orientation judgements but orientation variation does not interfere with hue judgements. Although the interference pattern is asymmetric for this one pair of line orientations, the general trend for adults with hue and line orientation seems to be interference in the texture segregation task. In Garner's (1974) terms interference implies an integral structure.

Since there are no available data in the literature, I have used a speeded sorting task modelled after Garner and Felfoldy's (1970) to assess the nature of perceived structure for these stimuli under conditions of more extensive processing (Callaghan, 1990a). Analysis of these data (see row 3, Table 5.1) confirm that the sorting times are equivalent for all types of decks. Thus, for adults the structure of hue/orientation stimuli changes from integral under less extensive to separable under more extensive processing conditions.

Next consider children's data from a texture segregation study (Callaghan, 1990b). Figure 5.3 presents mean RTs for correct segregation responses to hue/orientation arrays at three developmental levels. Note that the only pair of orientations that were varied in this study were horizontal/vertical, and the hue and orientation differences used were approximately equal. It is clear from the graph that there was strong symmetric interference for segregation judgements for both 5- and 7-year-olds. These results are consistent with adult data. Unexpectedly, the pattern is one of asymmetric interference for 11-year-olds. Hue interferes with orientation judgements (i.e., longer RTs for Hue-Low Similarity as compared to Hue-Control arrays) but orientation variation does not interfere (i.e., equivalent RTs for Orientation-Low Similarity and Orientation-Control arrays) with hue judgements. I suspect that these data are anomalous and will not be replicated in studies currently being run. Nevertheless, the general pattern across all age groups for hue/orientation stimuli appears to suggest an integral structure.

I have also obtained developmental data (Callaghan, 1990a) regarding perceived structure for these stimuli using a speeded sorting task modelled after Shepp and Swartz (1976). For comparison I have presented these data with the

Table 5.1. Mean Sorting Time (sec) as a Function of Type of Stimulus Deck.

	Type of Stimulus Deck				
Age	Control Hue	Orthogonal Hue	Control Orien-tation	Orthogonal Orientation	Correlated
5 yrs	59.85	74.03	58.97	57.30	56.93
7 yrs	43.07	45.69	45.49	49.81	39.76
A-dult	23.18	25.05	21.83	24.41	22.64

adult findings in Table 5.1. An interesting developmental trend emerges. For young children (5 years) there is asymmetric interference; orientation variation interferes with hue judgements (i.e., longer RTs for Hue-Low Sim compared to Hue-Control), but hue variation does not interfere with orientation judgements (i.e., equivalent RTs). This asymmetric interference is not evident for older children (7 years), or adults. Older children and adults show equivalent sorting times for all types of stimulus decks, which suggests separable structure.

To summarize the findings for hue/line orientation stimuli; with young children perceived structure changes from integral (symmetric) to integral (asymmetric) when processing becomes more extensive. For older children and adults the structure changes from integral to separable as processing becomes more extensive.

2. *Hue and Geometric Form*

First consider the adult data (from Callaghan, 1989) for hue/geometric form stimuli. Figure 5.4 presents mean RTs for correct segregation responses to hue/form arrays both when the forms that vary are circle/square (Figure 5.4A), and when they are curved/straight (Figure 5.4B) shapes. Regardless of the particular shapes that vary, the pattern for these stimuli is asymmetric interference. (Note that due to a scale difference between these two figures there may appear to be a symmetric interference effect in Figure 5.4B, however, the RT

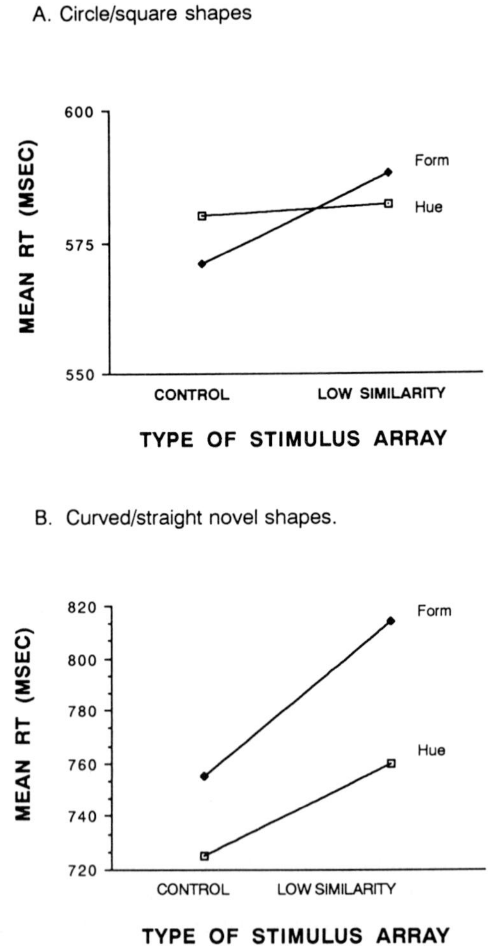

Figure 5.4. Mean Reaction Time (msec) for Correct Segretation Responses to Hue/Form a) Circle/Square and b) Curved/Straight Shapes Taken from Callaghan (1989).

for Hue-Low Sim arrays was not significantly longer than that for Hue-Control arrays.) Hue variation interferes with form boundary judgements (i.e., longer RTs for Form-Low Sim than Form-Control arrays), but form variation does not interfere with hue judgments (i.e., equivalent RTs). This suggests that the structure for these stimuli is asymmetric integral under less extensive processing conditions.

This is consistent with the structure reported by Garner (1983) for hue/form stimuli in experiments that employ speeded sorting tasks. However, as Garner has pointed out, for nonspeeded tasks the pattern of performance for these stimuli suggests a separable structure. Thus, the answer to the question of whether structure changes over levels of processing for adults with hue/form stimuli is: It depends on whether you compare the structure obtained for texture segregation (asymmetric integral) to that obtained for speeded (same structure) or nonspeeded (change in structure) tasks.

Figure 5.5 presents developmental data from a texture segregation experiment (Callaghan, 1990b) that varied hue and form (curved/straight). The older children (7- and 11-year-olds) replicate the adult pattern of asymmetric interference; hue variation interferes with form judgements (i.e., longer RTs for Form-Low Sim than Form-Control), but form variation does not interfere with hue judgements (i.e., equivalent RTs). In contrast, the young children show symmetric interference; hue variation interferes with form segregation and form variation interferes with hue segregation. Thus, under less extensive processing conditions the structure for hue/form stimuli appears to be integral for young children and asymmetric integral for older children and adults.

Shepp and Swartz (1976) investigated the perceived structure for these stimuli in a developmental study that used a speeded sorting task. They report that the structure for hue/form stimuli is integral for young children and separable for older children. When considering whether perceived structure changes across levels of processing for hue/form stimuli, it appears as though there is no change for young children, but there is a change from asymmetric integral to separable for older children. Recall that the structure for adults was asymmetric integral when speeded attentional tasks were used, and separable when nonspeeded tasks are used (Garner, 1983). I am currently replicating the Shepp and Swartz study using the same levels of hue and form that were employed in the texture segregation study in an effort to clarify what the structure is for children using these stimuli under conditions of extensive processing.

C. Conclusion: Structure Does Change in the Process of Selection

The trends reported here (summarized in Table 5.2) confirm the suggestion made earlier that in order to answer questions of structure one must consider the particular stimulus, the task demands or level of processing, and the developmental level of the observer. A change in structure was observed for older children and adults for both hue/orientation and hue/form stimuli. There

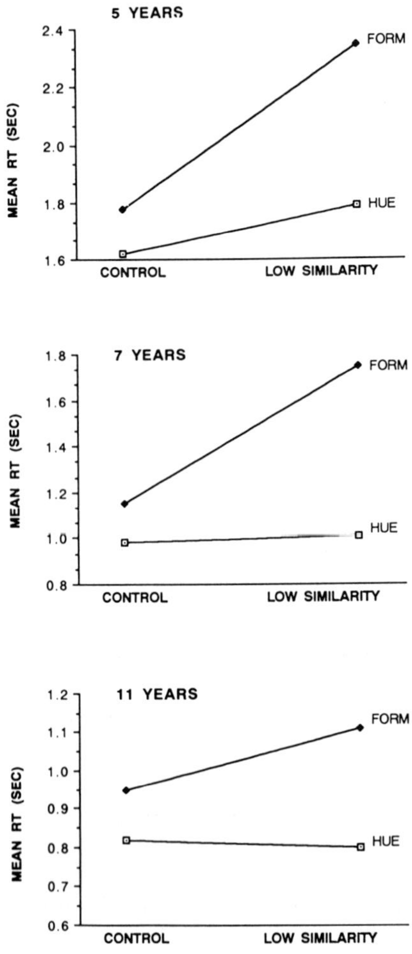

Figure 5.5. Mean Reaction Time (sec) for Children's Correct Segretation Responses to Hue/Form Arrays from Callaghan (in press).

was some change in structure for young children with hue/orientation stimuli, but no change for hue/form stimuli. Thus, it appears as though the structure available for access to older children and adults is different depending on extent of processing. For young children however, even more extensive processing does not guarantee that a new structure can be accessed.

Table 5.2. Summary of the patterns found when assessing perceived structure at differing levels of processing (early: texture segregation, late: speeded sorting) and of development (5, 7, 11 years and adult). Integral structure implies a wholistic, unitary percept, while separable structure implies a componential, dimensional percept.

		Structure		Change?
Age	Stimuli	Texture Segregation	Speeded Sort	
5 yrs	HO	integral	asymmetric	yes
	HF	integral	integral	no
7 yrs	HO	integral	separable	yes
	HF	asymmetric	separable	yes
11 yrs	HO	asymmetric	--	--
	HF	asymmetric	separable	yes
Adult	HO	integral	separable	yes
	HF	asymmetric	asymmetric	no
		(for nonspeeded:	separable)	yes

A number of researchers have reported (J.D. Smith & Kemler, 1984; L.B. Smith, 1984; 1989) that for very simple tasks and stimuli it is possible to reverse the usual trend such that even young children can gain access to dimensional structure. Likewise, for very difficult tasks and stimuli the typical pattern for adults can be altered so that even adults will not be able to gain access to dimensional structure. We have looked at two slices of task difficulty and confirmed the usual finding that young children are less flexible (i.e., have fewer structures to choose from at their disposal) than older children and adults. However, it appears that the effort demanded by the task can change a flexible processor into an inflexible one (Smith & Kemler-Nelson, 1984), or vice versa, depending on whether the task gets more or less difficult. This apparent plasticity of access to perceptual structure began to intrigue me at about the time I began to work on a project that explores the relationship between how one accesses structure in the perceptual world and how one forms symbols of that structure in drawings. Although there have been a number of attempts in the literature to isolate perceptual variables that may be important in the process of drawing (e.g., see Rosenblatt & Winner, 1988 for work on visual memory), there has not been a concerted effort to link together the knowledge we have about how humans access structure in their visual world with the study of how symbols, or drawings, of that world are formed. Preliminary findings from my own efforts in this direction will be discussed in the next section.

IV. STRUCTURE IN THE THE PROCESS OF DRAWING

Imagine that you arrived at the office very early one morning leaving your car in the large empty lot. At the end of the day, you leave a little early and the lot is jammed full. You happen to have a sky blue Aires wagon, a popular car in the area. After a hard day's work you can't remember exactly where you left your car so you begin to search from thc top of thc hill. When we engage in a typical analysis of the perceptual world like this, we are not conscious of the process of analysis nor do we usually try to improve our analysis; we simply look. When an artist looks at the same visual scene with the intention of drawing it (suppose the theme for this visual scene is 'Melancholy of the North American Workplace'), she is conscious of analyzing the visual details of light, colour and space. Furthermore, the artist is continually attempting to improve her skills of analysis. What are these skills of visual analysis that the artist uses, and are they different in quality from those that we all use so that we can get home at the end of the day?

I asked this question, among others, of a large group of Canadian artists. A preliminary analysis of their responses revealed that most artists believe that in addition to the 'ordinary' way of seeing, they also see things differently from nonartists. This difference is captured eloquently in the words of one artist, "I'm not so quick to categorize what I see, and therefore stop perceiving it's full reality. A house doesn't become a house. It's yellow, taller than the trees, dark windowed, lighter than the lawn, still against the moving clouds, has a red door, a dark roof, is deep not long, is ....all - instead of being just *a house* and therefore finished, categorized. Looking at the world like this takes time."

What is captured in this artist's description of how she looks at the world is a form of analysis that nonartists usually do not follow through with in their seeing the world. I would argue that it is not something that the nonartist cannot do, but rather is something that she does not choose to do, perhaps because it requires more effort to do so. Like the young child who prefers to access wholistic structure, and the older child and adult who prefer to access component dimensions of the stimulus (Kemler, 1983), the artist seems compelled to choose analysis that is based on sensory/perceptual attributes rather than conceptual categories.

The findings regarding flexibility of the selection process for perceptual structure mentioned earlier (Smith, 1984; Smith & Kemler-Nelson, 1984) suggest that adults are more flexible than older children, who are in turn more flexible than younger children in the process of selecting structure. From the artist's descriptions of their processes of seeing, it seems that it may be that artists are more flexible still than nonartists, and that this increased flexibility occurs throughout the development of the individual who is gifted in the visual arts. I decided to begin to explore this question by conducting a study, which asked whether children's drawings would be influenced by special training that promoted flexibility of structure selection.

A. General Methodology

1. Stimuli

The children in this study (5 and 10 years old) were asked to draw a simple infant's toy. The toy was mounted on a box so that it was at the child's eye level, and they were provided with markers having a range of primary colours and including all of the colours of the toy. Children made two drawings of the toy, one before and one after being exposed to one of four training conditions.

2. Training Conditions

The first, *perceptual flexibility training*, was modelled after the procedure used by Ivry (1981). In this condition children received special groups of four stimuli that varied on the dimensions of hue (seven Munsell levels were varied) and rectangle (seven levels of h/w ratio were varied). The stimuli were specially chosen so that observer's would have to shift their preferred classification (i.e., based on overall or dimensional similarity) in order to form two 'good' groups. Preferred strategy was assessed in a pretest, and this was followed by training that either promoted classifications based on overall similarity relations (if the observers preference was originally for dimensional structure), dimensional relations (if the preferred structure was integral), or both (if the observer was divided between the two structures). The object of training was to promote flexibility, not to shift selection to another structure, and so all training was ended by presenting a small number of stimulus groups that promoted the observer's original preference. Following this training experience children were asked to draw a second picture "now that they had experienced different ways of seeing things".

The remaining groups included a *tactile only* group, where children were blindfolded and given the toy to explore in their hands for three minutes, a *tactile plus visual* group, where children were given the toy to explore in their hands for three minutes while they could also view it, and a *control* group, where children were simply asked to draw another picture of the toy after a three minute break.

B. Results

The drawings collected in this study were rated by the experimenter on a scale of 1 to 10 as to how well they represented the toy. All drawings were grouped together and shuffled before rating, so the experimenter was unaware of the condition the drawings were produced in. The rating data was subjected to a repeated measures analysis of variance that revealed a significant interaction of age by training condition. Analysis of the interaction revealed that the only group that benefited from training was the 5 year old group who received perceptual flexibility training. Figure 5.6 presents some examples of the improvement found for some of the children in this group. Drawings before training are found in the left column, paired with the drawing completed after training in the right column.

These ratings provide a fairly global analysis of the potential change in drawings from before to after training and I am working along with others on more refined computer analysis of form in the drawings. Nevertheless, some interesting trends have emerged from the rating analysis. There are a number of ways that drawings improved from before to after training that included not only changes in form, but also in colour and perspective. Some common changes in form that occurred at both ages include improvements in the proportions of the parts of the object, including parts of the object that had been omitted on the first drawing, improving the accuracy of the line by either straightening it or curving it (e.g., bottom drawings Figure 5.6), and 'unscrambling' the parts of the object so that they are more clearly defined (e.g., middle drawings, Figure 5.6). Children at both ages also improved colour, usually by choosing the exact colours of the object on the second drawing after having used colours not found in the object in

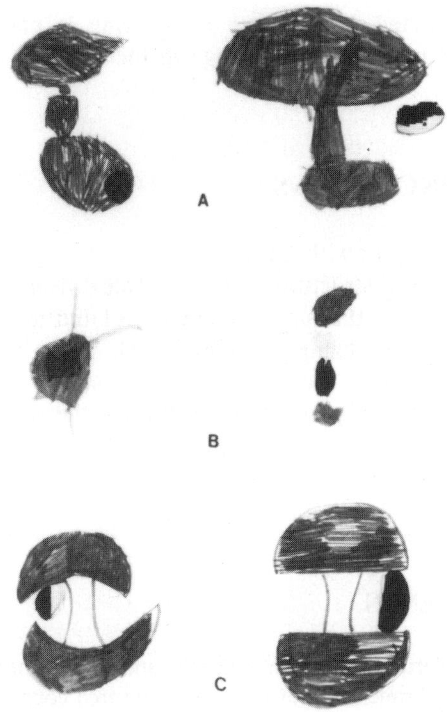

Figure 5.6. Samples of children's drawings before (left side) and after (right side) perceptual flexibility training. The drawings in A) and B) are from five-year olds and in C) are from a ten-year old.

their first drawing. Children who were producing very good drawings (usually older children scoring 8 or above on the rating scale) also made interesting perspective changes by choosing an unusual perspective (e.g., looking down at the object from an oblique angle) for their second drawing. If an unusual perspective was drawn well it received a higher rating than the usual side view.

C. Conclusions: Access to Structure is Important in Drawing

These findings suggest that experiences that encourage flexibility of access to perceptual structure may improve the ability to symbolize the structure of a simple object in drawings. This was especially true for the young children in this study and may be generally true for children who are typically not very flexible in their ability to access a variety of structures from the stimulus. That this simple training situation effected any change in drawing is encouraging that more extensive procedures may promote change in the drawing ability of even older children and adults.

V. GENERAL CONCLUSIONS

I began this chapter with a comparison of existing views on the nature of the process of perceiving structure in our complex visual world. I ended the chapter with a look at how that process may be related to something even more elusive, the process of symbolizing our visual world. There are some sophisticated answers to the first issue that have evolved as a result of a tremendous body of research that explores visual perception. Although the second area of research is only in it's infancy, perhaps it is even prenatal, it is a beginning of the process of understanding of how the mind then communicates, using a visual medium, about the information that was selected.

ACKNOWLEDGMENTS

This research was supported by a Social Sciences and Humanities Research Council of Canada grant, and a St. Francis Xavier University research council grant to the author. I am grateful to Jim Enns, Rich Ivry and Barbara Burns for their enlightened comments on an earlier draft of this chapter. I am also indebted to Tex Garner, who in his Seminar on Perception got me to take the first step in linking perceptual and artistic abilities. Thanks to Anne MacIsaac and Connie MacMullin for their help in running some of these experiments, and to the children for their wonderful productions.

REFERENCES

Callaghan, T. C. (1990a). Stimulus structure accessed for hue/line orientation stimuli using a speeded sorting task. Unpublished manuscript.

Callaghan, T. C. (1990b). Texture segregation in young children. In J. Enns (Ed.), The development of attention: Research and theory. North-Holland, 1990 Advances in Psychology Series.

Callaghan, T. C. (1989). Interference and dominance in texture segregation: Hue, geometric form and line orientation. Perception & Psychophysics, 46, 299-311.

Callaghan, T. C., Lasaga, M. I., & Garner, W. R. (1986). Visual texture segregation based on orientation and hue. Perception & Psychophysics, 39, 32-38.

Erikson, J. M. (1988). Wisdom and the senses: The way of creativity. New York: Norton.

Garner, W. R. (1974). The processing of information and structure. Hillsdale, N.J.: Erlbaum.

Garner, W. R. (1983). Asymmetric interactions of stimulus dimensions in perceptual information processing. In T. J. Tighe and B. E. Shepp (Eds.), Perception, cognition, and development: Interactional analyses. Hillsdale, N.J.: Erlbaum.

Garner, W. R., & Felfoldy, G. L. (1970). Integrality of stimulus dimensions in perceptual information processing. Cognitive Psychology, 1, 225-241.

Gibson, E. J. (1969). Principles of perceptual learning and development. New York: Appelton-Century-Crofts.

Gibson, E. J. (1987). Introductory essay: What does infant perception tell us about theories of perception? Journal of Experimental Psychology: Human Perception and Performance, 13, 515-523.

Gibson, J. J. (1966). The senses considered as perceptual systems. Boston: Houghton Mifflin.

Gibson, J. J., & Gibson, E. J. (1955). Perceptual learning: Differentiation of enrichment? Psychological Review, 62, 32-41.

Ivry, R. (1981). A developmental investigation of the flexibility of perceptual processing. Unpublished honor's thesis, Brown University, Providence, RI.

Kemler, D. G. (1983). Exploring and reexploring issues of integrality, perceptual sensitivity, and dimensional salience. Journal of Experimental Child Psychology, 36, 365-379.

Kahneman, D. (1973). Attention and effort. Engelwood Cliffs, N.J.: Prentice-Hall.

Rosenblatt, E., & Winner, E. (1988). Is superior visual memory a component of superior drawing ability? In L. Obler and D. Fein (Eds.), The exceptional brain. New York: Guilford Press.

Shepp, B. E., & Swartz, K. B. (1976). Selective attention and the processing of integral and nonintegral dimensions: A developmental study. Journal of Experimental Child Psychology, 22, 73-85.

Smith, J. D., & Kemler-Nelson, D. G. (1984). Overall similarity in adult's classifications: The child in all of us. Journal of Experimental Psychology: General, 113, 137-159.

Smith, L. B. (1984). Young children's understanding of attributes and dimensions: A comparison of conceptual and linguistic measures. Child Development, 55, 363-380.

Smith, L. B. (1989). A model of perceptual classification in children and adults. Psychological Review, 96, 125-144.

Treisman, A. M. (1986). Features and objects in visual processing. Scientific American, 255, 114B-125.

Treisman, A. M., & Gelade, G. (1980). A feature-integration theory of attention. Cognitive Psychology, 12, 97-136.

Treisman, A. M., & Gormican, S. (1988). Feature analysis in early vision: Evidence from search asymmetries. Psychological Review, 95, 15-48.

Ward, T. B., & Vela, E. (1986). Classifying color materials: Children are less holistic than adults. Journal of Experimental Child Psychology, 42, 273-302.

Commentary

Structure in the Process of Seeing and Drawing, T. C. Callaghan

LINDA B. SMITH
DIANA HEISE
Indiana University

> *"I am not so quick to categorize what I see, and therefore stop perceiving its full reality."*

With this comment, Callaghan's artist-subject emphasizes that perceived structure is the product of processing and inherently dynamic. The artist is clearly aware that there is no one-to-one map between stimulus and perceived structure. Instead, a single stimulus may yield many perceptions. This is not surprising since perceptual structures are not things but are the products of processing. The artist laments the fact that her learned categories sometimes cause her to see things in one way as opposed to other ways. This comment affirms the point that Diana Heise and I attempted to make in our chapter in this volume: that perceived structures and categories mutually influence one another. Sometimes theorists of categorization take the view that there is first a set perceptual structure out of which we make categories or concepts (or if these raw materials of perception are not enough, we then make concepts out of perceptual and nonperceptual stuff). But as Callaghan's artist-subject knows, this is wrong: what we see is both a step to categorization and a product of categorization. The fundamentally dynamic nature of perceptual structure, that it is a product of process, is seen in developmental changes in perceptual classification, in task and context effects, and in Callaghan's demonstration of the effects on processing time on what is perceived.

Percepts, Concepts and Categories
B. Burns (Editor)

6

Perceived Similarity in Perceptual and Conceptual Development: The Influence of Category Information on Perceptual Organization

BARBARA BURNS
University of Louisville

I. INTRODUCTION

These are particularly interesting times for those concerned with developmental changes in the perception of objects. First, recent hypotheses concerning the development of object perception have been more explicitly linked to models of adult representation and processing. Secondly, a variety of interesting links to perceived similarity have recently been made evident in descriptions of shifts in perceptual and conceptual development.

This chapter provides a framework within which to consider the importance of perceived similarity in attempting to characterize perceptual and conceptual development. In Section II, I review the literature that demonstrates developmental shifts in the perception of objects. I describe "traditional views" of the development of object perception as well as more circumscribed hypotheses such as the integral-to-separable developmental shift, the developmental shift in dimensional salience, and the syncretic and pointillistic views of the development of object perception. In Section III, I review literature that supports the notion of shifts in the conception of objects with an emphasis on the central importance of perceived similarity. These shifts include the perceptual-to-conceptual shift, the function-to-form shift, developmental studies of the vertical and horizontal structure of categories, the characteristic-to-defining shift and lastly, the thematic-to-taxonomic shift. The goal of this review is to highlight the role of perceived similarity among the proposed developmental shifts in perceptual and conceptual representation.

In Section V, I have reported a series of four experiments that explores the link between one characterization of the development of object perception, the integral-to-separable developmental shift, and the representation and processing of category information. The goal of these studies was to characterize across development the effect of category information on perceptual organization. A set of stimulus objects was developed that varied perceptually in form (h/w ratio) and size, but varied conceptually in category membership -- vases, cups and bowls. Also of interest in these studies was the influence of naming objects on classification performance.

II. SHIFTS IN PERCEPTUAL DEVELOPMENT

A. Traditional Views

Much of the current literature which examines shifts in the perception of objects across development refers to "traditional views" as proposed by Vygotsky (1962), Werner (1948; 1957), Wohlwill (1962) and Inhelder and Piaget (1964). These theories essentially proposed that objects are perceived by young children as undifferentiated wholes and that, with development, objects are perceived in terms of component parts or dimensions. A close examination of these traditional views reveals that such views described not only shifts in the perception of objects but also shifts in the "conception" of objects. For example, Werner described increasing differentiation and hierarchical organization as a general developmental process, characteristic of the child's perceptual *and* conceptual development (Werner & Kaplan, 1963). Similarly, Vygotsky (1962) supported a view of development of both perceptual *and* conceptual categories from organization based on overall, maximum similarity to analytic, principled organization. Wohlwill (1962), too, framed developmental changes in object perception within a continuum from perception to conception. Most generally, he argued that with development there was a decreasing dependence on "information in the immediate stimulus field" (1962, p. 87). Inhelder and Piaget (1964) proposed three developmental shifts in the perception of objects based on classification tasks. The first graphic collection phase emphasized configural aspects (e.g., spatial arrangement) as the basis of perceptual organization. This was followed by a nongraphic phase based on perceived similarity. Development proceeded to a mature organization based on logical, class inclusion hierarchies.

Thus, many of these "traditional" views described developmental changes in the perception of objects as moving from organization based on perceptual information to organization based on logical, conceptual information. Strong distinctions between the study of the perception of objects and the conception of objects, evident in contemporary literature, would likely seem quite puzzling to these "traditional" theorists.

B. Integral-to-Separable Developmental Shift

A substantial body of literature supports an integral-to-separable or holistic-to-analytic shift in perceptual development[1] (Kemler & Smith, 1978; Shepp, 1978; Shepp, Burns & McDonough, 1980; Smith & Kemler, 1977; 1978; Ward, 1980) based on converging tasks first employed by Garner to distinguish

[1]For the current purposes I have not described the varied perspectives and distinctions between Kemler Nelson's holistic to analytic shift (see Kemler Nelson, 1983) and Shepp's separability hypothesis (see Shepp, 1983).

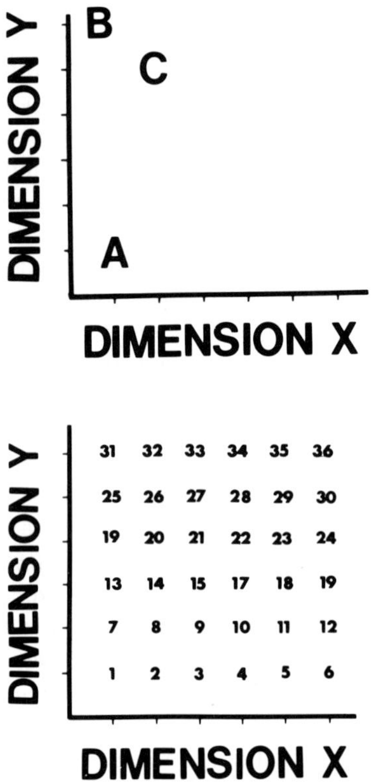

Figure 6.1. The distance relations of three objects presented for restricted classification (top). A matrix of objects varying in six levels on dimension X and six levels on dimension Y (bottom).

the perception of integral and separable stimulus combinations in adults (Garner, 1974). The core idea underlying the proposal for an integral-to-separable developmental shift was that early in development, children perceived objects in terms of similarity relations or as global integral wholes, and with increasing development, children perceived objects in terms of dimensional relations, that is, in dimensionally-organized psychological space. The strength of the integral-to-separable shift formulation rested on first, its explicit link to models and operations of adult perception and processing, and second, and relatedly, the specificity of "undifferentiated" perception as more than simply a lack of differentiation. Undifferentiated perception, characteristic of young children, was hypothesized to be similar to the holistic, but structured perception of integral dimensional combinations evidenced in adult performance on restricted classification, similarity judgments, and speeded sorting tasks.

The restricted classification task has been employed in a number of developmental investigations of the integral-to-separable shift in order to distinguish the dominant mode of representation of objects. The particular focus in the restricted classification task is on distinguishing similarity based on identity on a level of one dimension and considerable dissimilarity on a second dimension from similarity based on both of two dimensions composing the object. Figure 6.1 (top) shows three objects selected from a matrix of objects defined in two-dimensional psychological space[2] depicted at the bottom of Figure 6.1. Objects A and B share a level on Dimension X and differ considerably on Dimension Y. Objects B and C do not share a level on either Dimensions X or Y but are much closer in psychological space. Thus, the two kinds of similarity described above, dimensional identity and overall similarity, compete as available bases of classification.

Garner has shown that objects generated from some dimensional combinations which he termed "separable" (e.g., size and brightness) are consistently classified by adults in terms of identity on a dimensional level. The shared level on a component dimension overrides the direct-distance similarity relations in a two-dimensional psychological space. Other dimensional combinations termed "integral" (e.g., hue and brightness) are not classified in terms of component dimensions by adults, but are consistently classified by the direct distance similarity relations within two-dimensional psychological space.

The integral-to-separable developmental shift in object perception is supported by evidence from the restricted classification task. Young children base their similarity of *both* integral and separable dimensional combinations in terms of overall similarity relations. With increasing age, children shift their basis for classification of separable dimensional combinations towards the pattern evident in adults, that is, towards similarity based on identity of a level on a component dimension (Shepp et al., 1980; Smith & Kemler, 1977; 1978; Ward, 1980).

After this shift in object perception was first described, investigators questioned whether these developmental changes in object perception were best described as an emerging dimensional organization of objects, or a shift towards an increasing accessibility of an established dimensional organization. The answer to this question has been quite clear. Under certain task demands, even very young children have been shown to be able to access the dimensional structure of

[2]Selection of levels on each of the dimensions shown are presumed to be based on scaling data; hence the space is organized in terms of psychological distance relations.

objects (Kemler, 1982; 1983; Kemler & Smith, 1979, Smith & Kemler, 1978; 1979; Smith, 1979; 1983; 1984; Ward, 1980). The integral-to-separable shift has been more recently described as a *relative increase* in the tendency *to analyze* and a *relative decrease* in the tendency *to apprehend objects as integral wholes* (see Kemler, 1983).

C. Developmental Shifts in Dimensional Salience

Based on an extensive program of research, Odom and his colleagues have argued (Aschkenasy & Odom, 1982; Cook & Odom, 1988; Odom, Astor & Cunningham, 1975; Odom & Guzman, 1971; Odom & Corbin, 1973; Odom, 1978) that developmental shifts in object perception can be most properly characterized, not as shifts in perceived organization and structure, but rather as shifts in dimensional salience. Odom and his colleagues have documented "salience hierarchies" in classification, discrimination, problem-solving and memory tasks and have characterized perceptual and conceptual development in terms of changing salience values. Aschkenasy and Odom (1982) argued that the evidence for the integral-to-separable developmental shift can be accounted for in terms of developmental changes in dimensional salience. Their argument is that objects are represented in terms of dimensional components at all ages, but with development, the perceptual system becomes more sensitive to the differences between levels on dimensions composing objects. According to Odom, these developmental changes in perceptual sensitivity to differences among dimensional levels yield the pattern of developmental changes in perceived structure described within the context of the integral-to-separable shift. Kemler (1983) specifically addressed Odom's argument and closely examined Aschkenasy and Odom's assumptions and method for determining classification comparisons (see also Lane & Pearson, 1983). Of issue is Odom's inference that variations in classification performance as a function of variations in "dimensions defined by the experimenter" imply variations in sensitivity to *psychological* dimensions. Kemler described predictions based on the integral-to-separable shift for classification triads in which distance relations among objects were varied yet the dimensions underlying objects for classification were not represented as *psychologically independent* dimensions for the observer. She concluded that Aschkenasy and Odom's data were "exactly what the developmental form of the integrality-separability hypothesis predicts" (p. 365, 1983). Odom's conclusions necessitate the assumption of performance based on independent psychological dimensions. It appears clear that understanding the relation between the dimensional salience and integral-to-separable shift literature hinges on the identification and common specification of psychologically independent dimensions (Burns & Cohen, 1988; Burns, Schlewitt & Cohen, 1992).

D. Syncretism, Pointillism and Object Perception

There is a substantial and well-grounded literature supporting seemingly contradictory views concerning the developmental progression of object perception in terms of "parts" and "wholes". Vurpillot (1976) reviewed the literature (beginning in 1890!) for both the "syncretic" and "pointillistic" views of young children's object perception. Syncretism has been defined as the view that young children perceive whole objects more easily than parts of objects. In contrast, pointillism has been defined as the view that young children perceive the parts of an object more easily than the whole (Vurpillot, 1976, p. 126). The evidence for each view has come from a variety of similarity, identification, and naming tasks. Her extensive review led Vurpillot to speculate that both properties of the task and properties of perceptual organization determined which "level of organization" dominated perception.

For example, Elkind, Koegler and Go (1964) reported evidence for changing patterns of part-whole identification across development. They presented figures composed of easily-recognizable parts (e.g., an airplane constructed with vegetables, a bicycle constructed from candy) and measured developmental changes in part and whole identification. In contrast to previous work by Meili-Dworetzki (1956) who found that wholes were reported at an earlier age than parts, Elkind reported developmental increases in the recognition of *both* parts and wholes. When asked to identify objects, very young children (4-5 years) typically reported only parts of objects, five- to six-year olds predominantly reported wholes, and six- to eight-year olds reported both the parts and the whole in a sequential, nonhierarchically-integrated manner (e.g., candy canes, lolly pops, a bicycle). Children over eight years of age, however, reported both the parts and the whole *in a hierarchical organization* (e.g., an airplane made up of vegetables). Elkind pointed to properties of the stimulus, or "field effects", and secondarily to limited attentional abilities at different ages, as reasons for changing patterns across development in the perception of objects in terms of parts or global configurations.

Since Elkind's work, various patterns of whole and part perception have continued to be reported. Prather and Bacon (1986) questioned whether young children had the ability to perceive both parts of an object and a global configuration at the same time under identical task demands. Prather and Bacon manipulated the classification level of parts (basic, superordinate, subordinate), and the ambiguity of the wholes, and measured the frequency with which preschoolers identified both the parts and the whole. In a *cueing* condition in which children were prompted to elaborate on what they could see, preschoolers reported both the parts and the whole for 73% of simple and complex objects. In a *non-cueing* task, even young preschoolers reported both the whole and the parts on 40% of simple objects. Consistent with the previous conclusions of Elkind and Vurpillot, Prather and Bacon concluded that the critical issue concerning children's object perception is not *whether* children are able to perceive both the parts and the whole, but rather how to characterize the influence of object complexity and task demands on this ability.

Recent developmental work by Kimchi (1990) using hierarchical stimuli has helped illuminate the specific stimulus conditions and task demands under which parts and wholes are perceived. Kimchi's work also provides some explicit connections between the wholes and parts described in the context of syncretism and pointillism, and the wholes and parts previously described in the context of the integral-to-separable developmental shift. Kimchi reviewed the adult literature demonstrating a global-to-local progression in perceptual processing in adults (e.g., Broadbent, 1977; Navon, 1977; 1981), and pointed out that the thrust of most of the adult work has been to identify the order in which the two perceptual levels, the global whole and the local constituent parts, are processed. She has previously argued that this framing of the question is too simplistic (Kimchi, 1982; Kimchi & Palmer, 1985) and has shown that the relationship between the global and local level of perceptual processing is affected by two stimulus aspects: relative size and number of local elements. In her investigations with adults, Kimchi demonstrated that the global and local levels were perceptually separable (as defined within Garner's framework) when the local elements were relatively large in number, but were perceptually integral when the local elements were relatively few in number (Kimchi, 1982; Kimchi & Palmer, 1985).

In a recent developmental investigation, Kimchi (1990) presented triads of objects to children and determined the frequency of similarity judgments based on the global configuration of objects as compared to similarity based on the identity of individual elements. Her findings were clear. Similarity judgments by preschoolers were predominantly based on global configuration when the number of elements composing the objects was small and their relative size was large. (See Kimchi & Goldsmith, Figures 3.2, 3.3, current volume for examples of her stimuli.) However, when the number of elements was large and the relative size was small, similarity judgments were predominantly based on local elements. This pattern of findings mirrored the pattern found previously in adults. In a second experiment, Kimchi explored the contrast between similarity based on identity of global configuration, and similarity based on identity of local elements. Results showed that when the number of elements in a figure increased so did the likelihood of global configuration similarity. Again, this pattern was true for very young children as well as adults. It seems clear from Kimchi's findings that the proper characterization of the development of object perception must incorporate some characteristics of stimulus complexity that do *not* change across development but do influence whether objects are perceived in terms of their global configurations or individual elements at all developmental levels.

III. SHIFTS IN CONCEPTUAL DEVELOPMENT

A. Perceptual-to-Conceptual Shifts in Development

The examination of classification performance across development using complex pictures which contain perceptual as well as symbolic-linguistic forms of organization has provided support for a perceptual-to-conceptual shift in development. A body of literature supports the "traditional" characterization described earlier showing that young children prefer to sort complex pictures by perceptual attributes, and that older children shift their preference to a sort based on conceptual attributes (Bruner, Olver & Greenfield, 1966; Denny, 1975; Denny & Moulton, 1966).

A now-classic experiment by Olver and Hornsby (1966) examined the groupings of complex pictures by children from 6 to 11 years of age, and found that young children's similarity groupings were based on perceptual (e.g., "They are all red") properties but that older children's similarity groupings were based on functional (e.g., "They all make noise") or nominal (e.g., "They all are vehicles") properties. This perceptual-to-conceptual shift has been replicated with a variety of picture classification tasks (Denny, 1975; Denny & Moulton, 1966). More than twenty years ago, Flavell reviewed the existing literature on concept development and concluded that there was substantial evidence for a developmental shift "from equivalences based on the more concrete and immediately given perceptual situational and functional attributes of objects to equivalences of a more abstract, verbal-conceptual sort, in particular the use of class names ("animals", "fruit", etc.) as the basis for grouping" (p.996, 1970).

The perceptual-to-conceptual shift has continued to find support in the developmental literature. Gollin and Garrison (1980) demonstrated robust developmental changes in cognitive processing in a sorting task with training on perceptual versus nominal rules. Again, young children employed perceptual dimensions to sort objects whereas older children based their sorts on nominal or conceptual dimensions. A variety of grouping tasks was employed by Melkman, Tversky and Baratz (1981) in their examination of children's use of perceptual (color, form) and conceptual attributes. They found converging support for a developmental progression from perception to conception, and more specifically characterized this shift as a shift from "color" to "form" to "concept". Four-year olds preferred either color or form, five-year olds preferred form, and nine-year olds preferred conceptual attributes as the basis for grouping and classification.

Tversky (1985) has argued that the early reliance by young children on perceptual information is not due to either a lack of knowledge of conceptual features or an inability to disregard highly salient perceptual features. Tverksy (1985) manipulated the salience of perceptual features of objects by comparing grouping of basic-level objects in name-only and picture conditions. Preschoolers grouped objects consistently on the basis of perceptual features in both conditions. In contrast, older school-age children showed grouping patterns in both conditions consistent with more conceptual, superordinate organization. These findings supported her argument that young children's preferences for the perceptual organization of objects are not due simply to an inability to disregard salient perceptual attributes. Tversky proposed that the shift from perceptual to conceptual classification may be due to a combination of (1) knowledge acquisition (i.e., learning that perceptual features may be less useful than conceptual features) and (2) a generalized increase in active strategic processing.

B. Function-to-Form Shift

Relevant to a review of the development of object perception are two prominent language development theories, Clark's Semantic Feature Hypothesis (1973) and Nelson's Function-Core Concept Hypothesis (1973; 1974), which have proposed contrasting bases for linguistic development. Whereas Clark emphasized perceptual features as the fundamental organizing principles of meaning and concepts, Nelson emphasized equally fundamental functional features. Clark (1973) argued that word organization is built feature by feature and that salience predicted when and which perceptual features are acquired. Nelson (1973) similarly focused on the importance of independent functional properties in describing the developing organization of concepts. These two theories have prompted numerous investigations as to the relative importance of perceptual versus functional similarity using a variety of categorization and concept-learning tasks.

Support for the importance of functional similarity was garnered by Nelson (1973) who manipulated functional and perceptual (i.e., form) features and reported that functional features showed primacy over perceptual features. Nelson and Nelson (1978) described findings from an unpublished study by DeVos and Caramazza (1977) in which preschoolers were asked to label objects which resembled cups, glasses and bowls, adapted from a study by Labov (1973). (Labov (1973) had previously examined the boundaries of word meaning for cup-glass-bowl objects and had shown in adults that meanings of physically identical stimulus objects shifted with context.) The DeVos and Caramazza study examined the question as to the dominance of perceptual and functional features across development, and demonstrated that preschoolers labelled cups, glasses and bowls on the basis of function, whereas 8 year-olds relied on the features of form. Gentner (1978) further examined the existence of a function-to-form shift but created novel objects that permitted her to independently contrast functional similarity against perceptual similarity. It is important to note that in the previous work showing support for a function-to-form shift, the stimulus objects to-be-categorized differed perceptually as well as functionally. Gentner's results contrasted with the previous reports as she found that very young children and adults relied on perceptual features (i.e., form), and it was only the intermediate-aged children who labelled objects in accordance with functional features.

Prawatt and his colleagues (Prawatt & Wildfong, 1980; Anderson & Prawatt, 1983) returned to the more naturalistic Labov-type stimuli and examined the effect of various functional contexts (pictorial, verbal) on labelling of prototypic-cups, -glasses, and -bowls across development. In contrast to the initial DeVos and Caramazza data, Prawatt reported that older, or intermediate-aged children relied on functional features (as determined by context) to base their labels, whereas preschoolers focussed on perceptual features and ignored functional features.

Tomikawa and Dodd (1980) explicitly manipulated the salience of perceptual similarity, and contrasted perceptual and functional similarity in object sorting and concept-learning tasks with much younger children (2-3 years of age). Their findings showed that when perceptual and functional features are independently varied, perceptual features, in both high and low salience conditions, dominated performance. These results, similar to some of the previous work showing a perceptual to conceptual shift, provided support for the idea that perceptual rather than functional similarity dominates the organization of young children's concepts. Nelson (1983) reviewed many of the studies purported to test her "function-to-form" shift and argued that much of the research has been misguided. She claimed that she never posited a function-to-form developmental shift, but rather was describing the central importance of function in the conceptual core. More specifically, she stated her position regarding form, function and development is that "the way in which form and function interact may differ at differ points of development, but they will always interact" (p. 393). Of particular relevance for the current review, Nelson pointed out the complexity of contrasting perceptual against functional features, as an unconfounded contrast would require the *matching of the salience of form and function* from some independent test. A related problem, pointed out previously by Kemler (1983), is that a critical assumption underlying much of the analysis of the function-to-form shift is that the properties of function and form are *psychologically independent attributes* for young children. Kemler, drawing a parallel to the work supporting an integral-to-separable developmental shift in object representation, suggested that in the function versus form literature "questions about salience are quite possible reducible to questions about overall similarity relations. The young child who generalizes *ball* from his red rubber toy to the grape on the table may apprehend them as similar overall. We should resist jumping to the conclusion...that form is a psychologically real property for the child, and that, as such, it is particularly salient for him." (p. 376, 1983). It appears that the debate as to the relative importance of function and form has not adequately considered this underlying assumption concerning perceived structure and psychological independence.

C. Category Structure and Development

In the following the large literature on category structure has been first separated into issues related to vertical (i.e., subordinate, basic, superordinate) and horizontal (i.e., family resemblance, criterial attributes) category structure and the role of similarity has been highlighted. Second, the explict analyses of the role of perceived category structure in category learning by Kemler Nelson and Ward have been reviewed in some detail.

Vertical Category Structure. Eleanor Rosch's well-known work on category organization has established the primacy of basic-level categorization (Rosch, Mervis, Gray, Johnson, & Boyes-Braem, 1976). Basic-level categories, such as dog, apple and screwdriver, may be contrasted with superordinate-level categories, such as animal, fruit and tool, and subordinate-level categories, such as sheep dog, Rome apple, and Phillips screwdriver. The basic level has been described as fundamental because it is most cognitively efficient, most differentiated, the level at which the similarity of within-category members is highest relative to the similarity of objects from different categories, and the level at which objects have the most correlated attributes. In addition to basic-level objects having high within-category similarity, objects at this level can be described as having a very strong perceptual basis as members have similar overall shapes and members are interacted with using similar motor movements (Hoffman, 1982; Rosch et al., 1976).

Language learning has been shown to progress from basic-level to superordinate to subordinate categorization (Anglin, 1977; Daehler, Lonarda & Bukato, 1979; Horton & Markman, 1980; Mervis, 1987; Mervis & Rosch, 1981; Mervis & Crisafi, 1982; Rosch et al., 1976). Mervis and Crisafi's (1982) demonstration of the ease with which children form basic-level categories provided insight into the degree to which the relative degree of differentiation is a factor underlying this effect. Mervis and Crisafi showed that basic-level categorizations were more differentiated than objects from other levels of categorization. That is, basic-level objects (which were categorized first) showed the highest within-category similarity *and* the highest between-category dissimilarity. Mervis and Crisafi (1982) commented that children's difficulty in categorizing at the subordinate level may be related to the integral-to-separable shift in object perception, as subordinate classification requires that component dimensions be distinguished. Mervis has argued (1984, Mervis and Mervis, 1982) that the major source of difference between young children's basic-level categories and older children and adult's basic-level categories is that "children are attending to or emphasizing different attributes from adults" (1987, p.207). The reasons for weighting attributes differently include (1) not knowing the cultural significance of individual attributes, (2) having differing salience hierarchies for particular attributes, and (3) including incorrect attributes.

Tversky (1989) considered to why basic-level objects are most psychologically fundamental and examined the perceptual factors underlying young children's grouping of natural categories at the basic level. Tversky and Hemenway (1984) had shown in adults that the basic-level differed from other levels in that objects share "good parts" at the basic level. According to Tversky and Hemenway, these good parts, such as the leg of a pair of pants or the seat of a chair, appear to have both functional significance and perceptual salience. Extending this approach, Tversky (1989) demonstrated that young children were particularly sensitive to "good parts" of common objects. These results converged nicely with the argument that similar parts reflect similar functions, and suggested that it is this similarity that underlies and influences the previously described shift from perceptual to more conceptual (e.g. functional) classification.

Fenson, Camerson and Kennedy (1988) examined the role of perceptual similarity underlying the development shift toward taxonomic grouping of basic- and superordinate-level categories. Fenson et al. attempted to equate perceptual similarity of objects (as measured by adult dissimilarity judgments) with a basic- and superordinate-level match. The logic was that this procedure would allow perceptual similarity to *not* be confounded with conceptual similarity. For example, a standard object (e.g., a duck) was presented with a matching object which varied in perceptual similarity (e.g., another bird that varied in similarity to the duck). This matching object was embedded in a set of distractors (e.g., three other animals, that varied in low, medium or high similarity). Fenson et al.'s findings provided support for the idea that *when perceptual and conceptual similarity were varied independently and to the same degree, the difficulty of selecting a match was due to perceptual similarity, and not due to conceptual similarity.* Performance was not affected by whether distractor objects were at the basic- or superordinate-level. According to Fenson et al., "it is no easier to match a golfball with a football or a poodle with a collie (basic matches) than to match a sheep with a squirrel or a hammer with a saw (superordinate matches)" (p. 905, 1988). In addressing the contrast between these findings and previous work, Fenson et al. argued that "prior studies may have underestimated the role played by perceptual factors and overestimated the role played by categorical associations in children's decision making" (p. 905, 1988).

Fenson et al. argued that there is a developmental shift in the influence of perception on categorization but for young children, perceptual similarity is the organizing principle guiding *both* basic- and superordinate-level categories. In contrast with previous conclusions about similarity and category level, Fenson et al. suggested that for young children, early superordinate-level categories are not more abstract than basic-level categories. Clearly, the Fenson et al. work represents a novel examination of perceived similarity and category level.

Horizontal Category Structure. In addition to characterizing the vertical structure of categories, Rosch (see especially Rosch, et al., 1976) has provided the theoretical argument and experimental basis for a fundamental shift in thinking about the horizonal structure of natural categories. Rosch argued that natural categories are organized according to a family-resemblance structure in which single attributes are neither necessary nor sufficient for determining membership in a category. Each category is organized around a best example or a prototype, and the similarity of one item to this core object determines its degree of category membership. Thus, members of a category are most often similar to each other on a number of attributes. Membership in a category cannot be predicted on the basis of one criterial attribute but rather can be predicted as a probabilistic function of a number of attributes. This organization can be contrasted with the traditional view of concepts as having criterial attributes (see Smith & Medin, 1981, for a review and analysis of this distinction).

That category membership has a graded structure has been supported by a variety of tasks from similarity judgments (Tversky & Gati, 1978) to category-membership verification (Smith, Shoben & Rips, 1974). Rosch argued that a graded structure may function as a cognitive universal (Rosch, 1975b), and a large literature supports the idea that children structure their categories with a prototypicality gradient, although several differences have been shown between children's and adults' category structure (Barrett, 1982; Bjorklund, Thompson, 1983; Bjorklund, Thompson, & Ornstein, 1983; Duncan & Kellas, 1978; Kay and Anglin, 1982; Kossan, 1981; Posnansky & Neuman, 1976; Saltz, Soller & Sigel, 1976; Mansfield, 1977; White, 1982).

Horizontal Category Structure, Category Learning and the Integral-to-Separable Shift. Recently, Kemler Nelson has suggested that the integral-to-separable shift, one reported developmental shift in object perception, and children's category learning are linked. Kemler Nelson (1984) examined the category learning performance of kindergarten and fifth-grade children using categories that could be organized either on the basis of overall similarity in a family-resemblance structure, or on the basis of traditional criterial-attribute structure. Kemler Nelson argued that the holistic processing style of the young child would be mismatched with the learning of categories defined by criterial attributes. When the categories to-be-learned were distinguished on the basis of family resemblance structure, young children's holistic processing style would be well-matched, and would lead to highly successful performance in the task. Results supported her hypothesis that processing style was related to category-learning performance. (See also Kossan, 1981 for a similar pattern of findings.) Kemler Nelson showed that fifth-graders learned both types of category structure equally well, whereas kindergartners successfully learned the family resemblance categories but had difficulty in learning the criterial-attribute categories. Kemler Nelson (1984) provided converging evidence for this linkage between processing style and category learning by manipulating intentional and incidental learning conditions and examining the learning of family resemblance categories in adult subjects. These results showed that the holistic, or overall similarity, processing approach was successful only in the incidental learning condition. Kemler Nelson distinguished between the goal-oriented analytic task demands of the intentional learning task and the real-world characteristics of the incidental learning task, and argued that incidental learning task demands and family resemblance categories were more representative of real-world learning situations.

Ward and his colleagues (Ward & Becker, current volume; Ward & Scott, 1987; Ward, Vela & Hass, 1990; Ward, Vela, Perry, Lewis, Bauer & Klint, 1989) have also explored the links among representation, processing style and category-learning performance in children and adults. In contrast to Kemler Nelson, Ward has proposed that category-learning tasks with categories organized by a family resemblance structure are approached analytically by both young children and adults. Ward's results have supported this proposal that the differences in category learning performance between younger and older children were due to a differential flexibility in switching initial hypotheses concerning the dimensional attribute that distinguishes categories in such tasks. Ward argued that the young child's holistic processing mode previously found in classification tasks was not evident in category-learning tasks because children understand that a "different mode of processing is required when learning concepts." Ward argued was that the increased ability that children demonstrated in learning categories organized by family resemblance structure was not due to the overall similarity structure of family resemblance category, but rather due to the more informative nature of component attributes of categories organized within the family resemblance structure.

Recently, Ward et al. (1989) examined the underlying basis for category generalizations in preschoolers, second graders, and college students and attempted to specifically identify the individual dimensional attributes or clusters of attributes their subjects would employ to extend the gradient of a novel category label. A specific subset of the literature on category membership (some of which has been reviewed above) suggested that there were four types of manipulations that may provide the basis for membership: overall shape, size, number and type of parts. Overall shape was manipulated as it has been implicated as an important perceptual characteristic common to basic-level objects (Rosch et al., 1976). Number and type of parts was manipulated based on findings from Tversky and Hemenway (1984) showing the importance of "good parts" as a feature for basic-level category membership. Size was also manipulated as an example of a dimension not typically used to define category membership.

Ward et al.'s findings supported his argument (Ward, 1991; Ward & Scott, 1987) that young children's category generalizations are based on individual component attributes rather than on family resemblance or overall similarity structure. Young children generalized categories on the basis of single attributes, and with increasing age there was clear evidence of the use of multiple attributes or clusters of attributes.

The discrepancies in findings and interpretation between Kemler Nelson's and Ward's work have not gone unnoticed. Kemler Nelson has suggested some fundamental problems in the methodology and line of reasoning employed by Ward (Kemler Nelson, 1988, 1990). Ward has challenged Kemler Nelson's criticisms, isolated some differences in stimulus set characteristics used by Kemler Nelson and himself, and has effectively drawn attention to the utility of their different viewpoints in evaluating future research relating perceptual and conceptual development (Ward, 1990; Ward & Becker, current volume).

D. Characteristic-to-Defining Shift

Keil (1984; 1987; Keil & Batterman, 1984) has examined the structure of categories for children at differing ages and has provided strong evidence for a characteristic-to-defining shift in concept development. Most generally this shift can be described as a shift "from meanings based on bundles of characteristic attributes (overall similarity relations) to correct adult meanings based on criterial properties" (Kemler, 1983, p. 376). This shift is most simply understood by considering the contrast between defining a word on the basis of high characteristic/low defining or low characteristic/high defining features. Consider two meanings for the word uncle, first, a good friend of your father's who gives you presents at Christmas and your birthday, and secondly, your mother's two-year- old brother. The first set of descriptors has high characteristic features but lacks a critical defining feature. The second descriptors contain the critical defining feature but lack highly characteristic features. Keil has shown that children undergo a shift from describing words with no regard for a single critical feature and high regard for features that are characteristic or typical of that category as a whole, to describing words on the basis of specific central and critical features. Keil has argued that this shift does not support a gradual differentiation theory. This is not a shift from simple to more complex and elaborated information. Rather the shift is from weighting all of the features "equally" to weighting a few features as fundamental and critical to the organization of meaning. Keil has been careful to point out that this characteristic-to-defining shift is not in opposition to a prototype-based or other probabilistic representation. He has argued that the features that become central and fundamental to the organization of meaning may be organized around prototypes. This conceptualization of the characteristic-to-defining shift in word meaning parallels previous descriptions of the integral-to-separable shift in the perceptual classification of objects (see Smith, 1989). Keil and Kelly (1989) have elucidated this parallel explicitly:

"Early in the acquisition of knowledge about categories children may know so little about the dimensions or organizing principles of a conceptual domain that they (1) cannot perceive in a separable manner the different dimensions that adults use to organize the domain or (2) although they can perceive these different dimensions, they do not know which are more important than others. Thus they adopt the strategy of treating all dimensions as roughly equivalent and then they cluster items based on maximum similarity summed across all dimensions. With increasing knowledge and experience, they begin to perceive the dimensions less integrally or learn that a subset of those dimensions, perhaps only one, can be used as an efficient, or socially accepted, or theoretically useful means of organizing those items." (Keil & Kelly, 1989, p. 500-501).

E. Thematic-to-Taxonomic Shifts

Another developmental shift relevant to our understanding of the changing representations of objects across development has been described as the thematic-to-taxonomic shift. Young children have been shown to prefer to group objects on the basis of familiar thematic or scripted visual scenes (e.g., banana with monkey). In contrast, older children prefer to group objects which share common superordinate relations (e.g., banana with apple) (Smiley & Brown, 1979; Tenney, 1975). Analyses of the thematic-to-taxonomic shift have progressed along the same developmental path as several previously described shifts in perceptual development. Initially, the thematic to taxonomic shift was considered to be due to an increase in cognitive capacity (i.e., ability to recognize taxonomic relations), and only later was the shift described in terms of a cognitive preference. Smiley and Brown (1979) convincingly demonstrated that the thematic-to-taxonomic shift should be characterized as a change in conceptual preference and not as a change in conceptual capacity. Similarly, the argument posed by Markman (1981) was that both thematic and taxonomic relations can be used by the young child to classify but that thematic relations were more salient than taxonomic relations for preschool-age children. (But see Bauer & Mandler, 1989, Nelson, 1988 for alternative views). The thematic-to-taxonomic shift has been viewed as a subset of the previously described (and more general) perceptual-to-conceptual developmental shift. Clearly, objects in thematic groupings share relations among objects which are more concrete and perceptually-based than the shared relations among categories which have often been described as conceptually-based (Markman, 1981).

Factors shown to increase the use of thematic grouping when contrasted with perceptual grouping have included lowering the salience of color as an alternative dimension, employing free classification rather than triadic classification procedures, and varying the composition of stimulus materials (Bernt, 1988). Factors which have shown to increase the use of taxonomic grouping as contrasted with thematic grouping include decreasing the salience of the spatial arrangement of objects to-be-sorted (Markman, Cox, & Machida, 1981) and varying the familiarity of the stimulus sets to be classified (Horton, 1982, cited in Markman & Callanan, 1983). In studies which have demonstrated both thematic and taxonomic organization in young children, it has been shown that young children consistently show higher levels of both identification and verbal justification for thematic as opposed to taxonomic organization (Scott, Serchuck & Mundy, 1982).

A recent study by Fenson, Vella and Kennedy (1989) examined very young children's sensitivity to thematic and taxonomic organization and manipulated the category level (basic, superordinate), as well as the perceptual similarity underlying taxonomic organization. Fenson et al. did not pit thematic organization against taxonomic organization in these studies, but rather attempted to assess in very young children the ability to extract each type of organization as a function of perceptual similarity. Results showed that the recognition of thematic relations increased from 2 to 3 years of age. Children as young as 2 years of age were shown to be able to identify taxonomic relations based on *shared perceptual features*. Fenson et al. suggested that rather than asking which is harder, thematic or taxonomic grouping, it is more productive to ask what factors affect the difficulty of recognizing each type of organization. Fenson et al. found that (1) the recognition of thematic organization is constrained by experience of the child and familiarity of specific thematic relations, and that (2) the recognition of taxonomic organization is constrained by the perceptual similarity of items to-be-grouped. Fenson et al.'s basic conclusion, echoed by Markman (1989), was that the ability to classify thematically is not an indication of the inability to classify taxonomically.

Labelling and Classification Performance. One of the most interesting developments in the last few years related to the thematic to taxonomic shift was the finding that the procedure of naming an object before classification increased the use of taxonomic grouping (Markman, 1989). This has occurred even when the objects were given novel labels (Markman & Hutchinson, 1984) or labels from a second language (Waxman & Gelman, 1986).

Markman (in press; Markman & Hutchinson, 1984) argued that upon hearing the label for an object children "constrained the possible meanings of new words to refer mainly to categorical relations" (p.1, 1984). Naming objects was proposed to force the child to look beyond the thematic relations evident and to attend to possible taxonomic relations. Markman has referred to these as the *Whole Object* and *Taxonomic Assumptions.* "When children hear a novel label they assume that the label refers to the object as a whole, and not to its parts, or substance, or color, etc., and that it refers to other objects of the same kind or same taxonomic category and not to objects that are thematically related" (p.4, in press). Markman argues that children's tendency to extend labels categorically guides them in making correct judgments about novel objects. Markman and Hutchinson (1984) provided support for this assumption in a series of studies (see also Markman, 1989) of 4- and 5-year-old children who were either asked to "find another *one*"(no-label condition) or "find another *dax*" (label condition) in a grouping task. Thematic grouping dominated the no-label condition and taxonomic grouping dominated the label condition. Similar results have been reported by Au and Markman (1987), Landau, Smith and Jones (1988) and Baldwin (1989).

Waxman (1990) recently showed that this increase in taxonomic salience as a function of labelling interacted with level of categorization (i.e., basic, superordinate, subordinate). A second set of factors shown to mediate between labelling and taxonomic grouping involves the familiarity of particular labels and the child's existing vocabulary. Waxman, Shipley and Shepperson (1991) have reported that familiarity and existing vocabulary significantly modified the interaction of labelling and level of categorization reported by Waxman (1990).

IV. RELATING PERCEPTUAL AND CONCEPTUAL PROCESSES

The notion that psychologists could profit from the joint consideration of perceptual and conceptual processes is not, of course, new. Shepard and Podgorny (1978) detailed the similarity of results of cognitive tasks using "nonsymbolic" and "symbolic" stimuli and concluded that perceptual processes and cognitive function were highly interrelated and that understanding and characterizing their relationship was critical to the development of a comprehensive theory of cognition *and* perception.

Recently, two chapters in a volume on categorical perception (Harnad, 1990) have explicated the empirical and theoretical similarities of particular areas of perception and conception. These chapters, in fact, may provide the foundation for a qualitative change in thinking about the study of perceptual and conceptual processes. Medin and Barsalou (1990) reviewed the (currently) highly distinct areas of sensory perception categories ("categorical perception") and semantic categories. Medin and Barsalou critically analyzed a variety of important (and basic) issues in the study of categorization and described the high degree of empirical and theoretical similarity across these two areas. The issues analyzed included category structure, bases of classification, uses of categories, category acquisition, and category stability and flexibility. Their arguments for a change in thinking about the relation between sensory and semantic categories are highly compelling.

Keil and Kelly (1990) underscored the importance of relating perceptual and conceptual processes within the study of categories and development. Keil and Kelly provided an equally compelling case for considering parallels between shifts in perceptual representation with shifts in semantic and conceptual representation. As described earlier, Keil and Kelly make explicit the analogy between the integral-to-separable developmental shift and the characteristic-to-defining shift. Keil and Kelly go on to describe a general model which could incorporate common underlying mechanisms for these two shifts in development.

The current chapter distinguished the literatures on perceptual and conceptual development and suggested that "traditional theorists" would find the independent study of perceptual and conceptual processes across development quite puzzling. Medin and Barsalou and Keil and Kelly provide strong arguments to suggest that current researchers interested in perceptual and conceptual processes should also be puzzled by the apparently independent treatment they have received. In the next section, empirical work examining the effect of category information on perceptual organization is described. Findings from these studies provide support for the potential of this approach for advances in the mutual understanding of perceptual and conceptual processes and their development.

V. EMPIRICAL RESEARCH

A. Overview of the Present Experiments

The focus of the current research is on characterizing the influence of category information on the perceptual organization of objects across development. Experiment 1 examined the perceived structure of objects varying in size and form (six levels of h/w ratio) which corresponded conceptually to a continuum of vases, cups or bowls in preschool children and adults. Findings from Experiment 1 showed differences in the basis for similarity when objects were selected for classification from within a category as compared to when objects were selected from differing categories. Classifications based on identity of a level on form increased significantly on between-category triads as compared to within-category triads. Experiment 2 replicated the major findings from Experiment 1 with additional age groups using a nine-level continuum of vases, cups and bowls and examined additional manipulations of psychological distance among objects on within- and between-category triads. In addition, procedural changes in the naming task in Experiment 2 allowed more precise subject-defined within- and between-category designation. Results showed that for all ages, the basis for similarity differed for objects presented from within- and between-category triads. However, this distinction was significantly reduced or eliminated when the distance relations among objects on within-category triads were made more dissimilar. Experiment 3 examined the influence of differing naming procedures on the perceptual organization of objects when presented in within- and between-category classification triads for adult subjects. Findings demonstrated that naming objects immediately before classification further increased the distinction in basis for similarity on within- and between-category triads. Experiment 4 showed that this difference in the basis for similarity of objects from within- and between-category triads was also evident under speeded task demands. Although the overall level of dimensional classifications was reduced under speeded conditions within-category triads continued to produce fewer dimensional classifications than between-category triads.

Critical Stimulus Set Properties. The stimulus sets employed in each of the four experiments consisted of objects, similar to those described by Labov (1973), that lack well-defined category boundaries. Labov (1973) demonstrated various conditions that influenced the use of words and the location of boundaries of cup, mug, bowl, glass (with and without handles) and vase categories. Labov's conclusions have been further confirmed and extended by Kempton (1978) who completed cross-cultural investigations of category boundaries with highly similar drinking vessels.

The novel and potentially powerful aspect of the stimulus sets used in the experiments reported here is the construction of objects by crossing levels of two well-specified dimensions, h/w ratio and size. This yielded a set of objects in two-dimensional psychological space with approximately equal levels of dissimilarity on each dimension. Although the levels on the form dimension were selected to have equal steps of perceptual distances (based on pairwise similarity judgments), the resulting psychological space along the form dimension was interpreted by subjects as being composed of conceptually labelled zones with vase, cup and bowl regions. (See Shepard & Cermak (1973) for a similar stimulus set construction.)

B. Experiment 1. The Perceived Structure of VCB Across Development

In Experiment 1, objects resembling vases, cups and bowls were selected from a matrix of objects varying in form (h/w ratio) and size, and presented in restricted classification triads to preschoolers and adults. On each trial, classification based on shared identity of a level on form and considerable dissimilarity in size was pitted against classification based on overall similarity in both form and size dimensions. Three questions were addressed. First, what influence would category information have on perceptual organization? Second, would this relation between category information and perceptual organization differ for preschoolers and adults? Third, would naming objects (as vases, cups, or bowls) influence perceptual organization for both preschoolers and adults (see Markman)?

Method

Subjects. A total of 38 children (ages 37 to 55 months) and 24 adults participated. The children attended a university preschool and the adults attended a large state university. Adult subjects were paid $4.00 for participating.

Stimuli. Stimuli consisted of black outline drawings of objects varying in size and form which were mounted on white cards (13.0 X 7.5 cm.). Form varied in six levels of h/w ratio (.5, .7, 1.1, 1.6, 2.4, 3.4) and phenomenologically corresponded to a continuum of objects varying from bowls to cups to vases. The six levels of size were obtained by enlarging one set of six base figures in the following proportions: 20%, 40%, 60%, 80% and 100%. The dimensions of the base figures were approximately: 1.3 X .55, 1.1 X .65, .9 X .8, .7 X 1.0, .5 X 1.4, and .4 X 1.8 mm. The edges of the base and top of each rectangle were slightly rounded and at the top of each rectangle was a very narrow elipse (representing depth). The specific levels on each dimension were selected on the basis of a series of adult dissimilarity judgment and naming tasks.

Design. From the matrix of 36 stimuli constructed by combining the six levels of size and form (see Figure 6.1, bottom), triads were selected and presented to subjects for classification. Triads were selected such that classification by a shared level on one dimension (A + B in Figure 6.1, top) was pitted against classification by overall similarity relations (B + C in Figure 6.1, top). Haphazard classifications were defined as classifications of objects *not* based on overall similarity or shared dimensional relations (i.e., objects A + C in Figure 6.1). Objects in each triad had distance relations as shown at the bottom of Figure

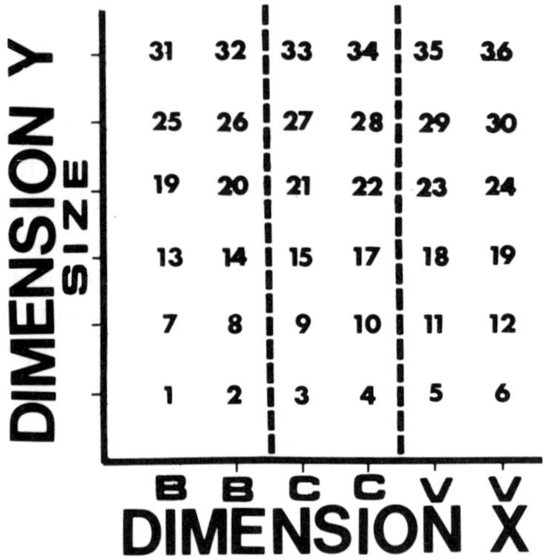

Figure 6.2. The matrix of objects varying in h/w ratio and size presented in Experiment 1. Note that the six levels on dimension X varied in category membership--bowl (B), cup (C) and vase (V).

6.1; objects A and B were five levels apart on dimension Y and shared a level on dimension X, objects B and C were one level apart on dimensions X and Y. Objects A and C were one level apart on dimension X and four levels apart on dimension Y.

1. Category-Triad Types. When the six levels of form are designated as bowl-1, bowl-2, cup-1, cup-2, vase-1, vase-2, two major types of triads, within- and between-category triads, can be distinguished. Within-category triads were defined as triads in which all three objects were selected according to the distance relations described above, and in which all three objects shared a level of form from one category (B1 and B2, C1 and C2, or V1 and V2). For example, objects 1, 31, 26, in Figure 6.2 constitute a within-category triad. Between-category triads were defined as triads in which A and B shared a level of form from one category and C was selected from a second category (e.g., objects 2, 32, 27, in Figure 6.2).

2. Size Triads. Triads were also employed which examined the basis for similarity when objects A and B shared a level in size and differed considerably in form. On size triads, objects A and B shared a level in size and were 5 levels apart on form. B and C were one level apart in size and form (e.g., objects 7, 12, 5 in Figure 6.2).

Procedure. A total of 60 triads was presented to each subject consisting of twenty-four within-category triads (four each of B, C, V presented twice), twenty-four between-category triads (three each of BBC, CCB, CCV, VVC presented twice) and twelve size triads (six from varying levels of the size continuum presented twice) in a randomized order. On half of the triads, the two objects that shared a dimensional level were adjacent to each other. On the other half the two objects that shared a similarity relation were adjacent.

For both age groups, subjects sat at a table facing the experimenter and three stimuli were placed in front of them. In the unlabelled condition the subject was asked "which two go together the best". In the labelled condition subjects were asked to name each object as a vase, a cup or a bowl. They were then asked "which two go together the best". Preschool children received a modified set of triads in the labelled condition consisting of nine within-category and nine between-category triads. Preschool children were tested in two 20-minute sessions which were held no longer than three days apart. Adults were tested in one 20-minute session.

Results and Discussion

Haphazard Classifications. In Experiments 1-3 there were no significant effects or interactions in the analyses of the percentage of haphazard classifications (defined as classification of objects A + C in Figure 6.1). Haphazard responses averaged from less than 6% to 13% across the experiments reported. All analyses reported in Experiments 1-3 were performed on the percentage of dimensional classifications out of the total number of classifications.

Size Triads. Triads in which a level on size formed the basis for dimensional classifications yielded a highly consistent pattern of performance and no developmental differences. Children and adults classified size triads by overall similarity on 94% and 99% of triads, respectively. Thus, according to terminology discussed by Garner (1974), these stimulus objects composed of size and form dimensions yielded a classification pattern that was asymmetric (at least for between-category triads). Levels on the dimension of form were used on between-category triads as the basis for classification whereas levels on the dimension of size were not used as the basis for classifications.

Category-Triad Types. The mean percentages of dimensional classifications for within- and between-category triads are shown in Figure 6.3 for pre-school children and adults. At the top are shown the mean percentages of

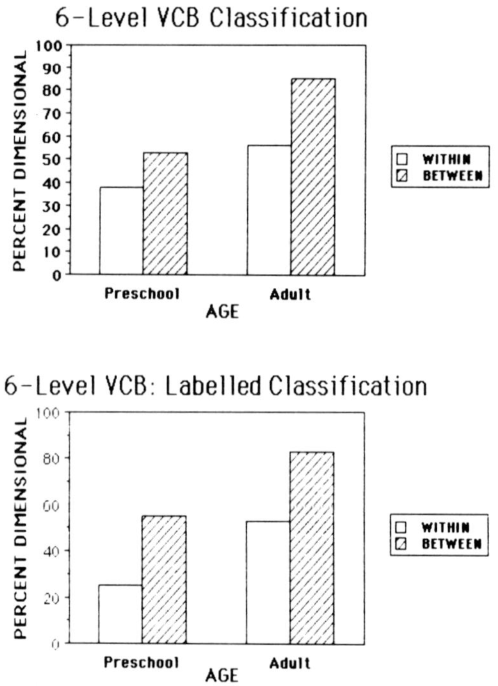

Figure 6.3. Mean percentage of dimensional classifications for within- and between-category restricted classification triads in preschool children and adults. The unlabelled condition is shown at the top; the labelled condition is shown at the bottom.

dimensional classifications for the unlabelled condition, at the bottom are shown the corresponding data for the labelled condition. An analysis of variance with the factors of Age, Labelling Condition and Category-triad type revealed a significant effect of Age ($F(1,58) = 56.46$, $p < .001$), a significant effect of Category-triad type ($F(1,58) = 89.98$, $p < .001$) and a significant Age x Category-triad type interaction ($F(1,58) = 6.69$, $p < .02$). There was no effect of Labelling Condition and no other significant interactions.

The first question of this study concerned the effect of category information on the perceptual organization of objects varying in size and form dimensions. The findings revealed that between-category triads were classified by shared dimensional relations significantly more often than within-category triads for both children and adults. The percentages of dimensional classifications for within-category triads were 29% (children), 51% (adults) in contrast to 43% (children), 81% (adults) for between-category triads.

The second question under investigation concerned developmental differences in the relation between category information and perceptual organization. The Age x Category-triad type interaction reflected that the difference in basis for classification for the two category-triad types was greater for adults as compared to children. Overall, children showed the same pattern of significantly fewer dimensional classifications for objects from within categories as compared to objects from between categories. It should be noted that the overall level of dimensional responding by preschoolers across all types of triads was considerably less than the level exhibited by adults.

The question as to the influence of naming objects on the relation of perceptual organization and category information was examined by comparing classification performance in the classification-only and naming-before-classification conditions. Results revealed no effect of the naming task on classification performance. Additional information from the naming condition comes from the following consideration of the naming functions for each age group.

Naming Patterns. Figure 6.4 shows the naming patterns for the six levels of form by subjects in the labelling condition. Children used two naming patterns. Six children (Figure 6.4, top) showed fairly sharp boundaries for bowls and vases and more graduated boundaries for cups. The six other children (Figure 6.4, middle) who employed only two labels (three children used only bowl and cup labels, three others used only bowl and vase labels) showed fairly sharp boundaries for the distinction between their two main categories. The naming functions for the adults are shown at the bottom of Figure 6.4. Adults had sharp boundaries for the bowl and cup distinction, agreed on one clear exemplar of cup, provided mixed responding for the second cup exemplar, and showed agreement on the two vase exemplars. The mixed responding for this second cup does not reflect individual differences in performance but rather individual differences among the adults. Three adults classified this object (level 4 of form) as a cup on 90% of the triads whereas the other nine subjects classified it as a vase on 74% of triads.

Comparing the preschoolers' and adults' naming functions revealed decreasing category fuzziness with age. Adult category boundaries were more consistent both within individuals and across individuals as a group than preschooler's category boundaries. These findings were not unexpected given

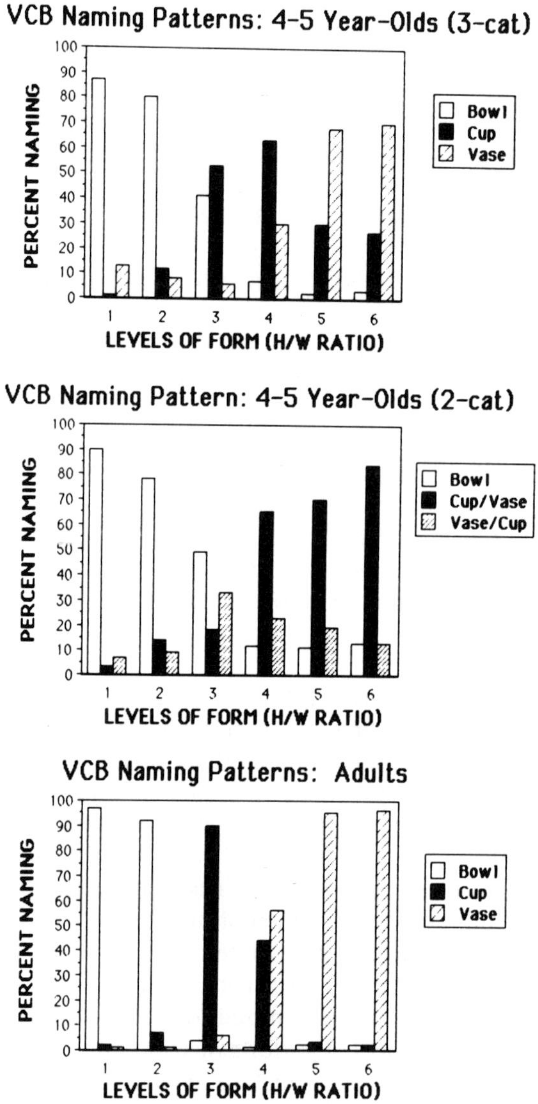

Figure 6.4. Naming patterns for the six levels of form for preschool children who employed three categories (top), preschool children who employed only two categories (middle), and adults who employed three categories (bottom).

previous discussions by Anglin (1985) and Nelson and Nelson (1978), as well as more recent studies of the development of boundaries of fuzzy categories by

Alexander and Enns (1988). (See Section VI. Conclusions for further comparisons.)

Summary of Experiment 1

The findings from Experiment 1 showed that perceptual organization is influenced by category information in both preschoolers and adults. Objects constructed from the same component dimensions (size and form) were classified differently depending on the category membership of the other objects presented on a triad. As in previous descriptions of perceptual development (i.e., the integral-to-separable shift), as measured by the restricted classification task, the overall level of dimensional classifications increased with age. However, *the important novel finding is that category information influenced the basis for similarity; preschoolers and adults switched their basis for classification from overall similarity relations on within-category triads to shared dimensional relations on between-category triads.*

C. Experiment 2. The Perceived Structure of VCB; Extensions and Replications

Experiment 2 employed children from three age groups, in addition to adults, such that the influence of development on the phenomenon described in Experiment 1 could be characterized in more detail. As in Experiment 1, the question under investigation was what was the influence of category information on the perceptual organization of objects. Triads of objects from an extended matrix of nine levels of form (h/w ratio) and six levels of size were selected such that perceptual organization based on identity on a dimensional level could be contrasted with perceptual organization based on overall similarity relations. In this experiment, in which the levels on the form dimension were extended to nine, an additional triad type (Type 2) was employed which allowed further manipulations of the overall similarity distance relations on both within- and between-category triads. Finally, in this experiment changes in naming procedures were implemented in order to allow a more precise subject-defined determination of which objects presented for classification constituted within- and between-category triads.

Method

Subjects. A total of 51 children attending a recreational summer camp participated as subjects and formed three age groups. The first age group contained 17 children with a mean age of 4 years 6 months. The second age group contained 18 children with a mean age of 6 years 8 months. The third age group contained 16 children with a mean age of 9 years 4 months. In addition, 11 college students participated and received a stipend of $4.00.

Stimuli. Black outline drawings which varied in nine levels of h/w ratio (.32, .40, .59, .75, 1.0, 1.3, 1.9, 2.5, and 3.1) and six levels of size were constructed and mounted on white cards (12 X 18 cm.). Size levels were produced by reducing (73%, 85%) and enlarging (109%, 129%, 146%) the base objects (100%). The dimensions of the base objects were 6.6 x 2.1, 5.9 x 2.4, 2.9 x 4.9, 4.3 x 3.3, 3.8 x 3.8, 3.3 x 4.3, 2.8 x 5.1, 2.4 x 5.9 and 2.1 x 6.6 cm. As in Experiment 1, the edges of the base and top of each rectangle was rounded; in addition a narrow ellipse at the top of each rectangle was included to depict depth. The nine levels on the dimension of form were determined on the basis of pilot studies in which similarity judgments were obtained from 10 additional adults. In this task, all possible pairs of the nine base objects were shown to each subject four times each to produce a total of 144 trials. Subjects were asked to rate each pair of stimuli on a scale from 1 to 10 such that 1 would refer to a pair that was highly dissimilar and 10 would refer to a pair that was almost identical. Results from this task showed that the one-level intervals of h/w ratios did not differ significantly ($p > .05$).

Design

1. Naming Task. Before completing the classification task, subjects were asked to name each of the 54 objects constructed by the nine levels of h/w ratio and 6 levels of size. Objects were presented individually in a randomized order.

2. Distance-Triad Types. From this 9 X 6 matrix of objects which contained additional levels of form, a triad type was constructed that differed from those used in Experiment 1 in the distance relations for the overall similarity (B + C) classification. Type 1 triads in this experiment were identical to the triads described in Experiment 1. In contrast, in the new triads (Type 2), objects B and C were two levels apart on form and one level apart on size. On Type 2 triads, objects A and B shared a level on form and were five levels apart on size. Thus, both distance-triad types placed dimensional and similarity classifications in conflict, however the similarity between items B + C in Type 2 triads was less than on Type 1 triads.

3. Category-Triad Types. In Experiment 2, the designation of within- and between-category triads was determined on the basis on individual naming functions obtained from the naming task. For each subject, category names were assigned to each level of form if on at least 67% of the trials one label was consistently provided. Out of the total triads presented, only those triads which could be assigned as a within- or between-category triad were included in the subsequent analysis of category-triad types. The criteria for inclusion were as follows. Within-category triads were defined as triads in which all three objects were named as from within one category: vase, cup or bowl. Between-category triads were defined as triads in which objects A and B were named as from one category (e.g., both bowls) and object C was consistently named (on > 66% of trials) as from a different category (e.g., cup).

Size Triads. As in Experiment 1, an additional set of five size triads (presented three times each) was included which allowed the examination of object classification when the dimension on which objects shared a level was size.

Procedure. A total of 96 possible within- and between-category triads (determined by the criteria described above) composed of thirteen Type 1 triads, fourteen Type 2 triads, and 5 Size triads each appearing three times each in a random order was presented to adult subjects.[3] Adults completed both the naming and classification tasks in one 20-minute session. Children in the three age groups received the same category-triad, distance-triad, and size-triad types as adults but only received them two times each totalling 64 trials. For children, the naming and classification tasks were divided into two sessions with the first session lasting 20 minutes and the second session lasting 15 minutes.

Results and Discussion

Size Triads. Overall similarity relations dominated the basis of classification of size triads. The percentages of similarity classification for 4-5 year-olds, 6-7 year-olds, 8-11 year-olds, and adults were 85%, 98%, 96% and 94% respectively.

Naming Patterns. Figure 6.5 shows the naming functions for each age group. As in Experiment 1, the naming functions showed a gradual decrease in fuzziness of category boundaries with age. The consistency of naming patterns within individual subjects increased with age and the consistency across age groups also increased with age. Given previous reports by Labov and others concerning the ambiguity of category boundaries for these particular objects, it is noteworthy

[3]Note that the designation of within- and between-triads was calculated on the basis on individual subjects' naming data.

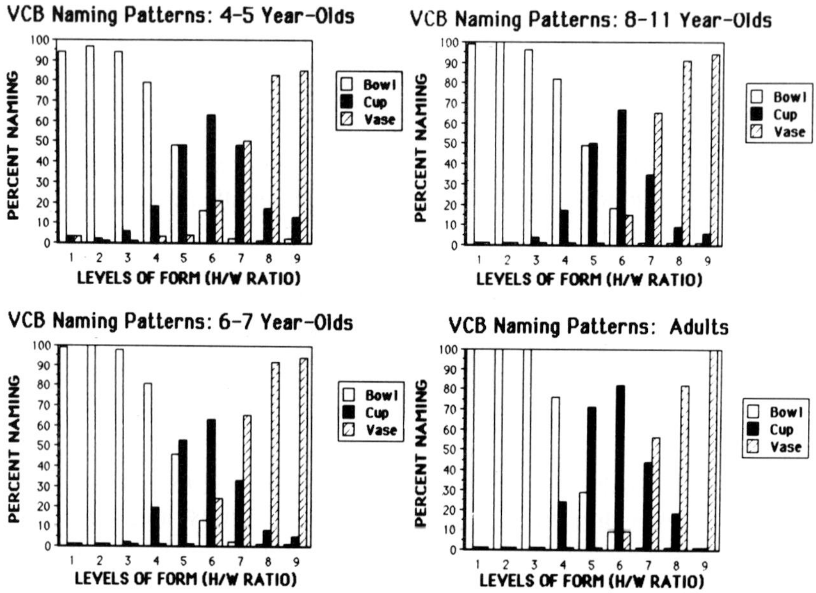

Figure 6.5. Naming patterns for the terms bowl, cup and vase shown across the nine levels of form shown for 4-5 year-olds (top, left), 6-7 year-old (bottom, left), 8-11 year-olds (top, right), and adults (bottom, right).

that consistency was quite high for each age group.[4]

Classification Patterns. The classification patterns of within- and between-category triads are shown separately for Type 1 and Type 2 triads at each age group in Figure 6.6. An analysis of variance of the mean percentage of dimensional classifications with the factors of Age, Distance-triad type (1, 2) and Category-triad type (within, between) revealed a significant main effect of Age ($F(3,58) = 7.43$, $p < .001$), Distance-triad type ($F(1,58) = 69.97$, $p < .001$), and Category-triad type ($F(1,58) = 5.83$, $p < .02$). There was also a significant interaction of Distance-triad type and Category-triad type ($F(1,58) = 7.53$, $p < .01$). There were no significant interactions with age and any other factor.

[4]However, it should be noted that selection of levels was based on a series of separate dissimilarity *and* naming tasks with adults. It is then not surprising that the adults showed consistency; the levels on form were selected initially to approximate such naming patterns.

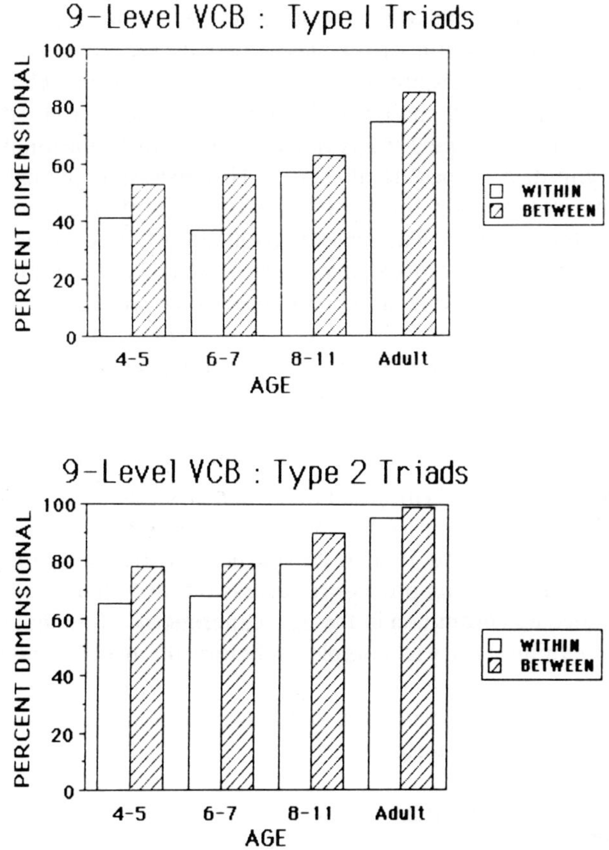

Figure 6.6. Mean percentage of dimensional classifications for within- and between-category triads across development separated by Distance-triad types.

The results from this experiment replicated and extended the findings characterized in Experiment 1. That similarity for objects from within- and between-category triads is organized differently was replicated with three age groups in addition to adults. There were no significant interactions of age and category-triad type or any other factor. As in Experiment 1, an increase in the overall level of dimensional classifications occurred with increasing age. The pattern of classifications did not differ for 4-5 and 6-7 year-old children ($p > .05$) but the 8-11 year-old group had significantly higher dimensional classifications across all triads than both younger groups (p's $< .05$). The adults had significantly higher dimensional responding than all other age groups (p's < .05). The introduction of an extended matrix of objects with nine levels of h/w ratio allowed the further examination of distance relations for both within- and between-category triads. Consistent with previous investigations which varied distance relations of the overall similarity classification (Burns et al., 1978; Burns, 1985; Shepp et al., 1980), Type 2 triads yielded increased dimensional classifications for both category-triad types at all ages. However, the increase in dimensional responding was greater for Type 2 triads for within-category triads than for between-category triads. The mean percentages of dimensional classifications for within-category triads across all ages were 53.4 for Type 1 triads and 80.5 for Type 2 triads. In contrast, the mean percentages for between-category triads across all ages were 64.3 and 80.9 for Type 1 and Type 2, respectively. Development did not interact with the manipulation of increased psychological distance of the overall similarity relation for Type 2 triads.

Although this study replicated the category-triad effect, additional questions were raised. With the nine-level matrix, the percentage of dimensional classifications was greater for both within- and between-category Type 1 triads as compared to Experiment 1. This was particularly true for adults. The possibility that the differing naming procedures across the two studies were responsible for this difference was examined in the next experiment. Note that in this experiment the entire naming task was completed before the classification task whereas in Experiment 1 naming trials were completed before each classification triad. Experiment 3 directly compared these two naming task procedures and their effect on classification performance.

Summary of Experiment 2.

Experiment 2 used a subject-defined designation of within- and between-category triads and an extended continuum of vase, cup and bowl objects. Performance by three age groups in addition to adults showed a consistent difference in the basis for similarity of within-category and between-category classification triads for Type 1 triads only. Category membership influenced perceptual organization for 4-5 year olds, 6-7 year-olds, 8-11 year-olds and adults for Type 1 triads only. The implementation of Type 2 distance triads did not yield the same distinction between basis for similarity of within- and between-category triads and showed the limits of this finding.

D. Experiment 3. Naming Procedures, Category Salience and Perceived Structure of VCB

Two questions related to category salience were pursued in this study. The first question concerned the limits of the influence of category salience on the perceptual organization of size vs. form objects. Specifically, we were interested in whether *naming objects immediately before each classification triad increased the distinction in basis for similarity for within- and between-category triads*. Experiment 1 employed a procedure in which subjects named each triad of objects immediately before classification; Experiment 2 required subjects to complete the entire naming task before initiating the classfication task. The current study contrast these two naming procedures to further understand the efffects of naming objects on perceptual organization. The second question concerned the role of the subjects' response set on classification performance. Can we account for the differing bases of classification for objects from within a category and from differing categories in terms of a higher-level strategy? For example, did subjects strategically attempt to maintain consistency across within-category and between-category triads? To address this issue, classification performance of within- and between-category triads in blocked conditions (all within, all between; or vice versa) was compared to classification performance in the standard mixed condition.

Method

Subjects. A total of 48 adults (15 males, 33 females) enrolled at a large state university participated in this experiment as part of an introductory psychology course credit.

Stimuli. The stimuli were identical to those used in Experiment 2.

Design

1. Category-Triad Types. Within- and between-category triads were selected from a matrix of objects constructed from a continuum of nine forms defined by the experimenter as three bowls (1-3), three cups (4-6) and three vases (7-9) and six levels of size. A total of 8 between-category triads and 9 within-category triads (both experimenter-defined, see footnote 3) was presented three times each in one of two conditions. In the blocked condition, all 24 between-category triads were presented before the 27 within-category triads (or vice-versa). In the mixed condition, the 51 triads were presented in random order. In both conditions, an additional 10 size triads was also included (5 triads selected from across the levels of size, presented twice).

Procedure. The name-during condition employed the procedure of Experiment 1. On each trial, subjects were asked to name each object as a bowl, cup or vase before being asked "which two go together the best". In the name-at-end condition, subjects were presented with each of the 54 objects for naming one at a time at the end of the classification procedure. Subjects were told they had to name each object as a bowl, cup or vase. Twelve subjects received one of the four conditions:blocked/name-at-end, blocked/name during, mixed/name-at-end and mixed/name during.

Results and Discussion

Size Triads. As in the previous experiments, size triads were predominately classified by overall similarity relations. Subjects did not organize similarity of objects by identity on a level on the size dimension. Overall similarity classifications on size triads averaged 92%.

Category-triad Types. An analysis of variance on the mean percentages of dimensional classifications with the factors of Presentation Order (blocked, mixed), Naming procedure (name-during, name-at-end) and Category-triad types (within, between) was performed. Determination of which triads were within- or between-category triads was made on the basis of individual naming functions using the criteria described in Experiment 2. The ANOVA revealed a significant main effect of Category-triad type ($F(1,43) = 16.0$, $p < .001$) and a significant interaction of Category-triad type and Naming procedure ($F(1,43) = 6.16$, $p < .02$). There were no other significant main effects or interactions.

Figure 6.7 presents the mean percentages of dimensional classifications for the within-category and between-category triads (as defined by individual subjects' naming procedures) in the two naming procedures. The name-during procedure yielded a higher percentage of dimensional classifications for between-category triads (74.1% vs. 65.4%) and a lower proportion of dimensional classifications for within-category triads (54.9% vs. 60.9%) than the name-at-end procedure (p's $< .05$). It appears that the procedure of naming objects immediately before classification further distinguished the basis of classification for objects from within and differing categories.

The finding that naming objects before each classification triad increased the difference in bases for classification for within- and between-category triads

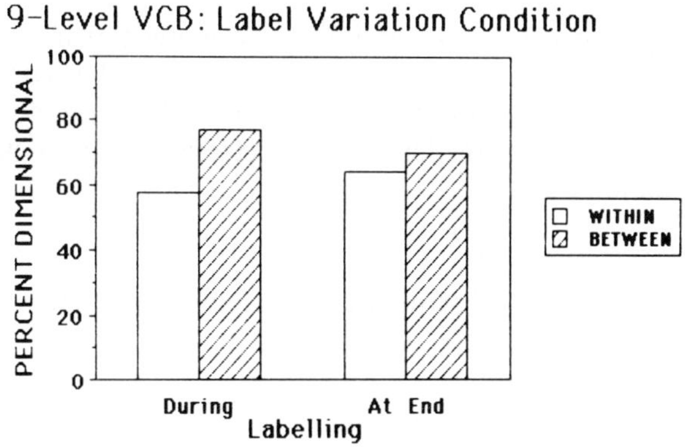

Figure 6.7. Mean percentage of dimensional classifications for within- and between-category restricted classification triads in adults. Performance shown separately for triads in the two label variation conditions.

parallel Markman's recent demonstration that naming influences attention as well as acts to improve the consistency of category judgments. (See Ward, 1990, for an enlightened discussion of the role of novel labels in directing attention to objects.) However, the results from this experiment are somewhat inconsistent given the findings from Experiment 1 in which no differences were found for subjects (both children and adults) who labelled during the classification procedure as compared to those who never labelled the objects. Although one might speculate that objects from the six-level matrix would have more defined category boundaries as compared to the nine-level matrix, the naming functions do not support this account, and this inconsistency remains a puzzle.

In this experiment, category salience was examined from an alternative perspective by comparing classification performance of within- and between-category triads in blocked or mixed conditions. Analyses revealed that the procedure of blocking within- and between-category triads had *no* significant effect on performance. There was also no interaction of blocking and naming procedure. This supports the idea that category information interacted with the distance relations of objects from individual triads. Subjects appear to be highly flexible in switching their basis for classification from identity on a dimensional component (on between-category triads) to overall similarity organization (on within-category triads). These findings do not support an explanation of this phenomenon based on any simple form of response set.

Summary of Experiment 3

Naming objects immediately before classification increased the distinction between within- and between-category triads and further confirmed Markman's suggestion that naming acts to mobilize attention. Performance differences do not appear to be due to a response set as classifications in blocked and mixed conditions did not differ.

E. Experiment 4. The Influence of Speeded Task Demands on the Perceived Structure of VCB

The results of the previous three experiments suggest that category information reliably interacts with perceptual organization such that objects from differing categories are more likely to be perceived in terms of identity on dimensional levels which define that category. Objects from the same category are more likely to be perceived in terms of overall similarity across all the dimensions composing the objects to be classified. In this experiment, the question as to the influence of speeded task demands on the basis for similarity of objects presented on within- and between-category triads for classification is addressed. Triads which varied in category-triad type (within, between) and in distance relations (Type 1, Type 2) were presented to adult subjects on a CRT for classification. The bases for classification in the speeded task as well as reaction times for each type of classification were examined. Our first question concerned whether the previously-described relation between category information and perceptual organization would be evident under speeded task demands. Second, we were interested in characterizing differences in the speed of classification for within- and between-category triads when the basis for classification was shared dimensional relations or overall similarity relations.

Method

Subjects. Twenty females undergraduates enrolled at a small liberal arts college participated in this study and received a stipend of $4.00.

Stimuli. The stimuli were identical to those used in Experiments 2 and 3.

Design. Before being presented with triads for classification, subjects were presented with each of the 54 objects (9 form levels x 6 size levels) on a CRT screen in a randomized order and asked to name each as a bowl, cup or vase. Classification triads were presented to subjects on the CRT in a triangular

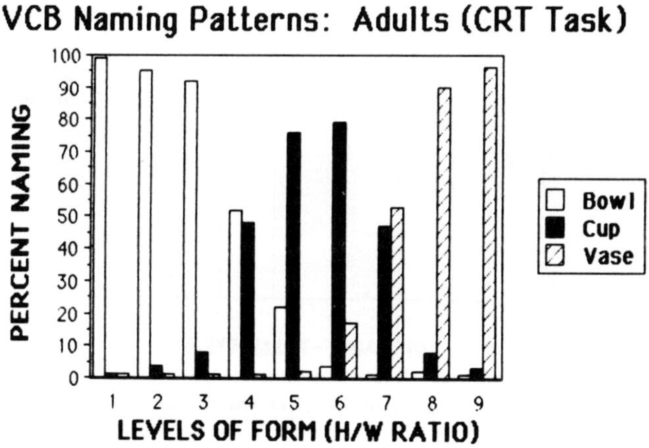

Figure 6.8. Adult naming patterns for the terms bowl, cup and vase shown across the nine levels of form presented individually on a CRT.

arrangement and subjects were instructed to select the object on the top-left ortop-right (objects A, C in Figure 6.1) that went best with the object on the bottom (object B in Figure 6.1). Thus, subjects were required to based their classifications on either overall similarity or shared dimensional relations. Haphazard classifications were not available as a basis for classification. Subjects were instructed that their response times would be recorded and they were to make their responses as quickly as possible.

1. Triad Types. Two types of distance-triads (Type 1 and 2, as described in Experiment 2 and 3) and two types of category-triads (within- and between-category, as described in Experiment 1, 2, and 3) were presented. Initial selection of category-triad types were experimenter-defined within the continuum of bowl 1-3, cup 4-6 and vase 7-9 for the nine levels of form.

After practice, a total of 178 triads were presented for classification. Half the subjects received Type 1 triads before Type 2 triads, half received Type 2 triads first. Within the block of Type 1 triads, 24 within-category triads and 12 between-category triads were presented twice each. The block of Type 2 triads was composed of 17 within-category triads and 36 between-category triads presented twice each. Within- and between-category triads were presented in a randomized order within each block.

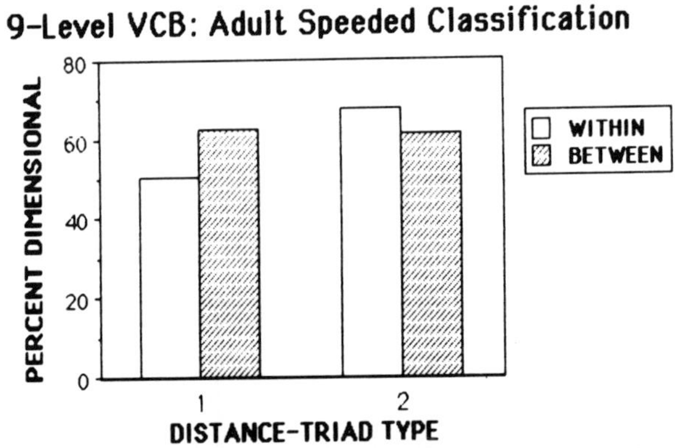

Figure 6.9. Mean percentage of dimensional classifications for within- and between-category (subject-defined) triads in adults. Performance is separated for Type 1 and Type 2 distance triads.

Results and Discussion

Naming Functions. The results of the naming task are shown in Figure 6.8. The three categories, vase, cup, bowl, are clearly identified with level 4 and 7 of form showing a fuzzy category boundary. Further analyses of naming data revealed that this fuzziness represents a decrease in individual consistency for the naming of these objects rather than overall group differences in the placement of category boundaries.

Basis for Classification. The mean percentages of dimensional classifications were entered into an ANOVA with the factors of Distance-triad type, Category-triad type and Order. As in Experiments 2 and 3, the identification of within- and between-category triads for this analysis was based on individual naming functions using the criteria described in Experiment 2. A significant effect of Distance-triad type ($F(1,18) = 9.33$, $p < .007$) was revealed. The mean percentages of dimensional classifications for Type 1 and 2 triads were 56% and 64%, respectively. The pattern of findings from this experiment differed from the previous experiments as there was no main effect of Category-triad type. However, there was a significant Category-triad x Distance-triad type interaction ($F(1,18) = 11.56$, $p < .004$). As shown in Figure 6.9, there was a significant difference in the proportion of dimensional classifications in within- and between-category triads for Type 1 distance relations (50.4% versus 62.5%) but no significant difference for Type 2 triads (67.5% versus 61.2%).

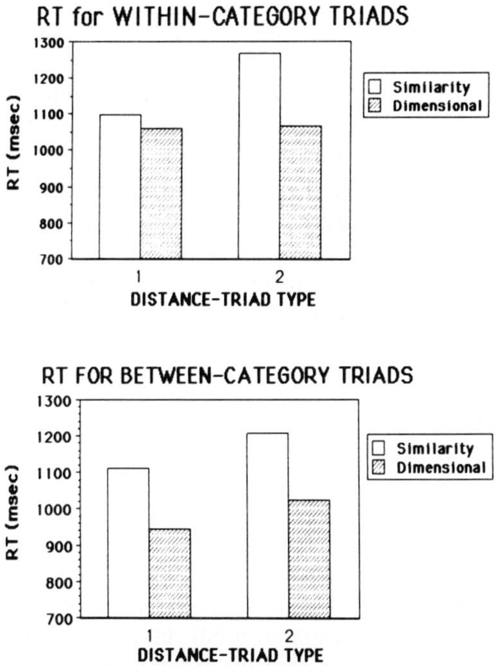

Figure 6.10. Reaction times for adult classifications based on identity on a shared dimensional and overall similarity relations. Performance is separated for within-category (top) and between-category (bottom) triads.

Thus, the overall level of dimensional responding was reduced under speeded demands but the phenomenon described in Experiments 1, 2, and 3 was replicated in Experiment 4 for Type 1 distance-triads. This suggests the limits of the effect of category information on perceptual organization under speeded classification task demands. Even if previously a subject consistently had labelled all three objects as cups, if the objects presented on a trial was such that the overall similarity basis for classification was considerably decreased, subjects were likely to classify on the basis of identity on a dimensional component. It is interesting to note that no effect was found for receiving the distance-triad types in differing orders ($F(1,18) = 1.82$, $p > .05$). Again, this suggests that subjects were not responding on the basis of an overall strategy or response set, but rather altered their basis of classification according to particular distance relations on individual triads.

RT for Classification. The top of Figure 6.10 shows the reaction times for within-category triads of each distance-triad type. The bottom of the figure shows the corresponding data for between-category triads. Analysis of variance of the mean reaction times for classification judgments with the factors of Distance-triad type (1,2), Category-triad type (within, between) and Basis for classification (similarity, dimensional) revealed three significant main effects and no significant interactions. Reaction times for classifications of Type 1 triads were significantly faster than for Type 2 triads ($F(1,19) = 18.72$, $p < .001$), reaction times for classification of between-category triads were significantly faster than within-category ($F(1,19) = 4.44$, $p < .05$) and reaction times for dimensional classifications were significantly faster than similarity classifications ($F(1,19) = 6.49$, $p < .02$) (see Figure 6.10).

Summary of Experiment 4

The pattern of classification performance with speeded task demands supported the general findings from Experiments 1-3 which employed nonspeeded restricted classification tasks. However, under speeded task demands, only Type 1 triads showed the pattern of fewer dimensional classifications on within-category triads than between-category triads. In addition, under speeded conditions, the percentages of dimensional classification were lower for all triad types. A similar effect has been previously reported by Ward (Ward, 1978; 1983; Ward et al., 1986) and J.D. Smith and Kemler Nelson (1984) who showed that subjects instructed to respond quickly, or instructed to provide their "first impression", produced more similarity classifications for separable stimulus sets than typically observed without speeded task demands. The pattern of similarity classifications slower than dimensional classification in this context is novel and we are currently replicating the finding. The pattern of dimensional classification taking less time than similarity classification for both within- and between-category triads does not support any simple holistic (i.e., first and fastest) to analytic (i.e., the result of secondary processes and slower) model of information processing. We are currently extending this work to examine the limits of subjects' flexibility and we are examining whether subjects have the ability to extract shared dimensional relations on within-category triads as easily as on between-category triads.

VI. CONCLUSIONS

The current experiments explored the relation between one characterization of the perceptual development of objects, the integral-to-separable developmental shift, and the representation and processing of category information. The goals of these studies were to characterize the effect of category information on perceptual organization across development and to examine the influence of naming on perceptual organization.

Our findings were that preschoolers, elementary-school children and adults were more likely to classify size vs. form objects (which had conceptual codes associated with them) in terms of shared dimensional relations when the objects were selected from two differing categories than when the objects were selected from the same category. Category information influenced the selection of basis for similarity in restricted classification tasks under both nonspeeded and speeded task demands. Naming tasks (especially when completed immediately before each classification triad) increased the distinction between basis for classification of within- and between-category triads.

The present findings support previous descriptions of perceptual development as an increasing predisposition to organize similarity in terms of identity on shared dimensional levels. Across all types of triads there was an increasing predisposition with age to classify objects on the basis of identity of a shared component dimension. *Our novel contribution is the demonstration that children have the flexibility to moderate their attention to different bases for classification as a function of category information within the same task and within the same set of objects. The current results demonstrate that even young children can access both a dimensional organization and a similarity organization from the same dimensional combinations in a consistent (and seemingly automatic) fashion.*

Relating Perceptual and Conceptual Development

What is clear from these findings is that judging similarity of objects which have symbolic-linguistic forms of organization associated with them is done in reference to category membership. Subjects at all ages shifted their basis for judging similarity when objects were presented in the same or differing categories. Shared membership in a category (e.g., cups) influenced subjects to attend to component size and form dimensions as wholistic objects and to judge similarity on the basis of overall similarity relations. When classification triads contained objects from two categories, they were classified on the basis of identity on the form dimension (which distinguished category membership), i.e., shared dimensional relations formed the basis for similarity. The current findings may be further characterized and understood within the attentional framework described by Treisman (Treisman & Gelade, 1980; Treisman & Schmidt, 1982) such that focal attention is more likely to be exhibited when objects are within a category, however such speculations await future research studies using these types of stimulus objects and employing feature integration paradigms (see Prinzmetal, 1981; Prinzmetal & Wright, 1984 for similar theoretical arguments regarding the influence of cognitive and linguistic factors on perceptual organization).

The current findings underscore the importance of considering the interrelations between perceptual and conceptual development. In the nearly-classic Gelman and Markman (1986) study in which children were asked to judge similarity of a bat, a crow and a flamingo, the overwhelming dominance of conceptual similarity over perceptual similarity was demonstrated. Children judged that the flamingo and crow were more similar in spite of low perceptual similarity (see Smith & Heise, current volume, for a related discussion). Our data support Gelman and Markman's finding that category information may be employed as a basis for similarity even in the context of low perceptual similarity. Our contribution has been to show how easily and quickly their bases for classification can be changed when such category boundaries are crossed. The complexity of deciding the level at which a category is considered to be a category, such that processes of perceptual organization yield wholistic perceptual units is another open research question.

The current findings also contribute to the literature on developmental changes in the characterization of category boundaries. These studies suggest that the acquisition of conceptual information is moderated by aspects of perceptual organization. Most relevant to the present work is a recent study by Alexander and Enns (1988) who examined categorization by preschoolers and adults of objects on the boundary between novel concepts. Alexander and Enns presented subjects with a continuum of objects from two differing and related novel concepts. Alexander and Enns supposed that studying children's assignment of objects on a fuzzy boundary between concepts would reveal the particular type and number of features used to include an object in a category.

In Alexander and Enns' work, the continuum of novel concepts varied along thirteen features which were varied in 2 to 7 levels from a "meagle" to a "borg". The features included levels on motion (speed, type), personality, and visual components (body color, height, width, mouth shape, eyebrow angle and shape, nose color and shape, upper and lower limb length). Alexander and Enns found that boundaries became less fuzzy and more consistent with age. They also found that the reasons for inclusion in each category became more conventional and perceptually-based with increasing age. This finding is consistent with some previous work by Nelson and Nelson (1978) who argued in support of a developmental shift in increasing category breadth and attributed this shift to changing bases for word organization from idiosyncratic rules to conventional perceptual features. Finally, Alexander and Enns reported that objects near the boundaries of categories became less variable and "less sensitive to contextual effects such as spatial configurations and the presence of verbal labels" (p. 1381, 1988).

The current set of objects which vary in only two dimensions represents a simplest-case stimulus set with which to examine some of the same issues of concern to Alexander and Enns. While this has some obvious advantages, the generalizability of the current findings for more complex real-world objects which vary in n-dimensional space is of serious concern. For example, Medin, Wattenmaker and Hampson (1987) have already demonstrated how persistent adults are in sorting cartoon animals (varying in head shape, number of legs, body markings, and tail length) on the basis of single dimensions (see also Martin & Caramazza, 1980; Ward & Becker, Chapter 12, current volume). Our findings support the idea discussed by Murphy and Medin (1985) that one cannot account for conceptual coherence in terms of simple attribute-matching procedures. Adults and children appear to be highly flexible and may manipulate their attentional grain, or level of attentional analysis, to employ wholistic and analytic strategies as a function of category membership.

Final Conclusion

In the introduction, literature was presented on the perceptual and conceptual development of objects such that a framework could be built with which to consider the experiments reported here. The findings from the current research have underscored the need to coordinate the investigation of perceptual and conceptual development. Recent work by Linda Smith and Diana Heise (Chapter 6, current volume) has supported the idea that conceptual structure is based on perceptual structure and has provided a compelling argument as to the relative importance of perceptual similarity in modelling cognitive development. The current results firmly establish the importance of relating category information and perceptual organization across development as well as suggest several different directions for research. However, within the study of attention, the idea that objects change similarity as a function of context has a long and rich history. Almost thirty years ago, Shepard (1964) showed that attention was influenced by the changing context in which objects were perceived. In a discussion of separable (or analyzable) stimuli Shepard concluded that "the psychological metric is strongly dependent upon the subject's state of attention" (p. 83), which is "influenced by the context of other stimuli presented in the experiment" (p. 84). A number of highly influential category-learning models (Medin & Shaeffer, 1978; Nosofsky, 1986, 1989) have focussed on changing distributions of dimension weights as a central tenet in category learning. The current findings add to a long list of support for the idea that similarity is context-dependent (Lockhead, current volume; A.F. Smith, current volume; Tversky, 1977; Tversky & Gati, 1981), and that similarity differences are due to differences in how selective attention operate on different dimensional combinations. We are convinced that the continued coordinated investigation of perceptual *and* conceptual processes across development may provide a framework such that we may make significant advances in understanding processes of attention and their development.

Medin et al. (1987) conclude their article on constraints on category construction saying "when people build family resemblances they may not only use bricks, but also mortar" (1987, p. 277). The current findings suggest some subtleties in the regulation of mortar use such that bricks for the same garden wall can be adhered but breaks in the wall can easily be made to allow clear passage.

ACKNOWLEDGEMENTS

This research was supported by a University of Louisville Graduate Research Council grant. I wish to acknowledge the research assistance of Nicolette Basile, Evelyn Farr, Lillian Ortiz, Julieta Macias, George Kim and the late Brian Lewis. I am grateful to A. F. Smith for providing comments and critique on an earlier version of this manuscript. Portions of this work were presented at the Psychonomic Society Meetings, Nov., 1989, and the Canadian Psychological Association, May, 1990.

REFERENCES

Alexander, T.M., & Enns, J.T. (1988). Age changes in the boundaries of fuzzy categories. Child Development, 59, 1372-1386.

Anderson, A.L.H. & Prawatt, R.S. (1983). When is a cup not a cup? A further examination of form and function in children's labeling responses. Merrill-Palmer Quarterly, 29(4), 375-385.

Anglin, J. (1977). Word, object, and conceptual development. New York: Norton.

Anglin, J.M. (1985). The child's expressible knowledge of word concepts: What preschoolers can say about the meaning of some nouns and verbs. In K.E. Nelson (Ed.), Children's language (Vol. 5). Hillsdale, NJ: Lawrence Erlbaum Associates, Inc.

Aschkenasy, J.R. & Odom, R.D. (1982). Classification and perceptual development: Exploring issues about integrality and differential sensitivity. Journal of Experimental Child Psychology, 34, 435-448.

Aslin, R.N. & Smith, L.B. (1988). Perceptual development. Annual Review of Psychology, 39, 435-473.

Au, T.K. & Markman, E.M. (1987). Acquiring word meaning via linguistic contrast. Cognitive Development, 2, 217-236.

Baldwin, D.A. (1989). Priorities in children's expectations about object label reference: Form over color. Child Development, 60, 1291-1306.

Barrett, M.D. (1982). Distinguishing between prototypes: The early acquisition of the meaning of object names. In S.A. Kuezaj II (Ed.), Language development (Vol. 1: Syntax and semantics. Hillsdale, NJ: Lawrence Erlbaum Associates, Inc.

Bauer, P.J. & Mandler, J.M. (1989). Taxonomies and triads: Conceptual organization in one- to two-year-olds. Cognitive Psychology, 21, 156-184.

Bernt, F. (1989). Children's use of schematic concepts in free classification tasks: A closer look. Journal of Genetic Psychology, 150(2), 187-195.

Bjorklund, D.F. & Thompson, B.E. (1983). Category typicality effects in children's memory performance: Qualitative and quantitative differences in the processing of category information. Journal of Experimental Child Psychology, 26, 328-340.

Bjorklund, D.F., Thompson, B.E. & Ornstein, P.A. (1983). Developmental trends in children's typicality judgments. Behavior Research Methods & Instrumentation, 15, 350-356.

Broadbent, D.E. (1977). The hidden preattentive processes. American Psychologist, 32, 109-118.

Bruner, J.S., Olver, R.R. & Greenfield, P.M. (1966). Studies in Cognitive Growth. New York: Wiley.

Burns, B. (1986). The relation of perceived stimulus structure and intelligence. Further tests of a separability hypothesis. American Journal of Mental Deficiency, 91, 196-200.

Burns, B. & Cohen, J. (1988, Nov.). The developmental shift in color-to-form preference: Differential dimensional salience or increasing separability? Paper presented at the Psychonomic Society Meetings, Chicago, IL.

Burns, B., Schlewitt, L., & Cohen, J. (1992). Is dimensional salience based on perceptual or conceptual factors? : An analysis of color-form difference thresholds and restricted classification performance. Manuscript submitted for publication.

Burns, B., Shepp, B.E., McDonough, D. & Wiener-Ehrlich, W. (1978). The relation between stimulus analyzability and perceived dimensional structure. In G.H. Bower (Ed.) The psychology of learning and motivation: Recent advances in theory. Vol. 12, New York: Academic Press.

Clark, E.V. (1973). What's in a word? On the child's acquisition of semantics in his first language. In T.E. Moore (Ed.), Cognitive development and the acquisition of language. New York: Academic Press.

Cook, G. & Odom, R.D. (1988). Perceptual sensitivity to dimensional and similarity relations is free and rule-based classification. Journal of Experimental Child Psychology, 48, 319-338.

Daehler, M.W., Lonardo, R. & Bukato, D. (1979). Matching and equivalence judgments in very young children. Child Development, 50, 170-179.

Denny, D.R. (1975). Developmental changes in concept utilization among normal and retarded children. Developmental Psychology, 11, 359-368.

Denny, D.R. & Moulton, P.A. (1976). Conceptual preferences among pre-school children. Developmental Psychology, 12, 509-513.

DeVos, L.F. & Caramazza, A. (1977, March). The role of form and function in the development of natural language concepts. Paper presented at the meeting of the Society for Research in Child Development, New Orleans.

Duncan, E.M. & Kellas, G. (1978). Developmental changes in the internal structure of semantic categories. Journal of Experimental Child Psychology, 26, 328-340.

Elkind, D., Koegler, R.R., & Go, E. (1964). Studies in perceptual development II. Part-whole perception. Child Development, 35, 81-90.

Fenson, L., Cameron, M.S. & Kennedy, M. (1988). Role of perceptual and conceptual similarity in category matching at age two years. Child Development, 59, 897-907.

Fenson, L., Vella, D. & Kennedy, M. (1989). Children's knowledge of thematic and taxonomic relations at two years of age. Child Development, 60, 911-919.

Flavell, J.H. (1970). Concept development. In P.H. Mussen (Ed.), Carmichael's manual of child psychology (3rd ed., Vol. 1). New York: Wiley.

Garner, W.R. (1974). The processing of information and structure. Hillsdale, N.J.: Lawrence Erlbaum Associates, Inc.

Gentner, D. (1978). What looks like a jiggy but acts like a zimbo? A study of early word meaning using artificial objects. Papers and Reports on Child Language Development, 15, 1-6.

Gibson, E.J. (1969). Principles of Perceptual Learning and Development. New York: Appleton-Century Crofts.

Gollin, E.S. & Garrison, A. (1980). Relationships between perceptual and conceptual mediational systems in young children. Journal of Experimental Child Psychology, 30, 325-335.

Hoffman, J. (1982). Representations of concepts and the classification of objects. In R. Klix, J. Hoffman & E. Von der Meer (Eds.). Cognitive research in psychology: Recent advances, designs and results. Amsterdam: Elsevier Science Press.

Horton, M.S. (1982). Category familiarity and taxonomic organization in young children. Doctoral dissertation, Stanford University.

Horton, M.S. & Markman, E.M. (1980). Developmental differences in the acquisition of basic and superordinate categories. Child Development, 51, 708-719.

Inhelder, B. & Piaget, J. (1969). The early growth of logic in the young child. New York: Norton.

Kay, D. & Anglin, J.M. (1982). Overextension and underextension in the child's expressive and receptive speech. Journal of child language, 9, 83-98.

Keil, F.C. (1989). Concepts, Kinds, and Cognitive Development. Cambridge, MA: MIT Press.

Keil, F.C. & Kelly, M. (1990). Developmental changes in category structure. In S. Harnad (Ed.), Categorical perception. New York: Cambridge University Press.

Keil, F.C. & Batterman, N. (1984). A characteristic-to-defining shift in the development of word meaning. Journal of Verbal Learning and Verbal Behavior, 23, 221-236.

Kemler, D.G. (1982). Classification in young and retarded children: The primacy of overall similarity relations. Child Development, 53, 768-779.

Kemler, D.G. (1983). Holistic and analytic modes in perceptual and cognitive development. In T. Tighe & B.E. Shepp (Eds.), Perception, cognition and development: Interactional analyses. Hillsdale, NJ: Lawrence Erlbaum Associates, Inc.

Kemler, D.G. (1983). Exploring and reexploring issues of integrality, perceptual sensitivity, and dimensional salience. Journal of Experimental Child Psychology, 36, 365-379.

Kemler, D.G. & Smith, L.B. (1978). Is there a developmental trend from integrality to separability in perception? Journal of Experimental Child Psychology, 26, 498-507.

Kemler, D.G. & Smith, L.B. (1979). Accessing similarity and dimensional relations: The effect of integrality and separability on the discovery of complex concepts. Journal of Experimental Psychology: General, 108, 133-150.

Kemler Nelson, D.G. (1984). The effect of intention on what concepts are acquired. Journal of Verbal Learning and Verbal Behavior, 23, 734-759.

Kemler Nelson, D.G. (1988). When category learning is holistic: A reply to Ward and Scott. Memory and Cognition, 16, 79-84.

Kemler Nelson, D.G. (1990). The nature and occurrence of holistic processing. In B.E. Shepp & S. Ballesteros (Eds.), Object Perception: Structure and Process. Hillsdale, NJ: Erlbaum.

Kempton, W. (1981). The folk classification of ceramics: A study of cognitive prototypes. San Diego, CA: Academic Press.

Kimchi, R. (1990). Children's perceptual organization of hierarchical visual patterns. European Journal of Cognitive Psychology, 2(2), 133-149.

Kimchi, R. & Palmer, S.E. (1982). Form and texture in hierarchically constructed patterns. Journal of Experimental Psychology: Human Perception and Psychophysics, 8, 521-535.

Kimchi, R. & Palmer, S.E. (1985). Separability and integrality of global and local levels of hierarchical patterns. Journal of Experimental Psychology: Human Perception and Performance, 11, 673-688.

Kossan, N.E. (1981). Developmental differences in concept acquisition strategies. Child Development, 52, 290-298.

Labov, W. (1973). The boundaries of words and meanings. In C.J.N. Bailey & R.W. Shuy (Eds.), New ways of analyzing variation in English. Washington, D.C.: Georgetown University Press.

Landau, B., Smith, L.B., & Jones, S.S. (1988). The importance of shape in early lexical learning. Cognitive Development, 3, 299-321.

Lane, D.M. & Pearson, D.A. (1983). Can stimulus differentiation and salience explain developmental changes in attention? A reply to Hagen and Wilson, Jeffrey and Odom. Merrill-Palmer Quarterly, 29(2), 227-233.

Mansfield, A.F. (1977). Semantic organization in the young child: Evidence for the development of semantic feature systems. Journal of Experimental Child Psychology, 23, 57-77.

Markman, E.M. (1981). Two different principles of conceptual organization. In M.E. Lamb and A.L. Brown (Eds.), Advances in Developmental Psychology. Hillsdale, NJ: Lawrence Erlbaum Associates, Inc.

Markman, E.M. (1989). Categorization and Naming in Children: Problems of Induction. Cambridge, MA: MIT Press.

Markman, E.M. (in press). The whole object, taxonomic, and mutual exclusivity assumptions as initial constraints on word meanings. In J.P. Byrnes & S.A. Gelman (Eds.), Perspectives on Language and Cognition: Interrelations in Development. Cambridge: Cambridge University Press.

Markman, E.M. & Callanan, M.A. (1984). An analysis of hierarchical classification. In R. Sternberg (Ed.), Advances in the Psychology of Human Intelligence, Vol. 2, Hillsdale, NJ: Erlbaum.

Markman, E.M., Cox, B. & Machida, S. (1981). The standard object sorting task as a measure of conceptual organization. Developmental Psychology, 17, 115-117.

Markman, E.M. & Hutchinson, J.E. (1984). Children's sensitivity to constraints on word meaning: Taxonomic versus thematic relations. Cognitive Psychology, 16, 1-27.

Martin, R.C. & Caramazza, A. (1980). Classification in well-deferred and ill-defined categories: Evidence for common processing strategies. Journal of Experimental Psychology: General, 109, 320-353.

Medin, D.L. & Barsalou, L.W. (1990). Categorization processes and categorical perception. In S. Harnad (Ed.), Categorical perception. New York: Cambridge University Press.

Medin, D.L., Wattenmaker, W.D. & Hampson, S.E. (1987). Family resemblance, conceptual cohesiveness and category construction. Cognitive Psychology, 19, 242-279.

Meili-Dworetzki, G. (1956). The development of perception in the Rorschach. In B. Klopfer (Ed.), Development in the Rorschach technique. New York: World Book.

Melkman, R., Tversky, B. & Baratz, D. (1981). Developmental trends in the use of perceptual and conceptual attributes in grouping, clustering, and retrieval. Journal of Experimental Child Psychology, 31, 470-486.

Mervis, C.B. (1987). Child-basic object categories and early lexical development. In U. Neisser (Ed.), Concepts and conceptual development: Ecological and intellectual factors in categorization. Cambridge: Cambridge University Press.

Mervis, C.B. & Crisafi, M.A. (1982). Order of acquisition of subordinate, basic and superordinate level categories. Child Development, 53, 258-266.

Mervis, C.B. & Mervis, C.A. (1982). Leopards are kitty-cats: Object labeling by mothers for their 13 month olds. Child Development, 53, 267-273.

Mervis, C.G. & Rosch, E. (1981). Categorization of natural objects. Annual Review of Psychology, 32, 89-116.

Murphy, G.L. & Medin, P.L. (1985). The role of theories in conceptual coherence. Psychological Review, 92, 289-316.

Navon, D. (1977). Forest before trees: The precedence of global features in visual perception. Cognitive Psychology, 9, 353-383.

Navon, D. (1981). The forest revisited: More on global precedence. Psychological Research, 43, 1-32.

Nelson, K. (1974). Concept, word and sentence: Interrelations in acquisition and development. Psychological Review, 81, 267-285.

Nelson, K. (1973). Some evidence for the cognitive primacy of categorization and its functional basis. Merrill-Palmer Quarterly, 19, 21-39.

Nelson, K. (1983). Concepts, words and experiments: Comment on "when is a cup not a cup?" by Anderson and Prawatt. Merrill-Palmer Quarterly, 29(4), 387-394.

Nelson, K. (1988). Constraints on word meaning? Cognitive Development, 3, 221-246.

Nelson, K.E. & Nelson, K. (1978). Cognitive pendulums and their linguistic realization. In K.E. Nelson (Ed.), Children's language (Vol. 1). New York: Gardner Press.

Nosofsky, R.M. (1986). Attention similarity and the identification-categorization relationship. Journal of Experimental Psychology: General, 115, 39-57.

Nosofsky, R.M. (1989). Further tests of an exemplar-similarity approach to relating identification and categorization. Perception and Psychophysics, 45, 279-290.

Odom, R.D. (1978). A perceptual-salience account of decalage relations and developmental change. In L.S. Siegal & C.J. Brainerd (Eds.), Alternatives to Piaget. New York: Academic Press.

Odom, R.D., Astor, E.C. & Cunningham, J.G. (1975). Effects of perceptual salience on the matrix task performance of four- and six-year-old children. Child Development, 46, 758-762.

Odom, R.D. & Cook, G.L. (1984). Perceptual similarity, integral perception and the similarity classifications of preschool children and adults. Developmental Psychology, 20, 560-567.

Odom, R.D. & Guzman, R.D. (1972). Development of hierarchies of dimensional salience. Developmental Psychology, 6, 271-287.

Olver, R.R. & Hornsby, J.R. (1966). On equivalence. In J.S. Broner, R.R. Olver & P.M. Greenfield (Eds.), Studies in cognitive growth. New York: Wiley.

Posnansky, C.J. & Neuman, P.G. (1976). The abstraction of visual prototypes of children. Journal of Experimental Child Psychology, 21, 367-379.

Prather, P.A. & Bacon, J. (1986). Developmental differences in part/whole identification. Child Development, 57, 549-558.

Prawatt, R.S. & Anderson, A.L.H. (1980). A reply to Nelson's comments on "when is a cup not a cup?". Merrill-Palmer Quarterly, 29(4), 395-397.

Prawatt, R.S. & Wildfong, S. (1980). The influence of functional context on children's labeling responses. Child Development, 51, 1057-1061.

Prinzmetal, W. (1981). Principles of feature integration in visual perception. Perception and Psychophysics, 30, 330-340.

Prinzmetal, W. & Millis-Wright (1984). Cognitive and linguistic factors affect visual feature integration. Cognitive Psychology, 16, 305-340.

Rosch, E. (1975). Universals and cultural specifics in human categorization. In R.W. Brislin, S. Bochner & W.J. Lonner (Eds.), Cross-cultural perspectives on learning. New York: Sage.

Rosch, E., Mervis, C.B., Gray, W., Johnson, D. & Boyes-Braem, P. (1976). Basic objects in natural categories. Cognitive Psychology, 8, 382-439.

Saltz, E., Soller, E., & Sigel, I. (1976). The development of natural language concepts. Child Development, 43, 1191-1202.

Scott, M.S., Serchuck, R. & Mundy, P. (1982). Taxonomic and complementary picture pairs: Ability in two- to five-year olds. International Journal of Behavioral Development, 5, 243-256.

Shepard, R.N. (1964). Attention and the metric structure of the stimulus space. Journal of Mathematical Psychology, 1, 54-87.

Shepard, R.N. & Cermak, G.W. (1973). Perceptual-cognitive explorations of a toroidal set of free-form stimuli. Cognitive Psychology, 4, 351-377.

Shepard, R.N. & Podgorney, P. (1978). Cognitive processes that resemble perceptual processes. In W.K. Estes (Ed.), Handbook of Learning and Cognitive Processes. V. HIllsdale, NJ: Erlbaum.

Shepp, B.E. (1978). From perceived similarity to dimensional structure: A new hypothesis about perceptual development. In E. Rosch & B.B. Lloyd (Eds.), Cognition and categorization. Hillsdale, NJ: Lawrence Erlbaum Associates, Inc.

Shepp, B.E., Burns, B. & McDonough, D. (1980). The relation of stimulus structure to perceptual and cognitive development: Further tests of a separability hypothesis. In J. Becker & F. Wilkening (Eds.), The integration of information by children. Hillsdale, NJ: Lawrence Erlbaum Associates, Inc.

Smiley, S.S. & Brown, A.L. (1979). Conceptual preference in thematic or taxonomic relations: A nonmonotonic age trend from preschool to old age. Journal of Experimental Child Psychology, 28, 249-257.

Smith, E.E. & Medin, D.L. (1981). Categories and concepts. Cambridge, MA: Harvard University Press.

Smith, E.E., Shoben, E.J. & Rips, L.J. (1974). Structure and process in semantic memory: A featural model for semantic decisions. Psychological Review, 81, 214-241.

Smith, J.D. (1990). Analytic and holistic processes in categorization. In B.E. Shepp & S. Ballesteros (Eds.), Object perception: Structure and process. Hillsdale, NJ: Lawrence Erlbaum Associates, Inc.

Smith, J.D. & Kemler Nelson, D.G. (1984). Overall similarity in adults' classification: The child in all of us. Journal of Experimental Psychology: General, 113, 137-159.

Smith, L.B. (1979). Perceptual development and category generalization. Child Development, 50, 705-715.

Smith, L.B. (1983). Development of classification: The use of similarity and dimensional relations. Journal of Experimental Child Psychology, 36, 150-178.

Smith, L.B. (1984). Young children's understanding of attributes and dimensions: A comparison of conceptual and linguistic measures. Child Development, 55, 363-380.

Smith, L.B. (1989). A model of perceptual classification in children and adults. Psychological Review, 96, 125-147.

Smith, L.B. & Kemler, D.G. (1977). Developmental trends in free classification: Evidence for a new conceptualization of perceptual development. Journal of Experimental Child Psychology, 24, 279-298.

Smith, L.B. & Kemler, D.G. (1978). Levels of experienced dimensionality in children and adults. Cognitive Psychology, 10, 502-532.

Tenney, Y.J. (1975). The child's conception of organization and recall. Journal of Experimental Child Psychology, 19, 100-114.

Tomikawa, S.A. & Dodd, D.H. (1980). Early word meanings: Perceptually or functionally based? Child Development, 51, 1103-1109.

Treisman, A. & Gelade, G. (1980). A feature integration theory of attention. Cognitive Psychology, 12, 97-136.

Treisman, A. & Schmidt, N. (1982). Illusory conjunctions in the perception of objects. Cognitive Psychology, 14, 107-141.

Tversky, A. (1977). Features of similarity. Psychological Review, 84, 327-352.

Tversky, A. & Gati, I. (1978). Studies of similarity. In E. Rosch & B.B. Lloyd (Eds.), Cognition and categorization. Hillsdale, NJ: Lawrence Erlbaum Associates, Inc.

Tversky, B. (1985). The development of taxonomic organization of named and pictured categories. Developmental Psychology, 21, 1111-1119.

Tversky, B. (1989). Parts, partonomies, taxonomies. Developmental Psychology, 25, 983-995.

Tversky, B. & Hemenway, K. (1984). Objects, parts, and categories. Journal of Experimental Psychology: General, 113, 169-193.

Vurpillot, E. (1976). The visual world of the child. New York: International Universities Press, Inc.

Vygotsky, L.S. (1962). Thought and language. Cambridge, MA: MIT Press.

Ward, T.B. (1978). Dimensional responding in children and adults as a function of stimulus and response variables. Dissertation Abstracts International, 39, 4565B. (University Microfilms No. 78-23,091).

Ward, T.B. (1980). Separable and integral responding by children and adults to the dimensions of length and density. Child Development, 51, 676-684.

Ward, T.B. (1983). Response tempo and separable-integral responding: Evidence for an integral-to-separable processing sequence in visual perception. Journal of Experimental Psychology: Human Perception and Performance, 9, 103-112.

Ward, T.B. (1991). The role of labels in directing children's attention. In J.T. Enns (Ed.), The development of attention: Research and theory. Amsterdam: Elsevier Science Publishers.

Ward, T.B., Vela, E. & Hass, D. (1990). Children and adults learn family resemblance categories analytically. Child Development, 61, 593-605.

Ward, T.B., Vela, E., Perry, M.L., Lewis, S., Bauer, N.K. & Klent, K. (1989). What makes a vibble a vibble? A developmental study of category generalization. Child Dev., 60, 214-224.

Ward, T.B. & Scott, J.G. (1987). Analytic and holistic modes of learning family-resemblance concepts. Memory and Cognition, 15(1), 42-54.

Waxman, S.R. (1990). Linguistic biases and the establishment of conceptual hierarchies: Evidence from preschool children. Cognitive Development, 5, 123-150.

Waxman, S.R. & Gelman, R. (1986). Preschoolers' use of superordinate relations in classification and language. Cognitive Development, 1, 139-156.

Waxman, S.R., Shipley, E.F. & Shepperson, B. (1991). Establishing new subcategories: The role of category labels and existing knowledge. Child Development, 62, 127-138.

Werner, H. (1948). The comparative psychology of mental development. New York, NY: International Universities Press, Inc.

Werner, H. (1957). Comparative psychology of mental development (revised edition). New York: International Universities Press.

Werner, H. & Kaplan, B. (1963). Symbol formation: An organismic-developmental approach to language and the expression of thought. New York: Wiley.

White, T. (1982). Naming practices, typicality and underextension in child language. Journal of Experimental Child Psychology, 33, 324-346.

Wohlwill, J.F. (1962). From perception to inference: A dimension of cognitive development. Monographs of the Society for Research in Child Development, 72, 87-107.

Commentary

Perceived Similarity in Perceptual and Conceptual Development: The Influence of Category Information on Perceptual Organization, B. Burns

ALBERT F. SMITH
State University of New York at Binghamton

Burns generated stimulus sets by crossing two attributes--shape and size--and collected data on restricted classification of stimulus triples. In each triple, the reasonable classifications were by level of shape and by overall similarity. However, Burns's stimulus sets varied in a third attribute--conceptual category--that was correlated with shape. Although steps of shape were equal, some adjacent levels spanned a category boundary whereas others did not. Thus, stimulus triples with identical relative arrangements in shape x size space differed according to whether all three items, or just two of them were from the same conceptual category. Burns found consistently that when all three items were from the same conceptual category, subjects grouped together the two items that were most similar overall, but that when two items were from one category and the third was from another, subjects grouped together the items that were the same in shape--that is, the items from the same conceptual category.

I make three observations about these results. The first concerns the attribute structure of stimulus sets that are used to investigate perceived interactions of stimulus attributes. When category was an attribute of the triples--that is, when the items within a triple varied in category--category served as a basis of classification (see A.F. Smith, Chapter 14, current volume). Otherwise, overall similarity was used. Had these experiments been carried out with rectangles rather than with rounded rectangles, so that there were no categories, it is plausible that no performance differences would have been observed between the analogues of what Burns has labelled the between-category and within-category triads. Although Burns's studies illuminate the potential impact of knowledge on classification, Burns may go too far in asserting that knowledge influences perception. Rather, her results may indicate simply that the functional attributes of the within-category triples are shape and size, whereas the functional attributes of the between-category triples are category and size.

Table 1. Stimulus Set for Speeded Classification.

Size	Shape Y1	X2	X3
p	C'	A	C
q	D'	B	D

This brings me to my second point, which concerns converging operations. Results from the restricted classification task have often been combined with results from other tasks to draw conclusions about perceived interactions of stimulus attributes. Burns's data suggest that shape and size interact integrally, but that category and size are separable attributes. It would be of considerable interest to know whether speeded classification performance would be consistent with Burns's restricted classification results.

Consider, for example, the stimuli represented in Table 1. These would be generated by crossing orthogonally three levels of shape with two levels of size. The letters in the labels of the levels of shape indicate category membership: Items A, B, C and D are members of Category X, whereas C' and D' are members of Category Y. Thus, discriminating between A and D is a within-category discrimination with correlated attributes, whereas discriminating between C' and B is a between-category discrimination between two stimuli that have the same relation in shape x size space as do A and D. The various speeded classification tasks used to make inferences concerning attribute interactions would be carried out with A, B, C and D, and also with A, B, C', and D'. If category membership has the same effect on speeded classification performance as it does on restricted classification, one should find the data pattern characteristic of integrally-interacting attributes with the single-category set of items (see Felfoldy, 1974), but the data pattern characteristic of separably-interacting attributes with the items from different categories. These results would not mean that the perceived interaction of shape and size is inconsistent. Rather, they would mean that shape and size interact integrally, but that category and size interact separably. Perhaps the most important conclusion to be drawn from extensive research on perceived attribute interactions is that the subject's attributes need not coincide with those of the experimenter. Although the experimenter may designate shape as an attribute in each set of items, the subject may perceive variation in shape in one set but variation in category-membership in the other.

Third, it is essential that the subject's ability to exhibit a behavior be distinguished clearly from his or her propensity to do so. In restricted classification of triples, the subject is instructed to group together the two items that go together the best. The subject's responses reveal his or her grouping criterion. When a subject classifies a between-category triad by category, this does not mean that he or she would deny the similarity of the two stimuli that are closest in shape x size space. Rather, it suggests that when category is available as an attribute, it assumes a primary role. A useful contribution to the understanding of the interaction of knowledge and perception, within Burns's framework, would involve assessing the role played by category information in varied experimental tasks. Research using stimulus sets in which category membership is not confounded with physical attribute would be an important step in this direction.

Felfoldy, G.L. (1974). Repetition effects in choice reaction time to multidimensional stimuli. Perception & Psychophysics, 15, 453-459.

Percepts, Concepts and Categories
B. Burns (Editor)

Perceptual Similarity and Conceptual Structure

LINDA B. SMITH
DIANA HEISE
Indiana University

I. INTRODUCTION

Many recent discussions of categories and concepts contrast the perceptual and conceptual bases for categories. These discussions pick up a persistent theme in developmental theory. This theme posits a trend from perception to conception and it has been played throughout the works of Piaget (1929), Vygotsky (1934/1962) and Werner (1948). It is echoed by Flavell (1970), Wohlwill (1962), and Bruner and Olver (1963) and more recently by Gentner (1989) and Keil (1989). The idea is that children shift from perceptually-bound representations of objects that are global and holistic to ones that are principled and articulated along abstract dimensions. So, for example, Flavell (1970) wrote that conceptual development grows from "equivalences based on the more concrete and immediately given perceptual, situational, and functional attributes of objects to equivalences of a more abstract, verbal-conceptual sort." Since development is directional, this theme implies that conception is in some way "better than" perception.

Many developmentalists argue against a shift from perceptual categories to conceptual ones (e.g., Brown, 1990; Mandler & Bauer, 1988; Wellman & Gelman, 1988). They argue that categories are conceptually structured from the very beginning. But even for these theorists, the research agenda is defined in terms of categories structured by "mere appearance" versus categories with a "rich theoretical structure that goes beyond superficial similarity." In these discussions, perceptual and conceptual categorization are presented as mutually exclusive processes. Moreover, conceptual categorization is presented as smart and mature and perceptual categorization is presented as deficient, and to be abandoned.

We believe that this perception-*versus*-conception organization of research works to the detriment of empirical and theoretical progress. This definition of the research agenda dismisses the contribution of perceptual structure to conceptual structure and does not study it seriously. Our purpose in this chapter is to defend perception. We counter the perception-*versus*-conception approach by arguing that the contribution of perceptual categories to conceptual categories is considerable and continuous throughout development. Conceptual structure does not replace or even override perceptual categories. Instead, conceptual structure is based on perceptual structure.

This view that perception is the grounding force for conception has been argued recently by Johnson (1983) and Lakoff (1987). This view is also gaining attention in the cognitive development literature (see Gelman, 1990; Gentner & Rattermann, 1990; Mandler, 1990). In this chapter, we specifically consider the role of perceptual similarity in category development. Our defense of perceptual similarity rests on three points:

1. Perceptual similarity is dynamic. It varies with the attributes attended to.

2. Experience in perceiving the relations in the world influences attention and perceptual similarity. Because of this, perceptual similarity moves about in meaningful ways.

3. Perceptual categories can be abstract and can convey -- indeed be the source of -- conceptually relevant knowledge.

Before presenting the defense, we assess the case against perceptual similarity.

II. THE CASE AGAINST PERCEPTUAL CATEGORIZATION

Keil (1989) building on the writings of Quine (1970) contrasts an "animal similarity space" and a "conceptual similarity space." Animal similarity according to Keil, is original similarity. It is uninterpreted feature counts and lists of correlations between perceptual features. According to Keil, this animal or original similarity is the *atheoretical* tabulation of information we get when we look at the world. It is constant and unchanging and therefore stupid. Fish and whales look alike regardless of what we know. But what objects look like is not always what they really are. Thus, development is away from "the immediate, the subjective, animal sense of similarity to the remoter sense of similarity determined by scientific hypotheses and posits and constructs" (Quine, 1977). In this view, the one we argue against, perceptual similarity is replaced by belief systems composed of causally related features and nonperceptual properties.

We offer three examples of how ideas such as these organize the current research on categories and concepts.

A. From Basic to Superordinate Categories

Medin and Ortony (1989; see also, Medin et al., 1990) suggest that perceptual similarity although not a very intelligent mechanism may precisely fit the needs of the infant in discovering her first categories. Research on perceptual development (e.g., Kemler Nelson, 1990; Smith, 1989) indicates that very young children often compare multidimensional objects wholistically across all dimensions at once. Medin and Ortony argue that this original similarity space fits the child's task of making a first partitioning of the world's objects into categories.

Medin and Ortony point to the fit between overall perceptual similarity and the basic level categories described by Rosch & Mervis (1975). The categories dog, cat, car, house, and bird are basic level categories and these categories do seem to be organized by perceptual similarity or in Rosch's terms by family resemblance. Dogs *look* alike, or at least most dogs seem to be mostly alike in most of their perceptual properties (see, Rosch, et al. 1976; Biederman, 1985).

In contrast to basic level categories, superordinate categories seem to be structured by a few general abstract properties. Animals, for example, differ widely and are not as a group perceptually like each other. Moreover, by the consensus view, there are no specifiable perceptual properties that all (or even most) animals possess. The consensus view is that superordinate categories are structured by nonperceptual properties (Carey, 1985; Gelman, 1988; Mandler & Bauer, 1988; Markman, 1989).

There is much developmental data consistent with a trend from basic to superordinate categories. In classification tasks, young children form spatial groupings of basic level categories such as shoes versus dogs before they form superordinate groupings (Rosch et al., 1976). Names for basic categories are learned fast and considerably before names for superordinate categories (MacNamara, 1982). These data and arguments suggest a developmental trend that proceeds from Quine's "animal" similarity which gives us basic categories to conceptual similarity which gives us superordinate categories.

However, in their full complexity, the developmental data do not squarely fit a *unidirectional* trend from basic to superordinate categories (see, Mervis, 1987). An early sensitivity to superordinate category structure can be seen in young children's overgeneralizations in word learning. Young children's overgeneralizations honor superordinate category distinctions in that children sometimes mistakenly call cows "doggie" but do not mistakenly call cars "doggie" (see Waxman, 1980, for relevant data). Mandler and Bauer (1988; see also, Mandler, Bauer & McDonough, in press) have shown that even prelinguistic children are sensitive to superordinate categories. They observed 12-month-olds' "categorizations" in a free play situation. Their measure of categorization consisted of the sequence of touching objects. Same category objects were touched in rapid succession more frequently than different category objects (see Sugarman, 1983). By this measure, Mandler and Bauer found that infants readily made superordinate classifications of the sort "dog and horse versus car." Indeed, in their task, 16-month-olds made superordinate classifications more readily than they made basic level ones (e.g., poodle and collie versus horse).

Mandler and Bauer take for granted that perceptual properties alone are not enough to organize superordinate categories. Under this assumption, their data provide support for early conceptually based categorization. Mandler and Bauer point to the categories formed by the children -- chicken, fish, cow, and turtle versus motorcycle, airplane, van, and train engine -- as proof that perceptual similarity was not involved. They point to the way the children played with the toys -- making the animals walk and talk and making the vehicles speed about with "vroom-vroom" sounds as evidence for nascent conceptual structure. Perceptual similarity, they conclude, does not control categorization even at its beginning.

We are not sure that this conclusion is warranted. Perceptual structure may well be sufficient to explain Mandler and Bauer's data. Specifically, there may be heretofore undiscovered perceptual properties that distinguish animals and vehicles, and babies might well be sensitive to these properties. Of course, to find these properties, we have to look. Later, in this chapter, we will present some evidence of such properties that we found when we did look.

B. Perceptual Categorization versus Essences

Keil (1989) recently argued that perceptual similarity is not sufficient to explain the psychological structure of even basic level categories. A thought experiment makes the point. Consider a cow. Is it still a cow if some psychologist covers it in sheepskin, paints it purple, cuts off a leg, and adds moose horns? Keil (1989) has shown that people strongly believe that naturally occurring objects do not change their identity with changes in their perceptual appearance -- no matter how severe. People possess beliefs about naturally occurring objects (or natural kinds) that attribute their identity to the processes of their origin or that imbue them with an undefined essence.

In this research, Keil affirms the distinction between the perceptual processes we use to identify objects in the world and our concepts of them (see Smith & Medin, 1981, for more on this important distinction). The idea is that when we come across some unknown object, we may use perceptual similarity to identify it and assign it to a category. But our categories -- our concepts of what it *means* to be a particular kind of object -- are much more than lists of perceptual or even functional properties. People have organized sets of causal beliefs, or theories, that distinguish kinds of objects according to whether they are naturally occurring or manmade, alive or not, terrestrial or water living and so on (see Carey, 1985; S. Gelman, 1988; & Keil, 1989). According to Keil, it is these theories about the causal connections between perceptual properties and their origins that form the conceptual core of a category. And thus, alterations in the "mere appearance" of an object in ways (e.g., painting) that do not violate central theoretical beliefs (e.g., an object is a cow if its mother was a cow), do not alter the object's identity. This theoretical core is sometimes referred to as the category's "essence" in recognition of the relation between these views and nominal essentialism in philosophy (see Keil, 1989; and also Murphy & Medin, 1985 and Medin & Ortony, 1989).

Causal theories increase in complexity and scientific accuracy with age (e.g., Carey, 1985; Gelman, 1988). Consistent with this age-related growth in conceptual structure, Keil (1989) has shown that the belief in "category essences" that transcend appearance also increases with age. Three- and 4-year-olds often maintain that an object's identity does change with changes in appearance. Thus, again, the developmental trend is described as moving from perceptual similarity to conceptual structure.

There are a number of ongoing disputes in this literature. They all concern whether perceptual appearance is enough to explain performance for any stimuli at any developmental level. For example, Wellman and Gelman (1988) argue that very young children believe in "essences" too and ignore appearance in domains in which they have sufficient knowledge. Carey, Gelman, and Keil each argue that essences that transcend perceptual appearance are characteristic of peoples' beliefs about naturally occurring objects but not manmade objects. By these arguments, artifacts are characterized by an impoverished theoretical structure and thus their category structures are more controlled by superficial perceptual similarity than are the category structures of natural kinds. Greer and Sera (1990; see also, Mandler, Bauer and McDonough, in press) argue in contrast that there is no principled distinction between artifacts and natural kinds. They report that people have causal theories about the internal parts and workings of complex artifacts such as computers and radios and will maintain that these objects' identities go unchanged despite radical changes in outward appearance.

The research agenda, the theoretical disputes, are all defined in terms of perception versus conception. Is this the most useful definition of the research question? When we come upon an object in the world, when we come face-to-face with a dog or a chair, we use sophisticated perceptual processes to determine what it is (e.g. Biederman, 1985). But if we are to take the data and claims about essences seriously, our concepts of objects are largely unrelated to and largely unaffected by the perceptual systems that have evolved to recognize real objects in the real world.

C. Category Induction

When asked to make inductive inferences from what is known about objects in one category to objects in another, conceptual structure and not perceptual similarity is again said to be the preeminent force. The typical empirical study again pits perceptual and conceptual solutions. The state-of-the field is cogently summarized by a thought experiment derived from the work of Carey and Gelman. Imagine a real monkey, a mechanical monkey, and a real snake. We tell you that the real monkey has a duogleenan inside. Is the mechanical monkey or the real snake more likely to also have a duogleenan inside? A real monkey and a mechanical monkey look alike, but the real monkey and real snake share "deeper" properties that place them in the same superordinate categories.

Carey (1985) and Gelman and Markman (1986, 1987, see also, Gelman, 1988) have shown that people (sometimes as young as three years of age) make inductive inferences about the nonperceptual insides of objects in accord with superordinate category structure. In two important studies, Gelman and Markman (1986, 1987) presented children with triads of pictures as in Figure 7.1. The crow and bat were black and the flamingo was colored pink. By Gelman and Markman's analysis, the crow and bat were similar overall whereas the flamingo and the crow belong to the same conceptual category. The children were told the names of each object (bird, bird, and bat) and then were told that the crow laid eggs. They were asked which other object, bat or flamingo, also laid eggs.

Previous research has shown that young children often freely classify objects like these by perceptual similarity (Tversky, 1985). Gelman and Markman found, however, that children made category inductions along conceptual lines. Gelman and Markman concluded from these results that words access conceptual structure and that categories -- even for very young children -- are organized by a rich theoretical structure that goes beyond appearance. We have no quarrel with this conclusion.

We do, however, question the ancillary assumption that perceptual properties have *no* role in conceptual structure. Gelman and Markman's assumption that perception does not matter is highlighted by their design of experiments and choice of stimuli. Gelman and Markman chose stimuli that distorted the similarity relations that actually exist in the world. The bat and crow are drawn and colored to emphasize the few features that real bats and crows have in common and to de-emphasize the many real differences that exist between

Figure 7.1. A sample triad of stimuli used by Gelman & Markman (1986).

bats and crows and to also de-emphasize the many real similarities that exist between crows and flamingos. Gelman and Markman selected their stimuli for a good reason -- to provide a strong test of the hypothesis that children would use conceptual structure even in the context of strong countering perceptual similarity. Their goal was to choose between two hypotheses -- conceptual structure controls performance or perceptual structure controls performance. Our question is whether this is a theoretically sensible goal.

Consider two more thought experiments that are variants on Gelman and Markman's study. In the first thought experiment, the triad of stimuli looks like that in Figure 7.2. Objects A and B are clearly most similar overall. The "conceptual" information that is provided is that inside A is a bird, inside B is a bat, and inside C is another bird. Now, if the object in A lays eggs, which other object, B or C, is also likely to lay eggs? Given the impoverished character of the perceptual information, one would be wise to generalize according to the labels provided. This experiment if actually conducted, would provide a perfectly good test of whether children can ignore perceptual information. Like Gelman and Markman's original studies, perceptual similarity is pitted against conceptual similarity; and also like Gelman and Markman's stimuli, *the perceptual stimuli provide little information about how the objects really look.* Although the stimuli in Figure 7.2 might provide an adequate test of the perception-versus-conception

Figure 7.2. Three stimuli for a thought experiment.

hypothesis, they provide a poor test of the uses of perceptual and conceptual information in real category and concept formation.

Now consider a second version of Gelman and Markman's study. In this version, the stimuli consist of a real living flamingo, a living crow, and a living bat. The objects are not named. Nothing is said about them except the fact that one (the crow) lays eggs. In the richness of the real perceptual information, there seems little doubt that egg-laying would be generalized from one bird to the next. Of course, these stimuli are inappropriate to the task of determining whether children use perceptual *or* conceptual information in category induction because the rich perceptual structure and the conceptual structure converge. However, these stimuli might be useful if our research goal is to determine how children form the categories they do. The point is that determining how real categories are formed requires an understanding of perceptual similarity.

III. IN DEFENSE OF PERCEPTUAL SIMILARITY

Defending the role of perceptual similarity in conceptual structure requires a clear statement of just what perception and conception are. We take perception to refer to the structure of immediate experience. We take conceptual structure to refer to knowledge that is (or can be made to be) explicit -- that is knowledge that is easily talked about and thought about. These definitions appear to the same as those of Keil (1989) and Mandler (1990). However, where we differ is that we believe perceived similarity embodies and reflects much implicit knowledge. Many theorists (e.g. Keil, 1989) write about perceptual similarity as if it were given in the stimulus. This naive realism assumes that the perceived similarity between two objects is constant and unchanging just as the physical measurements of the properties of objects are constant and unchanging. But perceived similarity is the result of psychological processes and it is highly variable. The central process that changes perceptual similarity is attention. The perceived similarity of two objects changes with changes in selective attention to specific perceptual properties.

The importance of attention in categorization has been recognized by other investigators. Murphy and Medin (1985) persuasively pointed out that when people categorize objects and reason about categories, they shift attention in principled ways among sets of perceptual features. For example, people seem to use different perceptual properties when they classify an object as an animal versus when they classify that same object as a pet. Murphy and Medin (see also, Keil, 1989) suggest that people's explicit causal theories or beliefs about objects organize and drive the selection of relevant perceptual features. The idea is that although perception is involved, perception alone -- animal similarity -- is not enough. According to Murphy and Medin, the real force in categorization is the conceptual structure that organizes and interprets perception. That is, by their view, conceptual structure is the cause and changing attention weights on specific dimensions is the effect.

Our view is quite the opposite: the dynamic nature of perceptual similarity is a causal force in the development of conceptual beliefs.

A. Perceptual Similarity is Dynamic

Among those who study perceptual similarity, there is one agreed upon fact, the perceptual similarity between any two objects varies considerably. Perceptual similarity varies with the attributes attended to (Shepard, 1964; Nosofsky, 1984). The dynamic nature of similarity is evident in the empirical research on perception and perceptual categorization and in formal theories of similarity (Goldstone, Medin, & Gentner, in press; Nosofsky, 1984; and Tversky, 1977). In formal theories, similarity is some function of some weighted combination of features and attributes. As the feature weights change, so similarity changes.

For example, in Smith's (1989) model of perceptual classification as in Nosofsky (1984) and Shepard (1987), similarity is calculated as an exponential decay function of the distance between stimuli in the psychological space. The similarity between two objects, O_i and O_j, then is

$$S_{ij} = e^{d_{ij}}$$

Distance, d_{ij}, is defined as the sum of the weighted dimensional differences

$$d_{ij} = \sum_{K=1}^{N} W_K[O_{ik} - O_{jk}]$$

where O_{ik} minus O_{jk} is the difference between objects i and j on dimension k, N is the number of dimensions, W_k is the weight given dimension K and $O \leq W_k \leq 1.00$ and

$$\sum_{K=1}^{N} W_k = 1.00.$$

The present point is that this formula *defines* perceptual similarity; perceptual similarity is *just this* (or some calculation like it). And if perceptual similarity is some weighted combination of dimensional similarities, then perceptual similarity necessarily varies with the magnitude of the difference between stimuli on the dimensions *and with the dimension weights.*

Consider the triad of stimuli on the left hand side of Figure 7.3. These stimuli are represented in terms of their coordinates on two varying dimensions. Object A is identical to Object B on dimension X but differs from B considerably on dimension Y. Object B differs from Object C by a small amount on both dimensions. Using the equations above, the right hand side of Figure 7.3 shows the *perceived similarities* between objects A and B, objects B and C, and objects A and C as the dimension weights change from perfect selective attention to

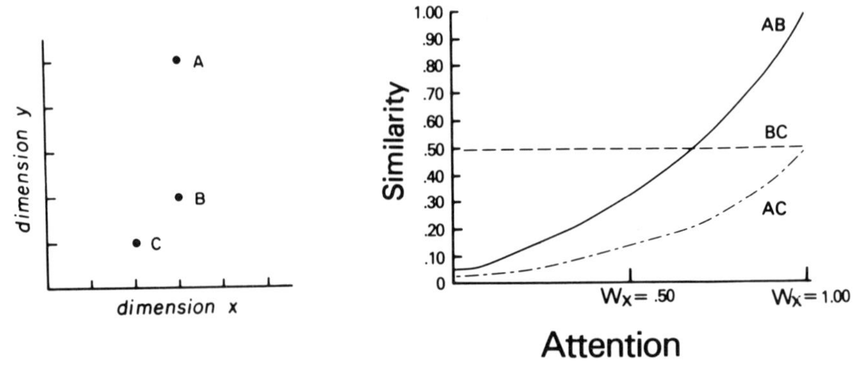

Figure 7.3. Left -- A triad of stimuli represented in terms of their coordinate values in a 2-dimensional space. Right -- the similarity between pairs in the triad as a function of selective attention to dimension x.

dimension Y (W_y = 1.00, W_x = 0) through equal weighting of the two dimensions, (W_x = .50, W_y = .50) to perfect selective attention to dimension X given dimension X (W_x = 1.00, W_y = 0). As is apparent, the absolute value of the similarity of Objects A and B changes considerably with changes in the feature weights. Moreover, there is a dramatic change in the relative similarities of AB, AC and BC. Which two objects in the triad are the most similar depends on which features are attended to.

If we extend these notions to Markman and Gelman's stimuli in Figure 7.1, a perceptual interpretation of their results is possible. Under one set of feature weights, say one that emphasized overall shape and color, the crow is perceptually more similar to the bat than the flamingo. Under another set of feature weights, for example, one that emphasized beaks or feet, the crow and the flamingo might be the *perceptually* more similar pair. Importantly, then, a demonstration that children perceive a bat and crow to be similar in a classification task and a crow and flamingo to be similar in the category induction task *need* not mean that children shifted from perceptual similarity to conceptual similarity when asked to make inductions. They may only have shifted the perceptual feature weights. Perceptual similarity may have played the key role in both judgments.

Nosofsky (1984) represents the effects of shifting feature weights on perceptual similarity in terms of the stretching of the psychological similarity space in one direction or another. Borrowing this idea, we represent hypothetical

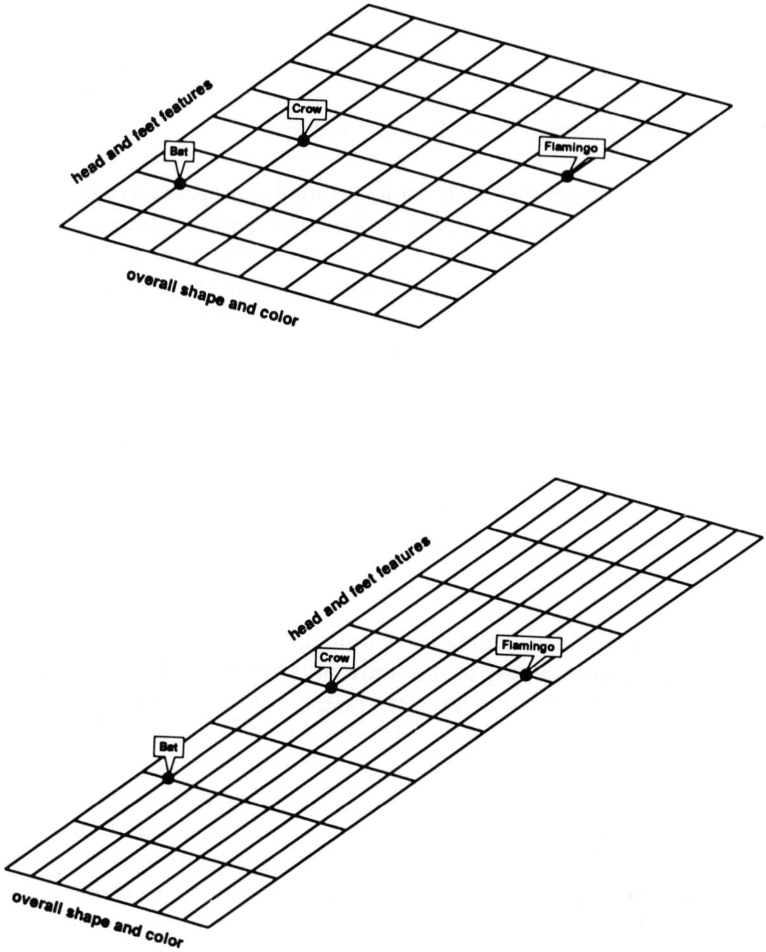

Figure 7.4. A theoretical depiction of the similarity relations between Gelman and Markman's stimuli under two different sets of feature weights.

similarity relations between Gelman and Markman's stimuli in Figure 7.4. For ease of discussion, the many-dimensional space is compressed into two dimensions. The X axis represents some combination of variation along overall shape and color and the Y axis represents variation on some combination of head and feet features. The three dots represent the location in that space of Gelman and Markman's crow, bat, and flamingo. The distances between the dots represent their perceptual similarity to each other. Figure 7.4 top illustrates the case for an equal weighting of all the dimensions; for these particular drawings, the bat and crow are the most perceptually similar objects among the three. Figure 7.4 bottom represents the similarity relations when attention is focussed on the head and the feet. The space is stretched so that the crow and flamingo are now closer, that is, more similar than the bat and crow.

One might argue that such stretched similarity spaces are not the original similarity or "animal" similarity meant by Keil and Quine. These theorists meant "raw" similarity unconfounded by knowledge or whatever mechanism pushes feature weights and similarity around. Consistent with these views, Smith (1989) defined raw similarity as the similarity relations that result from the equal weighting of all perceptual features. This overall similarity seems rightly raw in that it is the *default* similarity, the one that dominates perception whenever limits are placed on performance and/or there is no previous experience. Even for adults, an equal weighting of dimensions dominates when processing time is limited, when stimuli are complex, and when there is lack of relevant knowledge (see J. D. Smith & Kemler-Nelson, 1984; Smith, 1981). This is also the similarity that dominates 2- and 3-year-olds' perceptual categorization in most task circumstances (Smith, 1989).

Thus, the equal weighting of all dimensions does seem to be the point of origin for perceptual similarity in the sense of being the zero state of the system. However, overall similarity is neither mechanistically nor developmentally distinct from a stretched similarity space. There is but one system, one manner of calculating perceptual similarity -- the one that is mathematically described by the equations (see Smith, 1989, for an empirical demonstration of this claim). The possibility of changing feature weights is not an add-on to some more primitive form of similarity. Separate features and changing feature weights are inherent in the very nature and process of perceptual similarity. This claim is supported by the facts of animal perception and by the facts of perceptual similarity in human infancy.

Shifts in feature weights are characteristic of perceptual similarity in non-human animals - i.e., in real "animal similarity" (Mackintosh, 1965; Sutherland, 1964). The phenomenon of cue-blocking (or overshadowing) provides a particularly good example. Consider an illustrative cue-blocking experiment. In original learning, the organism is trained to respond to red objects for a reward but not to respond to blue ones. The shapes of the objects vary and do not matter in the response-reward contingencies. If animals formed conceptual rules, a reasonable one would be "red wins, blue loses, and shape doesn't matter." If we

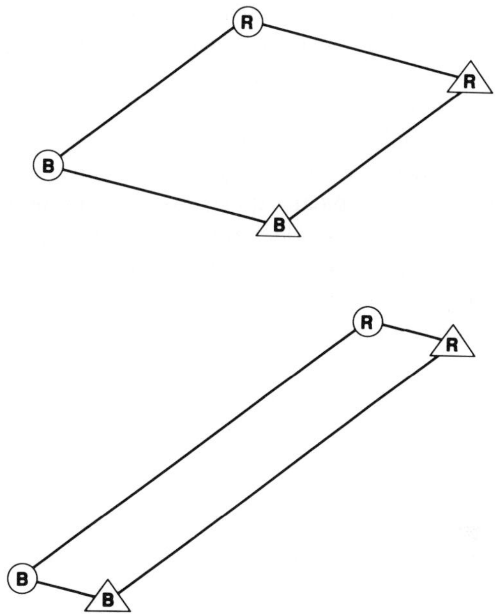

Figure 7.5. An example of a stretched similarity space in which color is weighted more than shape in the calculation of similarity.

stopped the experimental procedure here and tested the organism with novel shapes and colors, the pattern of response would fit the rule. The organism would respond to all and only red objects and ignore their shape. Figure 7.5 depicts the effects of learning on the perceptual similarity space. Original learning takes the organism from a similarity space in which colors and shapes matter equally (top) to one in which color is (virtually) all that matters (bottom).

The phenomenon of cue-blocking shows that this interpretation is correct. In a cue-blocking experiment, a redundancy phase is added. In this phase, shape is correlated with color. For example, all the red objects are now triangles and all the blue ones are circles. The organism is given some lengthy set of these redundancy trials. The question is whether, given the previous original learning, the animal notices the redundancy. If during original learning, the organism learned to selectively attend to color and ignore shape such that the similarity space has been severely stretched in one direction, then the organism should not notice the redundancy. The results show that in tasks such as these animals do not learn about the added redundancy. After the redundancy phase, the organism responds to all and only red objects and ignores shape just as if the redundancy training had not occurred. Selective attention is clearly a force in "animal similarity." The similarity space for pigeons and rats is easily stretched.

Another literature that suggests that "raw" similarity may rarely be realized is the infant habituation literature. Consider what happens to perceived similarity in a prototypic habituation study (for real and similar studies, see, e.g., Bornstein, 1985). In a typical experiment, infants might be shown repeated examples of objects that are alike on some dimension but vary on a second dimension. For example, they might be shown red squares, red circles, and red triangles. Looking time is recorded until the babies no longer look much at the red squares, circles, and triangles. Then, the infants are shown new objects -- either one that differs on the previously constant (or relevant) dimension (e.g., a green square) or one that offers a new value on the previously varying (or irrelevant) dimension (e.g., a red cross). The standard result, given repeated examples of the same color during habituation, is that infants look more to the novel color (green square) than the novel form (red cross) during dishabituation. Importantly, if infants were shown repeated examples of the same shape but varying colors during habituation, they would look more at the novel shape than the novel color.

Presumably, the mechanism behind such habituation-dishabituation effects is short-term changes in features weights. If the habituation stimuli are all red but with varying shapes, the common redness of the habituation stimuli must somehow stretch the similarity space along the color dimension (or the varying shapes cause a shrinkage of the space along the shape dimension). The habituation phase causes a change in the perceived similarity of objects; in our example, the habituation phase made differences in color count more in the calculation of perceptual similarity than differences in form. Original similarity, the perceptual similarity of infancy, is inherently dynamic; it moves about because of the mechanism through which it is computed. The mechanism is one in which feature weights may vary from context to context. The use of habituation as a method to study infant perception depends on the reality of malleable feature weights in psychological similarity.

If perceptual similarity shifts so readily in the artificial laboratory contexts of operant conditioning and habituation experiments, might not perceptual similarity shift systematically and *meaningfully* in response to the contingencies and correlations that exist between perceptual features in the real world?

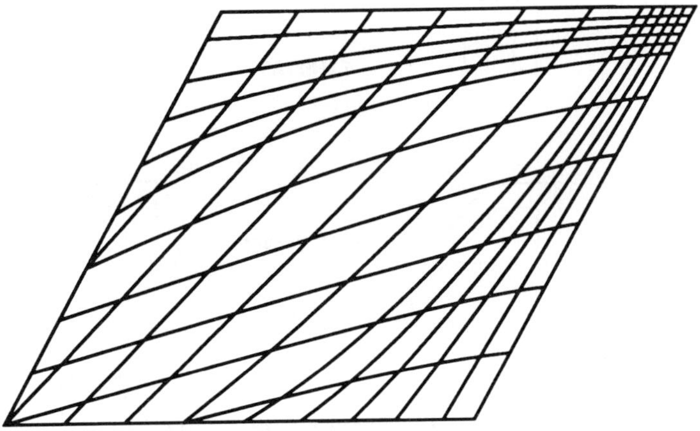

Figure 7.6. A distorted region of a similarity space as might result from increased attention to a particular conjunction of features (the relevant conjunction is the compressed area).

B. Knowledge and Perceptual Similarity

The relevance of perceptual similarity for conceptual structure stems from the role of real world experience on feature weights. One kind of knowledge that pushes feature weights around and stretches the similarity space is implicit knowledge about relations between perceptual features. As Rosch (1973) argued, perceptual features do not vary orthogonally in the world. They come in causally related clusters. Birds with web feet tend to have bills. Objects with dog-like feet tend to have dog-like heads. Evidence from laboratory experiments indicates that both adults (Medin et al., 1982) and older infants (Younger & Cohen, 1983; Younger, 1990) are sensitive to such correlations. This empirical evidence indicates that experience with correlations causes increased attention to the *combinations* of features that enter into correlations.

Figure 7.6 represents how such a correlation might organize a similarity space. The grid lines represent "equal" physical distances. Perceptual similarity is represented by real Euclidean distance in the space. In the depicted space, objects that possess a particular conjunction of values are closer together or perceptually more similar than objects that possess other particular conjunctions of values. Thus the compacted upper right of the figure might be the region of the stimulus space in which objects with dog-like feet *and* dog-like heads fall. If this figure were an accurate depiction of perceived similarity, it would mean that the similarity between objects with both dog-feet and dog-heads is greater than the

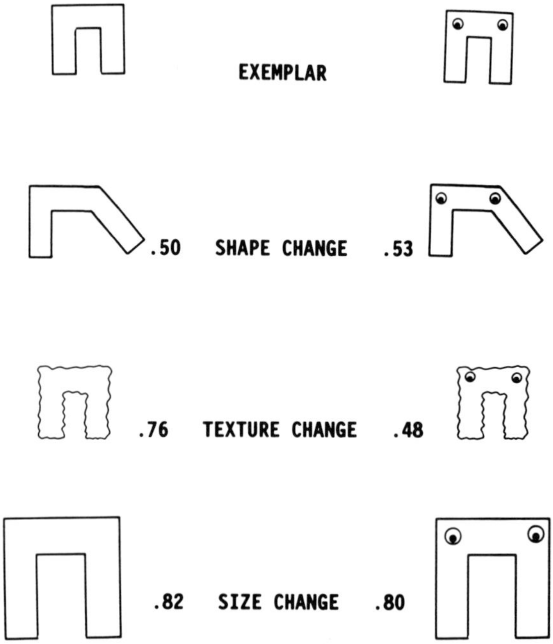

Figure 7.7. Stimuli and mean propostion name extensions to test stimuli in Jones, Smith & Landau (1990).

similarity between objects with both dog-feet *and* pig-heads. Do young children, in their everyday interactions in the world, learn about such correlations in ways that systematically and to good purpose distort the similarity space?

1. The Case of Eyes and Texture

Recently, Jones, Smith, and Landau (1990) discovered a co-relation effect in children's novel word extensions that suggests a powerful role for combinations of perceptual features in natural category formation. Jones et al. found that the perceived similarity of objects for 36-month-olds changed dramatically with the addition of a constant perceptual property. In the experiment, children were shown a novel object and it was named. For example, they might be shown the object at the top left of Figure 7.7 and told that it was a Dax. Children were then asked what other test objects were also Daxes. As shown in the left column of the figure, one test object differed from the exemplar only in shape, one only in texture, and one only in size. While the exemplar (the experimenter-named) Dax remained in view, children were asked separately whether each one of the test objects was also a Dax. The proportion of times the children agreed that the exemplar's name was also the name of the test object is given next to each test object. As the results attest, these children called the test objects by the same name as the exemplar when *the object was identical in shape to the exemplar object.* Shape changes from the exemplar mattered considerably more than changes in texture and size.

The right side of Figure 7.7 shows the stimuli for a second (between-subjects) condition. In this condition, toy eyes were affixed to each object. The addition of a constant stimulus property (the eyes) radically changed children's judgments as is evident in the proportions of name extensions. When eyes were added, the exemplar's name was extended to a test object only if the test object was the *same shape and same texture* as the exemplar.

These results demonstrate the synergistic relation that must exist between perceptual properties and knowledge. The presence of eyes activates children's knowledge about kinds of categories. That knowledge then guides attention, changing the weights accorded to the various dimensions, and thus shifting the perceptual similarities among the objects.

The kind of co-relation that exists between eyes and texture is different from the correlations between properties that have been studied by Medin et al. and Younger and Cohen. The kind of correlation studied by previous investigators is between specific properties, for example, between feathers, webbed feet, and bills. The kind of co-relation that underlies 3-year-olds novel word extensions does not seem to be between eyes and a *specific* texture.

In the world, if having eyes is correlated with a specific texture, it is with one that is soft and pliable. In the Jones et al. study, the Dax exemplar was made of wood. In one experiment, the texture-change test item was a sponge, in another, it was a soft cloth. If the presence of eyes had signalled the importance of particular textures, then the children should have been more likely to accept these softer Daxes in the Eyes condition than in the No-eyes condition even though they had a different texture than the exemplar. However, the results are just the opposite of this specific-feature-correlation hypothesis. Children in the No-eyes condition readily called these soft objects Daxes. However, children in Eyes condition rejected these items. They required objects with eyes to have the same wooden texture as the exemplar to be called a Dax. Apparently, the presence of eyes indicated that texture -- whatever that texture might be -- is criterial in categorization.

This kind of co-relation -- between a specific property (eyes) and a *dimension* of variation (texture) -- should prove a powerful aid in learning new category structures, one that would supplement the use of correlations between specific properties (e.g., wings and feathers). A learned correlation between the specific properties of wings and feathers is a local distortion of the similarity space as shown in Figure 7.6. Such a local distortion would help the categorizer distinguish objects that do or do not belong to an already-learned category.

A learned correlation between a cue and dimension of variation (like that between eyes and texture), however, causes a more widespread distortion of similarity. The entire similarity space would be stretched in one direction exaggerating texture differences in the context of eyes relative to the context of no eyes. Such an expansion of the space should help the child organize multiple categories of birds, and bats and walruses, and of real monkeys, mechanical monkeys, and snakes because the objects in these categories differ from one another in the texture of their surfaces. The idea, then, is that the similarity space reflects the fact that surface texture differences matter more for categorizing all objects with eyes than such differences matter for categorizing objects without eyes. A general expansion of the similarity space along the texture dimension for objects with eyes would increase the similarity of real crows and flamingos (which both have feathered surfaces) and decrease the similarity of crows and bats (since crows are feathered and bats are furry) relative to eyeless counterparts.

If we imagine multiples of local and dimension-wide distortions of the similarity space -- distortions resulting from real-world correlations between specific properties, from co-relations between material kind and kind of motion, eyes and texture, eyes and kind of motion, shape and motion, and so on -- then what emerges is a bumpy and irregular similarity space that *organizes itself into multiples of categories at multiple levels* in context dependent ways.

We suggest that the result of spending time in the world -- of looking at it, hearing it, and feeling it -- is a structured and context dependent similarity space. The result is a dynamic similarity space that fits the structure of the world. Rosch (1973) suggested that many natural categories are given in the world in correlated features sets. Her proposal seems right. The reason that the thousands of categories named by concrete nouns emerge so rapidly in the first three years of life and are conceptually well understood by four or five years of age is that they are given in a context dependent perceptual similarity.

This idea that perceptual similarity *reflects* the co-relations between perceptual properties as they exist in the world presumes a particular mechanism.

We presume what most formal theories of perceptual similarity and perceptual categorization posit -- that learned *atheoretical* associations shift the features weights and the computed similarities in systematic ways that reflect the structure of the world (see Nosofsky, 1986; Gluck & Bower, 1988; Medin & Schaffer, 1978).

There is another explanation of the effect of eyes on 3-year-olds' categorizations -- one built on represented causal theories. For example, 3-year-olds' judgments could be organized by knowledge structures of the sort: objects with eyes are natural kinds, members of the same natural-kind class have the same genetic structure, the same genetic structure produces the same material substance; the same material substance produces the same surface texture; therefore, texture matters for objects with eyes. Such a complicated set of beliefs seems implausible to us but more to the point, we have evidence that such beliefs are not necessary for the co-relation effects reported by Jones et al.

We (Heise & Smith, 1990) have demonstrated that the same phenomenon observed in children's novel word extensions occurs in adults' learning of arbitrary categories. Billman (1989) also has reported compelling evidence that makes the point. In our study, adults learned two different classification systems in a single experiment. For example, subjects might be given red bugs and blue bugs and learn to classify the objects into four groups such that the red bugs were partitioned into two subgroups by shape (circular vs. dropshape) and the blue bugs were partitioned into two subgroups by number of legs (2 versus 4). Given adults proclivity for selective attention and the formation of criterial property categories (Medin, Wattenmaker and Hampson, 1987), one might expect adults in this task to learn only to attend to shape and/or number of legs. Transfer tasks, however, showed that adults learned a more general co-relation between color and the other dimensions. In the transfer task, adults were given new sets of bugs and asked to freely classify them into groups. Unlike typical transfer tasks, then, we did not ask how well adults put new items into the just learned categories, we asked how well they formed new categories of new items -- on their own -- and without instruction.

Adults' transfer performance in this free classification task suggests they had learned an implicit "rule" of the sort: red bugs are distinguished into subkinds by their shape but blue bugs are distinguished into subkinds by their number of legs. Given the training described above, adults classified red bugs by their shape even when the novel bugs had shapes never seen before and they classified the blue bugs by their number of legs even when the numbers of legs were different from those in the learning phase. This acquisition of a context dependent shift in feature weights was obtained with various transpositions of the cuing and transfer dimensions.

These results suggest that context-dependent distortions of the similarity space are easily set up. Moreover, the adults in these experiments had no pre-experimental naive causal theories about how one dimension should be related to another in classifying cartoon bugs. The subjects also do not appear to have developed such causal theories in the course of the experiment. Clearly, causal theories or well-developed belief systems are not necessary for context dependent and useful shifts in attention to perceptual properties.

2. *The Case of Lexical Form-Class*

The case of eyes and texture indicates that one perceptual property may influence attention to another perceptual property and systematically alter the perceived similarities of objects. But knowledge of perceptual properties is only one force on perceptual similarity. Other *non*perceptual forces (though probably not explicit knowledge) also play a role. For example, young children, shift their attention among dimensions in novel word extension tasks as a function of the syntactic form class of the word.

In the Jones, Smith, and Landau (1990) study, (see also, Landau, Smith, & Jones, 1988), young children's selective attention to shape with eyeless stimuli and their attention to shape and texture with eyed stimuli occurred only in a word extension task. When children were asked to freely classify the same stimuli, they showed no differential attention to the dimensions of size, shape, and texture. The presence of a novel word, therefore, organized the feature weights and shifted them from the default value of equal attention to all dimensions. In those studies, the novel word was a count noun.

Recently, Smith, Jones, and Landau (1990) examined 36-month-olds' shifting attention to shape and color when the novel word was an adjective as well as when it was a count noun. In one study, half the children were presented with an exemplar object and told it was "a dax" and half the children were presented with the same exemplar object but were told it was "a dax one." The exemplar was made of wood, possessed a zig-zag shape and was colored a glittery (and highly reflective) combination of silvery gold. The critical test items consisted of (1) two unique objects that were the same shape but different color and (2) two unique objects that were the same color but different shape. Children were shown each test object one at a time and asked whether it was a "dax" in the noun

Table 7.1. Mean proportion extensions to critical test trials in the Adjective and Noun conditions in Smith, Jones & Landeau (1990). (Standard deviations are in parentheses.)

	Noun		**Adjective**	
	Same Color Diff. Shape	**Same Shape Diff. Color**	**Same Color Diff. Shape**	**Same Shape Diff. Color**
No Cave	.20 (.23)	.85 (.11)	.51 (.31)	.78 (.29)
Cave	.29 (.22)	.94 (.08)	.73 (.18)	.27 (.20)

condition or a "dax one" in the adjective condition. The exemplar was in view through out the entire procedure.

The results are given in Table 7.1 in the row labelled "No cave". Children in the noun condition attended to shape calling same shape items a *Dax*; children did *not* call same color but different shape items a *Dax*. Children also showed a weak shape bias in the adjective condition. However, in the adjective condition, individual children's performances differed from each other. Most children maintained that the same-shape objects were "dax ones." But some children believed same-color objects were "dax ones" and some children responded haphazardly. The shiny color of the exemplar was highly noticeable and indeed was spontaneously commented on by *all* the children. Nonetheless, children did not systematically interpret the novel adjective as referring to this novel and salient feature. These results show that the presence of a novel word, particularly a count noun, organizes attention to shape.

The results of a second experiment showed that children's attention in the adjective condition could be as highly organized as their attention in the noun condition -- if additional forces directed their attention among the varying dimensions. The procedure in this second experiment was identical to that in the first with the exception that the stimuli were presented inside a small dark cave. The child and experimenter looked through the opening of the cave to view the exemplar and each test item. A spotlight in the cave illuminated the stimuli. The cave and spotlight had the effect of heightening the already salient glittery color.

The results from this second experiment are also given in Table 7.1 in the row labelled "Cave". Again, in the noun condition, children extended the novel word to new items according to shape alone. But in the adjective condition, children attended to color and called objects "dax ones" only if they possessed the same glittery color as the exemplar.

These results show that both lexical knowledge and local forces conspire to control children's attention to dimensions and the perceived similarity of objects. Presumably, in the course of learning language, children have learned that shape is the principle determiner of membership in the categories labelled by count nouns (see Landau et al., 1988). The act of naming objects thus may serve as a cue to stretch the similarity space along the shape dimension. Apparently, however, adjective categories are not well constrained by kind of property and thus the context of an adjective does not systematically distort the similarity space. But local, context-specific, *ad hoc* forces can work synergistically with the adjective context to forcefully organize attention and thereby constrain possible interpretations of the novel adjective.

We do not find these results surprising. They fall right out of the mechanism that underlies perceptual similarity. However, since it seems that many kinds of knowledge shift feature weights -- associative connections between perceptual properties, knowledge about syntactic form classes, local *ad hoc* effects, some might question whether distortions of similarity are truly *perceptual* phenomenon. Is the shifty nature of perceptual similarity too contaminated by (implicit) knowledge to be rightly considered perceptual? We believe there is little to be gained from pursuing this question. The critical point is that perceived similarity is dynamic and shifts in meaningful ways.

C. Conceptually Relevant Perceptual Properties

The power of perceptual similarity and shifting attention among features becomes more impressive when we consider abstract perceptual properties. There are lots of perceptual properties that are not easy to talk about but that may shape our conceptions of the world from very early in development. We are thinking of the perceptual properties that distinguish biological from nonbiological motion (e.g., Berthental, Proffitt, Spertner & Thomas, 1985; R. Gelman, Spelke, & Meck, 1983; Bullock, R. Gelman, & Baillageon, 1982), that allow us to predict what novel objects can and cannot move alone (Massey & R. Gelman, 1988), and to determine whether an action by one object on another was intentional (Leslie,

1984; 1988; Spelke, 1982). There is a growing research effort to discover such perceptual properties that must underlie, indeed organize, conceptual categories.

Briefly, we would like to consider one perceptual property that may underlie the contrast between naturally occurring objects and manufactured objects. Our hypothesis is that manufactured and naturally occurring objects differ *in a perceptible way* in their surface gradients. Consider the two stars shown at the top of Figure 7.8. They have the same global shape. But if one of these is a manufactured holiday star and the other a living starfish, it is the one on the right that was made in a factory and the one on the left that was found at the beach. Naturally occurring objects tend to have an intricate surface gradient that is distinct from that of manufactured objects.

Mandelbrot (1983) described the surfaces of natural objects as fractal curves. These surfaces are self-similar across changes in scale. Thus, if we

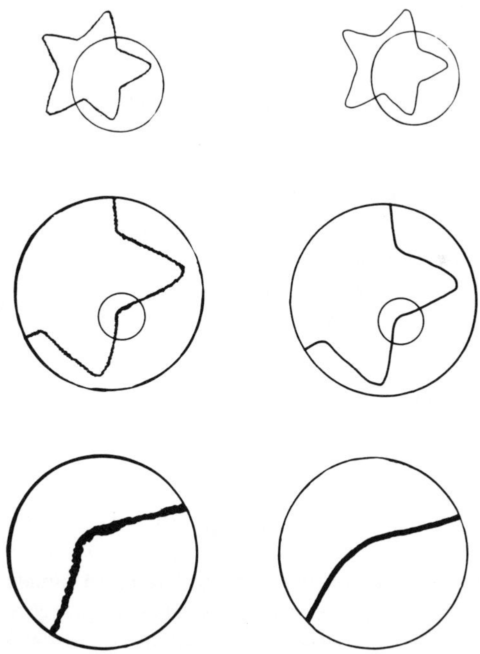

Figure 7.8. Two objects with the same global shape but different surface gradients and enlargements of portions of those objects.

magnified the starfish, we would see bumps on the surface. If we magnified those bumps, we would see bumps on the bumps and so on. We may speculate, following Mandelbrot, that the complex and seemingly irregular surface gradient of natural objects stems from the fact that they grow and are caused by a multiple of converging forces whose effects accrue in time. In contrast, the scale-specific structure of manufactured objects presumably derives from the fact that they were made at a particular scale level and at a particular point in time.

Are children sensitive to the surface gradient differences between naturally occurring and manufactured objects? Diana Heise, Susan Rivera and I have preliminary evidence that 12-month-old children are. In this ongoing research, the task is preferential looking. In the baseline experiment, the stimuli were toy vehicles and life-like model animals. Model animals, of course, are manufactured and so it might be argued that they are poor stimuli for a study of the perceptual properties afforded by naturally occurring objects. However, the perceptual properties that make one model animal seem life-like and another toy-like or robot-like are presumably the same ones along which naturally occurring and (typical) manufactured objects differ.

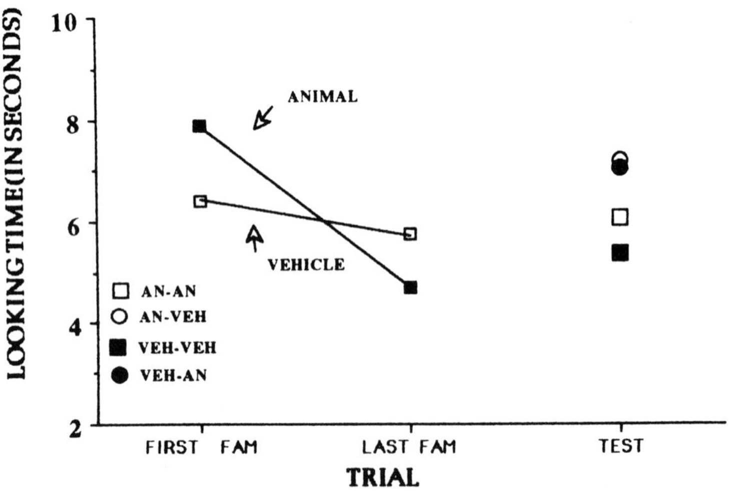

Figure 7.9. Mean looking time on the first and last of 14 familiarization trials showing Animals or Vehicles and on the test trial in the Between-category (AN-VEH/VEH-AN) and Within-category (AN-AN/VEH-VEH) conditions when the stimuli were intact toy animals and vehicles.

In the experiment, the babies were repeatedly shown in alternation two toys and then after familiarization were shown a novel test toy. In two Between-Category conditions, the novel test toy was from a different superordinate class than the familiarization toys. In the Within-Category conditions, the novel test toy was from the same class as the familiarization stimuli. For example, children in the Animal/Between-Category condition might see a goat and cow during familiarization and a boat during test. Children in the corresponding Animal/Within-Category condition would see a goat and cow during familiarization and an elephant during test. The results in Figure 7.9 show clear evidence that infants discriminate animals from vehicles. Looking time during test goes up more in the Between-Category conditions than in the Within-Category conditions.

There are a number of perceptual properties along which babies could (and probably do) distinguish toy animals and toy vehicles that include eyes, mouths, global shape, and by our hypothesis, surface gradient. We examined the infants' ability to use surface gradient cues in a subsequent experiment by removing all potential cues except surface gradient and asking if babies could still distinguish animals from vehicles. Specifically, we cut up the toys into approximately 2" x 2" pieces, removing any pieces with eyes, mouths, or parts of

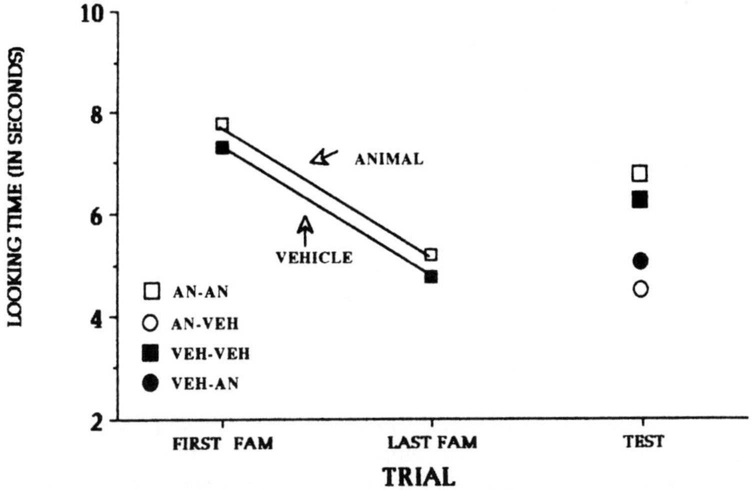

Figure 7.10. Mean looking time on the first and last of 8 familiarization trials showing Animals or Vehicles and on the test trial in the Between-category (AN-VEH/VEH-AN) and Within-category condition (AN-AN/VEH-VEH) conditions when the stimuli were cut-up toy animals and vehicles.

faces (and all wheels from the vehicles). We separately mounted the remaining pieces of each toy in a circle for viewing by the infant and repeated the baseline experiment with these cut-up stimuli. The results shown in Figure 7.10 indicate that children discriminated cut-up animals from cut-up vehicles. Contrary to the first experiment, children looked *more* at the Within-category test stimulus than the Between-category stimulus. Although a preference for the familiar as opposed to novel was shown with the "parts," the results, nonetheless, show babies discriminate animal textures from vehicle textures. Thus, there do appear to be perceptual properties that distinguish animals from vehicles. Along at least one perceptible property, surface gradient, animals and vehicles do not look alike. Perhaps, then, the babies in Mandler & Bauer's study were simply classifying the objects by what they looked like.

Our finding that babies are sensitive to a texture cue that may distinguish animals from vehicles has implications beyond Mandler & Bauer's results. The findings provide a useful starting point for thinking about the relation between perceptual and conceptual structure. The results show a perceptual dimension along which many naturally occurring and manufactured objects may be distinguished and to which infants are sensitive. Moreover, it is the nature of the human apparatus for perceptual similarity to link up this dimension with co-occurring properties and in this way to selectively enhance attention to this dimension in the stimulus contexts in which it most matters -- for example, in the context of stimuli with eyes, or stimuli that move in a certain biological way.

Our claim here is not that surface gradient is a perfect or fool-proof cue by which to distinguish naturally occurring from manufactured objects. There are stones so smooth from the continual action of the waves that they are as perfect spheres as ball bearings and there are talented artists who can fool the eye with their creations. Our idea, though, is that surface gradient is an important force in starting, maintaining, and grounding conceptual structure to reality.

This concludes our defense of perceptual similarity as a major player in category development. Perceptual similarity is context dependent. It is shifted about by implicit knowledge of relations between perceptual properties. And, the perceptual properties that can be emphasized or de-emphasized in similarity judgments may, like the presence of eyes and like the surface gradient differences between animals and vehicles, be highly conceptually relevant -- of a kind worth developing a causal theory to explain.

IV. PERCEPTUAL SIMILARITY AND CAUSAL THEORIES

Perceptual and conceptual similarity are not, by our view, the same thing. Conceptual similarity does not reduce to perceptual similarity. But perceptual and conceptual similarity are also not independent. We will consider first how perceptual and conceptual similarity are different. Then, we will consider the causal dependencies between the two.

A. Perceptual and Conceptual Similarity are Not the Same

When we look at objects, what we *perceive* reflects much implicit knowledge --- a myriad of associations between perceptual features, linguistic, social, and physical context. Perceptual similarity is thus dynamic and smart but it is not smart in the same way that conceptual structure is. Perceptual similarity is not a set of causal beliefs and explanations; it embodies causal relations but it does it not represent them. Thus, when children look at objects with eyes they perceive the similarities and differences in texture and their perception is smart and adaptive. But there is in the perception no causal understanding of why textures matter for objects with eyes.

Rozin (1976) in his landmark paper on intelligence distinguished between welded implicit knowledge of the sort that allows birds to migrate and transportable explicit knowledge such as that used by navigators on ships. Rozin characterized implicit knowledge as rigid and inflexible. Our view of the implicit knowledge that is embodied in the dynamics of perceptual similarity does not fit this description. Rozin also characterized implicit knowledge as being welded to particular contextual factors that are necessary for the knowledge to show itself. The knowledge behind perceptual categorization would seem to be welded in the sense of being tightly determined by the actual perceptual properties present, past specific learning, and context. Explicit knowledge, true intelligence according to Rozin, is knowledge that is more context free -- that may be *voluntarily* transported across contexts to new problem spaces. Conceptual structure may be smart in this way.

This distinction between implicit perceptual structure and explicit, transportable, conceptual knowledge is attested to by the difficulty one has in *consciously* controlling perceived similarities. Whereas conceptual similarity may be strongly influenced by being told some new fact, perceptual similarity may be moved along only a little by "mere talk." Being told that a purple, three-legged beast with moose horns is a cow, doesn't make the unfortunate animal *look* like a cow (although one might, if given that information by experts, claim to believe the odd beast to be a cow despite its looks). The separateness of perception and our easily talked about conceptual beliefs is evident in linguistic hedges (Lakoff, 1972) and statements such as, "It looks like a fish but it's really a mammal." This obvious distinction between perceptual knowledge, which is not transparent to conscious thought processes, and explicit conceptual knowledge probably underlies the temptation to empirically pit perception against conception in developmental experiments.

B. Perceptual and Conceptual Structure are Causally Related

The separateness and different status of perceptual and conceptual structure does not mean they are not causally related. There are causal relations that go in both ways. Explicit conceptual knowledge can push perception only so far (no amount of information will convince a person looking at a spoon that it is a tomato). However, conceptual knowledge can cause individuals to seek out perceptual dimensions that make sense of conceptual distinctions. This effect of conceptual knowledge on perception is aptly demonstrated in Heise's ongoing dissertation research. In that research, children are presented with real world correlates of the picture stimuli used by Gelman and Markman (1986). One triad consists of a ball of yellow wool yarn, a ball of yellow acrylic yarn, and raw unwoven wool. The two balls of yarn look very much alike. However when told that the wool yarn and the raw wool are both wool, 5 year old children (and adults) often look more closely and say such things as "Yes, these two are less shiny, that's how you can tell"). This example is telling on two counts. First, it shows how a conceptual distinction may invite the search for a perceptual distinction. Second, it shows how the human cognitive system attempts to keep conceptual and perceptual structures in line with each other. Conceptual growth does not cause us to abandon perception but it may cause us to take another look.

Although explicit conceptual knowledge may sometimes be a causal force in perceptual comparison, the causal relation between perception and conception goes mostly one way --- with perception the cause and conception the effect. Perceptual structure *ties* our conceptual beliefs to reality; it is the mapping function that takes our representations and beliefs and gives them meaning. Perceptual structure is the *data* that our conceptual theories are about and just as the data are the final arbiter of truth in science so is perception the unassailable constrainer of concepts in cognitive development.

Perceptual structure is what conceptual beliefs are about. The atheoretical, uninterpreted associations and correlations between perceptual features makes a bumpy terrain of the perceptual landscape. Our explicit conceptual beliefs try to make conscious sense of this terrain. *Because* our conceptual beliefs are about making sense of the world, perceptual and conceptual structure will generally agree.

However, our explicit conceptual beliefs, even at maturity, probably do not do a very good job of making explicit the implicit interrelations and co-relations that structure and organize perception. Perceptual knowledge is deep, highly entrenched, highly embedded and probably much smarter than conceptual knowledge (see Gentner & Rattermann, 1990, for a similar argument) this is so despite the fact that the knowledge embodied in perception is severely limited in its transportability across cognitive domains. The evidence on how people readily catch thrown balls without understanding the physics is but one example of the smart but welded nature of perceptual knowledge (see, e.g., Bingham, Schmidt, E. Rosenblum, 1989). We expect the same contrast between "smart perception" and "less smart" conception is true for categories. It is perceptual knowledge that is deep and it is the conceptual structure that is superficial and inaccurate as it depends on cultural and perhaps mistaken scientific beliefs (see Jeyifous, 1986; Lakoff, 1987; Putnam, 1975 for examples).

C. How Conceptual Structure Depends on Perceptual Structure

In his recent book, Keil (1989) examines Boyd's proposal about causal homeostasis as a model for human conceptual structure. The idea is that categories are not just clusters of correlated properties. Instead, there are sets of causally important properties that are contingently clustered. The correlated properties are not the results of happenstance. The laws of physics, biology, and behavior (the laws scientists seek to discover) are such that properties co-occur (and vary together) *because* of causal "homeostasis." The presence of some properties favors or causes the presence of others.

How is this structure in the world related to the psychological structure of categories? It is informative here to contrast our position with Keil's proposal. Following Quine, Keil considers an initial "original similarity" state that is represented as in Figure 7.11A. The perceptual primitives are represented by the circles, triangles, squares, and ellipses. These primitives are connected by pretheoretical associations that are formed when specific features co-occur in the world. What happens with development, according to Keil, is that causal beliefs are overlaid on this perceptual structure. These beliefs are shown by the heavier arrows in Figure 7.11B. (Keil suggests that 7.11B may in fact more closely represent the starting point for development than 7.11A since some domain-specific causal theories may be "innate".) With increasing age, theories are elaborated as in 7.11C. Early in development, Keil suggests, causal beliefs may interpret the perceptual associations between primitives but with development, they become more systematic, elaborate, and differentiated.

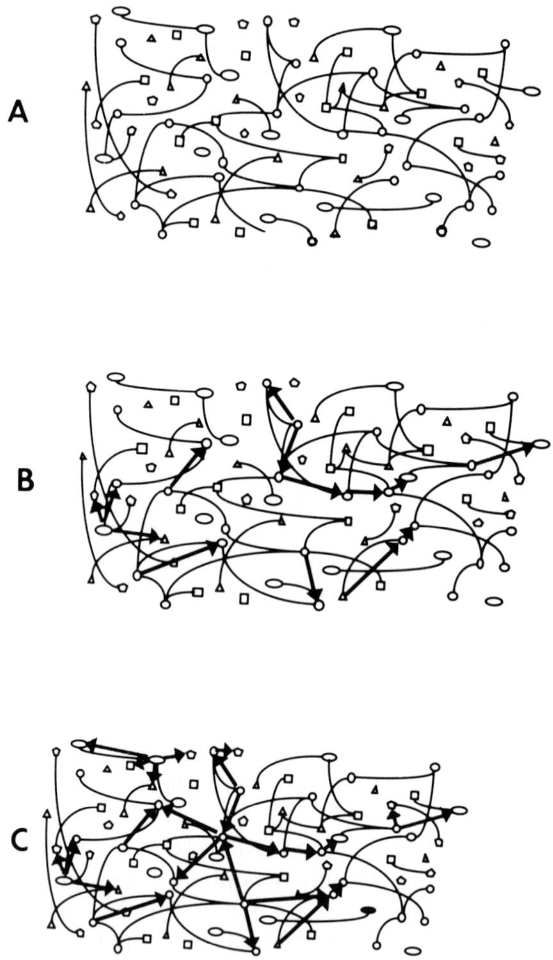

Figure 7.11. Taken from Keil (1989). A: Networks of features linked by associations; B,C: The emergence of theory indicated by arrows.

We concur with the main points of this outline of a theory. Yet we find its portrayal of perceptual structure disquieting. Again, static perceptual structure provides at best a minor constraint on conceptual structure. In Keil's description, perceptual structure only weakly reflects (in its associative connections) the causal homeostasis that exists between properties in the world. The causal homeostasis

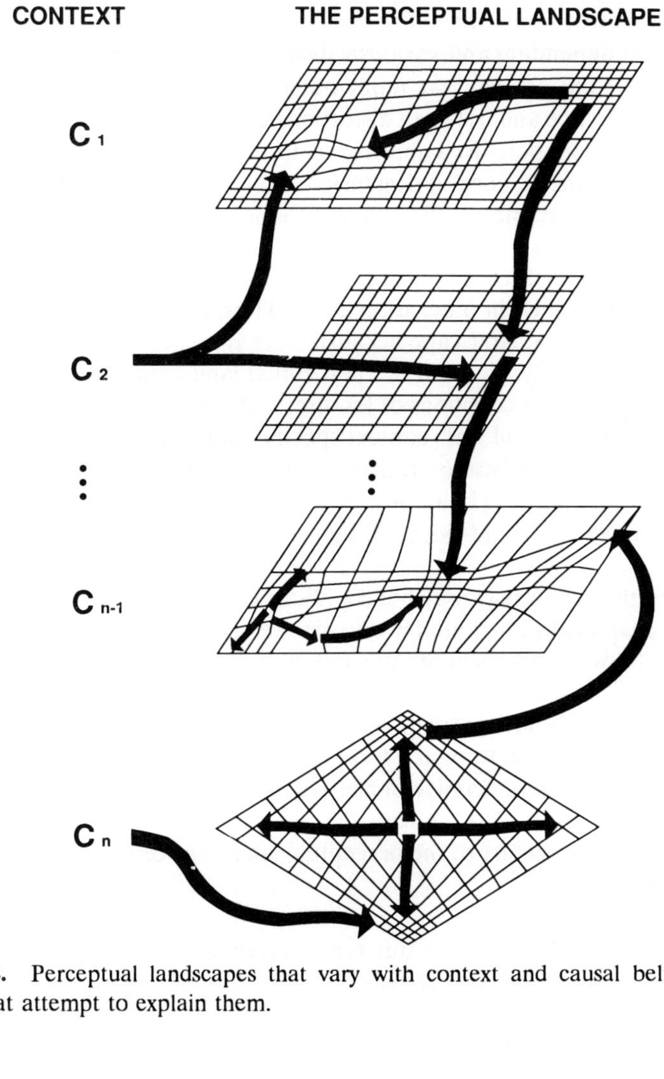

Figure 7.12. Perceptual landscapes that vary with context and causal belief systems (arrows) that attempt to explain them.

that Boyd writes about finds its psychological home in Keil's view in people's elaborated and systematic causal beliefs.

We offer as an alternative to Keil the outline of a theory of perceptual context C_1, is shown the perceived similarity of objects along some set of and conceptual structure depicted in Figure 7.12. What this figure attempts to depict is the changing perceptual landscape with changing context and people's theories about the perceived similarities. The dynamic perceptual landscape is represented by the context dependent and variously distorted grids. Conceptual theories are represented by the arrows that point out relations between distortions in the perceptual landscape and between distortions and context. So, next to dimensions in that context. The perceived similarities in the space are not uniform because of learned correlations, context-dimension associations, and language-dimension associations. We assume all these associations are acquired from being in the world. Context C_2 shows the distortions of *this same similarity* space in another context that activates another set of associations that shift the dimension weights and perceived similarity in other directions. Thus, in contrast to Keil, we do not present one perceptual landscape because there is no one landscape. There is no static original similarity. Instead the perceptual landscape emerges in particular contexts with particular coalitions of features. Morever, because of the nature of perceptual similarity and the context-specific shifts in feature weights, these dynamic perceptual landscapes reflect the causal structure of the physical and social world. Boyd's causal homeostasis finds its psychological home in this theory in the bumps and holes and ridges of perceived similarity.

People's explicit causal beliefs about categories and relations between objects are graphically illustrated by the arrows. By our view, people build causal connections between points in the perceptual landscape --- incorporating cultural and scientific knowledge --- in order to interpret and make sense of the complex perceptual terrain. Some causal beliefs may be local and about interdependencies in a single contextually determined terrain. Other naive theories may be grander and try to capture the dynamics of the changing emphasis and de-emphasis of features across contexts. It is unlikely that these naive theories adequately reflect or explain the rich structure of the perceptual landscape or the causal homeostasis in the world.

The difference between our representation and Keil's, then, is that the perceptual structure is richer and far more complicated. Perception provides the nutrients on which conceptual structure grows. Perception is the reality on which the conceptual system operates. Indeed, it is because we are perceptually in contact with the complex world that we need an elaborated conceptual structure that simplifies that complexity.

V. STRUCTURE AND PROCESS

The chapters of this book have all grappled with the question of the relation between structure and process. Although, we have not directly confronted the issue, the distinction between structure and process may be the root of the problem in understanding perception-conception interactions. The idea of the structure-process distinction is that cognition is composed of two parts: structure which is the represented knowledge and process which operates on the structures. In this view of cognition, structures are the nouns of thoughts, *the things*, and processes are the verbs, the actions. The problem with this metaphor is that perceptual and conceptual structures are designated as things as if that they were static and unchanging. This metaphor is particularly troublesome when thinking about development because development is about change (see, Smith, 1990; Smith & Sera, 1990).

How might cognitive development go forward if knowledge is a set of "things" --- if we acquire pieces of knowledge like new pieces of clothing? One possibility is that children just keep adding new pieces to their knowledge set. Another possibility is that children abandon or ("trade up") their use of one structure in favor of another. For example, they might abandon perceptual structure in favor of conceptual structure. We have argued that this particular solution to developmental changes in categorization does not fit the data.

There is an alternate to partitioning cognition into structure and process. The alternate view is that there is only process. By this view, perceptual and cognitive structures are not thing-like representations that reside in the child but are rather the emergent properties of complex processes. This alternative view is, of course, the one that underlies the theoretical endeavors known as connectionism (e.g., Rumelhart & McClelland, 1986) and dynamical systems theory (Thelen, 1989; Smith & Sera, 1990). It is the assumption that guides our work and forms our belief in perceptual similarity as dynamic. Perceptual similarity --- perceptual structure --- is dynamic because perceptual structure is not a thing; it is the emergent result of attentional processes.

VI. CONCLUSIONS

We have argued for a new research agenda -- one that does not pit perception against conception but instead asks how perceptual and conceptual structure are related and how they organize together to propel development forward. We have specifically argued for greater attention to the dynamics of perceptual similarity and perceptual structure. Perceptual similarity is a much maligned force in cognition. Philosophers such as Nelson Goodman (1972) have dismissed similarity as too slippery and variable a concept to have any explanatory power. By our view, the power of perceptual similarity stems directly from its (systematically) variable nature.

ACKNOWLEDGEMENTS

Portions of this chapter were presented under the title "In defense of perceptual similarity" at the Society for Research in Child Development, Kansas City, April 1989. The research reported in this chapter was supported by PHS grant S07RR7021L through Indiana University. We thank, Patricia Bauer, Barbara Burns, Susan Jones, Jean Mandler, Clay Mash, Doug Medin, and Maria Sera, for comments on earlier versions of this manuscript.

REFERENCES

Bertenthal, B. I., Proffitt, D. R., Kramer, S. J., & Spetner, N. B. (1987). Infants' encoding of kinetic displays varying in relative coherence. Developmental Psychology, 23, 171-178.

Biederman, I. (1985). Human image understanding. Computer Vision, Graphics, & Image Processing, 32, 29-73.

Billman, D. (1989). Systems of correlations in rule and category learning: Use of structured input in learning syntactic categories. Language and Cognitive Processing, 4, 127-155.

Bingham, G. P., Schmidt, R. C. and Rosenblum, L. D. Hefting for a maximum distance throw: A smart perceptual mechanism. Journal of Experimental Psychology: Human Perception and Performance, 15, 507-528.

Bornstein, M. H. (1985). Habituation of attention as a measure of visual information processing in human infants: Summary, Systematization, and Syntheses. In G. Gottlieb and N. Krasnegor (Eds.) Measurement of audition and vision in the first year of postnatal life: A methodological overview. Norwood, N. J.; Ablex, 253-300.

Brown, A. L. (1990). Domain-specific principles affect learning and transfer in children. Cognitive Science, 14, 107-134.

Bruner, J. S. & Olver, R. R. (1963). Development of equivalence transformations in children. Monograph of the Society for Research in Child Development, No. 28.

Bullock, M., Gelman, R., & Baillargeon, R. (1982). The development of causal reasoning. In W. J. Friedman (Ed.), The developmental psychology of time. New York: Academic Press.

Carey, S. (1985). Conceptual change in childhood. Cambridge, MA: MIT Press.

Flavell, J. H. (1970). Concept development. In P. H. Mussen (Ed.), Carmichael's manual of child psychology: Vol 1. New York: Wiley.

Gelman, R. (1990). First principles organize attention to and learning about relevant data: Number and the animate-inanimate distinction as examples. Cognitive Science, 14, 79-106.

Gelman, R., Spelke, E. S., & Meck, E. (1983). What preschoolers know about animate and inanimate objects. In D. Rogers and J. A. Sloboda (Eds.), The acquisition of symbolic skills. London: Plenum.

Gelman, S. A. (1988). The development of induction within natural kind and artifact categories. Cognitive Psychology, 20, 65-95.

Gelman, S. & Markman, E. M. (1986). Categories and induction in young children. Cognition, 23, 183-209.

Gelman, S. A. & Markman, E. M. (1987). Young children's inductions from natural kinds: The role of categories and appearances. Child Development, 58, 1532-1541.

Gentner, D. (1989). The mechanisms of analogical learning. In S. Vosniadou and A. Ortony (Eds.), Similarity and analogical reasoning (pp 199-241). New York: Cambridge University Press.

Gentner, D. & Rattermann, M. J. (1990). Language and the career of similarity. In S. Gelman (Ed.), Language and conceptual development.

Goodman, N. (1972). Problems and projects. Indianapolis: Bobbs-Merrill.

Goldstone, R. L., Medin, D. L., & Gentner, D. (in press). Relational similarity and the non-independence of features in similarity judgments. Cognitive Psychology.

Greer, A. E. & Sera, M. D. (1990). Artifacts vs. natural kinds: An empirical investigation. (Unpublished manuscript).

Gluck, M. A. & Bower, G. H. (1988). From conditioning to category learning: An adaptive network model. Journal of Experimental Psychology: General, 117, 227-247.

Heise, D. & Smith, L. B. (1990). Learning co-relations. (Unpublished manuscript).

Jeyifous, S. (1986). Antimodemo: Semantic conceptual development among the Yoruba. Doctoral dissertation, Cornell University.

Jones, S. S., Smith, L. B., & Landau, B. (1990). Object properties and knowledge in early lexical learning. (Under review)

Johnson, M. (1987). The body in the mind: The bodily basis of meaning, imagination, and reasoning. Chicago: University of Chicago Press.

Keil, F. C. (1989). Concepts, kinds, and cognitive development. Cambridge, MA: MIT Press.

Kemler Nelson, D. G. (1990). When experimental findings conflict with everyday observations: Reflections on children's category learning. Child Development, 61, 606-610.

Lakoff, G. (1972). Hedges: A study in meaning criteria and the logic of fuzzy concepts. In Papers from the Eighth Regional Meeting of the Chicago Linguistic Society. Also in Journal of Philosophical Logic (1973) 2, 458-508.

Lakoff, G. (1987). Women, fire, and dangerous things: What categories reveal about the mind. Chicago: University of Chicago Press.

Landau, B., Smith, L. B., & Jones, S. S. (1988). The Importance of Shape in Early Lexical Learning Cognitive Development, 3, 299-321.

Leslie, A. (1984). Infant perception of a manual pick-up event. British Journal of Developmental Psychology, 2, 19-32.

Leslie, A. (1988). The necessity of illusion: Perception and thought in infancy. In L. Weiskrantz (Ed.), Thought without language. Oxford: Clarendon.

Mackintosh, N. J. (1965). Selective attention in animal discrimination learning. Psychological Bulletin, 64, 124-150.

Macnamara, J. (1982). Names for things: A study of human learning. Cambridge, MA: MIT Press.

Mandelbrot, B. B. (1983). The fractal geometry of nature. San Francisco: Freeman.

Mandler, J. M. (1988). How to build a baby: On the development of an accessible representational system. Cognitive Development, 3, 113-136.

Mandler, J. M. (1990). How to build a baby: Part 2. (Unpublished manuscript)

Mandler, J. M. & Bauer, P. J. (1988). The cradle of categorization: Is the basic level basic? Cognitive Development, 3, 237-264.

Mandler, J. M., Bauer, P. J., & McDonough, L. (in press). Separating the sheep from the goats: Differentiating global categories. Cognitive Psychology.

Markman, E. M. (1989). Categorization and naming in children: Problems of induction. Cambridge, MA: MIT Press.

Medin, D. & Ortony, A. (1989). Psychological essentialism. In S. Vosniadou and A. Ortony (Eds.), Similarity and analogical reasoning. New York: Cambridge University Press.

Medin, D. L. & Schaffer, M. M. (1978). Context theory of classification learning. Psychological Review, 85, 207-238.

Medin, D. L., Ahn, W., Bettger, J., Florian, J., Goldstone, R., Lassaline, M., Markman, A., Rubinstein, J., & Wisniewski, E. (1990). Safe take offs - soft landings. Cognitive Science, 14, 169-178.

Medin, D. L., Altom, M. W., Edelson, S. M. & Freko, D. (1982). Correlated symptoms and simulated medical classification. Journal of Experimental Psychology: Learning, Memory and Cognition, 8, 37-50.

Medin, D. L., Wattenmaker, W. D., & Hampson, S. E. (1987). Family Resemblance, conceptual cohesiveness and category construction. Cognitive 4, 19, 242-299.

Mervis, C. B. (1987). Child-basic object categories and lexical development. In U. Neisser (Ed.), Concepts and conceptual development: Ecological and intellectual factors in categorization. Cambridge, England: Cambridge University Press.

Murphy, G. L. & Medin, D. L. (1985). Role of theories in conceptual coherence. Psychological Review, 92, 289-316.

Nosofsky, R. M. (1984). Choice, similarity, and the context of classification. Journal of Experimental Psychology: Learning, Memory, & Cognition, 10, 104-114.

Nosofsky, R. M. (1986). Attention, similarity, and the identification-categorization relationship. Journal of Experimental Psychology: General, 115, 39-57.

Piaget, J. (1929). The child's conception of the world. New York: Harcourt Brace.

Putnam, H. (1975). The meaning of meaning. In H. Putnam (Ed.), Mind, language, and reality, Vol. 2. London: Cambridge University Press.

Quine, W. V. O. (1960). Word and object. Cambridge, MA: MIT Press.

Quine, W. V. O. (1977). Natural kinds. In S. P. Schwartz (Ed.), Naming, necessity, and natural kinds. Ithaca, NY: Cornell University Press.

Rosch, E. (1973). Natural Categories, Cognitive Psychology, 4, 328-350.

Rosch, E. & Mervis, C. B. (1975). Family resemblances: Studies in the internal structure of categories. Cognitive Psychology, 7, 573-605.

Rosch, E., Mervis, C. B., Gray, W. D., Johnson, M. D., & Boyes-Braem, P. (1976). Basic objects in natural categories. Cognitive Psychology, 8, 382-439.

Rozin, P. (1976). The evolution of intelligence and access to the cognitive unconscious. Progress in Psychobiology and Physiological Psychology, 6, 245-280.

Rumelhardt, D. E., McClelland, and the PDP Research Group (1986). Parallel Distributed Processing. Volume 1: Foundations, Cambridge, MA: MIT Press.

Shepard, R. W. (1987). Toward a universal law of generalization for psychological science. Science, 237, 1317-1323.

Shepard, T. W. (1964). Attention and the metric structure of the stimulus space. Journal of Mathematical Psychology, 1, 54-87.

Smith, E. E. & Medin, D. L. (1981). Categories and concepts. Cambridge, MA: Harvard University Press.

Smith, J. D. & Kemler-Nelson, D. G. (1984). Overall similarity in adults' classification: The child in all of us. Journal of Experimental Psychology: General, 113, 137-159.

Smith, L. B. (1981). The importance of the overall similarity of objects for adults' and children's classification. Journal of Experimental Psychology: Human Perception and Performance, 1, 811-824.

Smith, L. B. (1989). A model of perceptual classification in children and adults. Psychological Review, 96, 125-144.

Smith, L. B. & Sera, M. (1990). A developmental analysis of the polar structure of dimensions. (Under review)

Smith, L. B., Jones, S. S., & Landau, B. (1990). Nouns, adjectives, and perceptual properties in very young children's lexical strategies. (Under review).

Spelke, E. S. (1982). Perceptual knowledge of objects in infancy. In J. Mehler, M. Garrett, and E. Walker (Eds), Perspectives on mental representation. Hillsdale, NJ: Erlbaum.

Sugarman, S. (1983). Children's early thought. Cambridge, England, Cambridge University Press.

Sutherland, N. S. (1964). The learning or discriminations by animalls. Edeavour, 23, 148-152.

Thelen, E. (1989). Self-organization in developmental processes: Can systems approaches work; In M. Gunnar & E. Thelen (Eds.), Systems and Development: The

Minnesota Symposia on Child Psychology, vol.22, pp. 11-117. Hillsdale NJ: Erlbaum.

Tversky, A. (1977). Features of similarity. Psychological Review, 84, 327-352.

Tversky, B. (1985). Development of taxonomic organization of named and pictured categories Developmental Psychology, 21, 1111-1119.

Vygotsky, L. S. (1934/1962). Thought and language. In E. Hanfmann and G. Vakar, trans. Cambridge, MA: MIT Press.

Waxman, S. R. (1990). Linguistic biases and the establishment of conceptual hierarchies: Evidence from preschool children. Cognitive Development, 5, 123-150.

Wellman, H. M. & Gelman, S. A. (1988). Children's understanding of the nonobvious. In R. Sternberg (Ed.), Advances in the psychology of intelligence, Vol. 4. Hillsdale, NJ: Erlbaum.

Werner, H. (1948). Comparative psychology mental development, 2nd Ed. New York: International Universities Press.

Wohlwill, J. F. (1962). From perception to inference: A dimension of cognitive development. Monographs of the Society for Research in Child Development, 27, 87-112.

Younger, B.A. (1990). Infants' detection of correlations among feature categories. Child Development, 61, 614-620.

Younger, B. A. & Cohen, L. B. (1983). Infant perception of correlations among attributes. Child Development, 54, 858-867.

Commentary

Perceptual Similarity and Conceptual Structure, L. B. Smith & D. Heise

GREGORY R. LOCKHEAD
Duke University

Many writers distinguish structure from process and distinguish perception from cognition. Smith and Heise argue there is only process. Perceptual and cognitive structures (or similarities) that are observed in data are, by their view, momentary reflections of dynamic processes.

When data structures differ in different situations, it is because attention in the dynamic system is then different. Data depend on the weights applied by attention to the similarity space. These weights, which distort the similarity space, are determined by the stimulus subset and by what the subject has experienced. Attention determines similarity in terms of relations among attributes. This view provides a convenient way to examine many situations. It is not clear that one really can distinguish between this position, which combines structure and process into a single outcome (cf. their Figures 7.4 and 7.6), and the perhaps more frequent idea that responses taken from a fixed similarity structure are biased. Both may make the same predictions.

To support their view (and other views), Smith and Heise claim it is important to study perceptual similarity. Their arguments are welcomed. Similarity measures have been too successfully deprecated on the basis of too little data. There is still much to be learned by using similarity to evaluate models of perception and cognition. Smith and Heise demonstrate this in two important ways. First, they show how many prior studies can be reinterpreted in terms of similarity, if effects of attention and the stimulus subset are considered. Second, they demonstrate the richness of such measures. Without modification, similarity measures can address such issues as essence, form-class relations, basic categories, knowledge, biological texture, development, and causation.

One summary message of the paper is found in an embedded sentence: "People build causal connections between points in the perceptual landscape--incorporating cultural and scientific knowledge--in order to interpret and make sense of the complex perceptual terrain." (p. 41) Thus, although Smith and Heise never say so directly, this paper is about meaning.

Percepts, Concepts and Categories
B. Burns (Editor)
1992 Elsevier Science Publishers B.V.

8

Reflecting on Representation and Process: Children's Understanding of Cognition

SUSAN E. BARRETT
Lehigh University

HERVE ABDI
University of Texas at Dallas
University of Bourgogne at Dijon

JILL M. SNIFFEN
Lehigh University

I. INTRODUCTION

Questions about the nature of representations, the structure of knowledge, and the relationship among perceptual and cognitive processes figure prominently in cognitive theories. These same issues play a central role in the child's evolving understanding of the cognitive system. In this chapter, we will explore preschoolers' and young elementary school children's conceptions of representations and the distinctions they make among various perceptual and cognitive processes. As we review these topics, it will become clear that even three-year-olds are aware of the unique status of mental representations. Young children recognize that they can manipulate representational entities, such as images, simply through mental effort and concentration, but that mental powers are not sufficient for altering physical objects. However, three-year-olds are less clear about the relationship between process and representation. They seem to recognize that some representations are the products of mental processes, such as pretending and imagining, but they find it difficult to reflect on the relationship between supposedly veridical processes, such as perception, and the representations they produce. This latter insight seems to depend on an understanding that perceptual and cognitive processes actively construct and transform information to produce knowledge, and as we will see, young preschoolers do not really appreciate this constructivist aspect of the mind.

An important part of understanding how the mind works involves understanding the distinctions and interrelations that exist among various perceptual and cognitive processes. In this review, we will explore the extent to which children distinguish among these processes and whether children's metacognitive knowledge is organized around these distinctions. As we examine children's conceptions of mental activities, it will become clear that even preschoolers understand that each perceptual system can function independently and that they have at least a rudimentary understanding that perceptual acts can be distinguished from more cognitive activities. Young elementary school children also make some distinctions among cognitive processes and we will consider what these distinctions reveal about how children organize their knowledge about the mind.

A. Chapter Overview

In recent years, developmental psychologists interested in children's ability to reason about intentional behavior and intervening mental processes have posited that even preschoolers possess a "theory of mind." Theories are also thought to guide conceptual development in other areas. For example, preschoolers have been credited with biological theories and theories about the physical world (see, e.g., Carey, 1985; Keil, 1989). Obviously, this approach to cognitive development raises a number of very important questions. Perhaps the most central of these is what it means to say that knowledge is structured around a theory. In the opening section of the paper, we will explore in some detail the "theory approach" to cognitive development and we will consider why psychologists such as Wellman (1990) believe that it is appropriate to credit young children with a theory of mind.

Two- and three-year-olds' speech is peppered with references to the mental world, and this has sometimes been cited as evidence that the young child possesses a theory of mind. In the second section of this paper, we will examine the mental lexicon to see whether natural language might provide some insight into the child's growing awareness of the mental world. Research on children's use of mental verbs suggests that four-year-olds appreciate that these terms have distinctly mental referents, even though these children do not appropriately discriminate among the mental verbs. As we explain in this section, these findings might suggest that preschoolers find it difficult to distinguish among more central cognitive processes.

Preschoolers could, of course, fail to discriminate among mental verbs even though they understand the distinctions and interrelations that exist among perceptual and cognitive processes. In the third section, we will review some recent studies that focus on whether children distinguish among perceptual processes and how children conceive of the relationship between seeing and knowing. One topic that figures prominently in these discussions concerns the child's conception of representation. We will consider whether young children distinguish between percepts and knowledge-based representations as well as their ability to deal with contradictory representations. This work will be framed in terms of a proposed shift in the child's evolving theory of mind in which a passive copy theory of knowledge gives way to a more active constructivist view (see, e.g., Chandler & Boyes, 1982; Wellman, 1990).

Traditionally, studies of metacognitive knowledge have focused on children's understanding of task and situational factors that affect performance. These studies are important because if children are able to distinguish between relevant and irrelevant situational variables in the context of simple cognitive activities, it would suggest that they are able to reflect on the cognitive processes that are used to accomplish these tasks. In the fourth section, we will review a number of studies designed to uncover what children know about various factors that impede or facilitate performance. A second topic addressed in this section concerns whether the child's metacognitive knowledge is structured around classes of cognitive activities. Adults distinguish among tasks based on the cognitive requirements associated with each. For example, they consider various short-term memory tasks to be similar even when the items that must be held in memory differ dramatically. In this section, we will examine whether children's perceptions of task similarity also vary as a function of the cognitive requirements of the task.

Finally, in the fifth section we will present some of our recent work that focuses on children's understanding of stimulus variables that differentially affect performance. Our findings show that both young elementary school children and older preschoolers evaluate the effects of these variables by reflecting on the specific processing requirements of individual tasks, an accomplishment that we think is made possible by their newfound conception of the mind as an active information processing system.

II. THE ROLE OF THEORIES

A. Scientific and Everyday Theories

Various types of conceptual structures support children's and adults' understanding of the world. A central problem facing researchers in cognitive development involves specifying how these structures are acquired, how they change over time, and the extent to which these structures are knowledge-dependent and domain-specific. Following Carey's (1985) seminal book, an increasing number of developmental psychologists have begun to adopt the theory metaphor as a way to characterize cognitive change. More specifically, researchers working in this tradition have posited parallels between conceptual change in the child and theory change in the history of science. Clearly, there are limits to these parallels and these limits involve both processing and structural considerations.

Consider first the process of theory building in science. Empirical testing, critical reflection, and rigorous theorizing are often considered the very essence of science and yet these are not the processes that characterize the young child's theory building. It is only in pre-adolescence that the child begins to carefully distinguish between theory and evidence and is able to coordinate the two explicitly (Kuhn, 1989). Thus, the process of theory building in the young child is not like scientific theorizing in either the novice or expert practitioner because the child lacks the metaconceptual skills necessary to reflect on both theory and the process of theory building. But it is also true that even adults and older children who are capable of critical and reflective theorizing rarely do so. Instead, the everyday theories of both child and adult are likely to reflect the same limitations and biases that color knowledge acquisition in general (Kahneman, Slovic & Tversky, 1982; Nisbett & Ross, 1980).

If the young child, and even the adult, lack or at least rarely use the skills most central to the scientific enterprise, what is gained by adopting the theory metaphor? Researchers working in this tradition believe that even though important processing differences exist between scientific and everyday theory building, there may be some important structural similarities that justify the metaphor. In discussing these similarities, Wellman (1990) focuses on an important distinction between framework theories and specific theories that has appeared in the philosophy of science. Wellman believes that the relationship between framework and specific theories in scientific thought parallels the relationship found in intuitive knowledge domains, and that it is this property that adds the most credence to the metaphor. In both scientific and everyday knowledge, framework theories serve three functions. Framework theories define the ontology of the domain, that is, they specify the kinds of entities and processes that are encompassed by the theory and in doing so define theoretical domains. Framework theories are also characterized by their causal-explanatory mechanisms: what counts as a satisfactory explanation differs across theoretical domains. Finally, framework theories constrain process in the sense that they may place limits on the methods that can be used to construct specific theories.

In contrast to framework theories which serve a definitional role, specific theories provide more detailed explanations for particular observations within the confines of the theoretical domain. As Wellman (1990) notes, it is at the level of specific theories that differences between everyday and scientific theories are most striking. In science, specific theories need to be formulated precisely so that they can be subjected to the appropriate empirical tests. Specific everyday theories, on the other hand, do not require such precision. Specific theories, as well as framework theories, are subject to change and revision. Specific everyday and specific scientific theories often change as a result of new observations. These changes, however, do not affect the existing framework theory. When framework theories do change it is typically a product of internal inconsistencies or changes in the scope of the framework theory.

Framework theories house a multitude of conceptual structures in addition to specific theories. Isolated hypotheses and theory fragments are a few examples of the structures housed within scientific framework theories, and scripts, schemas, mental models, associative nets and practical lore are among the variety of conceptual structures that our everyday framework theories encompass (Wellman, 1990). These more specific structures depend to varying degrees on framework theories for their sensibility. For example, explaining why paying for our food is part of the restaurant script involves an explanation that draws on our framework theory of economics, and explaining the courteous behavior of the waiter or waitress would draw on our everyday psychology (see Carey, 1985). It may be the case that certain conceptual structures, e.g., scripts and schemas, are more likely to be informed by multiple framework theories whereas others, e.g., specific theories, may be more closely tied to a single framework theory.

In the present chapter, we will focus on how the child's understanding of the human mind, and in particular, how his or her understanding of specific cognitive processes, changes over time. Following Wellman (1990), we argue that these specific acquisitions are housed under and constrained by the child's framework theory of mind. At the heart of the older preschooler's and adult's theory of mind is a shared commitment to what Wellman has termed a belief-desire psychology. Both the child and the adult predict human behavior on the basis of an actor's needs and desires and the actor's beliefs about how these might be satisfied (cf. Forguson & Gopnik, 1988). Wellman believes that even three-year-olds understand that mental representations or beliefs guide human action although they fail to comprehend the interpretative quality of these representations. Others (e.g., Flavell, 1988; Gopnik & Astington, 1988; Leslie, 1988; Wimmer & Perner, 1983) are not willing to credit children younger than four or four and a half with a representational theory of mind. Both camps, however, believe that only older preschoolers possess an active constructivist view of the mind.

Young preschoolers conceive of the mind as a storehouse of beliefs, images and ideas (Wellman, 1990). They do not focus on the process of knowledge acquisition, but instead focus on the endproducts of processing. Following Chandler and Boyes (1982), a number of investigators (e.g., Flavell, 1988; Pillow, 1988; Taylor, 1988; Wellman, 1990) have proposed that the young child holds a copy theory of knowledge in which the mind passively receives its beliefs from information that is copied or recorded by the senses. This copy theory is not long lasting, however, and soon gives way to an active constructivist view of knowledge acquisition. According to this more mature theory, the mind actively interprets the perceptual input and uses this information to construct beliefs and ideas which may or may not mirror reality. As the child moves toward an active constructivist view of the mind, he or she begins to focus more on the cognitive processes involved in knowledge acquisition. We believe that this focus on process makes it possible for the child to consider how situational factors might affect processing. In the present chapter, we will examine whether preschoolers and elementary school children recognize that situational factors can differentially affect an assortment of perceptual and cognitive processes. Evidence that children can distinguish the effects of stimulus variables would not only show that they are capable of focusing on process but might also suggest that they hold specific theories that are unique to different cognitive acts.

B. Theories and Conceptual Development

Framework theories constrain, or at least shape, how we acquire and reason about the concepts within each theoretical domain. Wellman (1990) has argued that preschoolers' reasoning about mental phenomenon, as well as their predictions of human behavior, are guided by their framework theory of mind. Two separate but related topics are encompassed by this theory. The first, which is not the focus of this chapter, centers around the child's understanding of intentional behavior. Wellman proposes that two-year-olds understand that people act to fulfill their desires but that at this age the child fails to appreciate that these actions are mediated or "framed" by the individual's beliefs about the world. The acquisition of this insight marks the three-year-old's commitment to a belief-desire framework theory. This framework theory undergoes revision during the preschool period as the child comes to realize that beliefs are not simply copies of the external world but instead are mental representations that are constructed on the basis of the available information which may or may not be sufficient for achieving an accurate representation. A second area subsumed by this framework theory concerns the child's understanding of the nature of mental representations and the processes that construct these representations and our analysis will focus on this topic.

Wellman's (1990) claim that three-year-olds conceive of the mind as a container for mental entities presupposes that children are capable of making some sort of principled distinction between mental and physical entities and, in fact, there is evidence that they can do so. By three years of age, children willingly acknowledge that they can form a mental image of an object and, more importantly, demonstrate some understanding of the special status of these representational entities. When asked to explain why they cannot touch their image of an object or why someone else is not able to see it, three-year-olds typically refer to the image's mental status, for example, by describing it as a thought or in their imagination or alternatively justify their answer by claiming that the image is not real (Wellman & Estes, 1986). In contrast, when asked to explain why they cannot touch an absent object, children's explanations center on the fact that the object is no longer present. An even more compelling finding is that children recognize that mental images are intangible in a different way than physical phenomenon such as shadows or smoke. When asked to explain their inability to touch these later items, children frequently allude to physical constraints whereas mental explanations dominate their discussions of images (Estes, Wellman & Woolley, 1989). Young preschoolers also realize that mental images can be manipulated through mental effort and concentration but that some form of physical contact is needed to transform physical objects. These findings suggest that three-year-olds have at least the beginnings of a representational theory of mind (Wellman, 1990). They recognize that mental entities, such as images and thoughts, are distinct from the things that they represent and that these mental representations have unique properties.

The work done by Wellman and his colleagues makes a compelling case that preschoolers recognize that mental phenomenon constitute a special class of intangible entities that fall within the domain of a framework theory of mind. Presumably, children draw on this theoretical framework to guide their reasoning about mental entities and intentional behavior. We may gain some insight into how the child's theory of mind guides his or her reasoning about mental phenomenon by considering how theories influence conceptual development in other domains. It is now fairly clear that which properties are considered most central to a concept as well as the types of inferences one draws on the basis of category membership depend in part on the encompassing framework theory. This has been shown most clearly in studies that have examined how children generalize properties when category membership is pitted against perceptual similarity (e.g., Carey, 1985; S. Gelman & Markman, 1986). For example, in one study (S. Gelman & Markman, 1986), children were taught that a particular bird feeds its young mashed up food. When later asked to predict how other animals nourish their young, preschoolers credited a very different looking bird with the same feeding habit, but did not predict that a bat that appeared very similar to the first bird would nourish its young this way. Children's willingness to extend biologically important attributes (e.g., feeding habit) across category members, coupled with their tendency to deny that other attributes (e.g., weight) generalize across category members, shows that even preschoolers make theoretically motivated distinctions among properties.

Studies which have looked at how children classify novel objects have revealed a complementary pattern of results (e.g. Keil, 1986, 1989; Mervis, 1987). In Keil's experiments, children were presented with stories about objects that undergo transformations dramatically altering their appearance. For example, in one story, a porcupine was made to look like a cactus. When asked to identify the transformed object, even kindergartners judged that the object was still a porcupine even though it shared more properties with a cactus. These results show that children draw on their framework theory of biology, and more specifically, their belief that natural kinds possess essential properties that cannot be altered by simple transformations, when making difficult decisions about potential category members.

More recently, R. Gelman (1990) and Brown (1990) have posited that one reason knowledge is acquired so rapidly by toddlers and young children is that they are sensitive to the different types of "skeletal causal principles" that structure specific knowledge domains. These principles are thought to constrain how the child filters information from the environment. One example of this is that when learning to pull an object across a table using a stick, children tend to focus on the properties of the stick that have the most functional significance (e.g., its rigidity and length). This tendency to focus on causally relevant properties makes it possible for the child to predict whether other implements can also be used to accomplish this task (Brown, 1990). In much the same way, skeletal properties having to do with locomotion are thought to guide the child in distinguishing between animate and inanimate objects by directing the child's attention to whether or not the object is capable of self-generated movement (R. Gelman, 1990; Massey & R. Gelman, 1988). More generally, it would seem that the child's sensitivity to the skeletal causal principles that structure specific knowledge domains enables the child to place the phenomenon within the appropriate framework theory and to use this theoretical framework to guide his or her thinking. Children's ability to distinguish between animate and inanimate objects and their understanding that desires only guide the behavior of animate beings[1] together with their understanding that mental phenomenon are not subject to the same constraints as physical objects shows that they possess the requisite knowledge to place these phenomenon within their theory of mind. They are then free to draw on this framework as they draw inferences or reason more generally

[1]This is not to deny that at times children borrow from this desire framework to explain the behavior of inanimate objects (cf. Inagaki & Hatano, 1987)

about mental events and intentional behavior.

III. THE MENTAL LEXICON

A. Early References to the Mental World

Part of what holds a theory together is a sense of agreement about what it refers to, and consequently, one function of a theory is to define the ontology of a domain (Carey, 1985). At least by the age of three, children seem to have some grasp of the ontological distinction between mental and physical entities. This suggests that they have separated out certain phenomena as falling within the scope of a theory of mind. What other evidence is there that young preschoolers possess a framework theory of mind? More specifically, is there any evidence that children are aware of the processes that construct these representations? Most children begin to speak about mental states before their third birthday and mental state utterances increase dramatically during the next year (Bretherton & Beeghley, 1982). The child's growing proficiency with the mental lexicon might be taken as evidence that young children are capable of distinguishing between mental processes and external acts or events (but see Piaget, 1929).

There are many problems with using natural language as a measure of the child's growing awareness of the mental world. One problem is that mental terms often serve conversational or "pragmatic" functions. For example, phrases like "I think" are used to soften indirect commands while expressions like "know what" may be used to get the listener's attention. A second, and more challenging, problem is that many mental verbs are correlated with specific behaviors or outcomes in the physical world. For example, "knowing" and "remembering" are almost invariably associated with successfully performing an act, and early studies supported Piaget's claim that preschoolers are unable to distinguish between these mental states and their behavioral correlates (Misciones, Marvin, O'Brien, & Greenberg, 1978; Wellman & Johnson, 1979). But as Johnson and Wellman (1980) note, in these studies the discrepancy between the mental and physical states may not have been sufficiently salient, and as a result, children may have relied more heavily on the readily observable behavioral evidence than they would in other circumstances.

To test this claim, Johnson and Wellman highlighted the discrepancy between the child's belief and the actual event by including a trick condition in their experiment. Children were instructed to watch as an item was hidden in one of two boxes. Then, after a very brief delay, the child was asked to retrieve the object. On trick trials, the boxes were surreptitiously altered so that the item was not found in the box in which it was hidden; instead it was discovered in the other box. As might be expected, this proved to be a very captivating situation for the child: He or she *knew* where the item should be but performed incorrectly nonetheless. If children simply equate mental verbs with particular outcomes in the physical world, or more specifically, if children believe that verbs such as "know" and "remember" mean that an individual has successfully accomplished something, then in these circumstances children should not claim to know or remember where the item was hidden. But, in fact, they did. Children in the trick condition protested that they knew or remembered where the item was but that the experimenter had tricked them. In short, children's performance under these circumstances demonstrates that they do not simply equate mental terms with behavioral outcomes. Instead, at least by the age of four, children realize that these terms have a distinctly mental referent.

Evidence that even younger children also appreciate this distinction comes from a study by Shatz, Wellman, and Silber (1983) which uses two-year-olds' spontaneous speech to assess their understanding of the mental world. Because the simple occurrence of a mental verb is not sufficient for establishing its reference to a mental state, Shatz et al. used the sentences preceding and following the target utterance as well as the general context of the interaction to guide their interpretation of these expressions. Only utterances which unambiguously referred to the thoughts, memories or knowledge of the speaker, listener or a third person were classified as mental state utterances. This analysis revealed that a sizeable number of children begin to use mental verbs to express mental states in the later part of their third year. What is perhaps an even more compelling finding is that approximately one-fifth of these mental state expressions involved explicit contrasts between reality and mentality, action and intention, or fact and belief. As an example, consider the following comment made by one of their three-year-olds: "Before I *thought* this was a crocodile; now I *know* it's an alligator." Statements such as this provide convincing evidence that the child is not simply equating mental verbs with behavioral events.

B. Mental Verbs

By the end of the fourth year, over 40% of the child's utterances that contain a mental verb are classified as referring to a mental state. But does children's use of these terms indicate that they are capable of distinguishing among cognitive processes? Johnson and Wellman (1980) tackled this question by examining whether or not preschoolers and young elementary school children differentiate the cognitive acts associated with remembering, knowing, and guessing. These acts can be distinguished along a number of dimensions. For example, both knowing and remembering can be used to indicate the presence of knowledge, but guessing cannot. Knowing also contrasts with remembering, however, in that it encompasses a wider range of phenomena. Knowing can be appropriately applied to inferences, deductions, generalizations, and present apprehensions, as well as acts of remembering.

Johnson and Wellman carefully devised a set of situations that would make it possible to evaluate whether children understand the distinctions between knowing, remembering and guessing. Their results revealed an interesting developmental pattern. In Johnson and Wellman's study, preschoolers were willing to say they knew and remembered where an object was hidden even when they had simply guessed correctly. And when the object was "hidden" under a transparent cover, these same four-year-olds claimed to remember where it was although they had not been provided with prior information about its location. Under these conditions, they also reported that they guessed where the object was hidden even though it was sitting in plain sight. So although four-year-olds were able to distinguish between mental and physical states, at least in "trick" situations in which their personal expectations were violated, they did not differentiate among knowing, remembering and guessing. The data from the kindergartners suggests that children begin to make some progress in this area during the next year. Kindergartners seemed to be starting to differentiate among these terms although they clearly lacked a full understanding of these distinctions.

First graders appeared to have a somewhat better grasp of these differences, but it was not until third grade that children had clearly worked out these distinctions. Third graders only claimed to guess in situations where they lacked the relevant information and only reported remembering when they had prior information about the object's location. These children also appreciated the distinction between knowing and remembering. They said they knew where the object was when it was clearly visible and when they could easily infer its location because the alternative location was an empty transparent container. Moreover, these third graders, in contrast with younger children, denied remembering in these situations. These findings suggest that third graders distinguish between cognitive processes that recruit prior knowledge and processes that depend primarily on the currently available information. It is not clear if these children also distinguish between processes that depend on simply reporting prestored information and those that depend on actively manipulating this information to

asked of the child with reference to him- or herself. The four-year-olds in this study correctly reported whether they had seen what was in the box, and if they had looked inside the box, they said they knew its contents. Children also recognized when their partner had looked inside the box but were not willing to say that their partner knew what was in the box. For these children, then, seeing was not sufficient for attributing knowledge to someone else. This might suggest a growing awareness that although beliefs are informed by perception, perception is not always sufficient for knowing. There are, however, a number of reasons for doubting this. First, the objects used in this study were fairly unambiguous (e.g., a coin, a piece of chocolate) and so there is no reason to think that additional information would be needed for identification. Second, the competitive nature of the task may have made some children more inclined to deny knowledge to their partners. And finally, the linguistic complexity of the question may have posed some difficulty (see Pratt & Bryant, 1990), although it is worth noting that the questions about looking and knowing were presented in the same form. At this point, the general conclusion we draw from these studies is that four-year-olds typically appreciate that seeing leads to knowing. The difficulties children experienced in Wimmer et al.'s task are most likely due to specific aspects of the task situation; however, their findings do suggest that children sometimes respond differently to questions about seeing and knowing. This would suggest that children do not simply equate these two activities.

The three studies cited above all involve situations in which the perceptual information is sufficient for knowing. But given our interest in whether children distinguish between perceptual and knowing acts, we may want to explore preschoolers' understanding of situations in which the perceptual information is not sufficient for knowing. In one of the early experiments on this topic, children were given an opportunity to watch (and listen) to a film and were then asked to predict what a viewer would know after watching a silent presentation of the same film. Mossler, Marvin and Greenberg (1976) found that up until about four years of age, children believe that even when their mother watches a film with the volume turned off, she will know about specific events in the film that depend on audio information. Children are thought to err in this situation because they tend to view adults as "all-knowing" figures. (Wimmer et al. (1988) report that a similar finding emerged in their pilot work. When children played the hiding game with a puppet that shared the name of their teacher, they always credited "Monika" with knowing the contents of the box even when she had not looked inside.) The fact that knowledge attribution depends on the age of the perceiver in a way that perceptual attribution does not suggests that children realize that there are differences between seeing and knowing.

Although children may be aware of some distinction between seeing and knowing, Mossler et al.'s work suggests they do not understand how the completeness of the perceptual information affects what can be known in a given situation. In Mossler et al.'s study, the perceptual input was incomplete because the observer was denied auditory input. Studies by Chandler and Helm (1984) and Taylor (1988), which varied the informativeness of the input presented to a single modality, have produced complementary findings. In these tasks, children are first shown a complete picture. For example, they might see a picture of a giraffe sitting beside an elephant. The picture is then covered so that only a small nondescript part is visible. Next the child is asked whether a naive observer would know what is in the picture after seeing only this very restricted view (e.g., only the tip of the elephant's trunk). Both Chandler and Helm (1984) and Taylor (1988) have shown that up until about six years of age, children credit a naive observer with knowing the identity of a pictured object even though the observer has only limited and ambiguous visual input. Once again, this finding would seem to fit with Chandler and Boyes' copy theory--anyone who sees an object should immediately know what it is. But, as Taylor has shown, even four-year-olds recognize that there are limits to what an observer would know based on a restricted view. Older preschoolers recognize that the observer would not know personal information about the depicted animal (e.g., its name or what it likes), and they do not expect the observer to know what the animal is doing (e.g., whether it is sitting or running) even though they themselves know what the animal is doing because they have had the opportunity to view the complete picture. So children are aware that additional information may be necessary for certain types of knowledge, but they believe that the identity of an object is so immediately apparent that it can be gleaned from viewing a small nondescript part.

It should be pointed out, however, that in the Taylor study, children were asked whether an observer would *know* a particular piece of information after viewing the picture. They were not asked to describe what the observer would *see*. If children were asked this question, they probably would not give an objective description of the physical evidence. Rather their descriptions might be expected to reflect what they know about the picture. For example, they might say the observer would see either the giraffe or the giraffe's back when looking at an unidentifiable part of this drawing. Such responses might indicate that the child has difficulty distinguishing between perceptual and knowledge-based representations.

Sodian and Wimmer's (1987) work on children's understanding of inference also suggests that four- and five-year-olds find it difficult to distinguish what a naive observer would conclude based on different pieces of evidence. In their study, an observer had to report the color of a ball that was removed from either a transparent container filled with balls that were all the same color or a transparent container filled with a mixture of different colored balls. Although the observer was always shown which container the ball was taken from, the transfer sometimes took place behind a screen. Four- and five-year-olds were only willing to say that an observer knew the color of the target ball if the observer watched the transfer. They did not recognize that even when the transfer could not be seen, the observer would always know the color of the ball if it was taken from the container filled with identical balls. In contrast, the six-year-olds realized that in this situation, knowledge does not depend on perceiving the entire event but instead can be based on logical inference. This might suggest that children under six do not appreciate that the mind can actively manipulate prior perceptual information to construct new knowledge.

The available evidence, then, suggests that older preschoolers are able to make rudimentary distinctions among perceptual acts and that they do not simply conflate seeing with knowing. But these initial distinctions all seem to involve a type of level 1 knowledge, that is, whether or not a particular object or event can be perceived or known under various conditions. What the preschooler seems to lack is an appreciation of the constructive processes associated with seeing and knowing. Before the child can appreciate that the informativeness of the visual evidence varies as a function of prior knowledge or that people can sometimes infer information about events that they do not directly witness, he or she needs to understand that knowledge is not something that is given out to all who witness an event but instead must be shaped from the available evidence. As these studies make clear, even five- and six-year-olds do not fully understand how the nature of the perceptual input affects this process.

V. REASONING ABOUT COGNITIVE PROCESSES

Three-year-olds conceive of the mind as a repository for mental entities (Wellman, 1990). They recognize that some of these representations correspond to actual objects and events in the world, but, even more importantly, they distinguish between these mental representations and their referents. Three-year-olds also understand that mental verbs, such as knowing and remembering, refer to actual mental states and not the behavioral outcomes they are typically associated with. But despite these impressive accomplishments, there are clearly limits to what the young preschooler knows. In a broad sense, what the young child fails to appreciate is that perceptual and cognitive processes actively manipulate information to construct knowledge. Perhaps this limitation is a function of the young child's tendency to focus on the representations themselves and whether they are "reality-oriented" or "pretend" rather then the processes that produce these representations. Once children come to the important realization that knowledge is not simply given in the world but must be actively created by the mind, they are in a position to consider how various perceptual and cognitive processes contribute to this endeavor.

When adults consider how the mind works, they are likely to think about the different processes that manipulate information to construct knowledge. Within this conception, perceptual processes are distinguished from one another and may also be distinguished from more derived higher-level processing. Adults may even organize their understanding of the cognitive system around different types of processes (Fabricius, Schwanenflugel, Kyllonen, Barclay & Denton, 1989), for example, by distinguishing between classes of activities involving recognition memory and those involving problem solving skills. In the remainder of this paper, we will explore how children's metacognitive knowledge is structured and whether they too distinguish among classes of mental activities. Although young children are able to distinguish among perceptual processes and recognize that there is an important distinction between perceiving and knowing, they still may fail to distinguish among higher-level cognitive processes. Studies on children's use of mental verbs provide some support for this view. Recall that even young elementary school children have difficulty distinguishing between situations that involve remembering prior information and those in which prior knowledge does not play any role. In our analysis, we will focus on whether research on children's understanding of variables that can affect cognitive performance supports this claim that young children are unable to distinguish among cognitive processes.

A. Evaluating Stimulus and Task Variables

Task and situational variables often affect the efficiency of cognitive performance. For example, studying for an exam might be difficult because there is a great deal to memorize or because there is a lot of noise in the dorm. In this section, we will review what preschoolers and elementary school children know about factors that can limit cognitive performance and the extent to which this work may provide us with some insight into how the child's metacognitive knowledge is structured. In this review, we will explore what children know about global and specific variables. Global variables are those factors that tend to have a similar effect on a variety of cognitive processes whereas specific variables are factors that limit performance in one task but not others. If children do not distinguish among cognitive processes, we might expect that once a child recognizes that a global variable affects performance, he or she will consider it to have a similar effect on all cognitive activities. But even if children prove to have a solid understanding of global variables, it does not mean that they actually consider the processing requirements of particular tasks when evaluating their effects because children do not need to distinguish among cognitive processes to make these predictions. In the latter part of this section, we will explore what children know about factors that have task-specific effects because these judgments require the child to discriminate among the processing requirements of different tasks.

There is now a good deal of evidence that preschoolers and young elementary school children are aware of at least three global variables that can affect performance. For example, four- and five-year-olds understand that background noise can make learning (Miller, 1982) and remembering more difficult (Wellman, 1977; Yussen & Bird, 1979), and that it can hinder their ability to pay attention and communicate (Yussen & Bird, 1979). Young children also understand that the number of items can affect performance. Preschoolers recognize that a longer list is harder to remember (Wellman, 1977) and realize that it is harder to tell a friend about a day filled with many activities (Yussen & Bird, 1979). These children also understand that performance can be hampered by increasing the number of distractors in a visual search task (Miller, 1985) or the number of targets in a monitoring task (Yussen & Bird, 1979). A third factor that children recognize affects performance is motivation. Preschoolers realize that a motivated child is likely to learn more and to remember more items (Miller, 1982; Wellman, Collins & Glieberman, 1981), and kindergartners believe that someone who tries hard will be able to remember even long lists of items even though they recognize that list length may still have some affect on performance (Wellman et al., 1981).

There is also evidence suggesting that once children recognize that a global variable limits performance in one situation, they also appreciate its relevance in other situations. Yussen and Bird (1979) asked children between the ages of four and six how global variables would affect three different cognitive processes: memory, communication, and attention. As expected, even the four-year-olds were aware of the negative effects of list length and distracting noise.

The six-year-olds also recognized that adults usually perform better than children and that severe time constraints can hinder performance. What is especially important about Yussen and Bird's findings, however, is that children's judgments did not vary across these activities. If a child judged that a variable affects one cognitive task, he or she considered it to affect all three. As Yussen and Bird note, this pattern of results suggests "...that children's understanding of cognitive processes may evolve in a system-like or 'synchronous' fashion" (p. 311).

Whereas global variables have similar effects on a variety of cognitive processes, other factors differentially affect performance. If children are able to discriminate among variables that differentially affect cognitive activities, then it would seem appropriate to credit them with at least a rudimentary ability to reflect on the processing requirements associated with different tasks. More specifically, if, for example, children realize that more familiar names are easier to recall but harder to recognize, we would feel fairly confident that they are able to reflect on the different processing requirements of recall and recognition tasks. Before reviewing children's knowledge of specific variables, it may be helpful to point out that all the studies we will be reporting have examined children's understanding of these more specific variables in the context of a single cognitive act. Consequently, it is not clear whether children recognize that these variables are only relevant for certain kinds of activities.

Wellman (1977) has examined young children's understanding of a wide assortment of different memory variables and found that five-year-olds are aware of some specific strategies that can aid recall. These children recognized that it would be beneficial to draw pictures of the items in a memory list and to look at these pictures when asked to recall the items. Some five-year-olds also realized that associated cues could facilitate retrieval, for example, that looking at a leaf could make it easier to remember a picture of a tree (see Gordon and Flavell, 1977; Ritter, 1978; for other examples of what preschoolers know about retrieval cues).

Knowledge of other task-specific variables seems to emerge later. For example, children under nine or ten generally do not realize that a categorized list of items tends to be easier to recall (Moynahan, 1973). Similarly, whereas nine- and eleven-year-olds clearly recognize that pairs of opposites are easier to memorize than random word pairs, six- and seven-year-olds do not (Kreutzer, Leonard & Flavell, 1975). Younger children are not totally unaware of the benefits of organization, however. Moynahan found that seven-year-olds realized that it would be easier to recall a row of colored blocks if the blocks were grouped by color. Young elementary school children are also sensitive to another form of organization: Kreutzer et al. found that 75% of their first graders and all of their third graders thought it would be easier to remember a collection of pictures if they were connected together through a story (Kreutzer et al., 1975).

During the elementary school years, children are becoming aware of a variety of specific variables that influence memory performance. Their facility in making these judgments suggests they have constructed a more specific memory model in which the specific processing requirements associated with different task conditions can be evaluated. Children are also refining their understanding of some of the factors that affect attentional performance. Older children consider both the attentional demands of the task and the nature of the distracting stimuli when making predictions about performance, and they are more adept at judging how stimulus factors such as the separation of targets or the similarity of distractors will affect visual search (Miller, 1985).

B. Classifying Cognitive Activities

Although elementary school children sometimes look very adult-like in their assessment of situational factors that are likely to affect performance, it is not clear that the cognitive requirements of specific activities are especially salient to them. In a recent study (Fabricius et al., 1989), both adults and elementary school children were asked to rate a variety of tasks, such as trying to find the North Star or learning a board game, according to how similar they were in terms of the mental activities each required. Whereas adults, and to some extent ten-year-olds, tended to use the amount of memory each task required as a way of organizing these mental activities, there was no evidence that eight-year-olds distinguished these tasks in terms of the amount of memory each required. A more detailed analysis of the adults' data revealed that four categories of mental activities were distinguished: memory, comprehension, attention, and inference. The memory tasks were further divided into activities that involved list memory and those that involved remembering to do something in the future (i.e., prospective memory). It would seem, then, that the adults relied heavily on the cognitive requirements associated with each task in making their judgments. One implication of these findings is that adults may organize their understanding of mental activities in a way that would make it easy to associate cognitive variables with classes of activities.

Ten-year-olds were also sensitive to the cognitive requirements of these activities, but their judgments revealed that the context in which each task occurred was also considered important. For example, whereas adults subdivided activities according to whether they were primarily memory or comprehension tasks, ten-year-olds tended to organize these same activities according to whether they involved games or learning situations and whether the remembering party's primary responsibility was to adults or friends. Eight-year-olds also tended to rely on contextual features when distinguishing among mental activities. For example, they considered returning a permission slip for a field trip and understanding directions to a friend's house to be highly similar activities, presumably because both are connected with taking a trip. Neither age group evidenced a concept of attention. Instead, these children, especially the younger ones, tended to subdivide activities according to whether they were visual or verbal tasks. This suggests that the perceptual components of these acts were more salient than the more central cognitive components. Again, these findings are important for the present discussion because they suggest that children are capable of distinguishing among mental activities, and that they do so on the basis of whether the activity involves primarily memory or perceptual processing. The distinctions children make, however, are not identical to those made by adults. Children are less likely to recognize a distinct class of attention activities and do not readily distinguish between memory and comprehension tasks. This might suggest that children would have some difficulty sorting cognitive variables that are differentially associated with these classes of cognitive activities. We are currently exploring this topic in our laboratory.

VI. THE INFLUENCE OF TASK-SPECIFIC VARIABLES

In this section, we will explore in more detail children's understanding of specific variables that differentially affect cognitive performance. Our hope is that research on children's understanding of task-specific variables together with studies of children's concepts of mental activities may provide us with some insight into the structure and organization of children's metacognitive knowledge. At this point, we still have only a limited understanding of the distinctions children make among perceptual and cognitive processes. We know that they do seem to make a fairly clear distinction between perceptual and more conceptual processes. They also distinguish among individual perceptual processes, although, at least in the early preschool years, they find it difficult to consider multiple perceptual representations of the same object.

As for how their knowledge of cognitive processes is structured, Fabricius et al.'s work suggests that eight-year-olds do not organize mental activities in the same way as adults. For example, they attach less weight to the memory requirements associated with specific tasks and do not distinguish between prospective and short-term memory tasks. Instead, they tend to place more weight on the context in which these activities occur. These findings might suggest that young elementary school children would find it difficult to judge how task-specific variables affect different classes of mental activities because these decisions are likely to require the child to focus on both the perceptual and central processing requirements of the tasks rather than the broader context in which they occur. Once the child understands the processing requirements of the task, he or she would then have to evaluate how variations in stimulus conditions or task demands would facilitate or impede processing. In making these judgments, it is likely that the child would not only have to consider in some detail the processes that mediate performance, but he or she might also have to think about the mental representations that guide processing.

A. Experiment 1

In our first experiment, we examined children's understanding of two specific variables, color and size, that would have different effects on the memory and monitoring tasks we chose to focus on. Color is an interesting variable to consider because although it is always present, it does not always have the same effect on performance. For example, typing all words beginning with the letter "a" in red may make them easier to find in the text, but the same red color is likely to be distracting to someone trying to read the paper. An interesting developmental question, then, concerns whether children initially overgeneralize their knowledge of specific variables, for example, by believing that color always improves performance. Evidence that they do so might suggest that they are unable to evaluate the importance of these factors in the context of a specific task.

The memory and monitoring tasks we chose were introduced as jobs at a zoo. In the memory task, the zookeeper was presented with a cage filled with tiny chicken houses and he had to remember which house contained a sick chicken. When the target house was a distinct color, this memory task would be easier. In this first experiment, children judged how changes in the color of the houses and variations in the size of the cage would affect the relative ease of the memory task. As the size of the cage increased, the configuration of houses remained unchanged although the distance between the houses increased. Thus

Figure 8.1. A diagram of one of the large cages used in the monitoring task. The figure positioned in the foreground is the zookeeper. The goats in the monitoring task and the houses in the memory task were arranged in this same configuration in all the cages.

there was no reason to think that the demands of the memory task would change as a result of these size variations. When undergraduates were asked to predict how these color and size variations would affect the difficulty of the memory task, they judged that the task would be easier if the houses were different colors but that changes in the size of the cage would not have any effect on the relative difficulty of the task. (The predicted effects of these stimulus changes are summarized in Table 8.1 and an example of one of the cages used in the first experiment is presented in Figure 8.1).

Color and size variations were expected to have a different effect on another one of the zookeeper's jobs. Children were told that the zookeeper also had to watch a small herd of goats and were asked to judge how changes in the color and size of the cage would affect this task. As the size of the cage in-

Table 8.1. Predicted Effects of Color and Size Variations on the Memory and Monitoring Tasks.

	Color		Size	
Task	More Colorful	Less Colorful	Smaller Smaller	Larger Larger
Memory	Easy	Hard	No Effect	No Effect
Monitoring	No Effect	No Effect	Easy	Hard

creased, the goats were farther apart, and adults recognized that this would make it more difficult to monitor the animals. Adults also agreed that variations in the color of the individual goats would not affect this task. To summarize, subjects concurred with our judgment that these color and size variations would have different effects on the memory and monitoring tasks. The memory task would be easier when the houses were different colors, and variations in the size of the cage were not expected to affect memory performance. In contrast, the monitoring task would be easier when the goats were confined to a small cage so the observer would not have to continually scan back and forth across the cage. In this situation, variations in the color of the goats' fur were not expected to affect task difficulty.

Eighteen first graders (mean age: 6-11) judged how color and size variations would affect these two tasks. Model scenes were used to illustrate these variations. The focus of each scene was a rectangular piece of carpet that served as the base of the cage. A zookeeper doll was centered at the front edge of the cage with his back toward the child so that he appeared to be looking inside. The cages for the memory task contained six houses which were positioned so that each one could be seen in its entirety. The same configuration of stimuli was used in all cages. The gray house in which the sick chicken lived was always the third house from the left. Three red arrows were pasted on a small piece of styrofoam positioned above the target house so that the child would know where the sick animal was in each display. For the monitoring condition, six goats were placed inside each cage. The goats were configured like the houses so that each animal afforded an unobstructed view. These scenes were photographed and each scene was presented to the child on a color slide. Each child was asked to make judgments in both the memory and monitoring conditions. The order of the conditions was counterbalanced across subjects. At the start of each condition, the experimenter explained the zookeeper's job and illustrated the possible variations in the size and color composition of the cages. Nine stimulus cages were created by varying the size of the cage and the color composition of the items. There were three levels of the color variable: the animals or houses were all gray, three were gray and three were brown, or each was a different color. Three size variations were possible: the cages were small, medium, or large. Children were then shown pairs of cages and instructed "to point to the one that would make the zookeeper's job easier." Children were also permitted to say that the zookeeper's job would be equally easy in both cages. For each condition, children were shown each of the nine cages paired with the other eight resulting in a total of 36 trials per condition. The left-right position of the cages varied randomly across subjects.

For each of the 36 trials, the child judged if one cage was easier than the other. Each child's preferences were coded as a 72-dimensional vector. To do this, for each trial, we simply designated one stimulus cage *a* and the other *b*. Deciding that *a* is easier than *b* is equivalent to selecting the pair (a,b) and rejecting the pair (b,a). Thus each trial was coded as the child's decision about two pairs. With this convention in mind, the answers of a single child in one condition can be expressed as the pairs which were selected. If the 72 possible pairs are arbitrarily ordered, then the pattern of answers of a child in one condition can be coded as a series of binary values (0,1), where 1 indicates that the pair has been selected, and 0 indicates that the pair has been rejected. If the child declares that the two stimulus cages are equivalent, each pair can be given the value .5.

Correspondence analysis was used to describe children's preferences in the memory and monitoring conditions. A brief digression about correspondence analysis may be useful as the method is not yet well known in the American literature. A more complete treatment can be found in Benzecri (1973, for readers fluent in French) or Greenacre (1984); additional examples can be found in Abdi (1988). Correspondence analysis is used to describe multivariate qualitative data. The data tables are arrays of non-negative elements whose rows represent the characteristics describing the observations. In our analysis, each row represents the answer of a given child in a given condition, and each column represents a pair of stimuli. The aim of the analysis is to display the observations and the characteristics as points on a map. As such, correspondence analysis can be seen as a variety of multidimensional scaling, or as equivalent to principal component analysis for qualitative data. Points that describe similar rows (or columns) are positioned close to each other. Dimensions can be interpreted as in (classical) multidimensional scaling and principal component analysis. The origin of the dimensions is the centroid (i.e., the average vector). In principal component analysis, each axis explains a proportion of the total variance of the data table. In correspondence analysis, each axis explains a proportion of the total chi-square (or inertia) of the data table. An interesting property of correspondence analysis is that it allows the simultaneous display of rows and columns in the same common space (the inertia and the factors extracted are the same for both the set of rows and columns, and each display allows for the reconstruction of the other). The relative clustering of the column vectors provides a measure of the coherence of children's preferences, and an examination of these points revealed that children's judgments were systematically related to the color and size differences in the displays. Although both the row and column vectors were used to define the space and both could be included in the same display, we have chosen to include only the row vectors in our graphs to facilitate the interpretation of these displays.

One advantage of correspondence analysis is that it enables the researcher to compare the actual data with ideal or theoretical patterns of performance. This advantage is especially important in developmental studies where it is likely that no single child will perform in perfect accord with the ideal. In correspondence analysis, these ideal or theoretical performance patterns can be coded as supplementary rows or columns in the data matrix. These points are not used to compute the analysis. They are projected onto the axes after the solution has been computed for the active rows and columns. The use of supplementary points in correspondence analysis is similar to the use of computing the correlation between an external variable and an axis in multidimensional scaling (i.e., each serves as an aid for interpreting the analysis). To give a more intuitive explanation, supplementary points are located in the places these points would have been if they had been in the data set, but they have no influence on how the analysis is computed. Once these points are projected onto the space defined by the analysis, the spread of data points around these supplementary points can be evaluated.

In the present study, supplementary points were defined to represent "ideal patterns" of responding. For example, one point was coded to represent an "ideal" subject who always picked the most colorful member of each pair and considered cages with the same color composition to be equally difficult. (This is the pattern of responses we would consider correct in the memory condition.) Children's responses were evaluated by examining the extent to which their responses clustered around various "ideal patterns" that were represented by supplementary points. It is important to stress that these supplementary points are not used in computing the correspondence analysis, and accordingly, do not play any role in defining the principal axes.

The results of the correspondence analysis for the first experiment are presented in Figures 8.2 and 8.3. Each point represents a single child's pattern of preferences in either the memory or monitoring condition. The data are presented in the plane defined by the first two dimensions of the analysis. The first principal axis accounted for 35.5% of the total inertia and the second axis accounted for an additional 16.2%. Thus the two-dimensional solution accounts

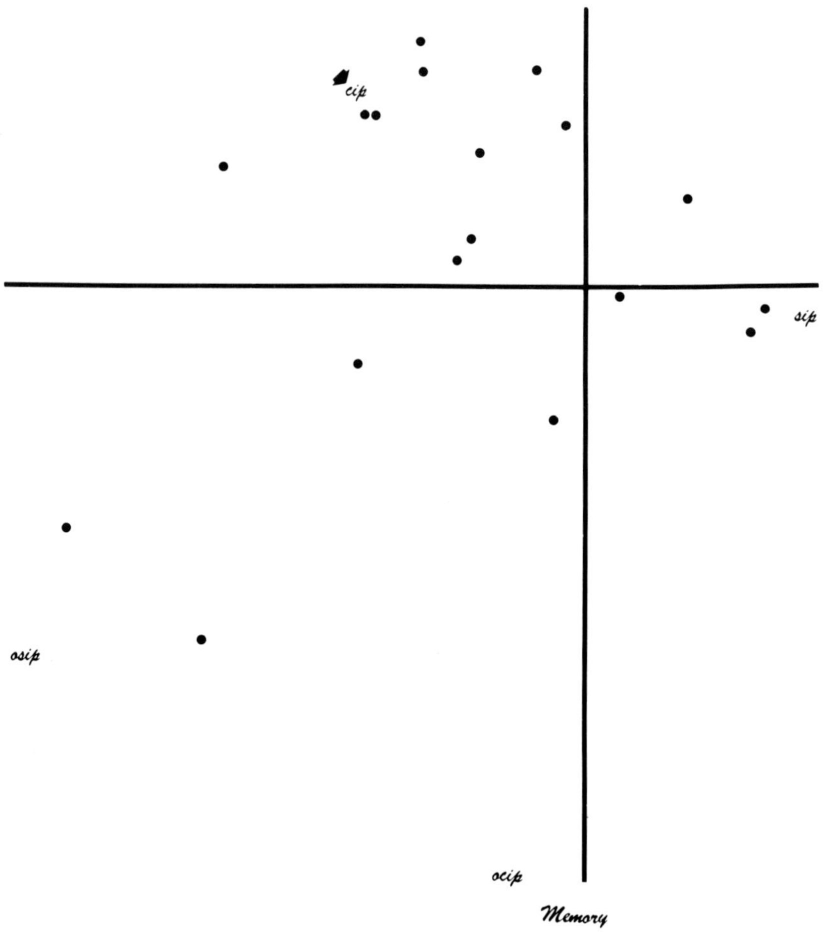

Figure 8.2. The memory data from the first experiment displayed in the plane defined by the first two dimensions of the analysis. The first dimension accounts for 35.5% of the inertia, and the second dimension accounts for 16.2% of the inertia. Children who consistently responded that color variations would facilitate performance would be characterized by the color ideal point, labeled *cip*. Ideal response patterns corresponding to the size ideal point (*sip*), opposite color ideal point (*ocip*), and opposite size ideal point (*osip*) are also included in this graph and subsequent graphs to help anchor the space.

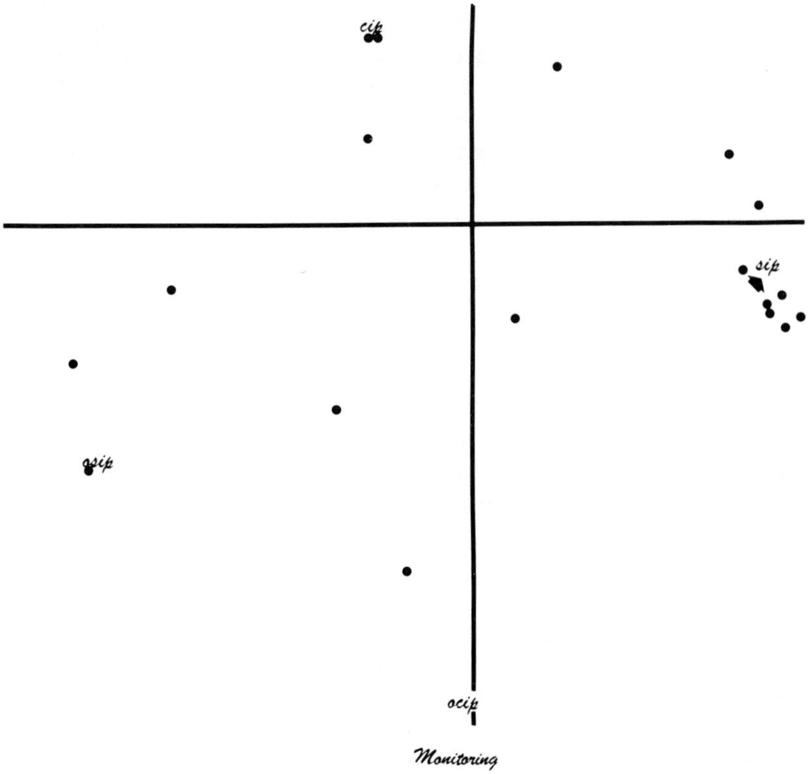

Figure 8.3. The monitoring data from the first experiment displayed in the same two-dimensional space. The first dimension accounts for 35.5% of the total inertia, and the second dimension for 16.2%. Perfect performance is marked by the size ideal point (*sip*).

for 52% of the total inertia (i.e., chi-square).[2] Four "ideal performance patterns" were coded as supplementary points and projected onto the display. Each

[2]It should be noted that because the analyses reported in this paper were performed on "doubled" matrices, the values reported in the text actually underestimate the amount of variability that is accounted for by each axis (see Greenacre, 1984, chapter 5).

performance pattern is "ideal" in the sense that it is perfectly consistent with a specifiable rule. In the color ideal pattern, all decisions were made on the basis of color: more colorful cages were considered easier, and whenever the color composition of both cages was the same, they were considered equivalent. Another ideal pattern was identified on the basis of an opposing rule: in the opposite color ideal pattern, all decisions were made on the basis of color but the less colorful items were preferred. Similarly, two additional ideal patterns were identified in which decisions were based completely on the size of the cage. In the size ideal pattern, the smaller cages were preferred, and in the opposite size ideal pattern, the larger ones were preferred. In both of these ideal patterns, two cages of the same size would be judged to be equally easy.

First consider the data from the memory condition which is presented in Figure 8.2. Slightly over half of the children clustered in the general vicinity of the color ideal point (labeled *CIP*) which suggests that most first graders realize that color can serve as a useful memory cue. A few children gave responses that were scattered around the size ideal point (labeled *SIP*) which means that they erroneously believed that the size manipulation was more important than the color manipulation in this task. But again, it is important to point out that most children recognized that the size manipulations would not affect the ease of the memory task.

Now consider children's judgments in the monitoring condition (see Figure 8.3). Twelve children were tightly clustered around the size ideal point (labeled *SIP*) which indicates that they correctly judged that it would be easier to watch the animals in the small cages. In fact, four children conformed perfectly to this ideal pattern. Moreover, none of the remaining children evidenced a strong preference for the larger cages and only one child considered the color variable relevant for this task. Thus, first graders generally understood that color variations are especially relevant for the memory task whereas size variations are important for the monitoring task.

The general results of this first experiment suggest that children clearly understood that it would be easier to watch the goats in the small cages and that color variations were not relevant for this task. Their judgments in the memory task, however, were less ideal. They seemed to consider both the relevant color and, to a lesser extent, the irrelevant size variations important for this task. The fact that children generally do not consider the size variations equally important for both tasks and their tendency to disregard the color variations when making judgments about the monitoring task, but not the memory task, suggests that first graders do distinguish between the processing demands of these two tasks. We decided to carry out a second experiment to see if a few minor changes in our stimuli might help children recognize that the size variations were not relevant for the memory task.

B. Experiment 2

In the second experiment, we decided to introduce two changes to see if we could improve performance in the memory condition. First, the houses were painted brighter colors so that the color differences would be more salient. A second change involved the cue that was used to remind the child which house contained the sick animal. Recall that in the first experiment a sign containing three arrows was suspended in the scene. Some children may have mistakenly thought this sign was present in the zoo even though they were explicitly told the contrary. To reduce the likelihood of this, a more artificial cue was used in the second experiment: a large red arrow was drawn directly on the slide. We hypothesized that these changes would produce even cleaner results. We expected to find that first graders' judgments would again be tightly clustered around the size ideal point in the monitoring task, but now we also expected to see a tight cluster of points around the color ideal point in the memory task.

In the first experiment, children performed extremely well in the monitoring condition. We found this somewhat surprising because in pilot work variants of this task often proved difficult. One factor which we thought may have facilitated performance in the first experiment was that the zookeeper was positioned in the scene. In the second experiment, we decided to remove the zookeeper from the display to see if this would have any effect on performance. We hypothesized that if children evaluate the effects of these stimulus variables by mentally simulating the zookeeper's job, their performance might vary as a function of whether or not he is physically present in the scene. We should point out that a family of dolls was positioned at the far edge of the cage in both the first and second experiment so that the child would have a visual guide to the scale of the scene even when the zookeeper was absent.

Eighteen first graders (mean age: 6-9) took part in the second experiment. The slides for the monitoring conditioning were exactly the same as those used in the first experiment except that the zookeeper was not photographed in the scene. The stimuli for the memory task also resembled what was used in the first experiment except that both the zookeeper and the sign were removed from the scene, and the houses were painted more distinct colors. After

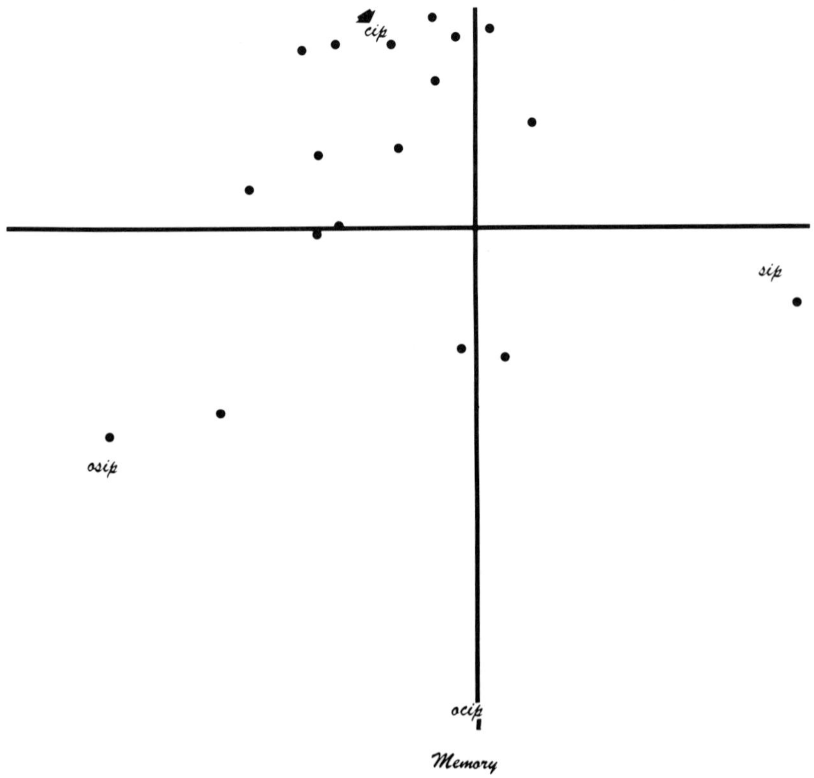

Figure 8.4. The memory data from Experiment 2 presented in the two-dimensional subspace defined by the analysis. The first dimension accounts for 36.6% of the inertia, and the second dimension accounts for an additional 16.6%. Points near the *cip* point represent children who realized that color variations would affect memory performance wheras changes in the size of the cage would not affect performance.

the houses were photographed, a large red arrow was drawn on each slide so that it pointed to the target house. The procedure was identical to that used in the first experiment.

A correspondence analysis was performed on the data from the second experiment. The first axis accounted for 36.7% of the total inertia and the second axis accounted for an additional 16.6%. Thus the 2-dimensional solution accounts for 53% of the total inertia. The results of the correspondence analysis are presented in Figures 8.4 and 8.5 along with supplementary points. The data from the memory condition, which are displayed in Figure 8.4, resembled the results of the first experiment: Children generally believed that the more colorful cages would facilitate memory, but their judgments were not tightly clustered around the color ideal point (labeled *CIP*). As can be seen in Figure 8.5, children's performance in the monitoring task also fell somewhat short of the ideal. Although half of the first graders were in the vicinity of the size ideal point (labeled *SIP*), the remainder were fairly spread out. It would seem, then, that first graders' performance is attenuated when the zookeeper is not present in the scene. The fact that this simple manipulation affects performance suggests that first grade may be a transitional period for this task. These children are able to judge how size variations will affect performance when the zookeeper is present in the scene to anchor the scale change but these judgments become more difficult when the zookeeper is not included in the scene. Along these lines, it is interesting to consider the relative positions of the outliers. In the first experiment, the zookeeper was depicted in the scene, and no child thought that it would be easier to watch the goats in the large cages. In the second experiment, however, a few children gave responses that were generally consistent with this opposite size ideal pattern (labeled *OSIP* in the figure). So when the zookeeper was not included in the scene, some children erroneously believed it would be easier to monitor the animals in the larger cages. It is possible that such judgments reflect an egocentric bias: Children may have based their judgments on how crowded the array appeared to them without considering the scale change.

The results of these two experiments suggest that first graders appreciate that remembering and monitoring are decidedly different cognitive activities. Children recognized that color variations would affect memory performance but that size variations would be unlikely to have much of an effect. In contrast, the size of the visual field was judged to be an important factor in the monitoring task, whereas color variations among the stimuli were considered to be of little importance. It seems that young school age children are able to evaluate the effects of stimulus factors in the context of particular cognitive acts. Even the way in which children arrived at their decisions suggests that they were not inclined to overgeneralize the effects of specific variables. In the memory task, many first graders appeared to understand immediately that it is easier to remember an item when it is a distinct color; they may even have had a general rule linking the distinctiveness of the target with improved memory performance. Although children's judgments diverged somewhat from the ideal, this may not be all that surprising given that in many pairs the target was not a distinct color. In pilot

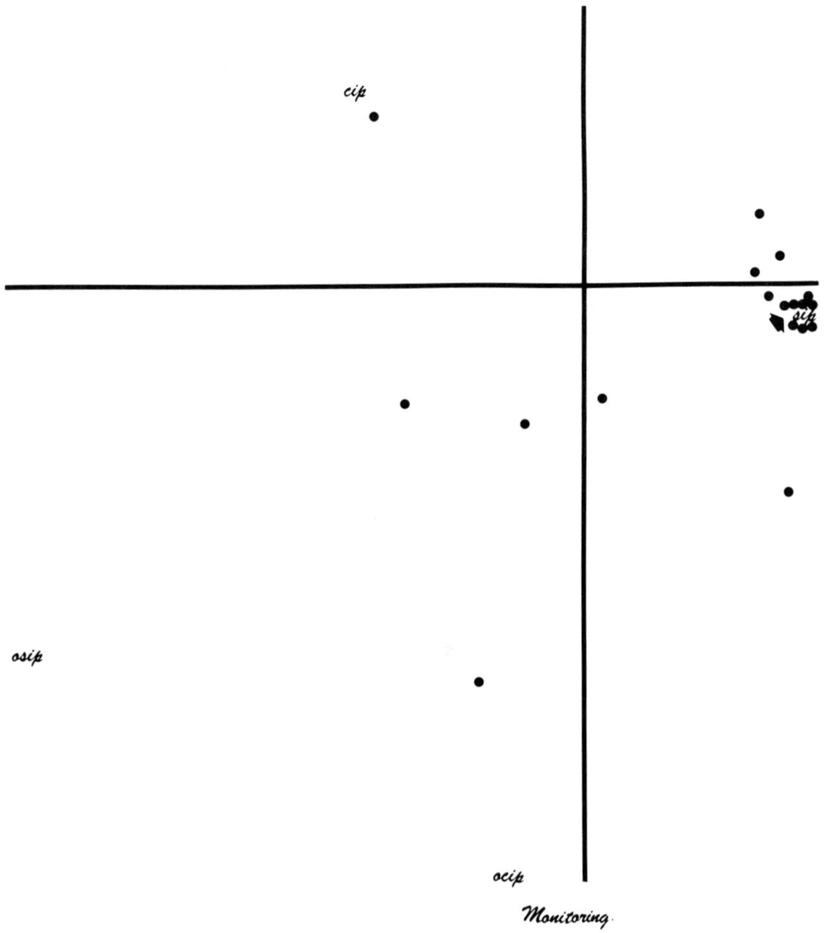

Figure 8.5. The monitoring data from Experiment 2 graphed in the same two-dimensional subspace. The inertia accounted for by the first dimension is 36.6% and the inertia accounted for by the second dimension is 16.6%. Here children who fall in close proximity to the *sip* performed optimally in the task.

work, we found that older children and adults tended to use a probability-based rule in these situations, judging the memory task to be easier when there were fewer items the same color as the target. That is, they thought the memory task would be easier if one of three gray houses was the target rather than one of six gray houses. First graders may have been less likely to consider this factor in their judgments.

Children have probably had numerous opportunities to observe how the distinctiveness of a target affects memory performance. In contrast, the attention task they were presented with in the monitoring condition was somewhat unusual, and it is unlikely that first graders possess a general rule linking monitoring ease with the size of the visual field. We suspect that when asked to make decisions about this task, children may have attempted to mentally simulate the zookeeper's activities. Including the zookeeper in the display probably facilitated such enactments and helped children understand how size would affect performance. Through these simulations, children may have come to realize that the zookeeper would have to rotate his head farther when watching the animals in the larger cages. At the end of the experiment, when asked how they decided where the zookeeper's job would be easier, some children commented that the zookeeper would have to turn his head back and forth when watching the goats in the big cages. (Many children even demonstrated this action as they gave their explanations.) A few first graders, however, commented that the larger cages would be more difficult because the distant animals would be harder to see. These children may have based their decisions on the general rule that an observer who is closer to an object will see it better (see Flavell, Flavell, Green & Wilcox, 1980).

The results of these first two experiments suggest that by the end of first grade, children are aware of many variables that affect cognitive performance and that they are able to evaluate the effects of specific variables with reference to particular cognitive activities. Some of the studies that we reviewed earlier (e.g., Miller, 1982; Wellman, 1977; Yussen & Bird, 1979) suggest that preschoolers are also aware of a variety of factors that can influence performance, but it is not clear if they too can evaluate the effects of stimulus variables in the context of specific activities. Our third experiment was designed to address this issue.

C. Experiment 3

In the third experiment, we asked preschoolers to judge how changing the number of items would affect mental and physical activities. We selected two tasks that had a strong motor component and two that were more cognitive in nature because we wanted to see if children first work out the effects of number variations in the context of physical activities. One reason we suspected this might be true is that motor tasks often provide direct feedback about how task manipu-

Table 8.2. Predicted Effects of Number Variations in Experiment 3.

	Task	Number of Items 3	12
Topple			
	Memory	Easy	Hard
	Tipping	Hard	Easy
Doors			
	Hiding	Hard	Easy
	Opening	Easy	Hard

lations affect performance. It is interesting to note in this context Markman's finding that five-year-olds can predict their performance in certain motor tasks more accurately than they can predict the amount of information they can recall (cited in Flavell & Wellman, 1977).

We chose number as the relevant variable in this experiment because differences in number are likely to be salient to young children (R. Gelman & Gallistel, 1978) and because number is likely to be one of the first stimulus factors that children recognize has an effect on performance (Wellman, 1985). In many situations, tasks become more difficult as the number of items increases. For example, a longer shopping list is harder to remember, and as the number of blocks increases, it becomes harder to build a stable tower. But increasing the number of items does not always have a negative impact on performance. For example, it is easier to hide in a room that affords many potential hiding places. It is not clear whether preschoolers appreciate that increasing the number of items can hamper performance in some situations while facilitating performance in others. Yussen and Bird (1979) report that in their pilot studies, children tended to believe that any task could be made more difficult by including more numerous or larger items. This suggests children might initially believe that number is a global variable and that increasing the number of items will hinder performance in all task domains.

Table 8.3. Probability of a Correct Answer as a Function of Age and Type of Task.

	Cognitive		Physical		
Subjects	Memory	Hiding	Tipping	Opening	Mean
Older Pre-schoolers (n = 9; mean age = 5-3)	.89	.89	.78	.78	.83
Younger Pre-schoolers (n = 10; mean age = 3-10)	.50	.40	.50	.50	.53
Mean	.68	.63	.63	.74	.67

The purpose of this final experiment was to see if children initially overgeneralize the effects of stimulus variables, for example, by believing fewer items will always make a task easier, or alternatively, whether they first consider the nature of the task before evaluating the effects of stimulus variables. If children's judgments prove to be task-dependent, we will consider whether they are initially more adept at judging the relative difficulty of physical, as opposed to cognitive tasks.

In our experiment, children were presented with four tasks. Two of these were easier when fewer items were included in the display and the other two were easier when the number of items increased. For each task, children were presented with two displays that only differed in the number of items each contained: one display had three items whereas the other had twelve items. Two of the tasks used pieces from a children's game called Topple. Preschoolers were presented with two unstable platforms perched on a base. One platform contained three yellow pieces, all positioned on the same half of the platform, whereas the other platform contained nine yellow pieces that were also clustered together on the same half of the platform. In the motor version of the task, children were asked which tower would be easier to tip over by adding another piece. Here the platform with the extra pieces would be easier to tip because of its greater unbalanced weight. (The predicted effects of these number variations are summarized in Table 8.2.) The same towers also served as stimuli for a cognitive task. In this task, children were asked to indicate which tower would make it easier to remember where the yellow pieces go. If children understand how this number variation affects memory, they should choose the tower with three pieces.

The stimuli for the other two tasks were pieces of cardboard containing either three or twelve little doors. In the motor task, children were asked where it would be easier to open the doors, and the display with the fewer doors was the right answer. Children were also asked where it would be easier to hide a penny, and for this more cognitive task, the correct response would be that hiding the penny would be easier when the display contained more doors.

As can be seen in Table 8.3, our older preschoolers (mean age: 5-3; range: 4-11 to 5-7) performed extremely well on all four tasks. These children were correct on 83% of the trials, and the few errors that did appear were scattered across all four tasks. In striking contrast, our younger subjects (mean age: 3-10; range: 3-6 to 4-6) found these tasks extremely difficult and were only correct on 53% of the trials. Perhaps more importantly, these children's error patterns did not suggest that they were uniformly adopting a rule that less is easier. If they had, they would have passed two of the tasks and failed the other two. Nor did children's performance vary as a function of the type of task. These general findings were confirmed by a logistic regression using age, type of task (motor vs. cognitive), type of motor task (tipping vs. opening), and type of cognitive task (memory vs. hiding) as the independent variables, and the probability of success on the task as the dependent variable. (We should note that a logistic regression was used rather than classical regression because the dependent variable is discrete rather than continuous, see Abdi, 1987; Hosmer & Lemeshow, 1989.) The effect of age is clearly significant (Wald's χ^2 (1) = 7.08, $p < .01$). The effects of task, type of cognitive task, and type of motor task were not significant (Wald's χ^2 (1) = .07, .13 and .54, n.s., respectively).

These results, then, suggest that children do not assume that stimulus factors will have the same effect on all tasks. Instead, children seem to treat each task as unique and attempt to evaluate the importance of stimulus factors in the context of each activity. This pattern of results, together with the findings of Fabricius et al. (1989), seem to suggest that children are quite sensitive to the differences between various tasks. In fact, it now seems likely that noticing the similarity among different examples of memory tasks, problem solving activities, or fine motor acts may be the real developmental achievement, and we are currently exploring whether children generalize their understanding of stimulus variables across classes of activities.

VII. CONCLUSIONS

In the course of this chapter, we have reviewed a wide range of studies that focus on how children conceive of the human mind. We began by discussing why many cognitive developmentalists have adopted the theory metaphor as a way to characterize cognitive change and have framed our review around Wellman's (1990) view of the child's developing theory of mind. Wellman (e.g., Wellman & Woolley, 1990; Wellman, 1990) has argued that the child's framework theory of mind undergoes major revisions during the preschool years. The child's first theory of mind is characterized by a desire psychology; he or she understands that individuals act to fulfill their desires but does not recognize that these actions are mediated by beliefs. Somewhere around their third birthday, children recognize that an individual's actions are informed by his or her beliefs. For Wellman, this achievement marks an initial commitment to a primitive belief-desire psychology. Not only do three-year-olds understand that mental representations play a role in intentional behavior, they also understand that different kinds of mental representations are possible. Three-year-olds distinguish among dreams, images, and reality-oriented representations, and in their view, reality-oriented representations are veridical copies of the external world. However, at the same time children make these distinctions, they also recognize that these mental entities all share the same ontological status and that these entities all fall within the scope of a theory of mind.

Four-year-olds are beginning to recognize that knowledge-based representations are not simply copied from the outside world. They recognize that beliefs are constructed and that even though beliefs are reality-oriented, they do not always mirror reality. Unlike the younger child, the four-year-old is able to reflect, at least to some extent, on both representations and the processes that construct them. Even three-year-olds, however, have some understanding of process. They recognize, for example, that perceptual processes are independent in the sense that one perceptual modality can be activated while another is not. In our review, we have focused on whether children also grasp more subtle distinctions, such as the difference between seeing and knowing or remembering and attending, and we have concluded that children do distinguish among these processes even though they may lack a complete understanding of how different factors influence processing.

Older preschoolers have been credited with a constructivist view of the mind. Wellman argues that once children reach four years of age, their theory of mind embodies the metaphor of a homunculus that actively interprets and manipulates information. Specific theories can develop within the confines of this constructivist framework. Elementary school children have been credited with both specific theories of self (Damon & Hart, 1988; Harter, 1985; Wellman, 1990) and specific theories of intelligence (Cain & Dweck, 1989; Wellman, 1990). At this point, it is not clear whether children also have specific theories about individual cognitive processes. As Wellman points out, metamemory might be conceived of as a collection of loosely organized pieces of practical lore, general rules, and thoughts about one's own mnemonic abilities. Alternatively, metamemory might be thought of as a specific naive theory of memory. This specific theory would be distinct from other specific theories, such as those focused on attention or other cognitive processes. But given that many complex cognitive activities recruit a variety of cognitive processes, does it really make sense to think that an individual would organize his or her metacognitive knowledge around specific theories that focus on each process in isolation? With some qualifications, we expect this is true.

Philosophers (e.g., Hesse, 1966) and psychologists (e.g., Gentner, 1982) interested in the development of scientific theories have posited that root metaphors guide theorizing and research activities in scientific domains, and that different metaphors lie at the heart of competing theories. Our intuitive understanding of the human mind is undoubtedly rooted in various metaphors. Perhaps, as Wellman claims, homunculi (or even demons, see Selfridge, 1959) underlie our everyday framework theory. It is also likely to be true that the specific everyday theories we hold about individual cognitive processes are rooted in metaphors.

Metaphors surely play a role in formal psychological theories. As Marshall and Fryer (1979) point out, the dominant models of memory have at their core metaphors that can be traced back to the ancient Greeks. For example, variations of the storehouse metaphor have played a central role in memory theories. Plato likened memory to an aviary, and many of the key distinctions this metaphor embodied were elaborated in a subsequent metaphor that likened memory to a well-laid out and well-indexed library. The library metaphor nicely captures many of the organizational issues that were neglected by the aviary metaphor, and Marshall and Fryer argue that most modern accounts of semantic memory build upon specific aspects of this library metaphor. Our understanding of other cognitive processes may be grounded in different metaphors. For example, modern theories of attention make reference to filters, spotlights, and bottlenecks.

It may be significant that each cognitive process has been described by a variety of metaphors, each of which seems to capture certain aspects of the process while glossing over others. For example, the library metaphor is particularly well-suited to questions about long-term memory whereas questions concerning recognition memory may be better addressed by variations of Plato's wax tablet metaphor (see Marshall & Fryer, 1979). The data we presented in the latter part of this chapter suggested that elementary school children distinguish among cognitive processes but the exact nature of these distinctions remains unclear. It is worth noting that our memory and monitoring tasks can each be conceptualized in terms of different metaphors. In the monitoring task, attention might be likened to a spotlight that must be continually moved across the visual field. The memory task can be framed in terms of the library metaphor; the task becomes easier when color can be used to index the target. Perhaps the distinctions children make among cognitive and perceptual processes mirror the different metaphors they use to reason about various aspects of the cognitive system. Thus children may consider activities similar if they can be conceptualized in terms of the same metaphor. Different metaphors highlight different task and situational variables, and children might be expected to generalize their knowledge about the effects of specific variables across activities that share the same root metaphor.

Metaphors play a central role in our intuitive psychology because they make it possible for us to construct mental models that can be used to reason about performance in different tasks and situations (see, e.g., Gentner & Stevens, 1983; Holland, Holyoak, Nisbett & Thagard, 1986). It is interesting to note that the metaphors that we suspect underlie specific theories within our everyday psychology typically make a distinction between representation and process. Representations are likened to tangible things, such as birds or books, that physical acts can be performed on whereas processes are likened to the actions that can be performed on these objects. Perhaps this distinction between representation and process is just as central to our everyday theory of mind as it is to theories in cognitive psychology.

ACKNOWLEDGEMENTS

Preparation of this chapter was supported in part by a Spencer Fellowship and HHS/NIH funds (2S07-RR07173-09) awarded to the first author. Portions of this paper were presented at the Biennial Meetings of the Society for Research in Child Development, Kansas City, 1989. We would like to thank Barbara Burns, Tara Callaghan, and especially Barbara Malt, for their helpful comments on an earlier version of this paper.

REFERENCES

Abdi, H. (1987). Introduction au traitment statistique des donnes experimentales. Grenoble: Presse Universitaire de Grenoble.

Abdi, H. (1988). A generalized approach for connectionist auto-associative memories. In J. Demongeot, T. Herve, V. Rialle & C. Roche (Eds.), Artificial intelligence and cognitive sciences (pp. 149-165). Manchester: Manchester University Press.

Benzecri, J. P. (1973). L ' analyse des donnes. Tome (Vol.) 1: La taxinomie. Tome 2: L' analyse des correspondences.
Paris: Dunod.

Bretherton I. & Beeghley, M. (1982). Talking about internal states: The acquisition of a theory of mind. Developmental Psychology, 18, 906-921.

Brown, A. L. (1990). Domain-specific principles affect learning and transfer in children. Cognitive Science, 14, 107-133.

Cain, K. M. & Dweck, C. S. (1989). The development of children's conceptions of intelligence: A theoretical framework. In R. J. Sternberg (Ed.), Advances in the psychology of human intelligence (Vol. 5, pp. 47-82). Hillsdale, NJ: Erlbaum.

Carey, S. (1985). Conceptual change in childhood. Cambridge, MA: MIT press.

Chandler, M. & Boyes, M. (1982). Social-cognitive development. In B. Wolman (Ed.), Handbook of developmental psychology (pp. 387-402). Englewood Cliffs, NJ: Prentice-Hall.

Chandler, M. & Helm, D. (1984). Developmental changes in contribution of shared experience to social role taking competence. International Journal of Behavioral Development, 7, 145-156.

Damon, W. & Hart, D. (1988). Self-understanding in childhood and adolescence. Cambridge, England: Cambridge University Press.

Escofier-Cordier, B. (1965). L ' analyse des correspondances. Doctoral dissertation, Universite' des Rennes, France. Later published in Cashiers du bureau univeritaire recherche operationelle. No. 13. (25-39).

Estes, D., Wellman, H. M., & Woolley, J. D. (1989). Children's understanding of mental phenomenon. In H. Reese (Ed.), Advances in child development and behavior (Vol. 22, pp. 41-87). New York: Academic Press.

Fabricius, W. V., Schwanenflugel, P.J., Kyllonen, P. C., Barclay, C. R., & Denton, S. M. (1989). Developing theories of the mind: Children's and adults' concepts for mental activities. Child Development, 60, 1278-1290.

Flavell, J. H. (1978). The development of knowledge about visual perception. In C. B. Keasey (Ed.), Nebraska symposium on motivation (Vol. 25, pp. 43-76). Lincoln: University of Nebraska Press.

Flavell, J. H. (1988). The development of children's knowledge about the mind: From cognitive connections to mental representations. In J. W. Astington, P. L. Harris, & D. R. Olson (Eds.), Developing theories of mind (pp. 244-267). Cambridge, England: Cambridge University Press.

Flavell, J. H., Flavell, E. R., Green, F. L., & Wilcox, S. A. (1980). Young children's knowledge about visual perception: Effect of observer's distance from target on perceptual clarity of target. Developmental Psychology, 16, 10-12.

Flavell, J. H., Green, F. L., & Flavell, E. R. (1990). Developmental changes in young children's knowledge about the mind. Cognitive Development, 5, 1-27.

Flavell, J. H. & Wellman, H. (1977). Metamemory. In R. V. Kail, Jr. & J. W. Hagen (Eds.), Perspectives on the development of memory (pp. 3-33). Hillsdale, NJ: Erlbaum.

Forguson, L. & Gopnik, A. (1988). The ontogeny of common sense. J. W. Astington, P. L. Harris & D. R. Olson (Eds.), Developing Theories of Mind (pp. 226-243). Cambridge, England: Cambridge University Press.

Gelman, R. (1990). First principles organize attention to and learning about relevant data: Number and animate-inanimate distinction as examples. Cognitive Science, 14, 79-106.

Gelman, R. & Gallistel, C. R. (1978). The child's understanding of number. Cambridge, MA: Harvard University Press.

Gelman, S. A. & Markman, E. M. (1986). Categories and induction in young children. Cognition, 23, 183-208.

Gentner, D. (1982). Are scientific analogies metaphors? In D. S. Miall (Ed.), Metaphor: Problems and perspectives (pp. 106-132) Atlantic Highlands, NJ: Humanities Press.

Gentner, D. & Stevens, A. L. (1983). Mental models. Hillsdale,N.J.: Erlbaum.

Gopnik, A. & Astington, J. W. (1988). Children's understanding of representational change and its relation to the understanding of false belief and the appearance-reality distinction. Child Development, 59, 26-37.

Gordon, F. R. & Flavell, J. H. (1977). The development of intuitions about cognitive cueing. Child Development, 48, 1027-1033.

Greenacre, M. J. (1984). Theory and applications of correspondence analysis. New York: Academic Press.

Harter, S. (1985). Competence as a dimension of self-evaluation: Toward a comprehensive model of self-worth. In R. Leahy (Ed.), The development of the self (pp. 55-121). New York: Academic Press.

Hesse, M. B. (1966). Models and analogies in science. Notre Dame, In: University of Notre Dame Press.

Holland, J. H., Holyoak, K. J., Nisbett, R. E., & Thagard, P. R. (1986). Induction: Processes of inference, learning, and discovery. Cambridge, MA: MIT Press.

Hosmer, D. W. & Lemeshow, S. (1989). Applied logistic regression. New York, Wiley.

Inagaki, K. & Hatano, G. (1987). Young children's spontaneous personification as analogy. Child Development, 58, 1013-1020.

Johnson, C. N. & Wellman, H. M. (1980). Children's developing understanding of mental verbs: Remember, know, and guess. Child Development, 51, 1095-1102.

Kahneman, D., Slovic, P., & Tversky, A. (1982). Judgment under uncertainty: Heuristics and biases. Cambridge, England: Cambridge University Press.

Keil, F. C. (1986). The acquisition of natural kinds and artifact terms. In W. Demopoulos & A. Marras (Eds.), Language learning and concept acquisition (pp. 133-153). Norwood, NJ: Ablex.

Keil, F. (1989). Concepts, kinds, and cognitive development. Cambridge, MA: MIT Press.

Kreutzer, M. A., Leonard, C. & Flavell, J. H. (1975). An interview study of children's knowledge about memory. Monographs of the Society for Research in Child Development, 40, (Serial No. 159).

Kuhn, D. (1989). Children and adults as intuitive scientists. Psychological Review, 96, 674-689.

Leslie, A. M. (1988). Some implications of pretense for mechanisms underlying the child's theory of mind. In J. W. Astington, P. L. Harris, & D. R. Olson (Eds.), Developing theories of mind (pp. 19-46). Cambridge, England: Cambridge University Press.

Marshall, J. C. & Fryer, D. M. (1979). Speak, Memory! An introduction to some historic studies of remembering and forgetting. In M. M. Gruneberg & P. E. Morris (Eds.), Applied problems in memory (pp. 1-25). London, NY: Academic Press.

Massey, C. & Gelman, R. (1988). Preschoolers' ability to decide whether pictured unfamiliar objects can move themselves. Developmental Psychology, 24, 307-317.

Mervis, C. B. (1987). Child-basic object categories and early lexical development. In U. Neisser (Ed.), Concepts and conceptual development: Ecological and intellectual factors in categorization (pp. 201-233). Cambridge, England: Cambridge University Press.

Miller, P. H. (1982). Children's and adults' integration of information about noise and interest levels in their judgments about learning. Journal of Experimental Child Psychology, 33, 536-556.

Miller, P. H. (1985). Metacognition and attention. In D. L. Forrest-Pressley & T. G. Waller (Eds.), Metacognition, cognition, and human performance: Vol 2. Instructional practices. (pp. 181-221) New York: Academic Press.

Misciones, J. L., Marvin, R. S., O'Brien, R. G., & Greenberg, M. T. (1978). A developmental study of preschool children's understanding of the words "know" and "guess". Child Development, 48, 1107-1113.

Mossler, D. G., Marvin, R. S., & Greenberg, M. T. (1976). Conceptual perspective taking in 2- to 6-year old children. Developmental Psychology, 12, 85-86.

Moynahan, E. D. (1973). The developmental of knowledge concerning the effect of categorization upon free recall. Child Development, 44, 238-246.

Nisbett, R. & Ross, L. (1980). Human inference: Strategies and shortcomings of social judgment. Englewood Cliffs, NJ: Prentice-Hall.

Piaget, J. (1929). The child's conception of the world. London: Routledge and Kegan Paul.

Pillow, B. H. (1988). The development of children's beliefs about the mental world. Merrill-Palmer Quarterly, 34, 1-32.

Pillow, B. H. (1989). Early understanding of perception as a source of knowledge. Journal of Experimental Child Psychology, 47, 116-129.

Pratt, C. & Bryant, P (1990). Young children understand that looking leads to knowing (so long as they are looking into a single barrel). Child Development, 61, 973-982.

Ritter, K. (1978). The development of knowledge of an external retrieval cue strategy. Child Development, 49, 1227-1230.

Selfridge, O. (1959). Pandemonium: A paradigm for learning. In Symposium on the mechanization of thought processes. London: H M Stationary Office.

Shatz, M., Wellman, H. M., & Silber, S. (1983). The acquisition of mental verbs: A systematic investigation of the first reference to mental state. Cognition, 14, 301-321.

Sodian, B. & Wimmer, H. (1987). Children's understanding of inference as a source of knowledge. Child Development, 58, 424-433.

Taylor, M. (1988). Conceptual perspective taking: Children's ability to distinguish what they know from what they see. Child Development, 59, 703-718.

Wellman, H. M. (1977). Preschoolers' understanding of memory-relevant variables. Child Development, 48, 1720-1723.

Wellman, H. M. (1985). The child's theory of mind: The development of conceptions of cognition. In S. R. Yussen (Ed.), The growth of reflection in children. New York: Academic Press.

Wellman, H. M. (1990). The child's theory of mind. Cambridge, MA: MIT Press.

Wellman, H. M., Collins, J., & Glieberman, J. (1981). Understanding the combination of memory variables: Developing conceptions of memory limitations. Child Development, 52, 1313-1317.

Wellman, H. M. & Estes, D. (1986). Early understanding of mental entities: A re-examination of childhood realism. Child Development, 57, 910-923.

Wellman, H. M. & Johnson, C. N. (1979). Understanding of mental processes: A developmental study of remember and forget. Child Development, 50, 79-88.

Wellman, H. M., Hogrefe, J., & Perner, J. (1988). Children's understanding of informational access as a source of knowledge. Child Development, 59, 386-396.

Wellman, H. M. & Woolley, J. D. (1990). From simple desires to ordinary beliefs: The early development of everyday psychology. Cognition,35, 245-275.

Wimmer, H., Hogrefe, G. J., & Perner, J. (1988). Children's understanding of informational access as a source of knowledge. Child Development, 59, 386-396.

Wimmer, H. & Perner, J. (1983). Beliefs about beliefs: Representation and constraining function of wrong beliefs in young children's understanding of deception. Cognition, 13, 103-128.

Yaniv, I. & Shatz, M. (1988). Children's understanding of perceptibility. In J. Astington, P. Harris, D. Olson (Eds.), Developing theories of mind (pp. 93-108). Cambridge, England: Cambridge University Press.

Yussen, S. R. & Bird, J. E. (1979). The development of metacognitive awareness in memory, communication, and attention. Journal of Experimental Child Psychology, 28, 300-313.

Commentary

Reflecting on Representation and Process: Children's Understanding of Cognition, S. E. Barrett, H. Abdi, & J. M. Sniffen

DENNIS H. HOLDING
University of Louisville

Perhaps we do use metaphors in our theories, and in constructing the mental models that underlie our reasoning about interaction with the world. In a sense it could hardly be otherwise, since all of our abstract words are originally metaphors, even when disguised in a Latin form. The word "abstract," after all, is simply the participle from "abs-" and "trahere" and thus corresponds to something like "outdrawn" or perhaps "dragged away from." Whether or not something functions as a metaphor is a matter of degree, depending on how much of the literal meaning is still present for users of the language. Perhaps then, children are aware of fewer literal associations and might therefore be said to employ fewer metaphors than adults.

Despite the emphasis of their conclusions Barrett, Abdi and Sniffen do not direct their efforts toward addressing the problem of children's metaphors. What the experiments show, instead, is that children do make metacognitive distinctions between the mental Processes (but not Structures) demanded by different task requirements. By and large, their judgments concur with those of young adults, both groups assuming, for example, that (skin) color is irrelevant to herding goats but that increased size would be a disadvantage. On the other hand, color should help in remembering the location of a sick chicken, while size is treated as irrelevant. However, unlike the adults, the children are less than unanimous in discounting size as an aid to memory.

What worries me about this, as an experimentalist, is that all we know is what the subjects say they expect would happen. But what would a memory test show? Perhaps the children are right. Perhaps they would remember larger chicken coops better because they would discriminate them better. Large letters are an aid to reading, and possibly to memory, and one might predict that minuscule chicken coops would be quite difficult to remember apart. In general, one ought probably to be cautious about obtaining information on metacognition without corresponding information about the cognition itself. This said, there is no doubt that children's judgments do imply that different tasks are seen as requiring different processes.

PART C. Categories, Concepts and Learning

Percepts, Concepts and Categories
B. Burns (Editor)

9

Basic Levels in Artificial and Natural Categories: Are All Basic Levels Created Equal?

MARY E. LASSALINE
EDWARD J. WISNIEWSKI
DOUGLAS L. MEDIN
University of Michigan

I. INTRODUCTION

This chapter addresses the nature of the basic level of conceptual structure in both naturally occurring and artificially constructed categories. We will look closely at the relationship between studies of the basic level that involve natural stimuli and those that involve artificial stimuli, and examine the extent to which structural properties of natural categories have been incorporated into artificial categories. More specifically, we evaluate studies of the basic level involving artificial categories with respect to how well conclusions from such studies extend to natural categories. In this context, we will address an important question: how confident can one be that a phenomenon called the "basic level advantage" *emerges for the same reasons* in artificial category studies as it does in natural category studies? One can view the basic level advantage as the output of a process operating on a representation. The fact that an artificial category study and a natural category study result in the same output by itself does not guarantee that they reflect the same representation and/or the same process. Before pursuing this point, however, we need to provide a bit more by way of background.

The single most important issue in the study of concepts has been the nature of conceptual structure. In the last 20 years or so, the investigation of this issue has led to a number of perspectives on concepts that are quite different from those of earlier views. As one example, the idea that all members of a category have fundamental characteristics in common that define category membership (i.e., the Classical view) has been replaced by the idea that category members *tend* to share properties that are neither necessary nor sufficient for category membership (see Smith & Medin, 1981, for a review).

Another notion that has been developed and refined over the years is that natural categories have both a vertical and horizontal component (Rosch, 1978). Drawing upon arguments presented in philosophy and logic (Wittgenstein, 1953; Zadeh, 1965), Eleanor Rosch proposed that natural categories can be viewed in terms of a gradient of typicality or goodness of category membership. Some members are the clearest cases of the category or the most prototypical members. Other members vary in terms of how good an example they are of the category (Rosch, 1975; Rosch & Mervis, 1975). Rosch further claimed that this horizontal dimension of conceptual structure is governed by the principle of *family resemblance* (see Wittgenstein, 1953). Under the family resemblance principle, category members considered the most prototypical are those with the most attributes in common with other members and the fewest attributes in common with members of contrasting categories. In a series of studies using both natural and artificial categories, Rosch and Mervis (1975) found a positive correlation between family resemblance and ratings of how good a member was of a category.

There is also a vertical component to category structure, given that objects can be categorized at a number of levels of generality. An object being driven on a highway, with four wheels, and a top that folds back, can be called a *convertible, a car*, or *a vehicle*. The category *car* is more general than *convertible* because it includes other objects (e.g., station wagons, hard top sedans) as well as the members of *convertible*. The category *vehicle* is more general than *convertible* and *car* because it contains other objects (e.g., trucks, trains) as well as the members of these categories. Such categories form a taxonomy, or a class inclusion hierarchy. In a seminal paper, Rosch, Mervis, Gray, Johnson, and Boyes-Braem (1976) singled out one level in such a hierarchy, which they called the basic level, as playing a central role in many cognitive processes associated with categorization. For example, the category level represented by *chair, hammer, car,* and *dog* is typically considered the basic level. These concepts can be contrasted with more general "superordinate" concepts, *furniture, tool, vehicle,* and *animal*, as well as the more specific "subordinate" concepts, *recliner, ball-peen hammer, convertible,* and *labrador retriever.*

Studies using a variety of cognitive tasks show that basic level concepts have advantages or privileges over other concepts. Pictures of isolated objects are categorized faster at the basic level than at other levels (Jolicoeur, Gluck, & Kosslyn, 1984; Murphy & Brownell, 1985; Murphy & Wisniewski, 1989; Rosch et al., 1976; Smith, Balzano, & Walker, 1978). People almost exclusively use basic category names in free naming tasks (Rosch, et al., 1976). Children learn basic concepts sooner than other types of concepts (Anglin, 1977; Brown, 1958; Horton & Markman, 1980; Mervis & Crisafi, 1982; Rosch et al., 1976). Finally, different cultures tend to employ the same basic level categories, at least for living kinds (Rosch, 1974).

Researchers have examined a number of explanations for the basic level advantage. Some explanations have focused on differences in the structure of categories at different levels (Gluck & Corter, 1985; Jones, 1983; Murphy & Brownell, 1985; Rosch et al., 1976). For example, people have argued that, relative to other concepts, basic concepts are the most differentiated (Murphy & Brownell, 1985; Rosch et al., 1976). That is, their members have many common features which are also distinctive. Other explanations have suggested that differences in the *content* of categories are responsible for the advantage (Tversky & Hemenway, 1984; Murphy & Wisniewski, 1989; Rosch, et al., 1976). For example, Tversky and Hemenway (1984) argued that the basic category advantage is due to the fact that these concepts primarily represent parts of objects.

One prominent experimental strategy for analyzing the basic level is to employ artificially constructed categories. The general rationale is to identify some variable or structural property of interest, incorporate that property into artificially constructed categories, and then run experiments to evaluate the role and importance of that property in categorization. This strategy allows one to control for a variety of correlated variables which might affect performance with natural categories. For example, basic level category names tend to be both more frequent and shorter than the labels for subordinate or superordinate categories (e.g., *dog* versus *labrador retriever* or *animal*). Does the basic level advantage hinge on frequency and length of category name rather than structural properties? Using artificially constructed categories, Murphy and Smith (1982) controlled for these factors and still observed that basic level categories were easiest to learn and were associated with the fastest categorization times. Therefore, neither word length nor frequency is necessary for one to observe a basic level advantage.

This paper is concerned with the basic level in both artificial and natural categories. We will pay particular attention to relationships between artificial and natural categories. Two critical issues concern the extent to which relevant structural properties of natural categories have been successfully incorporated into artificial categories and the related issue of just what conclusions concerning basic level categories are licensed by studies involving artificial categories. To anticipate, we will argue that the link between artificial and natural categories in analyzing basic levels has not been developed carefully enough to support strong inferences from one to the other. That's the bad news. The good news is that work using artificial categories is interesting in its own right because it can be successfully linked to theories and other metrics aimed at specifying structural properties which determine the relative difficulty of different levels of categorization.

This chapter is organized in the following way. We first take a closer look at the history and rationale for research with artificially constructed categories. We then describe some initial work on basic levels and theories and metrics relevant to them. Next, we critically examine the relation between artificial and natural basic level categories and describe some recent research in our laboratory on artificial basic levels. Finally, we try to summarize the current state of affairs both with respect to theories and metrics for the basic level and with respect to mapping between the natural and the artificial.

II. ARTIFICIAL CATEGORIES

At first thought the idea of condensing the years of experience associated with the learning of natural categories into an hour or so of training in a laboratory seems very, very tenuous. Probably the strongest argument for believing that important properties of categorization of natural objects can be captured in laboratory studies with artificial categories is the remarkable parallels between the two domains. For example, in their initial pioneering studies of the family resemblance principle, Rosch and Mervis (1975) included two experiments using artificial categories. Specifically, the stimuli were strings of letters conforming to a family resemblance structure. Rosch and Mervis found that degree of family resemblance affected ease of learning, identification response latency, and typicality rating, just as it does in natural categories.

These parallels have continued. For example, one striking phenomenon is that the prototype or best example of a category may be classified more accurately in transfer tests than previously seen examples that were used during original category learning (e.g., Homa & Vosburgh, 1976; Medin & Schaffer, 1978). Furthermore, people are sensitive to correlated attributes within artificial categories (Medin, Altom, Edelson & Freko, 1982) and within natural categories (Malt & Smith, 1983). And most relevant to our purposes, it appears that a basic level advantage can be obtained with artificial materials (e.g,. Murphy & Smith, 1982).

These successful parallels may have undermined natural conservatism about mapping between natural and artificial categories. Indeed, 15 years ago the position of devil's advocate would have been that findings from artificial and natural categories are closely linked. Now, the opposite stance is required for that role. Our present strategy will be to apply fairly strict standards in evaluating parallels.

The relationship between artificial and natural category studies has important consequences for models that have addressed basic level phenomena. These models typically have simulated the results of artificial category studies. Therefore, it is important to evaluate how well such studies reflect the "true" basic level phenomena involving natural stimuli. If the basic level emerges for the same reasons for artificial and for natural categories, then a theory that explains the basic level in one domain necessarily explains the basic level in the other.

A second, independent goal of the chapter is to evaluate concept learning models with respect to how well they account for basic level phenomena. We will examine a number of classes of models, whether or not they have explicitly addressed the basic level. Given the significance of the basic level, any theory should be held accountable for the cognitive phenomena associated with it. We will evaluate four major classes of models: probability-based models that capture feature distributions within and between categories; prototype models that integrate information across category instances to form a summary representation; exemplar-based models that preserve the feature structure of individual category members; and theories based on the use of explicit rules to determine category membership. In this chapter, the classes of models mentioned above are held accountable for the vertical organization of conceptual structure, that which characterizes categorization between taxonomic levels. Before describing these models in detail, however, we take a closer look at the basic level.

III. THE BASIC LEVEL IN NATURAL CATEGORIES

Psychological research on the basic level primarily began with the work of Eleanor Rosch and her colleagues (Mervis & Rosch, 1981; Rosch, 1975; Rosch, 1978; Rosch et al., 1976). In a series of very influential studies, Rosch et al. (1976) showed that the basic level is the most general level that is *informative* and the most general level that shows processing advantages in a variety of cognitive tasks (e.g., acquisition of concepts, identification of an object from an average shape, and identity judgment). Virtually all researchers have used one or more of these tasks to study the basic level, in either natural or artificial categories. Therefore, we will describe the Rosch et. al findings in some detail.

The basic level is the most general level that is informative in the sense that category members at this level share a large quantity of information. Rosch et al. (1976) provided evidence for this claim, using a number of converging measures. In one task, designed to examine the co-occurrence of attributes in common taxonomies of natural objects, subjects listed the attributes of superordinate, basic, and subordinate categories. The nine taxonomies used as stimuli are presented in Figure 9.1. For the nonbiological taxonomies, members of basic categories shared significant numbers of attributes. Categories at a more general level (i.e., superordinates) had few attributes in common, whereas categories more specific than the basic level (i.e., subordinates) shared only a few additional attributes. In a second task, subjects described the kinds of motor movements that they made in interacting with category members (e.g., bending one's knees in a particular way to sit on a chair). Basic level categories were the most general categories that shared a significant number of motor movements that could be made toward their members. Superordinates had few common move-

Superordinate	Basic Level	Subordinates	
		Nonbiological taxonomies	
Musical instrument	Guitar	Folk guitar	Classical guitar
	Piano	Grand piano	Upright piano
	Drum	Kettle drum	Bass drum
Fruit	Apple	Delicious apple	Mackintosh apple
	Peach	Freestone peach	Cling peach
	Grapes	Concord grapes	Green seedless grapes
Tool	Hammer	Ball-peen hammer	Claw hammer
	Saw	Hack hand saw	Cross-cutting hand saw
	Screwdriver	Phillips screwdriver	Regular screwdriver
Clothing	Pants	Levis	Double knit pants
	Socks	Knee socks	Ankle socks
	Shirt	Dress shirt	Knit shirt
Furniture	Table	Kitchen table	Dining room table
	Lamp	Floor lamp	Desk lamp
	Chair	Kitchen chair	Living room chair
Vehicle	Car	Sports car	Four door sedan car
	Bus	City bus	Cross country bus
	Truck	Pick up truck	Tractor-trailer truck
		Biological taxonomies	
Tree	Maple	Silver maple	Sugar maple
	Birch	River birch	White birch
	Oak	White oak	Red oak
Fish	Bass	Sea bass	Striped bass
	Trout	Rainbow trout	Steelhead trout
	Salmon	Blue back salmon	Chinook salmon
Bird	Cardinal	Easter cardinal	Grey tailed cardinal
	Eagle	Bald eagle	Golden eagle
	Sparrow	Song sparrow	Field sparrow

Figure 9.1. Taxonomies used in Rosch et al. (1976) Experiment 1.

ments whereas subordinates had about the same number of common movements as basic categories. In a third task, Rosch and her colleagues computed the ratio of the area of overlap to nonoverlap of the normalized shapes of category members as an index of within category similarity. They showed that in going from the superordinate level to the basic level, there was a large increase in the similarity of basic category members. In going from the basic level to the subordinate level, there was a significant but significantly smaller increase in similarity of subordinate category members.

Basic categories also were the most general categories that had processing advantages in a number of cognitive tasks. In a fourth task, subjects were given shapes that were averages of the shapes of superordinate, basic, or subordinate category members. Averages of superordinate members could not be identified significantly better than chance. On the other hand, basic categories were the most general categories for which averages of their members could be readily identified. Averages of subordinate members were no more identifiable than the averages of basic category members. In a fifth task, the most general names that aided the detection of objects presented in noise were those of basic categories. In a sixth task, basic category names were the most general names which facilitated judgments about whether two objects were physically identical.

In the cognitive tasks described above, performance at the subordinate level actually was as good or (sometimes) better than performance at the basic level. Rosch et al. (1976) also showed that for some tasks, the basic level had advantages over both the superordinate and subordinate levels. In a categorization task, subjects more quickly identified objects as members of basic categories than as members of superordinate or subordinate categories. In a naming task, subjects overwhelmingly labeled objects with the names of basic level categories. This result occurred even though in one stimulus set, a superordinate name was sufficient to distinguish one object from another, and in another set, a subordinate name was necessary to distinguish an object from other objects.

Finally, Rosch et al. (1976) examined developmental differences in the acquisition of concepts at different levels of generality. Two studies showed that children can sort objects into basic level categories earlier than they can sort objects into superordinate categories. In a task involving oddity problems, nursery and elementary school children and adults were shown sets of three color photographs. The sets varied on whether two of the three pictures came from the same basic level category or from the same superordinate category (the "odd" picture out always belonged to a different superordinate). Subjects were instructed to "put together the two that go together, the two that are alike." At all ages, sorting into basic level categories was almost perfect, but sorting into superordinate categories showed a significant improvement with age. In the other study, elementary school children and adults were given a set of 16 pictures, either from four basic level or from four superordinate categories, and asked to sort the entire set into "the ones that go together, the ones that are the same kind of thing." Results of this second, more traditional, sorting study were similar to the first. In a third developmental study, Rosch and her colleagues examined protocols of spontaneous speech during a child's initial language acquisition. They found that virtually all of the child's first utterances of concrete nouns were at the basic level.

Rosch explained the importance of the basic level in terms of a general principle called *cognitive economy*. This principle reflects a compromise between two somewhat conflicting reasons for having concepts. On the one hand, it would seem advantageous for a concept to be as informative as possible. As a result, we could predict many properties of an object by knowing its category. Carried to the extreme, however, this tendency would favor the formation of many specific concepts--one for *each* object encountered in the world. This extreme view conflicts with another reason for having concepts: they allow us to ignore differences among objects so that we can treat them equivalently as members of the same category. Carried to the extreme (i.e., ignoring all differences among objects), this tendency would favor the formation of a *single*, very general concept--one corresponding to *all* objects encountered in the world. Rosch suggested that the basic level results from the combination of these tendencies: basic level categories are both informative *and* general.

More specifically, Rosch implied that concepts at the basic level were more *differentiated* than those at other levels. In particular, she suggested that the basic level was the most general level with categories whose members had many attributes in *common* which were also *distinct* from the attributes of other categories at that level. In contrast, the members of the more general, superordinate categories have few features in common. Furthermore, although members of more specific, subordinate categories have slightly more features in common than those of basic level categories, many of these features are not distinctive. That is, they are shared by the members of other subordinate categories (e.g., members of *kitchen chair* share a number of features with the members of other categories of *chair*).

In addition to this *structural* explanation of the basic level, Tversky and Hemenway (1984) suggested that differences in the *content* of categories were responsible for the importance of the basic level. They argued that the features of basic categories which are both common and distinctive are primarily parts of objects. Reanalyzing the attribute listing data of Rosch et al. (1976), they found that a majority of the attributes that were shared by the members of basic categories were parts (e.g., "keys," "legs," "footpedal," and "strings" for the members of *piano*). In contrast, the proportion of attributes that were parts was less for subordinate categories (and was a minority for biological categories). Subjects listed only a few parts for superordinate categories. Furthermore, in two other studies, they showed that parts were the distinctive attributes of basic categories. Specifically, different basic categories tended not to share the same parts but instead, tended to share features that were not parts (e.g., "makes sound" for musical instruments and "sweet" for fruit). In contrast, subordinate categories tended to share the same parts but to differ on features that were not parts (e.g., subordinates of the basic category "chair" have legs, a seat, a back and typically arms, but "living room chair" is distinguished from "kitchen chair" by being large and soft).

Tversky and Hemenway (1984) suggested that parts and their configurations underlie many of the cognitive tasks that converge at the basic level. Parts and the way that they are configured primarily determine the shape of an object. Shape in turn, would seem to play an important role in a number of tasks described above. Recall that in going from the superordinate level to the basic level, there was a large increase in the overlap of the shape of basic category members. Basic categories also were the most general categories for which an average shape of their members could be readily identified. These results are consistent with the finding that basic level category members have many parts in common. Furthermore, because basic level categories have distinctive parts, they may also have distinctive shapes. This factor may account for why categorization of pictures was superior at the basic level. Finally, basic level categories were the most general categories that shared a significant number of motor movements that could be made toward their members. Intuitively, it seems that many such movements involve parts (e.g., we sit on the seat of a chair, we grasp the handle of a hammer, and so on).

The findings of Rosch et al. (1976) and Tversky and Hemenway (1984) applied to object taxonomies of biological and nonbiological categories. Researchers have found evidence consistent with a basic level in other domains, including environmental categories (Tversky & Hemenway, 1983), computer programs (Adelson, 1988), personality types (Cantor & Mischel, 1979) and events (Morris & Murphy, 1990; Rifkin, 1985). We describe two of these other domains in detail. Tversky and Hemenway (1983) suggested that there was a basic level for categories of scenes. Two of the three tasks that they used were originally employed by Rosch et al. In an attribute listing task, the number of features that subjects listed for scenes described at an intermediate (basic) level of generality (e.g., home, school, beach, and mountain scenes) was substantially greater than the number they listed for scenes described at a more general level (e.g., indoors and outdoor scenes). There was only a small increase in the number of attributes that subjects listed for more specific scenes (e.g., apartment, high school, lake beach, and Rocky mountains scenes). In a naming task, subjects primarily labeled photographs of scenes with their basic level names. In a third task, subjects primarily completed sentences describing activities performed in scenes with basic level names. The results of the last two tasks occurred even though in some cases, superordinate names were sufficient to distinguish scenes from others and in other cases, subordinate names were necessary to distinguish scenes from others. These findings for scenes parallel those of Rosch et al. (1976) involving objects.

Morris and Murphy (1990) suggested that there was a basic level for events (see also Rifkin, 1985). Three of the four tasks that they used paralleled those employed by Rosch et al. (1976). In one task, subjects listed the actions of events. The number of actions listed for events described at an intermediate (basic) level of generality (e.g., breakfast, dinner, classes, and tests) was substantially greater than the number listed for events described at a more general level (e.g., meals and school activities). There was only a small increase in the number of actions that subjects listed for more specific events (e.g., quick breakfast, family dinner, English class, multiple choice exam). In another task, subjects verified whether an action expressed as a verb phrase was part of either a subordinate, basic, or superordinate event category. So, for example, subjects verified whether the action "scream during the scary parts" was true of "horror movie" (a subordinate event), "movie" (a basic event) or "entertainment" (a superordinate event). Subjects were fastest to verify that actions were parts of basic events, slightly less faster to verify that they were parts of subordinates, and substantially slower to verify that they were parts of superordinates. In a third task, subjects were asked to choose appropriate names for events presented in story contexts. In most cases, they preferred to label the events with basic level names. One notable exception involved stories about events that could be classified in the same basic category (a horror movie and a comedy). Here, subordinate terms were necessary to distinguish these events. Unlike Rosch et al.'s findings with objects and Tversky and Hemenway's findings with scenes, subjects preferred subordinate terms slightly more than basic level terms.

A fourth task used by Morris and Murphy is interesting because it presents an alternative way of measuring the differentiation of a category. Recall that Rosch et al. viewed differentiation in terms of the *number* of attributes that were common and distinctive for a category. Such a measure does not take into account the possibility that some features are more important than others. An alternative way to measure differentiation, first suggested by Mervis and Crisafi (1982), is to subtract between-category similarity from within-category similarity. Morris and Murphy had subjects rate the similarity of category members at various levels of generality. Interestingly, they found that superordinate events were more differentiated than basic category events. They explained this apparent paradox by suggesting that functional features were more central to event concepts than to object concepts, since events are structured by goals rather than by perceptual shape. These features apparently carried a lot of weight in judging the similarity of superordinate members.

IV. THE BASIC LEVEL IN ARTIFICIAL CATEGORIES

Given the set of correlated measures that converge on the basic level, it is far from easy to identify which factors are central and which are derivative. Therefore, it is not surprising that investigators have turned to artificial category learning situations to disentangle these factors.

Murphy and Smith (1982) examined the basic level advantage for picture categorization found by Rosch et al. (1976). Looking at that study, Murphy and Smith noted that a variety of factors other than number of distinctive features may have accounted for why pictures were categorized faster at the basic level than at either the subordinate or superordinate level. In comparing subordinate categories to basic categories, subordinate names were considerably longer, some of the subordinates may have been unfamiliar (e.g., *cross-cutting hand saw*), and some of the differentiating features of subordinates may have been difficult to perceive (e.g., as in *green seedless grape*). In addition, there are some general characteristics of natural basic categories that may have accounted for their superior performance: basic categories are learned first, their names occur more frequently, their members occur more frequently than those of subordinates, and there may be a higher conjoint frequency between basic category names and objects (e.g., cars may be called *car* more frequently than they are called *vehicle* or their appropriate subordinate name).

In several studies, Murphy and Smith generally replicated the findings of the Rosch et al. picture categorization task, using artificial stimuli and removing the confounds described above. Those factors either were held constant (e.g., using CVC labels to equate category name length) or systematically varied (e.g., order of learning the different category levels was counterbalanced). In addition, Murphy and Smith explicitly designed the artificial stimuli of their first experiment so that they resembled natural subordinate, basic, and superordinate categories. Superordinates had a single attribute common to their members (namely, function), basic categories had many common features (which gave them distinctive shapes), and subordinate category members had only one attribute that could differentiate subordinates of the same basic category (e.g., there were two subordinate categories of "knife," differentiated by whether or not the blade was serrated). Figure 9.2 presents a sample of the stimuli (modeled after real-world

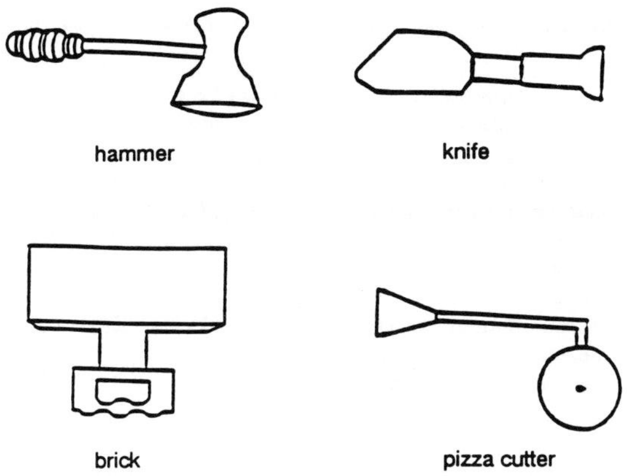

Figure 9.2. Examples of four basic tools used in Murphy and Smith (1982) Experiment 1.

tools) used in that experiment.

Although Murphy and Smith generally replicated the Rosch et al. findings, there also were important differences between their results and those of Rosch et al. In the experiment that most closely resembled that of Rosch et al., Murphy and Smith found that people categorized pictures into subordinate categories almost as quickly as they did into basic categories. In contrast, Rosch et al. found that categorization at the subordinate level was slowest. Murphy and Smith suggested that the Rosch et al. results may have resulted from some of the factors described above, including name length and perceptibility of distinguishing features.

While supporting the claim made by Rosch et al. that basic categories were associated with more distinctive attributes, Murphy and Smith argued that only perceptual features were critical to basic level superiority. The importance of perceptual features is most readily seen in comparing basic and subordinate categories to superordinate categories. Murphy and Smith (also Rosch et al., 1976) suggested that the advantage of basic and subordinate categories over superordinates arises because people typically do not have a single, perceptual representation for a superordinate. When deciding that an item belongs to a superordinate category, people must activate a *number* of representations. Such a decision involves extra capacity and more matching of features. To support this claim, Murphy and Smith conducted a study in which the members of superordinates shared a distinctive perceptual feature (size). After the experiment, subjects were divided into two groups, based on whether or not they reported attending to the size cue. For those subjects that did attend to the size cue (8 of

Table 9.1. Abstract category structure of Corter et al. (1988) Experiment 1.

Stimulus	Categories Top	Middle	Bottom	Feature Dimensions Gums	Eyes	Rash
1	queritism	burlosis	jirenza	1	1	1
2	queritism	burlosis	malenza	1	2	1
3	queritism	cretosis	gilenza	2	3	2
4	queritism	cretosis	surenza	2	4	2
5	philitism	midosis	habenza	3	1	2
6	philitism	midosis	kelenza	3	2	2
7	philitism	nitosis	tumenza	4	3	1
8	philitism	nitosis	valenza	4	4	1

12), categorization times were faster for superordinates than for intermediate level categories, which were not defined perceptually.

To test Murphy and Smith's claim that only perceptual features are critical for the basic level advantage, Corter, Gluck, and Bower (1988) conducted two experiments using artificial categories that were defined in terms of verbal or conceptual features rather than perceptual ones. The categories were diseases, characterized by three symptoms (gums, eyes and rash), each having four possible values (swollen, discolored, bleeding, or sore gums; puffy, sunken, red, or burning eyes; and blotchy, spotted, itchy, or scaly rash). Table 9.1 shows the diseases, symptoms, and abstract category structure of the experiment.

The category structure closely paralleled that used by Hoffman and Ziessler (1983), but substituted verbal or conceptual features for the perceptual features that they used. Based on the table, one can see that top-level categories are defined by a disjunction of values of the first symptom, middle-level categories are defined by a single value of the first symptom, and bottom-level categories are defined by a conjunction of values of the first and second symptoms. Given this structure, Corter et al. argued that the middle level should be the preferred level because it is the highest level for which category members share values on one or more symptoms. Like Hoffman and Ziessler (1983), they found that people categorized items faster at the middle level than at other levels, and learned the middle-level categories the fastest. These results suggest that perceptual features are not necessary for a basic level advantage.

In a developmental study, Mervis and Crisafi (1982) tested the claim made by Rosch et al. that children first learn to categorize at the basic level before learning to categorize at other levels. Like Rosch et al., Mervis and Crisafi

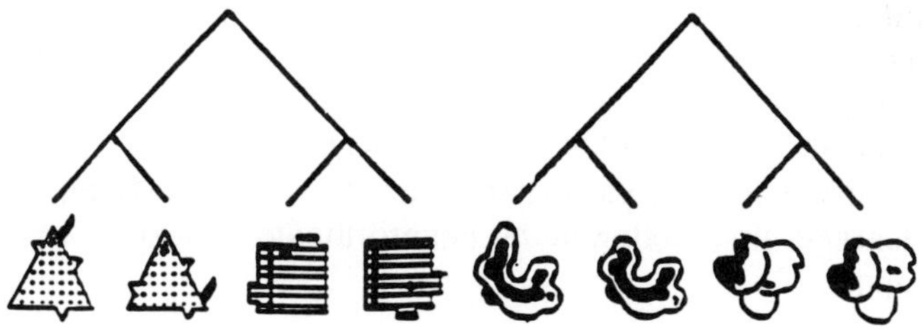

Figure 9.3. Examples of stimuli used in Mervis and Crisafi (1982) Experiment 1.

used a task involving oddity problems (see above). Figure 9.3 presents a sample of their stimuli.

Specifically, children were shown a picture of an object (the standard) and asked which of two other pictures went with the standard. The standard and one of the other pictures either came from the same subordinate, basic, or superordinate category. However, their task provided a more stringent test of the Rosch et al. claim in three ways. First, for the sets involving basic categories, children only could solve the problems by attending to cues at the basic level. In basic level sets used by Rosch et al., the "odd" item out came from a different superordinate, so it was possible to put the other two items together because they belonged to the same superordinate and not because they belonged to the same basic category. So for example, in the set *beagle, cocker spaniel,* and *jet*, one can put *beagle and cocker spaniel* together because they are both animals, rather than because they are both dogs. Mervis and Crisafi removed this possibility by using sets of pictures in which two items came from the same basic category but all of the items came from the same superordinate (e.g. *beagle, cocker spaniel and deer*). Second, Mervis and Crisafi used artificial stimuli which were never labeled, thereby ruling out some of the same alternative explanations noted by Murphy and Smith (see above). Third, this study also examined acquisition of subordinate categories (it was not considered in the Rosch et al. study).

Like Murphy and Smith, Mervis and Crisafi designed their artificial stimuli to mimic the structure of natural categories (see Figure 9.3). Basic category members had very similar overall shapes and shared a number of distinctive attributes. Superordinates shared very general abstract attributes. Subordinate category members shared more features than members of basic categories, but their features were not as distinctive.

As predicted, children were able to categorize at the basic level at a younger age than at other levels. For the basic-level sets, all children (ages 2-6,

4, and 5-6 year-olds) performed significantly more accurately than would be expected by chance. For the superordinate level sets, 4 and 5-6 year-olds performed significantly more accurately than chance. For the subordinate-level sets, only the 5-6 year-olds performed significantly more accurately than chance. These results imply a particular order of category learning. Basic level concepts are learned first, followed by superordinate concepts, and then subordinate concepts.

The stimuli used may have been problematic, though, on two counts. First, it is unclear that members of subordinate categories were discriminable from members of their contrast categories. For natural categories, subordinates may be much more distinct from one another. More generally, experimenters have the freedom to make subordinate categories arbitrarily similar to each other, and it is not clear how similar subordinates "should" be. This factor weakens the primary conclusion drawn from this work, namely that categorization at the basic level is acquired first.

Recently, Murphy (1991) has investigated whether the importance of parts in basic level categories (as detailed by Tversky and Hemenway, 1984) is caused by a psychological principle that requires basic concepts to represent such information. That is, the importance of parts could reflect an intrinsic bias of the human conceptual system. Alternatively, the importance of parts may reflect the structure of object categories in the world, rather than a special property of the conceptual system. By this account, people generally are sensitive to the common and distinctive information of categories, whether or not it is associated with the parts of objects. It just happens to be true of objects that this information is based on their parts. Murphy (1990) provides evidence that parts are neither necessary or sufficient for a basic level advantage. He found a basic level advantage for artificial categories whose members did not have parts in common but rather shared distinctive, nonpart information such as size, texture, and type of border. He also found that the reaction time advantage of basic categories (containing *distinctive parts*) could be increased even further by adding *distinctive nonpart* information to them. On the other hand, this advantage could be eliminated when the distinctive nonpart information was associated with another category.

Although none of the studies so far described are flawless, they encourage the view that one can reliably obtain an advantage for an intermediate level of categorization. Regardless of whether the parallels to the basic level with natural categories are precise, these findings do have implications for theories of categorization. As we shall see, it is nontrivial to develop a model that favors an intermediate level of categorization. Let's take a closer look at some of these metrics and models.

V. METRICS, THEORIES, AND THE BASIC LEVEL

There have been two theoretical approaches to analyzing the basic level. One strategy has been to specify what types of information or structural properties might be maximized at the basic level. The other strategy has been to develop and evaluate categorization models in terms of their ability to account for basic level advantages. In some cases, the metrics and theories are linked in that the theory may explicitly embody the metric of interest. We first describe some metrics and then turn to associated theories.

A. Metrics

Cue validity. One possibility is that the basic level maximizes the cue validity of features. The cue validity of a feature *f* is the probability that an item belongs to a category *c*, given that it has the feature, that is, P (c | f). In fact, Rosch et al. (1976) suggested that the most differentiated categories (i.e., basic categories) maximized a related measure, which they called *category cue validity*. This measure is the sum of the cue validities of features for the category. Carried to the extreme, maximizing either measure would favor the formation of a single, very general concept of all objects encountered in the world. Furthermore, a number of researchers (e.g., Medin, 1983; Murphy, 1982) have shown that for nested categories, category cue validity increases with the generality of a category, or stays the same. A simple example illustrates this point. Consider the nested categories *sparrow, bird,* and *animal*, and the feature "flies." Let *v* be the probability that an item with the feature "flies" is a *sparrow*. The probability *w* that an item with the feature "has wings" is a *bird* must be greater than *v* because there are more birds than sparrows that fly. Furthermore, the probability *x* that an item with the feature "flies" is an animal must be greater than *w* because there are more animals than birds that fly (e.g., bats, some insects, perhaps some fish, etc).

Category validity. Another possibility is that the basic level maximizes the category validity of features. Category validity is the probability that an item has a particular feature *f*, given that it belongs in some category *c*, that is, P (f | c).This measure emphasizes how well unknown features of an object can be predicted from knowledge of its category membership. Carried to its extreme, maximizing category validity would favor the formation of a concept for each object encountered in the world. Therefore, it always will pick out the subordinate or least inclusive level as the most advantageous (Medin, 1983), a problem directly opposite from that faced by cue validity (favoring the most inclusive level). Consider again the nested categories *sparrow, bird, and animal*, and the feature "flies." Let v be the probability that given that an item is a sparrow, it has the feature "flies" and let w be the probability that given that an item is a bird, it has the feature "flies." Because there are more birds than sparrows that do not fly (e.g., penguins, ostriches), w must be less than v. Furthermore, the probability x, given that an item is an animal it will have the feature "flies," must be less than w because there are more animals than birds that do not fly. Weighting category validity by a category's base rate avoids the extreme consequence of favoring the least inclusive level. Anderson (1990) uses this metric in his rational theory of categorization and has successfully modeled the basic level findings of Murphy and Smith (1982) and Hoffman and Ziessler (1983).

The product of cue and category validity. Jones (1983) suggested that the product of cue and category validity (which he called *category-feature collocation*) might determine the preferred level of categorization. Jones suggested applying this measure in the following way to predict which level should be basic. For each feature in each category of a nested hierarchy, compute its collocation score $P(f \mid c)*P(c \mid f)$. Then, "assign" each feature to that category in the hierarchy which has the highest collocation score for that feature. Basic categories should be those categories which have the greatest number of features assigned to them. Corter and Gluck (1990) have suggested an alternative way of applying this measure. For each category, one might compute the average of the collocation scores for all features. Basic categories should be those categories with the highest average collocation. Using a number of hierarchies involving natural and artificial categories, Corter and Gluck (1990) evaluated both measures along with a third metric called category utility (described below). They found that only category utility reliably predicted which categories should be basic.

Category Utility. Gluck and Corter (1985) developed a context sensitive metric analogous to H_t (information transmitted) of information theory, which they call *category utility*. It is the increase in the expected number of features that can be correctly predicted given knowledge of category membership versus no such knowledge. Specifically, the utility of a category C_k is the difference between the sum of squared category validity probabilities of its features, $\Sigma_i P(f_i \mid C_k)^2$, and the sum of squared base rates of those features, $\Sigma_i P(f_i)^2$, weighted by category's base rate, $P(C_k)$:

$$(1)\quad \text{Category Utility } (C_k) = P(C_k)\,[\Sigma_i P(f_i \mid k)^2 - \Sigma_i P(f_i)^2]$$

Gluck and Corter assume that category utility should be calculated over a universal set of features, including those that do not occur in one category but do occur in a contrast category (that is, over any feature that appears in any category). For example, if feature x appears only in Category B, but not in Category A, when calculating the category utility for Category A, x is included as a feature. This assumption prevents the addition of a single instance from radically changing the feature domain. In addition, though, this assumption raises the problem of determining the relevant universe of features, a problem which has not been fully addressed.

Category utility predicts that a novel item will be placed into the category whose utility increases the most when that item is considered a member of the category. Furthermore, it is assumed that the level whose categories have the highest utility is the basic level. Corter and Gluck (1990) showed that the level predicted by category utility to be basic corresponded to the empirically determined basic level in the experiments of Rosch et al. (1976), Murphy and Smith (1982), and Hoffman and Ziessler (1983).

B. Theories of Categorization

Computational models of the basic level. Fisher (1987) developed an incremental conceptual clustering program, called COBWEB, that uses category utility. Conceptual clustering is a task in which items are partitioned into categories based on some criterion and a concept is formed for each category. It differs from other concept formation tasks in which people explicitly are given feedback on whether or not an item belongs in a category. Using category utility as its clustering criterion, COBWEB adds a novel item to an existing category (i.e., one that it has previously constructed) or places the item into a new, singleton category. Specifically, COBWEB tentatively places the item into an existing category and calculates the average category utility for that partition of categories:

$$\frac{\Sigma_k P(C_k)\, [\Sigma_i\Sigma_j P(f_{ij} \mid k)^2 - \Sigma_i\Sigma_j P(f_{ij})^2]}{n} \qquad (2)$$

In this formula, $P(f_{ij} \mid k)$ is the probability of attribute i possessing value j in category k, $P(f_{ij})$ is the base rate of attribute i possessing value j, and n is the number of categories in the partition. Note that this is simply category utility summed across categories and divided by the number of categories.

This process is repeated, placing the item into a different category each time. In addition, COBWEB calculates the category utility for the partition that results when the item is placed in a new category by itself. COBWEB then classifies the item into the category that results in the partition with the highest category utility.

COBWEB incrementally constructs categories by examining novel items sequentially. Eventually, it constructs a hierarchy of categories, related to each other by class inclusion. Fisher (1988) modified the program to account for basic level phenomena. Specifically, the program augments the category hierarchy with *collocation-maximizing indices*: pointers from the root of the hierarchy to those categories that maximize the collocation of an attribute value f_{ij}. Collocation, again, is the product of cue and category validity (see Jones, 1983). An object is classified into the category with the greatest sum of collocation. Fisher (1988) has shown that his system categorizes items at the level corresponding to the empirically determined basic level in the experiments of Murphy and Smith (1982) and Hoffman and Ziessler (1983).

Anderson (1990) has developed and extensively evaluated a clustering model which, like Fisher's COBWEB, embodies the idea that the organism's goal is to maximize the inferences that can be drawn from category membership. In the model, a novel item is placed into its most *probable* category, which either may be an existing category (i.e., one that previously has been constructed) or a new, singleton category. The probability P_k that an item belongs to a category k is $P(k \mid F)$, or the conditional probability that the item belongs to the category k, given its set of attribute values F. This probability can be expressed as the following equation:

$$(3) \qquad Pk = P(k \mid F) = \frac{P(k)P(F \mid k)}{\Sigma k P(k)P(F \mid k)}$$

The probability P(k) in turn, has the following form:

$$(4) \qquad P(k) = \frac{cn_k}{(1 - c) + cn}$$

where n_k is the number of items assigned to category k so far, n is the total number of items seen so far, and c is the *coupling probability*. This probability (a fixed value) indicates how likely two objects come from the same category (for most of Anderson's simulations, this probability was set to .30). For large n, P(k) approximates n_k/n, strongly biasing the system to place items into large categories.

The probability that an item displays a set of attribute values F, given membership in category k, is computed by the following equation:

(5) $P(F \mid k) = \Pi\ P(ij \mid k)$

where $P(ij \mid k)$ is the probability of an item displaying value j on attribute i, given that it comes from category k. This probability in turn, has the following form:

(6) $$\frac{n_{ij} + 1}{n_k + m_i}$$

where n_k is the number of items in category k that have a value j on attribute i, n_{ij} is the number of items in the category that have the same value as the item to be classified, and m_i is the number of values on the dimension to be classified. Notice that the *normative* value for $P(ij \mid k)$ is calculated by n_{ij} / n_k (and this equation approximates that value for large n). However, the term m_i is included in the equation above to deal with problems of small samples. As Anderson notes (page 105), if a system has just seen one item in a category and it was red, one would not want to assume that all items in the category were red (as would be the case if the probability was simply n_{ij} / n_k. Assuming seven colors, equally probable on prior grounds, P (red|k) equals .25 by direct substitution into equation 6.

Anderson has accounted for the basic level findings of Murphy and Smith (1982) and Hoffman and Ziessler (1983) in the following way. He presented the model with random sequences of the stimuli used in those experiments, encoded as bit patterns (with unique subpatterns corresponding to different values on the various feature dimensions). The system forms categories at various levels by varying the value of the coupling probability.

Prototype and exemplar models of categorization. In general, these models have not been held accountable for hierarchical categorization. Rather, they usually characterize such within-level categorization phenomena as typicality effects and new/old recognition. Much of this work contrasts the predictions of prototype versus exemplar-based theories of categorization. (e.g., Medin & Smith, 1984; Nosofsky, 1988). To our knowledge, however, these models previously have not been compared with respect to their predictions concerning levels of categorization. We will examine a pair of models that are examples of general classes of models: the adaptive network model (Gluck & Bower,1988), which is an example of *multiplicative-similarity prototype* models (see also Massaro's fuzzy logical model, Massaro & Freidman, 1990 and Nosofsky, 1990), and the context model (Medin & Schaffer, 1978; Nosofsky, 1986), which is an example of *multiplicative-similarity exemplar* models (see also the array model of Estes, 1986, and Hintzman's MINERVA, 1986).

The adaptive network model. Prototype models combine information about individual examples into a summary representation. In the case of the adaptive network model, this information is represented as a set of weights that connect input nodes of the system to output nodes. The system learns categories in the following manner. A category item is presented as a pattern of activity across the input nodes. The pattern of activity corresponds to a set of attribute values that describe the item. The system responds by generating a pattern of activity across its output nodes. Each output node represents a category. The response of a given output node is:

$$(7)\quad o_j = \Sigma_i\, w_{ij}\, a_i$$

which is simply the sum of the activations of the input units to which the output node is connected, the a_i, multiplied by their weights, the w_{ij}. An input node has an activation value of 1 if the attribute value that it represents is present in the item. Otherwise, the node has a value of 0. The system is taught to activate the node of the correct category and to inhibit activation of the other nodes, by adjusting its weights, using the delta or least mean squares learning rule (Duda & Hart, 1973; Rescorla & Wagner, 1972):

$$(8)\quad \Delta w_{ij} = \beta\, a_i\, (o_j - \Sigma_i\, w_{ij}\, a_i)$$

Here, the change in the weight between input unit i and output unit j, Δw_{ij}, is the difference between the desired output from unit j, o_j (which is 1 if the stimulus belongs to category j and is 0 otherwise), and the sum of the activations of the input nodes to which the output node is connected, the a_i, multiplied by their weights, the w_{ij}. This difference is multiplied by the activation of input node i, a_i and by the learning rate, ß, which is left as a parameter ($0<\beta<1$). This learning rule is a gradient descent procedure leading to minimum squared error over a set of learning items.

After some amount of learning, the system is typically tested on how well it has learned. An unclassified stimulus pattern is presented and the system generates a pattern of activity across its output nodes. The output is converted into a response probability for each of the possible categories, $P(R_j)$, by computing the ratio of the output for a category o_j, to the sum of the outputs for all categories i, $\Sigma\, o_i$:

$$(9)\quad p(R_j) = \frac{o_j}{\Sigma\, o_i}$$

There are several problems with models of these types. First, it is well known that these relatively simple models only can learn *linearly separable* categories (e.g., Hinton, 1987; Minsky & Papert, 1969). Categories are linearly separable if their members can be correctly classified using a weighted, additive function of their features. Nosofsky (1991) also has shown that the adaptive network model is a special case of a class of multiplicative prototype models which have the same limitation. Apparently the constraint that categories be linearly separable does not apply to human category learning (e.g. Medin & Schwanenflugel, 1981). Second, the multiplicative prototype model also is not sensitive to correlated attributes within categories and in general, does not fare well in contrast with exemplar models that use multiplicative similarity (see Nosofsky, 1991 for a review).

To address some of these limitations, Gluck, Bower, and Hee (1989) slightly modified the adaptive network model by adding input nodes to the model which correspond to the *conjunctions* of single features. For example, a model with nodes for the features *a*, *b*, *c*, and *d*, would also have nodes for the conjunctions of those features (i.e., a+b, a+c, a+d, b+c, b+d, and c+d). By encoding the input in this way, the model is sensitive to correlated attributes. Furthermore, such an encoding may transform a learning problem that is not linearly separable with single features into one that is linearly separable in terms of combinations of features. Gluck, Corter, and Bower (1990) have shown that this model (which they call the configural-cue network model) reproduces the basic level advantage observed by Murphy and Smith (1982), Hoffman and Ziessler (1983), and Corter et al. (1988).

The context model. In contrast to prototype models, standard similarity-based exemplar models explicitly represent information about individual category members. Exemplar models assume that category items are stored in memory and retrieved in a probabilistic fashion. Also, an item is classified into the category to which it is most similar. For the Medin and Schaffer context model, the classification rule takes the form of Luce's choice rule (Luce, 1959). In this case, the probability of classifying an item as a member of a target category is equal to the sum of the similarities of that item to each category member, divided by the sum of the similarities of the target item to all items (both in the target category and in the contrast categories). Furthermore, the similarity of an item to a category member is a *multiplicative* function of the similarity of the item's attribute values and to those of the category member (this is not specified by Luce):

(10) P(classifying item in category A) =

$$\frac{\Sigma_a \Pi s_d}{[\Sigma_a \Pi s_d + \Sigma_b \Pi s_d]}$$

where a represents the items in category A, b represents the items in a contrast category B, d represents the attributes that make up each item, and s represents the similarity of the item's attribute values to those of a category member. If the target item and the comparison item have a matching value on attribute d, then s for that attribute equals 1; otherwise, the similarity parameter takes on the best fitting value between 0 and 1. Matches between the target item and comparison items in memory increase similarity; mismatches decrease similarity.

Exemplar models of this form, which combine information about category size with similarity information, will always derive a higher probability of classification for the category with the most exemplars, as long as the categories are nested, because within category similarity increases (or at least remains constant) with increases in category size. *Average* within-category similarity (dividing the sum of the similarities between the target item and each of the category items by the number of category items) may decrease as one moves up a nested hierarchy but *summed* similarity (simply summing the similarities between the target item and each category item across all category items, without dividing by the number of category items) can only increase.

Some provisos are in order concerning the predictions of exemplar models. First of all, exemplar models have almost exclusively been applied to transfer tests given after learning. They have not been developed as computational models of learning. It may be that some form of exemplar model that incorporates competitive learning could predict a basic level advantage. Our predictions hold only for the class of models in which learning is a monotonic function of the classification probabilities, as indexed by the equation above. One also must keep in mind that the context model permits selective attention to constituent dimensions. Therefore, a second hedge is that the equation above should only hold when one allows the similarity parameters to adjust in the direction of optimizing performance. With nested category structures, selectivity at a lower level will also benefit performance at higher levels of categorization. Therefore, the prediction that higher levels will yield better performance ought to hold even

Table 9.2. Similarity analysis of multiplicative exemplar model using average rather than summer similarity applied to Corter, Gluck and Bower (1988).

	Average Similarity		
Level	Within	Between	
Superordinate	$\frac{1+S_2+2S_1S_2S_3}{4}$	$\frac{S_1S_2S_3+S_1S_3+2S_1S_2}{4}$	
Basic	$\frac{1+S_2}{2}$	$\frac{3S_1S_2S_3+S_1S_3+2S_1S_2}{6}$	
Subordinate	1	$\frac{S_2+3S_1S_2S_3+S_1S_3+2S_1S_2}{7}$	
with $S_1 = 0$:			
Superordinate	$\frac{1+S_2}{4}$	0	
Basic	$\frac{1+S_2}{2}$	0	
Subordinate	1	$\frac{S_2}{7}$	
with $S_1 = S_2 = S_3 = .30$			Ratio
Superordinate	.338	.052	.867
Basic	.650	.058	.918
Subordinate	1.00	.093	.915

S_1= similarity parameter on first dimension (gums)
S_2= similarity parameter on second dimension (eyes)
S_3= similarity parameter on third dimension (rash)

when selective attention is allowed. The sole exception that comes to mind is the possibility that selective attention might operate more effectively at one level than others. For example, errors may drive the adjustment processes faster than correct guesses, and correct guesses should occur more frequently for the superordinate level than for the basic level. Still, it appears that the most natural application of exemplar models fails to predict a basic level advantage. At the very least, this represents a serious challenge to exemplar models that rely on summed similarity.

Alternatively, one could formulate an exemplar model based on average rather than summed similarity (with both the average and the sum taken over all members of a category). As a result, the model would favor the basic level over the superordinate in experiments such as those of Corter et al. (1988). To see this, consider the design in Table 9.1 in terms of average within-category similarity versus average between-category similarity. This is summarized at the top of Table 9.2. The special case in which S_1=0 is clear, as shown in the bottom half of Table 9.2. One immediately sees that the basic level will be favored over the superordinate level. Furthermore, as S_2 increases, the basic level will be favored over the subordinate (the exact point of the shift depends on whether one assumes performance is a function of differences versus ratios). Another way to see this is to let $S_1 = S_2 = S_3 = .30$ and note that the ratio of average within-category similarity to average within plus between category similarity is maximized at the basic level. This case is shown in the bottom third of Table 9.2. For at least some values of similarity, the average similarity exemplar model picks out the basic level.

We will return to the multiplicative exemplar model using average or normalized similarities a bit later. For the moment we simply note the following problem. The average similarity exemplar model implies an insensitivity to category size or frequency but the data suggest that people's categorization is, in fact, at least partially sensitive to frequency (e.g., Medin & Florian, 1991; Nosofsky, 1988). The asymptotic performance of the adaptive network model is also insensitive to category frequency. The summed similarity exemplar model handles frequency effects but does not pick out the basic level as easiest. Overall, we are in the uncomfortable situation of having no model which simultaneously gives an account of category frequency and basic levels effects. Rather than dwell on these difficulties, we introduce a few further ones. First, we take a second look at natural categories and then describe some additional results growing out of this analysis.

VI. FURTHER COMPLICATIONS

A. Natural categories reconsidered

Although both natural and artificial categories have yielded results favoring a basic level of categorization, one cannot be sure that the underlying basis for these results is the same. In fact, a closer analysis of the category structures used with artificial stimuli gives cause to fairly strong reservations. Consider again Table 9.1, summarizing the Corter et al. (1988) design. Note that for the middle level of categorization there is always a feature that only and always is associated with it. In short, a defining feature is present at the middle level. But we began this chapter by noting the strong evidence and near uniform consensus that natural concepts do not have defining features (let alone a single defining feature). The same problem arises for the Hoffman and Ziessler (1983)

Table 9.3. Abstract category structure of Murphy and Smith (1982) Experiment 1.

Stimulus	Feature Dimensions			
	Handle	Shaft	Head-Blade	Size
1	2	2	0	0
2	2	2	0	1
3	2	2	1	0
4	2	2	1	1
5	0	3	4	0
6	0	3	4	1
7	1	3	4	0
8	1	3	4	1
9	3	4	2	0
10	3	4	2	1
11	3	4	3	0
12	3	4	3	1
13	4	0	5	0
14	4	0	5	1
15	4	1	5	0
16	4	1	5	1

studies as their design also maps onto that of Table 9.1. It is not entirely clear how the abstract structure of the Murphy and Smith (1982) study should be represented but one plausible scheme, taken from Corter et al. and shown in Table 9.3, reveals that numerous defining features were available at the basic level and not at the subordinate or superordinate level. It also appears that there were defining features in Mervis and Crisafi's stimuli (see Figure 9.3).

The basic level advantage for artificial categories may be related to defining features. However, we have noted that natural categories are not characterized by defining features. As a result, the explanation for the basic level advantage in natural categories may have little to do with the explanation for this advantage in artificial categories. It is possible that a measure such as category utility picks out the basic level for both artificial and natural categories but the presence of defining features undermines the parallels. It is difficult to assess the seriousness of this defining feature issue. We turn attention now to some recent studies of levels effects in which defining features were not present. Having adopted a skeptical frame of mind, however, we should note that these studies are themselves far from immune to criticism.

B. Fuzzy artificial categories and the basic level

Recently, the first author studied levels effects using artificial categories without defining features (Lassaline, 1990). The studies were also designed to evaluate concept learning models with respect to the basic level. We decided that there was no absolute need to have three levels of categorization, as the theories and metrics we have described require only two levels for a contrast. Therefore the artificial categories were organized into a two-level hierarchy with two General categories and four Specific categories, two nested in each General category. In addition, training level was varied between rather than within subjects. One reason for this choice is that it has proven surprisingly difficult for subjects to learn multiple levels of categorization simultaneously. Previous studies with artificial categories have used the procedure of training people on one level at a time and only later integrating across levels.

The first experiment was an exploratory one, designed to show that a levels effect can be obtained for category structures that do not contain defining features. The design of the first study is summarized in Table 9.4. The stimuli were sixteen objects, each consisting of a value on four different dimensions, D1, D2, D3, and D4 (two shape dimensions and two texture dimensions), modeled after the tool-like stimuli of Murphy and Smith. The General level consisted of two categories (Categories A and B). There were four categories for the Specific level (Categories C, D, E, and F). Unlike previous work, there was no single feature or combination of features at any level that was necessary and sufficient to determine category membership.

Subjects were trained to verify category membership upon presentation of a picture of one of the objects and a category name. After training on the categories, subjects were tested with a speeded version of the category verification task. Average response latencies and error rates across subjects, by level, were used to evaluate predictions for ease of category learning and verification made by three theories: category utility, the adaptive network model, and the exemplar similarity model.

Category utility, the adaptive network model, and the summed similarity exemplar model predict an advantage for the General level of categorization. The general level partition has a category utility value of .266 whereas the specific level has a value of .211. The predictions of the average similarity exemplar model depend on values of the similarity parameters--large values favor the Specific level whereas smaller ones favor the General level. The predictions of the adaptive network model were derived through simulations assuming a configural cue

Table 9.4. Lassaline (1990) Experiment 1 Stimulus Structure.

Example	D1	D2	D3	D4	Category General	Category Specific
1	1	1	1	0	A	C
2	1	0	1	0	A	C
3	1	0	0	0	A	C
4	0	1	1	0	A	C
5	1	1	0	1	A	D
6	0	1	0	0	A	D
7	0	1	0	1	A	D
8	1	0	0	1	A	D
9	2	2	2	0	B	E
10	2	0	2	0	B	E
11	2	0	0	0	B	E
12	0	2	2	0	B	E
13	2	2	0	2	B	F
14	0	2	0	0	B	F
15	0	2	0	2	B	F
16	2	0	0	2	B	F

representation and setting any negative output values to zero before computing the response probabilities. Across variations in the simulations, these predictions did not vary. Specifically, whether or not negative outputs were transformed to zero, whether or not the stimulus representation was supplemented with configural cues, and across various learning rates, the adaptive network model predicts a General level advantage.

The results of this first experiment were consistent with the General level advantage predicted by category utility, the adaptive network model, and the summed and average similarity exemplar models. As Figure 9.4 indicates, the General level proved easier to learn than the Specific level. Although this experiment does not distinguish between models (since all of them predict a general level advantage), it did demonstrate that a levels effect can come about with fuzzy artificial categories.

The second experiment was somewhat more diagnostic in that the various models did not all make the same prediction. The abstract structure of the experiment is shown in Table 9.5. Each of the twelve stimuli was constructed from three dimensions, D1, D2, and D3 (two shape dimensions and one texture dimension), each of which had three possible values. Again, no defining feature is present. In addition, no single feature is even sufficient to determine category membership at either level, although the Specific categories do have a feature with perfect category validity. That is, all members of Category C have a value

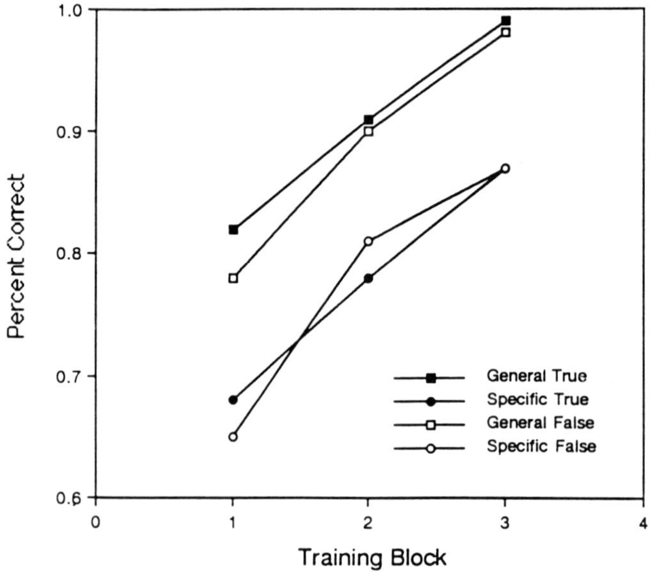

Figure 9.4. Training accuracy for General and Specific categories (Lassaline, 1990, Experiment 1).

of 1 on D2; all D's have a value of 1 on D3; all E's have a value of 2 on D2; and all F's have a value of 2 on D3.

The category utility value is .118 for the General categories and .267 for the Specific categories. Thus, category utility predicts a Specific level advantage. The average similarity exemplar model also predicts a Specific advantage. The summed similarity exemplar model predicts, of course, an advantage for the General level, as within-category similarity increases with category size. In our simulations, the adaptive network model also predicts an advantage for the General level.

The Specific level proved to be much easier to learn than the General level. These learning differences carried over to speeded categorization trials given after training, as shown in Figure 9.5. In brief, the data support category utility and the average similarity exemplar model over the summed similarity exemplar model and the adaptive network model.

The most challenging data come from a third experiment whose design is summarized in Table 9.6. Structure (One Dimension or Four Dimensions) was varied between subjects, such that each subject learned the twelve items in either

Table 9.5. Lassaline (1990) Experiment 2 Stimulus Structure.

Example	D1	D2	D3	Category General	Category Specific
1	1	1	2	A	C
2	1	1	0	A	C
3	2	1	0	A	C
4	1	2	1	A	D
5	1	0	1	A	D
6	0	0	1	A	D
7	2	2	1	B	E
8	2	2	0	B	E
9	1	2	0	B	E
10	2	1	2	B	F
11	2	0	2	B	F
12	0	0	2	B	F

the One Dimension or the Four Dimensions structure. Each of the sixteen items was constructed from four dimensions, D1, D2, D3, and D4 (two shape and two texture dimensions) each of which had four possible values. Note that we have introduced a defining feature for the Specific level of categorization, and thus a disjunction of two features is defining at the General level. Of primary interest was the effect of the distribution of the defining feature across dimensions. For the One Dimension case, the defining feature for each of the Specific categories consisted of a value on the same dimension (values 1, 2, 3, and 4 on D1 for Categories C, D, E, and F, respectively). For the Four Dimensions case, the defining feature was taken from a different dimension for each of the Specific categories (a value of 1 on D1, D2, D3, or D4 for Categories C, D, E, or F, respectively).

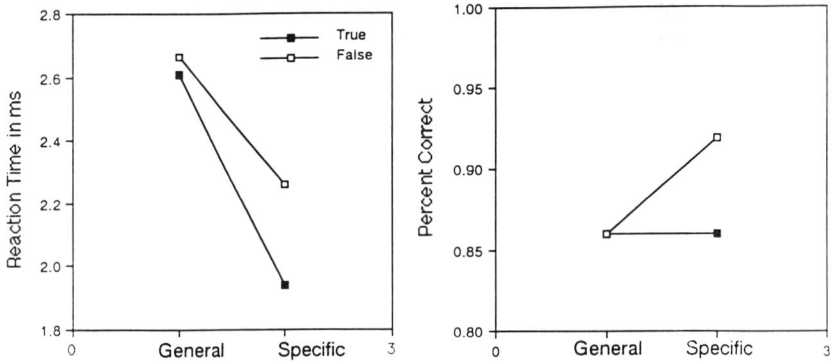

Figure 9.5. Category verification times and classification accuracy for General and Specific categories (Lassaline, 1990, Experiment 2).

Table 9.6. Lassaline (1990) Experiment 3 Stimulus Structure.

	Structure								Category	
	One Dimension				Four Dimensions					
Example	D1	D2	D3	D4	D1	D2	D3	D4	General	Specific
1	1	1	3	4	1	2	4	3	A	C
2	1	2	4	1	1	3	2	4	A	C
3	1	3	1	2	1	4	3	2	A	C
4	2	4	2	1	2	1	3	2	A	D
5	2	1	3	2	3	1	4	2	A	D
6	2	2	4	3	4	1	2	3	A	D
7	3	3	1	3	4	3	1	2	B	E
8	3	4	2	4	2	4	1	3	B	E
9	3	1	3	1	3	2	1	4	B	E
10	4	2	4	2	3	4	2	1	B	F
11	4	3	2	3	4	2	3	1	B	F
12	4	4	1	4	2	3	4	1	B	F

Neither the category utility measure nor the adaptive network model is sensitive to how the defining features are distributed, and thus predict no difference in category learning or verification between the One-dimension and the Four-dimension cases. Versions of the summed similarity exemplar model that allow selective attention can predict that the One-dimension case will be easier than the Four-dimension case. Given that the average similarity exemplar model also allows for selective attention, it can also predict that the One-dimension case will be easier than the Four-dimension case.

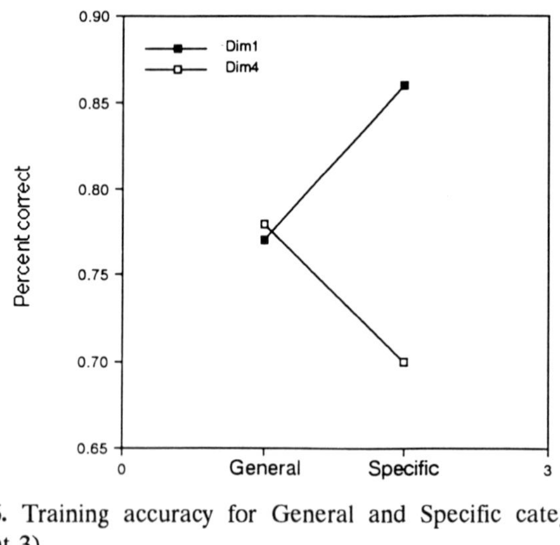

Figure 9.6. Training accuracy for General and Specific categories (Lassaline, 1990, Experiment 3).

With respect to the contrast between levels, the category utility is .167 for the General level and .250 for the Specific level, and thus category utility predicts a Specific advantage. Both the summed similarity exemplar model and the adaptive network model predict that the General level will be easier for both the One-dimension and the Four-dimensions cases. The average similarity model, like category utility, predicts that the Specific level will be easier than the General level in either case.

Somewhat surprisingly, levels interacted strongly with how the defining feature was distributed. As Figure 9.6 indicates, for the One-dimension case, the Specific level was much easier than the General level both in learning and at test, but for the Four-dimension case the General level was easier than the Specific level.

C. Summary

The first two experiments show that one can obtain clear levels effects for category structures that do not contain defining features. In these experiments, the favored level was picked out by the category utility measure. The third experiment, however, yielded an interaction of level with feature distribution that is outside the scope of category utility. Table 9.7 presents a summary of the predicted and obtained results of these experiments.

Table 9.7. Summary of predicted and obtained level advantage from Experiments 1, 2, and 3.

	Predicted Advantage				Obtained Advantage
	Category Utility	Context Model	Adaptive Network	Rule Account	
Experiment 1					
Level	General	General	General	General	General
Experiment 2					
Level	Specific	General	General	?	Specific
Experiment 3					
Level(1D)	Specific	General	General	Specific	Specific
Level(4D)	Specific	General	General	?	General
Structure	No Diff	One Dim[a]	No Diff	One Dim	One Dim

[a] One Dimension structure.

To our knowledge, no current model can account for the full pattern of Lassaline's results. Although the multiplicative average similarity exemplar model can predict these results, these predictions are parameter specific. We have been unable to find a principled reason for varying the parameters in the manner necessary to accommodate this set of results.

D. Rule-Based Accounts

Going beyond the formal models and metrics we have presented thus far, we will next briefly discuss a class of models that use rules to define category membership. Many of the participants in these studies report using rules. It turns out that a rule-based account may well capture many of Lassaline's results, but it is difficult to evaluate such an account quantitatively, as there are no well-specified rule-based models.

Turning to a rule-based account, the ordering of levels with respect to difficulty of learning and verifying category membership was, for the most part, consistent with subjects' use of rules across the three experiments. Subjects' use of rules to learn categories was tested for each experiment by deriving the simplest logical rules that captured the category structure used. The general idea is that categories described by simple rules should be easier to learn than categories requiring complex rules. Of course, it is possible that there might be some simple rule that is difficult to discover. In Lassaline's experiments, however, the dimensions and values are all fairly salient so it is plausible that simple rules were easier to discover than complex rules.

The rule account can also be used to predict performance on individual items. For example, if a subject learning the Specific categories from Experiment 1 uses the rule "a value of 1 or 2 on D3 leads an item to be placed in Category C or Category E, respectively; a value of 1 or 2 on D4 leads on item to be placed in Category D or F, respectively". If subjects learning the Specific categories used this rule, then performance involving the one member of each Specific category that doesn't possess the feature specified in the rule (Items 3, 6, 11, and 14 from Categories C, D, E, and F, respectively) should be poorer than for the other items. The results are consistent with use of this rule. The only inconsistency between the data and the simplest-rule account came from Experiment 2 (see Table 9.5 for category structure). The simplest-rule account predicted an advantage for certain items in this experiment that was not obtained. The derivation of predictions for Experiment 2 for a rule-based account was not as straightforward as that for the first and third experiments, as explicit rules to capture category membership were not as obvious. For the General categories, a value of 1 on the first dimension for Category A and a value of 2 on the first dimension for Category B characterizes four of the six members of each category. It is also possible to capture General category membership by using a conjunction of the necessary features that characterize the Specific categories, i.e. for Category A, conjoining the necessary feature for Category C, a 1 on the second dimension, and the necessary feature for Category D, a 1 on the third dimension. For each Specific category, in contrast, its necessary feature is not sufficient as one nonmember possesses it. There is one exception to using a rule based on the necessary feature for each Specific category (i.e. Item 10 for Category C, Item 7 for Category D, Item 4 for Category E and Item 1 for Category F); whereas for each General category, there are two exceptions to using the rule based on a conjunction of the second and third dimensions (i.e. Items 7 and 10 for Category A and Items 1 and 4 for Category B), and three exceptions to using the rule based on the first dimension (i.e. include Items 3 and 6 but exclude Item 9 for Category A, and include Items 9 and 12 but exclude Item 3 for Category B).

To determine whether or not subjects used any of the rules described above, data from Experiment 2 were analyzed by item. Verifying category membership for Item 6 for the General level and Item 4 for the Specific level was significantly more difficult than for all other items. There were no other item differences. While these two differences are both consistent with use of the rules described above, the results of the item analysis as a whole do not strongly suggest that subjects used rules as a strategy to learn categories and verify membership in Experiment 2. This inconsistency is not terribly damaging to the rule-based account, though, as it is possible that subjects were simply employing different rules than those proposed by the experimenter. Until a model based on use of logical rules is formally specified and supplemented with a mechanism for exemplar retrieval, it is impossible to quantitatively evaluate the rule-based account. Yet the notion that subjects were employing logical rules in classification may plausibly characterize performance across the three experiments.

Martin and Caramazza (1980) suggest that previous research with artificial categories which has been used to support similarity-based models may actually reflect use of a logical rule system for classification. These authors claim that even using categories with no set of defining features, subjects attempt to learn categories by a sequence of feature tests evaluating logical rules for category membership. This claim is supported by a series of experiments using artificial categories in which the pattern of reaction times for category verification and typicality judgments was consistent with performance based on sequential feature testing. Martin and Caramazza emphasized the importance of analyzing individual subjects' data, as subjects may be using the same process to perform the categorization task (in this case the process would be developing rules to capture category membership), but may differ in the specific rules developed.

Mixed support for a rule-based account comes from experiments by Nosofsky, Clark and Shin (1989) testing the extent to which exemplar-based similarity models can account for classification of rule-described categories. The rule account formulated by these authors assumes that subjects adopt the simplest possible rule for partitioning categories (Shepard, Hovland & Jenkins, 1961). In two of the experiments by Nosofsky et al. (1989), when subjects were explicitly instructed to use specific rules, classification performance was better predicted by the minimal-rule account than by the exemplar model, but could be predicted by the minimal-rule account. In contrast, when subjects were not explicitly instructed to use rules, the exemplar model provided a better account of the findings.

Experiments involving category construction do provide support for a rule-based account (Ahn and Medin, 1989). These authors proposed a two-stage categorization model that accurately predicted when family resemblance sorting will occur. In the first stage, initial categories are created based on values on a single primary dimension, a defining feature for the initial categories. Remaining exemplars not classified by values on the defining feature are classified in the second stage according to overall similarity to members of the initial categories. This model integrates use of an explicit unidimensional rule with exemplar similarity computations to produce categories. Models based on category utility or similarity alone were unable to predict when family resemblance categories would and would not be constructed.

In formulating rule-based models, psychologists have tended to assume that simplicity governs the choice of rule derived to capture category membership. Medin, Wattenmaker and Michalski (1987) argue instead that human rule induction is constrained by a category validity bias and a bias toward rules that are more specific than necessary, perhaps to maximize inferences and protect the rule system from over-generalizing. By exploring such constraints, one might develop more precise rule-based models. In addition, some mechanism governing exemplar storage and retrieval is needed to supplement the rule system. As things stand now, we cannot conclude a great deal about the viability of a rule-based account of hierarchical categorization, as current rule models are imprecise.

E. Identification Versus Classification

A limiting case of levels effects is identification learning, or learning in which each example must be placed into a separate category. The classic studies of Shepard, Hovland and Jenkins (1961) looked at the relative difficulty of different types of category partitions. The stimuli were the eight possible combinations of three binary valued dimensions (111, 110, 101, 011, 001, 010, 100, and 000). For two equal-sized categories, there were six distinct types of category partitions (one corresponding to a single defining feature, one to correlated attributes within a category, one to a linearly separable category with characteristic features, and so on). Shepard et al. found that classification learning was easier than identification learning except for the worst category structure, where no difference was found. Nosofsky (1984) demonstrated that the multiplicative summed similarity exemplar model could account for the ordering of learning difficulty, if selective attention to or weighting of dimensions is allowed. Surprisingly, however, the category utility measure is completely unable to account for the Shepard et al. results. Category utility predicts that identification learning should be easier than classification learning for four of the category types and that there should be no difference for a fifth type. Only for the defining features case does category utility predict classification to be superior to identification.

F. Summary

The initial rosy picture of parallels between artificial and natural basic level categories has turned somewhat problematic. Studies with artificial categories have often used defining features. This undermines the idea that relevant structural properties of natural categories have been incorporated into artificial categories but leaves the hope that some metric or theory will correctly predict levels effects in both domains. To our chagrin, however, none of the metrics and theories hold up well, even within the domain of artificially constructed categories. Lassaline's finding of an interaction of level with feature distribution was not predicted by any metric or theory. Before attempting to draw out the implications of this series of challenges, we take one further look at linkages between artificial and natural categorization.

VII. ARTIFICIAL VERSUS NATURAL CATEGORIES: FURTHER OBSERVATIONS

There are a number of recently observed aspects of natural categorization that typically have not been examined in artificial category studies. We see these characteristics as challenges to any complete theory of categorization. Furthermore, we are fairly sure that these aspects of natural categorization play an important role in the basic level advantage. Below, we describe these aspects and speculate on how they might be relevant to basic categories.

A. Conceptual Functions

Certainly, our concepts have uses or functions. In almost all research on concepts, including that involving the basic level, two conceptual functions have been emphasized. First, we use concepts to *classify* or identify items. Classification is a prerequisite for many cognitive and behavioral tasks. For example, building a bookcase might require nails, a hammer, a saw, and wood. It seems necessary that one be able to identify such objects before carrying out such a task. Second, we use concepts to *predict* information about an item. For example, if one has classified an object as a bird, one might reasonably conclude that it will fly. The importance of prediction is fairly obvious. If we can predict things about an object or event, we can anticipate potentially harmful or beneficial consequences, and take appropriate steps to avoid or insure them.

As we have noted, models typically have accounted for the basic level advantage by suggesting that this level maximizes some function of the cue and category validity of features. Cue validity emphasizes the classification function of concepts. It measures how distinctive a feature is with respect to other categories. A feature with maximum cue validity allows one to classify with certainly an item possessing that feature. Such a feature would be unique to members of the category and therefore very distinctive. Category validity emphasizes the prediction function of concepts. It measures how common a feature is with respect to other categories. Given that an item belongs to a category, one can predict a feature of a category member with certainty, if the feature has maximum category validity (i.e., is common among all members of the category).

Recently, some researchers have emphasized that people use concepts for a *variety* of purposes (Matheus, Rendell, Medin, & Goldstone, 1989; Medin, 1983). Concepts have other functions besides classification and prediction. Below, we briefly describe other possible functions of concepts.

Learning. We use concepts to learn other concepts. There are probably a variety of ways that concepts facilitate learning. For example, one might learn about an unfamiliar concept (e.g., *electricity*) using reasoning by analogy to a familiar concept (e.g., *water flow*). Gentner and Gentner (1983) discuss this type of learning. Existing concepts might also provide expectations or biases that guide learning of new concepts. For example, when you first learned about computers you previously may have been told that they could be used for text processing and mathematical calculations. Learning about computers may have been a process in which you related these abstract expectations to more concrete features of the computer (particular keypress sequences, displays on the computer screen, etc). Researchers in the field of machine learning have called this type of learning *operationalization* (Dejong & Mooney, 1986; Mostow, 1983). In general, it involves translating abstract, high level expectations into more specific, low level information.

Explanation. We use concepts to explain and understand why things happen. For example, one might explain why a bottle of Coca-Cola shattered after being left in the freezer for several hours by noting that water expands when it freezes, Coca-Cola is composed of almost entirely of water, and so on. The role of explanation is related to prediction: if we can explain an event (for example), we generally know the conditions under which it will occur. Therefore, if we can identify these conditions in a new situation, we can predict the event.

Conceptual combination. We also combine existing concepts to create new ones (see Murphy, 1988; Wisniewski & Gentner, 1991 for reviews). For example, one might combine the concepts *elephant* and *box* to form the new concept *elephant box*, which might mean "a special type of box for transporting elephants." Conceptual combination is another way that we learn concepts. It is also an efficient way of adding new terms to a language (Downing, 1978).

Communication. Of course, we also use concepts to communicate with other people. One role of concepts in communication is to efficiently convey knowledge. There is a wealth of information associated with a concept (Barsalou, 1989). Furthermore, people believe that their concepts are similar to other people's (Rey, 1983; Miller & Johnson-Laird, 1976). Therefore, in uttering a phrase, a speaker may mean more than what is explicitly stated and believe that the listener will have access to the knowledge implicit in the utterance. For example, a speaker uttering the sentence, "She began eating the soup and it burned her mouth" assumes that the listener has a similar concept of *soup* and does not have to explicitly state why the soup burned the woman's mouth.

Theories of the basic level have assumed that this level best captures the functions of classification and prediction. The design of category structures involving artificial stimuli primarily has varied factors associated with these functions (e.g., commonality and distinctiveness of features). Little work has attempted to relate the basic level to other conceptual functions. Furthermore, one intriguing possibility is that concepts at different levels in a taxonomy may differ in their conceptual functions. One possible reason for the basic level advantage in so many cognitive tasks is that these tasks correspond to the functions ideally suited for the basic level, and not other levels. Further research needs to examine the functions of other concepts like superordinates and subordinates.

Some work has addressed the idea that superordinate and basic concepts have different functions. Based on a linguistic analysis of basic and superordinate terms, Wisniewski and Murphy (1989) argued that people primarily use superordinate terms to conceptualize related groups of objects and use basic concepts to conceptualize single objects. To test this hypothesis, Murphy and Wisniewski (1989) had people categorize objects that were isolated or in scenes. The scenes were naturalistic depictions of actual places that contained members of a superordinate category. Consistent with many studies of the basic level, when the object was isolated, people were faster at categorizing an object at the basic level than at the superordinate level. However, when the object was in a scene, superordinate and basic categorization took about the same amount of time. Furthermore, superordinate categorization was more disrupted than basic categorization when the object was in an improbable scene. These studies suggest that superordinate and basic concepts have different functions.

B. Selective Induction Versus Constructive Induction

Artificial category studies of the basic level have the virtue that, in general, the category structure created by the experimenter closely approximates the structure perceived by the subject. The features typically used are well-specified and unambiguous. As a result, the experimenter and subject probably agree on the features that are present in the stimuli. (This characteristic also is true of other artificial category studies that do not explicitly address the basic level.)

However, many researchers have suggested that the crucial problem in induction is determining the units of analysis or constituents upon which to perform induction (e.g., Medin, Wattenmaker, & Michalski, 1987). That is, induction may involve selecting features from a feature space (i.e., selective induction) but more importantly, it involves constructing that feature space (i.e., constructive induction). In artificial category studies of the basic level, the constructive induction problem has been "solved" for the learner by the experimenter, who typically uses stimuli with well-specified, unambiguous constituents. This is as true for our studies as for any others.

Some researchers have argued that constructive induction may result from an interaction of people's prior expectations or background knowledge and information delivered by the perceptual system (e.g., Wisniewski & Medin, 1991). One criticism of artificial category studies is that they create situations in which a person's use of background knowledge is unnecessary for category learning. For example, we are sure that people do not need to use prior knowledge to learn the categories shown in Table 9.1.

Some evidence for this view of constructive induction comes from studies involving children's drawings. Wisniewski and Medin (in press) had two groups of subjects learn about the same categories, but the groups were given different, meaningful labels for the categories. Specifically, one group of subjects was told that the drawings were done by creative and noncreative children. The other group was told that the same drawings were done by farm and city children. The purpose of giving the groups different labels for the categories was to activate different, prior knowledge in the two groups during learning.

This simple experimental manipulation had important effects on the *kinds* of features that people perceived in the drawings. Furthermore, it appeared that people's prior knowledge helped determine these features. A striking finding was that different people sometimes perceived the *same* part of a drawing as a *different* feature. For example, consider the two drawings shown in Figure 9.7.

Many subjects in the Creative/Noncreative group believed that creative children were likely to draw detailed pictures. In contrast, many people in the Farm/City group believed that city children would draw people that they were likely to encounter in the city. Accordingly, one person in the Creative/Noncreative group interpreted the circled part of drawing 1 as *buttons* (evidence for detail), whereas a person in the Farm/City group interpreted the same part as a tie (evidence for a business person). In drawing 2, a person in the Creative/Noncreative group interpreted the circled part as a *pocket* (again, evidence for detail), whereas a person in the Farm/City group interpreted the same part of the drawing as a *purse* (evidence of a woman from the city).

Of course, such features were not constructed solely from people's prior expectations. For example, consider again the circled line configuration in drawing

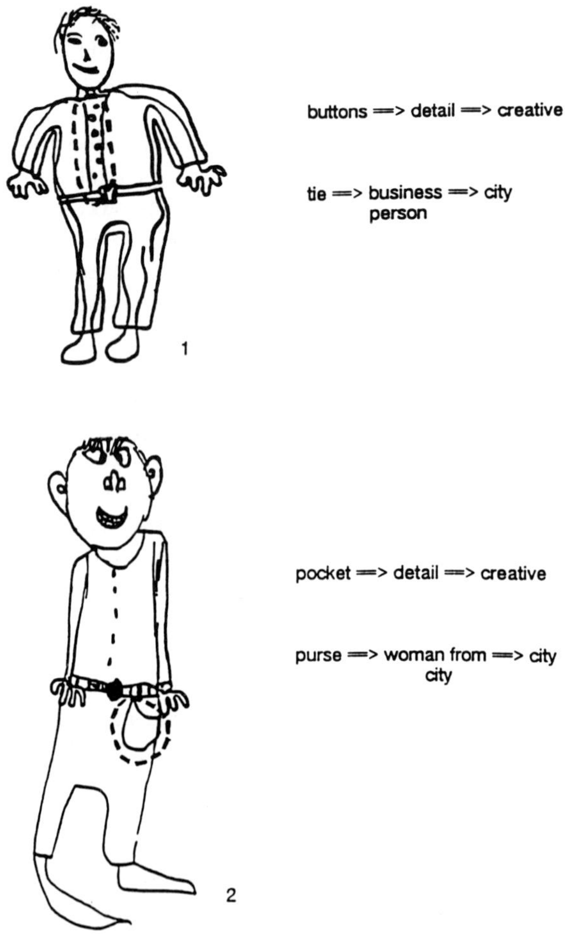

Figure 9.7. Constructive induction processes in classifying children's drawings (Wisniewski and Medin, in press).

2 of Figure 9.7. While a person interpreted the configuration as a *pocket*, therefore providing evidence of detail, the person did not interpret the configuration as *shoelaces, eyebrows, buttons*, etc., even though such features were mentioned by other subjects as evidence of detail. Most likely, the perceptual system provides important information that constrains which features are plausible, such as low-level shape descriptions and locations (e.g., Biederman, 1985; Marr & Nishihara, 1978).

In more natural learning situations, it may be more accurate to view features as *hypothetical entities* that people believe exist among the category members with varying degrees of certainty. As they experience more members and environmental feedback, they adjust these certainties. Sometimes the certainty of an hypothesized feature becomes very low and the hypothesis is abandoned. A new hypothetical feature may take its place. In fact, Wisniewski and Medin (1991) found that when people incorrectly categorized a drawing, they sometimes abandoned those feature interpretations that supported the incorrect category and reinterpreted features of the drawing in a way that was consistent with the correct category.

We should at least consider the possibility that determining the features of natural categories is a very important aspect of learning about them. Studies of the basic level using artificial stimuli have not addressed this aspect of category learning. There are at least two potential problems with ignoring constructive induction. First, constructive induction may be related in some way to the basic level advantage. Second, because of constructive induction processes, a person's perceived category structure could be different from the category structure assumed by the experimenter.

C. Features at Different Levels

A number of researchers have suggested that concepts at different levels in a taxonomy have different features, or similar features that differ in importance. (This suggestion also is consistent with the idea that conceptual function may vary with taxonomic level.) Rosch et. al (1976) suggested that the attributes of superordinate and basic concepts are qualitatively different. In particular, basic concepts contain many perceptually salient features whereas superordinate concepts contain more abstract or functional features (e.g., "plays music" for *musical instrument*). As previously discussed, Tversky and Hemenway (1984) argued that, compared to other concepts, basic concepts tend to represent the parts of objects, and that this difference accounts for the basic level advantage. Murphy and Wisniewski (1989) suggested that superordinate concepts not only contain abstract or functional features, but information about the scenes in which their members are likely to be found. For example, *furniture* might include a representation of a living room scene, linking the concepts *couch, lamp, coffee table, and chair* together, and a representation of a bedroom scene, linking the concepts *bed, dresser, and mirror* together. Furthermore, these basic concepts could be connected by relations. For example, *lamp and table* could be connected by the relation "sits on" and *chair and table* by the relation "oriented towards." On this account, superordinates are partly defined through their associated basic concepts and the relations between them.

Importantly, most artificial concept studies of the basic level have not varied the kinds of features that describe different levels. One notable exception is the Murphy and Smith (1982) studies. Basic level categories were described by parts and superordinates were described by their functions. However, as previously mentioned, basic categories in those studies could be defined by necessary and sufficient features.

D. Dimensions of Features

Artificial concept studies of the basic level typically have characterized features along the dimensions of commonality and distinctiveness. On the one hand, it seems clear that a model of concept learning should be sensitive to these dimensions. As we have suggested, they play an important role in the classification and prediction functions of concepts.

On the other hand, it seems unlikely that commonality and distinctiveness are the *only* important dimensions that characterize the features represented in concepts. At least in the case of artifacts, features can be characterized in terms of how relevant they are to the artifact's function. To take a simple example, consider the features "uses water" and "has a door that opens downward," associated with the concept *dishwasher.* It seems that "uses water" is much more relevant to the function of a dishwasher than "has a door that opens downward." Intuitively, this feature seems more important to our concept of dishwasher. This is the case even though "has a door that opens downward" is just as common among dishwashers as "uses water," and furthermore, is more distinctive of dishwashers than "uses water." Specifically, artifact concepts might include information about *which* features are related to the function and *how* they are related to the function. Among other things, such information would be crucial in understanding how to carry out the function associated with the artifact and in understanding how to repair an artifact that is not functioning properly.

Recent research supports the idea that the relevance of a feature to an artifact's function affects categorization. A study by Rips (1989) suggests that functional relevance is more important for categorization than other information, such as appearance. He gave subjects descriptions of artifacts that underwent two types of transformations. In one type, the artifact was changed so that its appearance resembled members of another category, while preserving its intended use. In another type, the artifact was changed so that its intended use resembled those of the members of another category, while preserving its appearance. In both cases, subjects tended to classify the artifact into the category whose intended use most resembled that of the artifact. More recently, in a study involving artificial artifact concepts, Wisniewski and Medin (1992) examined the classification of items composed of features that were relevant to the function of one category and features that were irrelevant to the function of another category, but more diagnostic of membership in that category. They found that subjects tended to classify the items into the category containing the functionally relevant features, even though those features were less diagnostic.

Besides functional relevance, features can be viewed at multiple levels of abstraction. For example, consider a drawing of a person that belongs in the category "drawn by a creative child." A low level feature such as a particular *configuration of lines* might be viewed as a *pocket*, which might be an example of *detail*, which in turn, might be an example of *creativity*. One consequence of this process is that features can be treated equivalently. So, the features *shoelaces* and *pocket* are different, but when viewed as examples of *detail*, they are similar.

As evidence for this claim, Wisniewski and Medin (1992) showed that when different features could be conceptualized as examples of the same, higher-level feature, people rated them as more similar than in a case in which they could not be conceptualized in that manner. For example, people were given two descriptions of objects, one containing the feature "uses a poisonous substance" and the other containing the feature "emits microwaves." In a neutral context, both of these features are very different. When told that both objects were "used for killing bugs," people rated the two features as more similar than when told that both objects were "used for transporting people underwater." Presumably, people inferred that "emits microwaves" and "uses a poisonous substance" were both examples of "methods for killing bugs." They were unable to conceptualize these features as examples of the same higher-level feature associated with "used for transporting people underwater" and thus rated them as dissimilar.

This dimension of features could have important effects on how concepts are learned. To take one simple example, consider complex stimuli such as drawings done by emotionally disturbed children. One indicator of emotional problems is the omission of body parts, such as hands, feet, and so on (Koppitz, 1964). Consider two conceptual clustering models that are shown drawings done by emotionally disturbed children. One model groups drawings based on category utility and uses selective induction. The second model views features at multiple levels of abstraction. (The second model also has prior knowledge about emotionally disturbed children.) Both models will cluster the drawings in different ways. The first model will group drawings into categories with common, overlapping features and maximize category utility. On the other hand, the second model will group drawings on the basis of whether or not they have missing body parts, even though such features may not overlap. For example, the feature "missing hands" in one drawing does not overlap with "missing feet" in another drawing. However, because the second model views features at multiple levels of abstraction, it will consider the features "missing hands" and "missing feet" equivalent and group these drawings together.

To summarize, it is clear that features have other important dimensions besides commonality and distinctiveness. Research on the basic level using artificial categories, ours included, has almost exclusively focused on these two dimensions. Exactly how other dimensions are related to the basic level is unclear. Nevertheless, we can speculate on some of the possible relations between the basic level and the dimensions described above. For example, with artifact categories, it may not be that the basic level is the one with the greatest number of common and distinctive attributes, *per se*. Rather, it may be that it is the level with the greatest number of common and distinctive attributes relevant to the category's function. Some evidence for this claim comes from Tversky and Hemenway's (1984) studies. They showed that categories at the basic level had the greatest number of part features, which typically are relevant to a category's function. Another possibility is that the commonality and distinctiveness of features depend *crucially* on the level of abstraction at which people view features. In the example above,it might be more accurate to view the features two members of the category "drawings of emotionally disturbed children" as having a common feature ("missing body parts") even though at a lower level, they possess different features (e. g., "missing hands" and "missing feet"). On the other hand, it might be more accurate to view the feature "round" as somewhat distinct in the categories basketball and cantalope, since at a higher level of abstraction they are different. One can dribble and shoot a basketball and "round" is important in these activities. It would be very difficult to do either activity with a cantalope.

VIII. SUMMARY AND CONCLUSIONS

We now return to two issues with which we began this chapter. First, we have been concerned with the extent to which parallels can be drawn between work using natural versus artificial categories in studying the basic level. It is clear from our analysis that most studies using artificial categories have not carefully incorporated structural properties of natural categories. In particular, experimenters have tended to employ artificial categories that possess defining features, which is inconsistent with what is typically assumed about natural categories. In addition, artificial categories have, for the most part, ignored differences in the types of features and in conceptual function across levels of categorical structure.

Do these differences truly matter? We do not know. A potentially fruitful strategy is to be more rigorous in bringing what's known about natural categories into the laboratory with artificial categories. For the moment, anyway, we believe that caution should be exercised in drawing conclusions about hierarchical categorization in general from studies involving artificial categories. What then, can be gained from the work done thus far using artificial categories?

The primary contribution made by studies of the basic level using artificial categories has been to make possible the evaluation of various classification theories with respect to their account of hierarchical categorization. This is the second concern that this chapter was intended to address. The metrics and theories considered here are constrained by simplifying assumptions about the representation of items to be classified, and are thus difficult to evaluate using natural categories, where the representation is less clear than with artificial ones.

Although category utility does an adequate job of predicting difficulty ordering in the set of basic level studies using artificial categories discussed here, it assumes that features are independent, and is thus insensitive to any nonindependence among features. This is clearly inconsistent with what is known about human categorization. In addition, category utility fails to predict difficulty ordering between identification and classification learning in all but the simplest case, the one in which categories possess defining features.

Exemplar models, in contrast, do well on capturing categorization within horizontal structure, including cases where categories possess correlated attributes. Exemplar models that compute similarity as a sum across items, like cue validity, predict that the most inclusive level of classification will be the easiest to learn. As exemplar models are extended to learning we may get a better picture of how and whether this generalization needs to be qualified. Average similarity exemplar models, in contrast, are not sensitive to category size, but the accuracy of their predictions depends on the particular best-fitting value of the similarity parameter. In addition, although an insensitivity to category size frees the average similarity exemplar model from the prediction that the most inclusive level is basic, it also makes it difficult for such a model to predict category size effects.

The adaptive network model, like the exemplar models, is sensitive to violations of feature independence. At least for Lassaline's studies, though, it did not correctly predict difficulty ordering between levels, but instead predicted a General advantage across experiments.

Finally, accounts based on use of logical rules to capture category membership remain too underspecified to be carefully evaluated. In addition, while the notion that humans learn categories by deriving the simplest logical rule by which to classify members has intuitive appeal and some empirical support with artificial stimuli, the derivation of rules governing natural category membership is not as straightforward.

In summary, the classification metrics and theories we have considered here do not fully account for the basic level advantage found with studies using artificially constructed categories. The prospects for an account of hierarchical categorization in natural categories are perhaps even more dim, as it is unclear how to apply these theories to natural categories. Not wishing to be harbingers of grim news, nor end this chapter on a completely pessimistic tone, we nonetheless find ourselves forced to conclude that our work in understanding the basic level is not done, and further, that without understanding a phenomenon so prominent in categorization as the basic level, we have a long way to go.

ACKNOWLEDGEMENTS

We gratefully acknowledge Woo-kyoung Ahn, Barbara Burns, Jim Corter, Judy Florian, Mark Gluck, Rob Goldstone, Evan Heit, Matt Kurbat, Joshua Rubinstein, David Thau, and Phil Webster for their helpful comments and suggestions on this chapter. We also thank Bob Dylan for inspiration.

REFERENCES

Adelson, B. (1983). Constructs and phenomena common to semantically-rich domains. Cognitive Science Tech. Report No. 14, New Haven CT: Yale University.

Ahn, W.K., & Medin, D.L. (1989). A two-stage categorization model of family resemblance sorting. In Proceedings of the Eleventh Annual Conference of the Cognitive Science Society, Ann Arbor, MI.

Anderson, J.R. (1990). The adaptive character of thought. Hillsdale, NJ: Erlbaum.

Anglin, J.M. (1977). Word, object and conceptual development. New York: W.W. Norton.

Barsalou, L.W. (1989). The instability of graded structure: Implications for the nature of concepts. In U. Neisser (Ed.), Concepts and conceptual development: Ecological and intellectual factors in categorization. Cambridge: Cambridge University Press.

Biederman, I. (1987). Recognition-by-components: A theory of image understanding. Psychological Review, 94, 115-147.

Brown, R. (1958). How shall a thing be called? Psychological Review, 65, 14-21.

Cantor, N., & Mischel, W. (1979). Prototypes in person perception. In L. Berkowitz (Ed.), Advances in experimental social psychology. New York: Academic Press.

Corter, J.E., Gluck, M.A., & Bower, G.H. (1988). Basic levels in hierarchically structured categories. In Proceedings of the Tenth Annual Conference of the Cognitive Science Society, Montreal, Quebec, Canada.

Corter, J. E., & Gluck, M. A. (1990). Explaining basic categories: Feature predictability and information. Unpublished Manuscript.

Dejong, G., & Mooney, R.J. (1986). Explanation-based learning: An alternative view. Machine Learning, 1, 145-176.

Downing, P. (1978). On the creation and use of English nominal compounds. Language, 53, 810-842.

Duda, R. O. & Hart, P. E. (1973), Pattern classification and scene analysis, New York: Wiley.

Estes, W.K. (1986). Array models for category learning. Cognitive Psychology, 18, 500-549.

Fisher, D. (1987). Knowledge acquisition via incremental conceptual clustering. Machine Learning, 2, 139-172.

Fisher, D. (1988). A computational account of basic level and typicality effects. In Proceedings of the Seventh National Conference on Artificial Intelligence, 233-238. Saint Paul, MN: Morgan Kaufman.

Gentner, D., & Gentner, D.R. (1983). Flowing waters or teeming crowds: Mental models of electricity. In D. Gentner & A.L. Stevens (Eds.), Mental models. Hillsdale, NJ: Erlbaum.

Gluck, M.A., & Bower, G. H. (1988). Evaluating an adaptive network model of human learning. Journal of Memory and Language, 27, 166-195.

Gluck, M.A., Bower, G. H., & Hee, M.R. (1989). A configural-cue network model of animal and human associative learning. In Proceedings of the Eleventh Annual Conference of the Cognitive Science Society, Ann Arbor, Michigan. Hillsdale, NJ: Erlbaum.

Gluck, M.A., & Corter, J.E. (1985). Information, uncertainty and the utility of categories. In Proceedings of the Seventh Annual Conference of the Cognitive Science Society, Irvine, CA.

Gluck, M.A., Corter, J.E., & Bower, G.H. (1990). Basic levels in the learning of category hierarchies: An adaptive network model. Unpublished manuscript.

Hinton, G. E. (1987). Connectionist learning procedures. Technical Report CMU-CS-87-115. Carnegie Mellon University.

Hintzman, D. L. (1986). Schema abstraction in a multiple-trace memory model. Psychological Review, 93, 411-428.

Hoffman, J., & Ziessler, C. (1983). Objectidentifikation in kunstlichen begriffshierarchien. Zeitscrift fur Psychologie, 16, 243-275.

Homa, D. & Vosburgh, R. (1976). Category breadth and the abstraction of prototypical information. Journal of Experimental Psychology: Human Learning and Memory, 2, 322-330.

Horton, M.S. & Markman, E. (1980). Developmental differences in the acquisition of basic and superordinate categories. Child Development, 51, 708-715.

Jolicoeur, P., Gluck, M.A., & Kosslyn, S.M. (1984). Pictures and names: Making the connection. Cognitive Psychology, 16, 243-275.

Jones, G.V. (1983). Identifying basic categories. Psychological Bulletin, 94, 423-428.

Koppitz, E. M. (1964). Evaluation of human figure drawings, New York: Grune & Stratton.

Lassaline, M.E. (1990). The basic level in hierarchical classification. Unpublished master's thesis, University of Illinois.

Luce, R.D. (1959). Individual choice behavior. New York: Wiley.

Malt, B.C., & Smith, E.E. (1983). Correlated properties in natural categories. Journal of Verbal Learning and Verbal Behavior, 28, 250-269.

Marr, D., & Nishihara, H.K. (1978). Representation and recognition of the spatial organization of three-dimensional shapes. Proceedings of the Royal Society of London, B200, 269-294.

Martin, R.C., & Caramazza, A. (1980). Classification in well-defined and ill-defined categories: Evidence for common processing strategies. Journal of Experimental Psychology: General, 109, 320-353.

Massaro, D.W., & Freidman, D. (1990). Models of integration given multiple sources of information. Psychological Review, 97, 225-252.

Matheus, C.J., Rendell, L.R., Medin, D.L., & Goldstone, R.L. (1989). Purpose and conceptual functions: A framework for concept representation and learning in humans and machines. The Seventh Conference of the Society for the Study of Artificial Intelligence and Simulation of Behavior. Sussex, England.

Medin, D.L. (1983). Structural principles in categorization. In Tighe, T.J. & Shepp, B.E. (Eds.), Perception, cognition, and development: Interactional analyses. Hillsdale, NJ: Erlbaum.

Medin, D.L., Altom, M.W., Edelson, S.M., & Freko, D. (1982). Correlated symptoms and simulated medical diagnosis. Journal of Experimental Psychology: Learning, Memory and Cognition, 8, 37-50.

Medin, D.L., & Florian, J.E. (1991). Abstraction and selective coding in exemplar-based models of categorization. In A.F. Healy, S.M. Kosslyn, & R.M. Shiffrin (Eds.), From Learning Processes to Cognitive Processes: Essays in Honor of William K. Estes, Vol. 2. Hillsdale, NJ: Erlbaum.

Medin, D.L., & Schaffer, M.M. (1978). Context theory of classification learning. Psychological Review, 85, 207-238.

Medin, D.L., & Schwanenflugel, P.J. (1981). Linear separability in classification learning. Journal of Experimental Psychology: Human Learning and Memory, 7, 355-368.

Medin, D.L., & Smith, E.E. (1984). Concepts and concept formation. Annual Review of Psychology, 35, 113-138.

Medin, D.L., Wattenmaker, W.D., & Michalski, R.S. (1987). Constraints and preferences in inductive learning: An experimental study of human and machine performance. Cognitive Science, 11, 299-339.

Mervis, C.B., & Crisafi, M.A. (1982). Order of acquisition of subordinate-, basic-, and superordinate-level categories. Child Development, 53, 258-266.

Mervis, C.B., & Rosch, E. (1981). Categorization of natural objects. In Rosenzweig, M.R. & Porter, L.W. (Eds.), Annual Review of Psychology, 32, 89-115.

Miller, G.A., & Johnson-Laird, P.N. (1976). Language and Perception. Cambridge, MA: Harvard University Press.

Minsky, M. & Papert, S. (1969). Perceptrons. Cambridge, MA: MIT press.

Morris, M.W., & Murphy, G.L. (1990). Converging operations on a basic level in event taxonomies. Memory and Cognition, 18, 407-418.

Mostow, J. (1983). Machine transformation of advice into a heuristic search procedure. In R.S. Michalski, J.G. Carbonell, & T.M. Mitchell (Eds.), Machine Learning: An Artificial Intelligence Approach, Vol. 1. Los Altos, CA: Morgan Kaufman.

Murphy, G.L. (1982). Cue validity and levels of categorization. Psychological Bulletin, 91, 174-177.

Murphy, G. L. (1988). Comprehending complex concepts. Cognitive Science, 12, 529-562.

Murphy, G. L. (1991). Parts in object concepts: Experiments with artificial categories, Memory and Cognition, 19, 423-438.

Murphy, G.L., & Brownell, H.H. (1985). Category differentiation in object recognition: Typicality constraints on the basic category advantage. Journal of Experimental Psychology: Learning, Memory and Cognition, 11, 70-84.

Murphy, G.L., & Smith, E.E. (1982). Basic level superiority in picture categorization. Journal of Verbal Learning and Verbal Behavior, 21, 1-20.

Murphy, G. L., & Wisniewski, E.J. (1989). Categorizing objects in isolation and in scenes: What a superordinate is good for. Journal of Experimental Psychology: Learning, Memory, and Cognition, 15, 572-586.

Nosofsky, R.M. (1986). Attention, similarity and the identification-categorization relationship. Journal of Experimental Psychology: General, 115, 39-57.

Nosofsky, R.M. (1988). Similarity, frequency and category representations. Journal of Experimental Psychology: Learning, Memory, and Cognition, 14, 54-65.

Nosofsky, R.M. (1991). Exemplars, prototypes and similarity rules. In A. Healy, S. Kosslyn, & R. Shiffrin (Eds.), From Learning Theory to Connectionist theory: Essays in honor of W. K. Estes, Vol. 1. Hillsdale, NJ: Erlbaum.

Nosofsky, R.M., Clark, S.E., & Shin, H.J. (1989). Rules and exemplars in categorization, identification, and recognition. Journal of Experimental Psychology: Learning, Memory, and Cognition, 15, 282-304.

Rescorla, R.A., & Wagner, A.R. (1972). A theory of Pavlovian conditioning: Variations in the effectiveness of reinforcement and non-reinforcement. In A.H. Black & W.F. Prokasy (Eds.), Classical conditioning II: Current research and theory. New York: Appleton-Century-Crofts.

Rey, G. (1983). Concepts and stereotypes. Cognition, 15, 237-262.

Rifkin, A. (1985). Evidence for a basic level in event taxonomies. Memory and Cognition, 13, 538-556.

Rips, L.J. (1989). Similarity, typicality, and categorization. In S. Vosniadou & A. Ortony (Eds.), Similarity and analogical reasoning. Cambridge: Cambridge University Press.

Rosch, E. (1974). Universals and cultural specifics in human categorization. In R. Breslin, W. Lonner, & S. Bochner (Eds.), Cross-cultural Perspective on Learning. London: Sage Press.

Rosch, E. (1975). Cognitive representations of semantic categories. Journal of Experimental Psychology: General, 104, 192-233.

Rosch, E. (1978). Principles of categorization. In Rosch, E., & Lloyd, B.B. (Eds.), Cognition and categorization. Hillsdale, NJ: Erlbaum.

Rosch, E., & Mervis, C.B. (1975). Family resemblances: Studies in the internal structure of categories. Cognitive Psychology, 7, 573-605.

Rosch, E., Mervis, C.B., Gray, W.D., Johnson, D.M., & Boyes-Braem, P. (1976). Basic objects in natural categories. Cognitive Psychology, 8, 382-439.

Shepard, R.N., Hovland, C.I., & Jenkins, H.M. (1961). Learning and memorization of classifications. Psychological Monographs, 75(13, Whole No. 517).

Smith, E.E., Balzano, G.J., & Walker, J. (1978). Nominal, perceptual, and semantic codes in picture categorization. In J.W. Cotton & R.L. Klatzky (Eds.), Semantic factors in cognition. Hillsdale, NJ: Erlbaum.

Smith, E.E., & Medin, D.L. (1981). Categories and concepts. Cambridge, MA: Harvard University Press.

Smith, L.B. (1989). From global similarities to kinds of similarities: The construction of dimensions in development. In S. Vosniadou & A. Ortony (Eds.), Similarity and analogical reasoning. Cambridge: Cambridge University Press.

Tversky, B., & Hemenway, K. (1983). Categories of environmental scenes. Cognitive Psychology, 15, 121-149.

Tversky, B., & Hemenway, K. (1984). Objects, parts and categories. Journal of Experimental Psychology: General, 113, 169-913.

Wisniewski, E.J., & Gentner, D. (1991). On the combinatorial semantics of noun pairs: Minor and major adjustments to meaning. In G. B. Simpson (Ed.), Understanding word and sentence. Amsterdam: North Holland.

Wisniewski, E.J., & Murphy, G.L. (1989). Superordinate and basic category names in discourse: A textual analysis. Discourse Processes, 12, 245-261.

Wisniewski, E.J., & Medin, D.L. (1991). Harpoons and long sticks: The interaction of theory and similarity in rule induction. In D. Fisher & M. Pazzani (Eds.), Concept Formation: Knowledge and Experience in Unsupervised Learning. San Mateo, CA: Morgan Kaufman.

Wisniewski, E.J., & Medin, D.L. (1992). Is it a mornek or a plapel? Prior expectations and functionally relevant features in category learning. Unpublished Manuscript.

Wittgenstein, L. (1953). Philosophical investigations. New York: Macmillan.

Zadeh, L.A. (1965). Fuzzy sets. Information and Control, 8, 338-353.

Commentary

Basic Levels in Artificial and Natural Categories: Are All Basic Levels Created Equal?, M. E. Lassaline, E. J. Wisniewski, & D. L. Medin

IRWIN D. NAHINSKY
University of Louisville

This chapter might be subtitled, "In search of the basic level." Although the author's confess frustration in the attempt to develop an integrated theoretical account of the basic level advantage within a hierarchical category structure, they arrive at a number of insights important for understanding information processing related to such structures in their search.

The basic level is defined as the conceptual level in a nested category structure at which elements are most easily categorized. Rosch introduced the concept in describing the finding that the category level above the most specific level but below the superordinate level is associated with optimum classification for natural categories. The author's turn to the category utility notion to help provide a framework for understanding the phenomenon. Thus, the basic level may be represented by categories for which the product of cue validity, P(category | feature), and category validity, P(feature | category), is a maximum. Large categories are not very prognostic of specific features, but specific features are not very prognostic of small categories. A level associated with some optimum joint values of the two measures might be expected to produce the greatest discrimination among categories. Various models for acquisition and representation of category structure may be assessed and understood using this notion.

There are tests of the ability of a number of models to predict the basic level effect. Adaptive network, prototype, exemplar-based, and rule-based models are considered. No one model was effective in predicting a level advantage in all situations, which embodied the complexities that might be involved in natural categories. Factors related to category structure were important in determining how various models fit the data. Category size, correlation of features, among other factors were related to information processing requirements, which, in turn, would affect strategies for sorting objects into categories.

The authors also point to considerations that affect the manner in which individuals organize the categorization process. The functional relevance of features in specifying a category, the role of explanation and prediction as a basis of organization, and the use of analogy to other concepts are among the factors considered. The authors have taken important steps to provide a framework within which to investigate the development of multilevel conceptual structures.

Percepts, Concepts and Categories
B. Burns (Editor)

10

Episodic Components of Concept Learning and Representation

IRWIN D. NAHINSKY
University of Louisville

I. INTRODUCTION

The nature of acquisition and representation of concepts has been of long standing interest in cognitive psychology. Early systematic investigations of concept learning focused upon the induction of rules in situations well defined for the subject. The resulting theories were based upon processes whereby information was extracted from a set of stimuli to form an abstract representation which embodied defining features. The classic work of Bruner, Goodnow, and Austin (1956) exemplifies the approach. The influence of this work upon theories of concept learning and representation was felt for many years.

Theories used to explain performance in laboratory tasks were "analytic" in nature. That is, they were based upon the notion that learning could be characterized by strategies in which information about stimulus attributes is abstracted from stimulus-category associations, with all information from stimuli not predictive of category membership left behind.

When interest turned to study of conceptual behavior in areas that extended beyond laboratory tasks, which used artificially derived stimuli to control acquisition variables, a corresponding interest in "nonanalytic" aspects of the process developed. Interest grew in variables associated with acquisition context. Variations in acquisition context have been shown to influence the induction process and the ultimate conceptual structure. In this chapter I will explore some of the major trends in conceptualizing the concept learning process in terms of the task stimuli and the context of acquisition. Section II concerns "classical" theories that postulate an abstractive process that allows for graded category membership. Sections III and IV deal with issues related to the influence of specific stimuli upon learning and representation. In Section V I will deal with evidence, mostly from my own research, that indicates a more central role for episodic information in both concept acquisition and representation than has generally been proposed. Sections VI and VII present a framework within which the semantic-episodic relationship in concept representation may be approached, with Section VII containing a processing model to guide relevant research in the area. It proposed that information within the context of category learning and representation is organized in an integrated structure that includes semantic and episodic information, with each type of information accessed by its own separate process.

In the course of the presentation I will attempt to distinguish the roles of semantic and episodic factors in concept learning and representation. There will be a discussion of the ways in which factors that affect mnemonic characteristics of stimuli associated with conceptual categories influence the learning process and, in turn, determine the nature of the category structure.

I will now briefly consider prominent theoretical notions that postulate a major role for abstractive processes in this area. Hypothesis testing theories and prototype theories are the major classes of these approaches.

II. THEORIES POSTULATING AN ABSTRACTIVE PROCESS

Theories based upon the notion that the concept acquisition and representation process are exclusively abstractive in nature dominated thinking in this area for many years. Experiments were done within the context of well-defined concepts with stimuli constructed for such research. Typically, subjects were presented with artificial stimuli, such as geometric figures, systematically varied along a number of dimensions. Individuals were observed in the learning of concepts in which rules were well defined. For example, a subject might have to discover the classification rule, "red figures", which would successfully classify all stimuli as members or nonmembers of the class defined by the experimenter. Results were used to develop models which could broadly be classified under the heading of hypothesis testing models (e.g., Bower & Trabasso, 1964; Falmagne, 1970; Gregg & Simon, 1967; Levine, 1966; Nahinsky, 1970; Restle, 1962; and Wickens & Millward, 1971). These theories in general contain a set of assumptions whereby individuals test hypotheses about potentially relevant dimensions or combinations thereof using feedback about correct stimulus classifications. The subject assumes the role of an "information filter" with no storage of specific stimulus information as part of the representation of the concept. That is, the subject extracts the attribute-category relationship from the stimulus representation and discards the other aspects of the event, including memory for the stimulus itself. The Bruner, et al. (1956) demonstration of the "conservative focusing strategy", a prime example of such an approach, set the stage for subsequent hypothesis testing models and illustrates the kind of processes embodied in hypothesis testing theories. In conservative focusing subjects observe a stimulus that is a category member, a "positive instance". They then vary only one of its attribute values per trial in choosing subsequent stimuli for hypothesis tests. If changing the value of the attribute (e.g., color) results in a negative instance, the characteristic is found to be relevant in defining the concept; otherwise the value varied is irrelevant. This inductive elimination process is abstractive, concentrating only upon attribute values rather than stimuli. As a matter of fact, when the concept and the stimulus domain are well defined, the stimulus selected for initial study should have no influence on the rule that is induced nor upon the course of solution. Memory of stimuli and their category membership status would play no role in the process and could hinder it by burdening working memory. Indeed, theorists viewed the stimulus as a vehicle for conveying category information and postulated that the stimulus did not enter into the induced category structure. Such theories were often successful in situations involving well-defined rules to be learned and a stimulus domain in which context is limited such that stimuli are not distinctive enough to be easily remembered.

As interest developed in the representation of "real world" and "natural language categories", it became apparent that class membership in many of these categories is not clearly defined. The distinction between category members and nonmembers is often "fuzzy". It appears that natural categories are often defined in terms of general resemblance to some representation not clearly specified. Insofar as a general resemblance depends upon a long term aggregation of stimulus information, individual stimuli implicitly assume some role. A major theoretical advance was the introduction of the concept of the prototype.

Posner (1969) and Rosch (1973, 1975, & 1977) represent major exponents of the prototype as a basis for understanding categorization in a number of domains. Posner and Keele (1968), generated ill-defined categories via random distortions of central random dot patterns, called prototypes, for each category. Subjects then were required to learn to classify the distortions correctly into their categories. After learning it was found that individuals could classify the central pattern with greater ease than other members not previously presented, even though this stimulus had not appeared in the learning phase. Since the prototype member was used to generate the category, it represented some kind of "average" of the category over dimensions of variation. Consequently, it was theorized that learning of ill-defined categories involves the acquisition of a prototype. The process presumably requires some internal averaging of presented stimulus dimensions to form the central representation. Such representation is to be distinguished from cumulative memory for individual stimuli which could serve as the basis for comparison to arrive at correct classification. Although old stimuli are classified with more confidence than new stimuli, this advantage tends to fade over time. Rosch and Mervis (1975) presented evidence favoring the prototype notion. For example, they found that more prototypical members of natural language categories shared more characteristics with other members of the category than did less prototypical members. Reed (1972) found that a prototype model predicted classification of schematic faces better than other models. Subjects tended to classify stimuli on the basis of distance from the prototype rather than on the basis of average distance from category members, distance from a "nearest neighbor", or average cue validity.

Smith, Shoben, and Rips (1974) examined representation of natural language categories from a related point of view. Semantic features serve as the basis for representing concepts in their approach. Some features are defining (i.e., common to all category members and necessary for inclusion). For example, "wings" may be a defining feature for the category, bird. Other features are characteristic (i.e., more or less typical of category members). For example, "undomesticatedness" is typical of birds, but not a characteristic necessary for class inclusion. To the extent that a member possesses characteristic features, it is a "typical" category member. Thus, category membership is graded, ranging from very typical members to very atypical members, with category boundaries "fuzzy". Categories may be represented by a composite of characteristic and defining features which assumes the role of prototype. Distance from the prototype in multidimensional feature space defines a member's typicality. Correlated measures of inter-stimulus similarity, judged typicality, and speed of category membership verification lend support to the conceptualization.

For the most part, theories discussed in this section posit a representational structure for a category based upon abstracted information in a semantic memory system. Correspondingly, a process for verifying category membership would depend upon tests for the presence or absence of key features of a given stimulus. Recent evidence suggests that the episodic frame of reference present during acquisition of category information may play a vital role in concept development and representation. Brooks (1978), for example, has explored situations in which nonanalytic approaches are prominent in arriving at a concept. In situations, such as learning artificial grammars, analysis into components often proves hard to apply. Brooks proposes that, in instances in which situational complexity defies analysis, the use of a salient exemplar as an analogy for the category provides an effective strategy. He found that subjects tended to find an exemplar of a category highly similar to a test stimulus such that a correct classification could be made. Thus, the experience with category exemplars provided a category structure that was not based explicitly upon abstraction of values but would still serve as the basis for correct classification. In this way, an episodic component could enter into the category structure. He presents examples in which results suggest that this approach is used. Note that a salient exemplar is to be distinguished from a prototype, which is usually a derived representation and not directly encountered in experience. Before considering approaches that incorporate the episodic frame of reference, it might be well to review major distinctions between episodic and semantic memory.

III. THE SEMANTIC-EPISODIC DISTINCTION IN CONCEPT LEARNING AND REPRESENTATION

The distinction between semantic and episodic memory proposed by Tulving (1972) has particular relevance for the study of concept representation in memory. Episodic memory refers to memory for stimuli in their spatial and temporal context. We cannot at this stage talk about a faithful representation of the stimulus event, free of encoding processes. However, insofar as an event is placed in memory within the context of its occurrence, it seems reasonable to label such a memory as episodic. In contrast to such memories, semantic memories involve "meaning", as defined in terms of derived interconnections between stored elements of information. Such memories would include concept representations in terms of classification rules along with related attribute information. I emphasize the notion that storage of event information within its spatial and temporal context should, strictly speaking, be qualifying criteria for inclusion in the category of episodic memory. Thus, a stimulus represented in episodic memory is tagged with respect to the context in which the individual encounters it, both in terms of the immediate context of its occurrence and in terms of events preceding and succeeding it. It is then retrieved within the framework of one's own life experience. The temporal consideration is important. For example, when an individual is expected to learn "horse" as a member of a stimulus list, it is clearly implied that the memorizer can distinguish between the occurrence of the word in its present context and prior knowledge of the word. To the extent that representation of a stimulus event enters into organization at a higher level, it loses its episodic quality. If a stimulus is stored with other stimuli according to some organizational requirements, be they associative, abstractive, or of some other sort, the memory loses its "event like" quality. However, if information is accessed directly from information about an event placed within the context of its occurrence, episodic memory is implicated. I believe the distinction will prove helpful in understanding issues raised in theories to be discussed. Before proceeding with the episodic-semantic distinction in the conceptual realm, it is well to consider issues related to the distinction.

Are Episodic and Semantic Memory Two Separate Systems? This issue is quite relevant for concept learning and representation. If systems are clearly separate, it is implied that abstract information about a class of objects is independent of event information encountered in the process. On the other hand, if the systems cannot be distinguished, then concept learning and representation are embodied in some internally recorded acquisition history. In turn, the processes operating upon these systems should reflect the nature of the representation. Certainly, degrees of interconnectedness ranging between these extremes are possible.

A major source of evidence on distinctiveness of the two systems comes from research on experimental dissociation. In such experiments, manipulations that involve semantic aspects of information processing are shown to affect tasks requiring semantic memory, but not tasks requiring episodic memory. The converse relationship seems to hold for manipulations that involve episodic aspects of information processing. As an example, Shoben, Wescourt, and Smith (1978) asked subjects to perform a sentence verification task involving attributions about known terms (e.g., "A canary is yellow"). They were also to perform a later recognition task involving presented sentences. Semantic relatedness between sentences containing given terms was varied. For example, "tigers have stripes", was a related true statement, while "Tigers have thighs" was an unrelated true statement, as determined by prior subjects' judgments. The number of sentences related to a given term (i.e., the amount of "fanning") was also varied, varying between four and eight sentences. The former manipulation produced commensurate variation in response latencies for sentence verification, a task related to semantic memory, while the latter manipulation produced only variation in sentence recognition, an episodic memory task.

In a task closer to categorization, Jacoby, Baker, and Brooks (1989) exposed one group of subjects to line drawings of objects to be named by the subjects. Another group was required to clarify incrementally these drawings of objects until they could be identified. Clarification was accomplished by the subject's pressing a key to remove "noise" incrementally from the drawing until it could be recognized. Upon test, subjects in the naming condition recalled more picture names then did clarification subjects. Conversely, clarification subjects took less time than naming subjects to clarify pictures previously presented for identification. The results indicated differential processing of stimuli, a result consistent with differentiation of the two memory systems.

The results of studies of experimental dissociation lend support to the notion of functional distinctiveness of the two types of memory. However, the question of whether or not the two systems can be represented as separate memory structures is unresolved. Indeed, results are not inconsistent with common storage of information, with differential retrieval of information dictated by task demands that activate access to different types of information. This possibility will be explored in the realm of categorization.

IV. EXEMPLAR-BASED THEORIES OF CATEGORIZATION

The role of the individual stimulus in concept acquisition has a more prominent role in exemplar-based theories than in hypothesis testing or prototype theories. In these theories the stimulus is mainly a means for conveying attribute-value information. As we will see, the stimulus in recent exemplar-based theories is a contextual reference point for the integration of attribute value information into the conceptual structure. The importance of a concept exemplar in learning and representation has been recognized for some time. Subjects perform better when given positive instances than when given negative instances, even when negative instances provide information needed for solution (Hovland & Weiss, 1953; Johnson, 1972). Despite this long standing evidence, emphasis on the role of individual stimuli is quite recent.

Several theories of category learning and representation are based upon the notion that memory traces for individual patterns are stored and enter into the development of the category structure. Major theories of this type may be labeled nontemporal-mediated. Nontemporal refers to the fact that the temporal context in which a learning stimulus is presented does not influence the way in which it enters into the representation of its category. Mediated refers to the fact that the stimulus-category relationship is mediated by its relationship to a composite representation of experienced members of the class. Thus, individuals do not base their assessment of a stimulus' category membership on a direct access to the stimulus-category link. Later, we look at an approach that is temporal and nonmediated.

Two major examples of current exemplar-based models are presented by Hintzman (1986) and Medin and Shaffer (1978). Both of these models, the former realized as a computer simulation, Minerva 2, characterize the acquisition process in terms of the aggregation of memory for stimulus events as episodic traces. In Hintzman' model, each stimulus trace is stored as a vector of features arranged with respect to dimensions of stimulus variation. Each position in the array is scored -1, 0, or +1 depending on whether it represents inhibition of a given property, is indeterminate, or represents excitation of the property, respectively. The trace for a stimulus to be classified is compared with the stimulus trace for each category member to arrive at a set of similarity values for each category member:

$$S(i) = (1/N_R)\Sigma_j P(j)T(i,j),$$

where S(i) is the measure of the similarity of the probe stimulus to stimulus i, P(j) is the score for the probe on attribute j in the array, T(i,j) is the corresponding score for stimulus i on attribute j, and the summation is over all attributes. The activation value, A(i), associated with stimulus i is a cubic function of S(i). In addition to producing a generalization gradient of appropriate form, this transformation implicitly introduces interaction among attribute values within the context of the stimulus, because the sign of the values enters into the product of terms, with positive activation scores for probe and stimuli for given attribute values producing positive values. The "echo" for a probe is a function of the activation level of each stimulus, i, and its value on each attribute. Thus, the echo for attribute j for the comparison is

$$C(j) = \Sigma_i A(i)T(i,j),$$

where the summation is over all category stimuli. This measure provides for contribution of information not in the probe but introduced through activation of other stimuli. The echo for a given probe stimulus is then the vector of C(j) values, with the strength of the echo determining the response.

Medin and Shaffer (1978) also provide for comparison of a stimulus with the set of category members to arrive at a classification decision. A probe stimulus is compared with the set of traces of known exemplars as in Hintzman's model. The probe is compared with each stimulus on each attribute, resulting in a similarity score for the attribute in the closed interval, [0,1], with 1 representing a perfect match. The similarity value for a given probe-stimulus comparison is the product of the resulting similarity values. For example, suppose stimuli vary with respect to shape, size, and color, with similarity parameters for nonmatching values p, q, and r, respectively. If the probe stimulus is a large, red, triangle and a known exemplar is a small, green, triangle, then the similarity score is q.r, with a value of one the multiplier for the matching value, "triangle". A category resemblance score for, say, category A is given by E_A, the sum of all similarity scores involving the probe and learned members of A. A corresponding score for category B in a two-category task can be derived for comparison of the probe with category B traces. The probability of an A response is then $E_A/(E_A+E_B)$, indicating that classification probability for a category is proportional to its overall similarity to the category.The multiplicative similarity measure requires that values do not act independently but, rather, interact within the context of the stimulus. Hence, the model is a context model, with stimulus providing the context. The original stimulus presentation then assumes an important role. However, each stimulus, old and new must be compared with all other traces (including an old stimulus with itself). In fact, this property is true of the Hintzman model. Neither model has an explicit provision for direct recognition of an old stimulus-category association. Therefore, the term, mediated, is used to describe the retrieval of stimulus -category associations. It is also true that the models are nontemporal insofar as the sequential context of presentation does not enter into the formulations.

These models have successfully predicted results in the category representation literature, including superior performance for the prototype, differential forgetting of old exemplars and the prototype, typicality effects, and category size effects. Nosofsky (1986) presented an extended version of this approach, and Nosofsky, Shin, and Clark (1989) provided support for the model with converging evidence from classification and recognition performance.

Insofar as comparison processes derive from overall representation of the stimulus set, the episodic-semantic distinction is not prominent. In fact, these theorists propose that a single memory system underlies the process.

V. EVIDENCE FROM THE AUTHOR'S LABORATORY

A. Evidence for Direct Access to Episodic Information in Category Representation

In this section I address the possibility that classification of stimuli into categories is influenced by direct retrieval of stored exemplars. This possibility implies a more extreme view of the relationship of episodic memory to categorization. However, it should be noted that such direct access to episodic information does not necessarily imply two separate memory systems. Indeed, it has not generally been claimed that episodic and semantic memory represent separate structures. Nonetheless, access to individual stimuli in concept representation would have important implications for conceptual processes. Even if individual stimuli are interconnected within a more general category representation, the related processes may have consequences not anticipated by exemplar-based theories discussed, as we will see.

We have done a series of studies that implicate direct access to individual stimuli in the learning and representation process (Nahinsky & Morgan, 1983; Nahinsky & Oeschger, 1975; Nahinsky, Oeschger, & O'Leary, 1977). Nahinsky and Oeschger (1975) presented a concept identification task in which the stimuli were well known personalities varied along four descriptive dimensions -- occupation, era, nationality, and real-versus-fictional. For example, one stimulus might be "Beetle Bailey -- a fictional, American, contemporary soldier". It should be noted that the task, explicitly defined for the subject, involved well-defined concepts in which the domain of possible solutions was limited to a finite number of attribute-value combinations possible for the sets of values given. For example, a solution might be "real politician" or "historical, American soldier". Only attribute-value information, given with each stimulus, was relevant for solution. The names and pictures, on the other hand, provided only episodic context. In one experiment, the experimental group was given a set of stimuli over trials of the type described which contained information sufficient for solution. Stimuli on remaining trials contained only the attribute-values. The control group subjects received the same problems, but with only the attribute-values for all stimuli. After solving three problems, both groups were presented a problem in which only the attribute-values were presented. The experimental condition showed initial superiority, but the control condition produced superior performance on the transfer problem. The results implicate the influence of episodic context in the development of a learning strategy. Insofar as individuals became dependent upon retrieval of contextual information during learning, they were not prepared for processing information without this episodic context.

Although the evidence favored a strong episodic component, it is still possible that analytic processes were associated with concept learning and representation. Consider a process in which stimuli are tested on each trial for proper classification and then for information content. If a stimulus contains information leading to elimination of previously plausible hypotheses or to addition of potentially correct hypotheses, it is encoded, possibly as a "chunk", and stored with some probability. On each trial, stimuli are retrieved and necessary hypothesis information is constructed. The comparison of such retrievals from stimuli rich in information can then proceed in a manner suggested by hypothesis sampling models. In this way, economies of memory may be achieved. A primitive mathematical model that embodied these assumptions was tested using data from one of the reported experiments. Comparison of predicted and observed learning probabilities for the first presentation cycle provided reasonably good fits.

Nahinsky et al. (1977) reported two followup experiments to explore the above notions. The number of stimuli needed to identify solution was varied over three values -- 2, 4, and 6, with strict control for number of trials needed to present solution-relevant information. Stimuli of the type used in the preceding experiments were used, and the two stimulus presentation conditions reported were also employed. In addition to these conditions, problem isomorphs using abstract symbolic stimuli were presented. There was a consistent increase in trials to last error with increase in number of stimuli needed to identify solution. In view of the fact that conditions were equated for rate of presentation of solution relevant information, the more plausible interpretation would involve the stimulus memory process suggested above.

Interestingly, the magnitude of the effect was monotonically related to the amount of stimulus context, with pictorial-name stimuli producing the greatest effect and abstract stimuli the smallest effect. Corresponding to this effect, it was found that memory for specific correct responses prior to the final error was greatest for pictorial-name stimuli and least for abstract stimuli. These converging findings support the notion that individuals incorporate memory for specific stimulus-category associations into the learning process, with context that distinguishes stimuli from one another facilitating the process. Hypothesis testing models would not predict that the variables studied would have an effect.

How is stimulus information incorporated into the learned category representation? Evidence just discussed suggests that a provision for direct access to stimulus-category associations is called for. Nahinsky and Morgan (1983) provided evidence favoring direct access to learned exemplar-classification associations, as distinguished from a process that involves overall similarity to the set of old exemplars. Subjects were presented with a cue learning task in which a single relevant attribute value defined solution. The stimuli were again the well known individuals described above. After the learning trials a series of test trials was run. Old stimuli were matched with new stimuli on the basis of task-relevant descriptive values. For example, an old stimulus might have been "William T. Sherman -- a real, American, historical, soldier", and a matched new stimulus might have been, "George A. Custer", who is associated with the same demographic description. Again, the differences between stimuli in the explicitly defined task should be only of mnemonic value for the subjects. Further, independently derived interstimulus similarity ratings indicated that matched old and new stimuli corresponded well in similarity to other old stimuli in the learning series ($\overline{X}$(old) = 4.28, $\overline{X}$(new) = 4.27). Further, the matched old and new stimuli were judged highly similar to each other ($\overline{X}$ = 7.57 on a 9-point scale). Despite the correspondence old positive instances elicited significantly lower latencies than matched new positive instances on test trials. In view of the fact that the two sets were informationally equivalent for correct classification, and were essentially mutually substitutable in terms of overall similarity within the learning set, episodic storage and access seem to be implicated.

A number of other investigators have found evidence for retrieval of individual stimuli in the concept representation process. Whittlesea (1987) independently varied distance of a test stimulus from a prototype and from a stimulus encountered in learning in a series of experiments with pseudoword categories. He found that performance depended more upon similarity to the previous instances than to the prototype. This classification on the basis of the "nearest neighbor" supports the storage and retrieval of event information as a strong component of category representation.

B. Influence of Learning History Factors upon Representation

As I have pointed out, episodic memory involves representation of information in a temporal as well as physical context. Models of categorization have not explicitly allowed for the effects of stimulus sequence in generating predictions. Hypothesis testing models have generally included some sampling process whereby learning is postulated to make transitions from stage to stage in a manner dependent only upon the learning state at a given time. Indeed, this assumption is one requirement of a Markov chain, which is used in deriving a number of hypothesis testing models. The fact that chains derived for such models postulate one absorbing state as a rule clearly implies that the system ends up in the same place, regardless of starting point. Such models de-emphasize the influence of variation of learning history in the learning process. Exemplar-based models also do not appear to provide for such variations in learning history.

It is interesting to note that the earlier work on concept identification reported by Bruner et al. (1956) included a detailed examination of how an initial positive instance, which serves as a focus, guides the learning process. However, examples examined were of clearly defined rules taken from a finite population, such that the outcome was indifferent to the starting point. In a larger stimulus domain involving less clearly defined concepts, the focus for testing hypotheses could have a profound influence upon the ultimate representation.

Some investigators have looked at the influence of temporal sequence upon the course of learning. Detambel and Stolurow, 1956; Hovland, 1952; Miller, 1971 found optimization of presentation sequences to provide solution information as soon as possible facilitated solution. Busemeyer and Myung (1988) presented evidence of serial position effects in categorization tasks.

We performed an exploratory experiment to examine the effect of stimulus sequence upon the learning of a category and its representation. The results also bear upon some of the other issues related to the semantic- episodic distinction in category learning and representation.

Method. Subjects were given the task of learning the predictive criteria for success in a hypothetical job training program. Stimuli were males, described with respect to four potentially relevant attributes: intelligence -- highly intelligent or of average intelligence, social orientation -- introverted or extroverted, athletic orientation -- unathletic or very athletic, and personality -- "uptight" or "laidback". The first value shown for each attribute occurred more frequently in the successful group than in the unsuccessful group.(Two additional values were used -- of low intelligence on the intelligence dimension and somewhat athletic on the athletic orientation dimension. However, these values, on instances to be indicated, did not result in violations of specified criteria for intercategory relationships.)

The two categories, "successful" and "unsuccessful", were linearly separable (i.e., all successful individuals possessed more relevant attribute values than the unsuccessful individuals). Wattenmaker, Dewey, Murphy, and Medin (1986) found that linearly separable categories were conducive to learning via a process involving analysis into individual stimulus components. The classification criterion can be based upon an additive function of relevant values. Each member of the successful category possessed at least two relevant values, while no member of the unsuccessful group possessed more than one relevant value. Table 10.1 shows the relationships between the two learning sets, with six stimuli shown for each category. The order of presentation for one condition is represented by the order of stimuli shown. The initial successful-category stimulus contains the two relevant values, highly intelligent and introverted, with no other relevant values. In contrast, the last stimulus in the category contains the two relevant values, unathletic and uptight, with no other relevant values. Therefore, the two stimuli complement each other with respect to qualifications for category membership. Further, there is a progression from the initial to the last stimulus, such that the first half of the successful stimuli resembles the first positive instance more than the last positive instance with respect to relevant attributes. However, the last half of the positive instance set resembled the last positive instance more than the first positive instance.

Each stimulus, mounted on a 4" by 6" card , was represented by a 2" by 2.37" photograph of a male taken from a college yearbook, with a distinctive name, e.g., Chester, typed to the right. The specific attribute values were typed below the name in a vertical array, reflecting the order of the attributes shown in Table 10.1.

For one half of the subjects, stimuli in both categories were presented in the order shown, with the first positive instance presented initially and all other stimuli presented such that positive and negative instances were randomly interspersed. For the other condition, presentation order of the series was reversed, save that the final positive instance was presented first.

Within each of the two conditions, one half of the subjects were presented only the sequence of 12 stimuli as described. For the other half of the subjects, each attribute-value combination was duplicated and represented by a different individual. However, the sequence of attribute value combinations described was preserved, with duplicated combinations following their mates, possibly with a member of the opposite category randomly interspersed.

Table 10.1. Stimuli for the Category Learning Experiment.

	Attribute			
Order-Stimulus[a]	A	B	C	D
Learning Stimuli for Successful Category				
1. Chester	1	1	0	0
2. Oswald	1	1	1	0
3. Delbert	1	1	0	1
4. Bruce	1	0	1	1
5. Lester	0	1	1	1
6. Percival	0	0	1	1
Learning Stimuli for Unsuccessful Category				
1. Sedgwick	1	0	0	0
2. Samuel	0	0	0	0
3. Samson	0	1	0	0
4. Melvin	0	0	1	0
5. Rubin	0	0	0	1
6. Wilbur	0[b]	0	0	0
New Test Stimuli for Successful Category				
1. Barnaby	1	1	0	0
2. Gregory	1	1	0[c]	0
3. Oscar	1	1	0	1
4. Bertram	1	0	1	1
5. Smedley	0[b]	0	1	1
6. Tobias	0	0	1	1
New Test Stimuli for Unsuccessful Category				
1. Zachary	1	0	0	0
2. Ezra	0	0	0[c]	0
3. Winston	0	1	0	0
4. James	0	0	1	0
5. Manuel	0[b]	0	0	1
6. Eli	0	0	0[c]	1

Table 10.1 (cont.)

Note. Attributes: A -- Intelligence, B -- Introversion-extroversion, C -- Athletic orientation, D -- Personality. Numerical values in the Table refer to relevance of attribute value for the successful category, with 1 relevant and 0 irrelevant. See text for explanation.

[a] Numerical value represents position in the one presentation order. Sequence was reversed for the other presentation order.

[b] "Average intelligence" was the attribute value used.

[c] "Somewhat athletic" was the attribute value used.

Subjects were explicitly instructed that the task was to discover the characteristics important in distinguishing between success and failure in the hypothetical training program for a business. It was emphasized that the descriptive attributes were potentially predictive, while the other characteristics, e.g., names and appearance, would not be relevant for classification. Each subject made a success-versus-failure judgment followed by a confidence rating on a 5-point scale ranging from "very little confidence" to "very confident". They were given feedback about correctness of their classification after each trial.

After the learning sequence was presented, subjects were told that the next set of stimuli would provide a test of their concept of the success category, and that no feedback would be given for their responses. Each subject was presented with the 12 learning stimuli common to all groups together with a set of 12 new stimuli, with the 24 stimuli presented in a random sequence. The 12 new stimuli were matched with the old stimuli on attribute-value combinations, with exceptions shown in Table 10.1. Subjects made classification and confidence judgments as in learning. In addition, recognition memory judgments for old-versus-new stimuli were elicited on a 5-point confidence scale following classification judgments. As these results were not central to the present issues, they will not be presented here.

Let us review some features of the experiment and their implications. First, the categories were linearly separable. In such a structure, class membership is determined by a linear weighting of independently operating attribute values. The additivity of the characteristics should serve to emphasize the abstractive nature of the task and, presumably, reduce the emphasis on individual stimuli. Thus, the appearance of an episodic component would be noteworthy.

Second, the task does not have an invariant component (i.e., there is no single defining feature). However, all descriptive attributes are correlated with category membership. In contrast, the other characteristics (i.e., pictures and names) constitute "noise" from an informational point of view (i.e., they contain no information about class inclusion criteria). Nonetheless, this lack of informational value may be contrasted with mnemonic value. As research previously discussed indicates (Nahinsky & Morgan, 1983; Nahinsky & Oeschger, 1975; Nahinsky et al., 1977), the mnemonic value of the stimulus may derive from its ability to use specific stimuli to retrieve associated components of task-relevant information and conserve working memory capacity for inference making. Since the task was defined for the subject, the distinction is important in considering the relative contribution of different components of the stimulus.

Related to the point just considered, old and new stimuli were matched for information relevant to category membership. Differences between the old and new set reside in the representations of the specified individuals. Since assignment of individuals to the old and new categories was not dependent upon characteristics related to appearance or name, there is no reason to believe that similarity relationships among stimuli would depend upon category membership or upon membership in the old or new stimulus set. Thus, superior classification performance for old stimuli over new stimuli should be attributable to an episodic component of retrieval.

Finally, the category structure is such that certain stimuli have greater overall similarity to members of their category than do other members. For example, the third stimulus in both the old and new sets, Delbert and Oscar respectively, is highly intelligent, introverted, and uptight, all three of which are values favoring the successful category. If we sum the number of times each value overlaps the old set (including the stimulus with itself), the grand total for all attributes is 12. In contrast, the first stimulus in each set possesses the relevant values, highly intelligent and introverted, which produces a corresponding overlap total of eight. Further, stimulus three has at least as high an overlap score on each attribute as stimulus one. Hence, any differential weighting of values will produce a superiority for the former stimulus, if an additive function is applied. This result implies that a model assuming independence and additivity of cues would predict superior classification performance for new stimulus three relative to old stimulus one. This inference is, of course, based upon the assumption that old exemplars have no privileged status with respect to the "noisy" information, which I have suggested is only mnemonic.

Overall Results. Classification responses on test trials were subjected to an ANOVA. Responses were recorded on a 10-point scale, ranging from 1 -- "very confident", for the unsuccessful category to 10 -- "very confident", for the successful category. Between-subject factors were learning presentation order and number of learning trials -- 12 or 24, and within-subjects factors were old-versus-new, category -- successful or unsuccessful, and items -- 1 to 6 (a collapsed variable for most comparisons). Number of learning trials had no significant effect upon test trial performance, $F(1,40) = .42$, $p > .50$. Consequently, the two learning trial conditions were combined for subsequent analyses. The category effect was highly significant, $F(1,40) = 34.84$, $p < .001$, $\overline{X}$(successful) = 6.76 and $\overline{X}$(unsuccessful) = 5.69. The old-versus-new variable also produced significance, $F(1,40) = 39.88$, $p < .001$, $\overline{X}$(old) = 6.71 and $\overline{X}$(new) = 5.73. The old-versus-new x category interaction was nonsignificant, $F(1,40) = .41$ $p > .50$. Thus, subjects were responding on the basis of category membership for both old and new stimuli. It is plain that there was bias for classifying stimuli as successful, possibly because of the relatively great within class variability and the fact that individuals in the unsuccessful category possessed traits that seemed desirable. In general, we can conclude that, despite the relatively short learning period, subjects developed a classification scheme based upon relevant criteria of category membership. The lack of difference between the 12 and 24 learning trials condition suggests that category differention occurred at an early stage. Data on proportion of correct classifications for stimuli appear in Table 10.2 and reflect classification results, as might be expected.

Influence of Episodic Factors. First, let us consider the effect of presentation order upon test trial classification. The crucial source of variation is centered around the order of success category stimuli. Table 10.2 shows mean classification scores for each of the six stimuli within the successful category for both old and new stimuli for each presentation order. The presentation order X old-versus-new X category interaction was significant, $F(1,40) = 4.21$, $p < .05$, which justifies an examination of the interaction of the first two variables within the successful category. An interaction contrast was performed for new stimuli, which partitioned the stimulus set into the subsets containing the first three stimuli and the last three stimuli, shown in Table 10.1, for the first presentation order. These two subsets represent the last three stimuli and the first three stimuli, respectively, for the second presentation order. The interaction of the two subsets with the presentation order variable constituted the approximate contrast and was highly significant, $F(1,200) = 11.19$, $p < .01$. Inspection of Table 10.2 shows that for new stimuli the first three stimuli had higher classification scores ($\overline{X}_{1\text{-}3}$=6.47) than the second three stimuli ($\overline{X}_{4\text{-}6}$=5.42) in the first presentation order. However, this relationship was reversed for the second presentation order ($\overline{X}_{1\text{-}3}$=5.74 and

Table 10.2. Test trial Responses for Category Learning Experiment for Successful Category.

	Table 1 Presentation Order		Reverse Presentation Order	
Stimulus	Mean Rating	Percent Correct	Mean Rating	Percent Correct
Old stimuli				
1.	7.86	82	7.68	82
2	7.50	82	6.64	68
3.	7.45	82	7.61	72
4.	7.68	82	6.68	77
5.	8.32	95	7.36	77
6.	5.64	50	7.09	73
New Stimuli				
1.	6.09	86	7.27	77
2.	6.05	59	5.86	55
3.	7.27	77	4.09	23
4.	5.77	55	7.86	81
5.	6.73	64	5.18	41
6.	3.77	18	8.50	91

$\overline{X}_{4\text{-}6}$=7.18), corresponding to the reversal of order of the sets for this condition. Thus, stimuli resembling an initially presented instance in potentially relevant characteristics were associated with a higher category membership score than stimuli not resembling such a focal stimulus. Therefore, for a given set of exemplars used to define a category, the learning history has a decided impact upon derived category structure. Insofar as temporal links in the acquisition process determine the derived concept representation, this representation incorporates the temporal interstimulus relationships, an important aspect of episodic memory I have noted. This result would suggest a significant modification of current exemplar-based models to account for such episodic effects.

The same analysis applied to old exemplars showed no corresponding effect, $F(1,200) = .03$, $p > .50$. These stimuli, then, obeyed different rules for accessing category membership relationships than did the new stimuli. The two results juxtaposed suggest that retrieval of episodic and semantic information involve two distinguishable processes. In the present context, semantic information refers to attribute value-category relationships induced in the learning process.

Next, the four old stimuli--1,3,4, and 6 were compared with the corresponding new stimuli within the success category. These represent subsets matched for attribute values. The resulting contrast proved highly significant, $F = 8.71$, $p < .01$ and favored the old stimuli. Thus, the difference can clearly be attributed to episodic variables.

The first and sixth stimuli in the set of old stimuli were contrasted with the third and fourth stimuli in the new set for average classification score, with a highly significant result, $F(1,77) = 8.71, p < .01$, $\overline{X}$(old) = 7.21 and $\overline{X}$new) = 6.33. Recall that for these stimuli the old stimuli had less overall similarity to the learning set than did the new stimuli. In fact, similarity values over the different attributes were such that no independent-cue model would generate an additive weighting scheme that would predict this result. However, an interactive cue model (e.g.,Medin & Shaffer, 1978) might deal with these data. In applying Medin and Shaffer's model, let p, q, r, and s be the similarity parameters for different attribute values on the intelligence, introversion-extroversion, athletic orientation, and personality attributes respectively. Then,

$$A_{s1} = 1+r+s+qrs+prs+pqrs,$$

$$A_{s3} = 1+s+rs+qr+pr+pqr,$$

where A_{s1} and A_{s3} are the similarity values for the successful category for stimulus 1 and stimulus 2 respectively. Further, the corresponding similarity values for the unsuccessful category are,

$$A_{u1} = q+pq+p+pqr+pqs+pqrs,$$

$$A_{u3} = qs+pqs+ps+pqrs+pq+pqs.$$

If we note that category association strength is a monotonic function of $A_{si}/(A_{si}+A_{ui})$, it is possible that some combination of similarity parameters would be compatible with the results. (A similar relationship holds for comparison between stimulus 4 and stimulus 6.) Insofar as an exemplar-based model relies on a comparison process in which cues interact within the context of the stimulus, it can at least partially explain the data. However, the general old-versus-new difference seems to require a further assumption about episodic effects. The context theory implicitly allows for a comparison of an old stimulus with itself. It seems reasonable that a recognition of the old association could be accommodated within the process, thereby dealing with the special status of old exemplars.

The temporal and physical acquisition context was clearly represented in the category structure induced by the subjects. It is important to note some implications for theories of category learning. First of all, the influence of temporal factors can have an important impact on how membership decisions are influenced over time. As an example, Brooks, Jacoby, and Whittlesea (cited in Jacoby and Brooks, 1984) found that presentation of category exemplars primed responses to later presentations of similar exemplars. So too, item-category associations that are temporally contiguous in memory may show some priming relationship. This possibility implies that categorization revealed at any time could be influenced by encounters with instances that have been associated with a category. New items, as well as other old items, could be influenced by such priming, insofar as attribute values associated with exemplars gain strength with presentation.

It is important to note that memory for old items in this context must be distinguished from mere recognition memory for item occurrence. It is the category-item association that is crucial to understanding the role of exemplars in category representation. Indeed, Nosofsky (1988) pointed out that the classification and recognition task are usually orthogonal (i.e., partition of stimuli is such that the category and its complement both have about an equal number of old and new stimuli during test trials). Thus, prior occurrence or nonoccurrence of a stimulus should have no impact upon the ability to classify a stimulus. Nosofsky presented evidence to support the distinction between classification and simple recognition memory. An old category exemplar could prime the correct classification for a similar or temporally contiguous old member of the category. However, the presentation of the stimulus might interfere with correct classification of an old nonmember that is associated with the former stimulus in time.

VI. CONDITIONS RELATED TO EPISODIC EFFECTS

I have reviewed a variety of data associated with category learning bearing upon the relationship between episodic aspects of the learning situation and ultimate category representation. Although there may be certain boundary conditions, which I will discuss, it seems that the original learning context, both temporal and physical, plays a larger role than many present viewpoints allow for. Exemplar-based models are based upon the notion of an integrated memory system in which stimulus information is stored in a cumulative manner and serves as the basis for classification decisions. Research discussed suggests that episodic information may play a more distinctive part in the process than these models provide for. However, consideration of a number of questions should precede introduction of a different approach.

Homa (1984) has presented a taxonomy of category types, which may help serve as a framework for considering a number of problems in the semantic-episodic interrelationship. He considers the following four criteria as a basis for classifying categories:

1. Is the category deterministic or probabilistic in nature?

2. Is the stimulus domain composed of a finite or an infinite membership?

3. Does each stimulus have an invariant component that can mediate classification?

4. If an invariant component exists, is the remaining (or complementary) information in the stimulus correlated with the invariant component?

The resulting 16 possible combinations associated with the variables form the taxonomic scheme. In general, if the concept lacks an invariant component and has a small stimulus domain, the episodic contribution may tend to be larger than an other cases. Indeed, exemplar-based models have received their support in such situations. Large stimulus populations with well delineated defining criteria would tend to minimize the contribution of memory for specific associations. However, consideration of another variable, the stimulus, may provide a different perspective. Three characteristics of the stimulus are important in assessing the likely strength of episodic components in category representation: uniqueness, encodability, and redintegrative power.

Stimulus Uniqueness. Tasks have included stimulus sets ranging from schematic geometric figures to representations of "real world situations". In stimulus populations constructed artificially by orthogonally varying some finite number of dimensions in discrete steps, each stimulus has at least one attribute value in common with a number of other stimuli. Stimuli are not readily discriminable, and the likelihood that a given stimulus and its classification can be retrieved without confusion with other stimuli is greatly reduced.

In contrast to the preceding situation, if stimuli are quite discriminable and embody unique aspects, their mnemonic value is great, thereby enhancing encoding and retrieval.

Encodability. A stimulus is encodable to the extent that it is compatible with the individual's repertoire of learned schema. A random dot pattern may be unique within a given context, but it will not resemble an available representation of a known pattern to allow for its encoding. As an example of encodable and unique stimuli, an individual with related physical characteristics is more likely to be represented in memory in a manner that sets her apart from other individuals.

Redintegrative Power. The retrieval of task relevant information via an individual stimulus will be related to the redintegrative value associated with stimulus components. That is, the given mnemonic context provided by a stimulus should have strong associations with task relevant information to facilitate retrieval of that information. Redintegration is said to occur when elements of a stimulus are integrated into a unit, such that retrieval of one element brings about the retrieval of the other elements. The unitization of stimulus elements, i.e., redintegration has been demonstrated by a number of investigators (e.g., Horowitz & Manelis, 1972; Horowitz & Prytulak, 1969), who studied conditions under which stimulus-response pairs are integrated into units. A stimulus consisting of arbitrary combinations of attributes not normally related to each other would not be expected to facilitate retrieval processes necessary for the task. To pursue the example described above, an individual can serve as a redintegrative cue provided that task relevant information consists of attribute values descriptive of individuals. Incidental cues associated with stimulus presentation might not serve such a role. When stimuli possess the characteristics discussed, it is possible that episodic factors will play an important role in representing a variety of categories, even those involving large stimulus populations and well defined concepts.

VII. A PROCESSING MODEL

This section will be devoted to discussing the outlines of a processing model that may help guide the investigation of the interaction between episodic and semantic aspects of category representation. The proposed approach is based upon the representation of the category structure as a network of interconnected units of knowledge which relate an instance to the category and to the attribute values that are relevant for category membership. The general scheme has much in common with recent connectionist models in the way in which information processing is characterized. One example is the framework provided by the

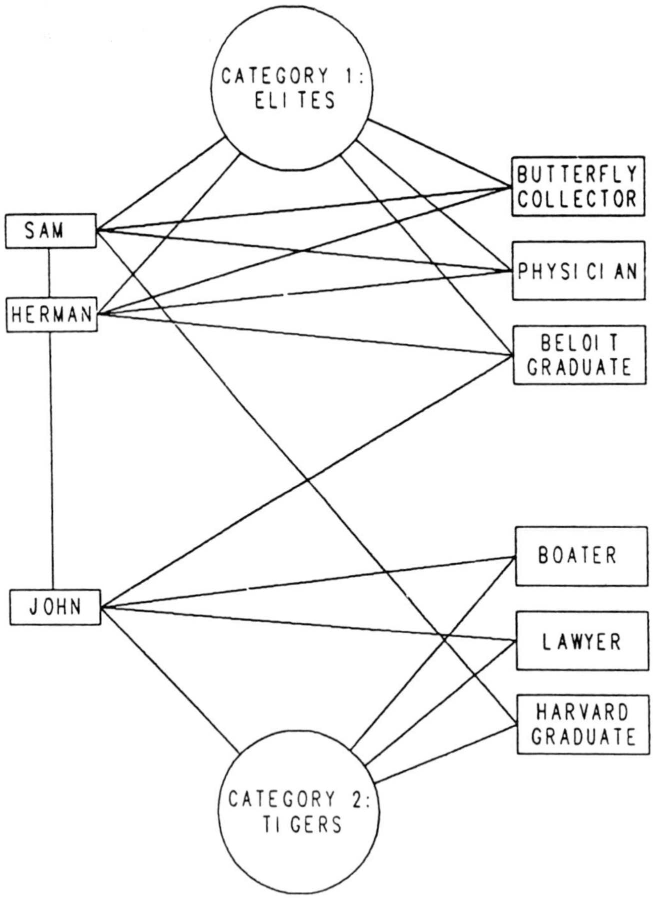

Figure 10.1. Attribute Value- and Exemplar-Category Links.

parallel distributed processing approach of McClelland, Rumelhart, and Hinton (1986), whose representational scheme is quite similar to the one I present, with allowances for specific characteristics of the knowledge structure to be described.

The major distinguishing feature of this model is in its separate representations of the stimulus and stimulus attributes, with corresponding links to a category of interest. The links represent episodic and semantic aspects of system respectively. Thus, episodic and semantic memory are conceived of as an integrated structure within this context. The two memories are distinguished by separate processes, one accessing stimulus-category links and the other accessing attribute value-category links. Insofar as certain mutual influences, to be discussed, operate, they may not be independent.

Figure 10.1 presents an example of how the stimulus-category relationships for stimuli in a classification situation would be depicted. Note that on the left stimulus nodes are represented as individual instances, realized as individuals, e.g., Sam, who is a member of the "Elites". On the right, the potentially relevant attribute values associated with Sam are shown as a series of nodes, each of which is linked to Sam. Two of Sam's attributes, butterfly collecting and being a physician, are weighted in favor of the Elites. Thus, the given nodes are connected to a category 1 node. Further, if Sam has already been encountered as an Elite, his node is connected to the category, Elite club member. In this way, two links to the category exist, the former "semantic" and the latter episodic. (Sam could have also been linked to a number of other characteristics to produce a more detailed representation. However, to the extent that, save for the potentially relevant attribute-values, individuals are distinct, this level of representation should suffice for characterizing the interconnections needed to describe links determining category decisions.)

The dual connections provide a basis for understanding the advantage that old instances may have in classification. Suppose the instance elicits a parallel search of relationships. Then the correct category response can result from a direct retrieval of the Sam-Elite link. It could also result from the retrieval of the attribute value-category connections, which should jointly have enough strength to produce the Elite response. In view of the Nahinsky and Morgan (1983) results, it seems likely that the former association would occur more rapidly than the latter. If we assume a parallel search from the stimulus node, Sam in this case, to other nodes associated with the representation, we might expect that the category node would be reached from Sam before it would be reached from the attribute-value nodes, if the process of evaluating attribute value-category connections takes longer than recognizing the stimulus-category link. Thus, old stimuli would have a time advantage over new stimuli. Elicitation of the correct classification on the basis of category information would depend upon cumulative strength of attribute value-category connections, with some threshold level required for a positive response. Note that attribute-value nodes are not connected directly. However, they may be linked by mutual connections to stimuli. Hence, stimulus context mediates interactions among attribute- values and can account for multiplicative relationships found in response patterns.

In general, activation of a node is assumed to spread activation to other associated nodes. Thus, presentation of Sam should not only activate values associated with him, but it should activate nodes of other stimuli associated with him in time. For example, Sam's presentation may activate a node associated with Herman, who appeared after Sam in learning. Since Herman is also an Elite, the association should be facilitated, if Herman follows Sam again. In contrast, John, a member of the "Tigers", follows Herman. Hence, the latter may interfere with the instance-category association of the former. Insofar as early stimuli may facilitate activation of associated attribute values, a mechanism exists for the advantage of these stimuli in determining the nature of the concept derived.

These hypothesized relationships can serve as the basis for a number of interesting tests of episodic effects upon concept representation. The formulation implies that category structure is dynamic, varying over time and context and capable of incorporating new information. New instances may change the representation, both in terms of reference instances and salience of attribute values.

This model implicitly incorporates a number of notions that have appeared in the literature on categorization. For example, context models imply previously encountered exemplars may vary in influence depending upon similarity to instances that are to be classified at a given time (Medin & Shaffer, 1978; Medin & Smith, 1981). Barsalou (1987) has pointed out the instability of graded categorical structures, with an emphasis on episodic components that operate over time. These examples of possible "local" influences over time can be dealt with by the model proposed in terms of activation of the various links associated with previously learned category associations.

Only a representational framework is presented here. A model for generating predictions must be derived. One example of an approach that might be adapted to this framework is the parallel distributed memory model presented by Murdock (1982, 1989). His model allows for the accumulation of featural information and interitem associations represented by vector products of corresponding item components. Item information is stored together with other informational components. Hence, a single integrated system is implicated, including featural, interitem association, and item information. Nevertheless, episodic and semantic components could be separated.

The proposed approach must be considered within a broader framework for conceptualizing the representation of general knowledge. The processing model envisions only the nature of an information structure and the way it may be accessed in representing concepts. Higher level decision processes can be hypothesized to control the use of the representation system. Ecological factors related to the one's needs in categorizing the environment would be involved. For example, if an individual wishes to select individuals for a specific mission and wants to avoid passing up any promising candidate, there would be an implicit requirement for great inclusiveness. In this case, general rules might be supplemented by salient exceptions, and the classifier would rely heavily on the episodic component of the system in the form of prominent past successes that have proved the exception to the rule.

Strategy decisions may be influenced by processing requirements of the stimulus populations, with distinctive stimuli containing well integrated information favoring episodic aspects of the system, as has been suggested above. Nonetheless, the system outlined suggests how strategies selected would process information in categorization tasks.

Research on the factors discussed is not only important for understanding the nature of category learning and representation. Such research provides an important opportunity for examining the interface between episodic and semantic memory.

VII. CONCLUSION

I have attempted to show that the individual instance may play a more prominent role in the representation of a category than has been assumed. I have presented evidence that individuals may store and access stimulus-category relationships directly, when they process information about a category they have learned. Further, these relationships may be accessed separately from information about attribute value-category relationships derived from stimuli.

The framework of a processing model was presented to deal with the evidence referred to. The proposal calls for a representation of both stimulus-category relationships and attribute value-category relationships. In this framework, episodic memory and semantic memory are part of an integrated memory structure, with episodic memory and semantic memory distinguished by processes activating the two types of links referred to.

ACKNOWLEDGEMENTS

I would like to acknowledge Herve Abdi and Susan Barrett for their extensive comments on an earlier draft of the chapter. I believe the final version is much better as a result. Teresa Richardson collected data for the experiment presented in section IV. B.

REFERENCES

Barsalou, L.W. (1987). Decentralized control of categorization: The role of prior processing episodes. In U. Neisser (Ed.). Concepts and conceptual development: Ecological factors in categorization. Cambridge, England: Cambridge University Press.

Bower, G. H., & Trabasso, T. R. (1964). Concept identification. In R. C. Atkinson (Ed.), Studies in mathematical psychology: Stanford: Stanford University Press.

Brooks, L. (1978). Nonanalytic concept formation and memory for instances. In E. Rosch & B. Lloyd (Eds.), Cognition and Categorization. Hillsdale, N. J.: Erlbaum.

Bruner, J. S., Goodnow, J.J., & Austin, G. A. (1956). A study of thinking. New York: Wiley.

Busemeyer, J. R., & Myung, I. J. (1988). A new method for investigating prototype learning. Journal of Experimental Psychology: Learning, Memory, and Cognition, 14, 3-11.

Detambel, M. H., & Stolurow, L. M. (1956). Stimulus sequence and concept learning. Journal of Experimental Psychology, 51, 34-40.

Falmagne, R. (1970). Construction of a hypothesis model for concept identification. Journal of Mathematical Psychology, 7, 60-96.

Gregg, L. W., & Simon, H. A. (1967). Process models and stochastic theories of simple concept formation. Journal of Mathematical Psychology, 4, 246-276.

Hintzman, D. L. (1986). "Schema abstraction" in a multiple-trace memory model. Psychological Review, 93, 411-428.

Homa, D. (1984). On the nature of categories. In G. H. Bower (Ed.), The psychology of learning and motivation. Vol. 18. New York: Academic Press.

Horowitz, L. M., & Manelis, L. (1972). Toward a theory of redintegrative memory: Adjective-noun phrases. In G. H. Bower (Ed.), The psychology of learning and motivation. Vol. 6. New York: Academic Press.

Horowitz, L. M., & Prytulak, L. S. (1967). Redintegrative memory. Psychological Review, 76, 519-531.

Hovland, C. I. (1952). A "communication" analysis of concept learning. Psychological Review, 59, 461-472.

Hovland, C. I., & Weiss, W. (1953). Transmission of information concerning concepts through positive and negative instances. Journal of Experimental Psychology, 45, 175-182.

Jacoby, L.L., Baker, J. G., & Brooks, L. R. (1989). Episodic effects on picture identification: Implications for theories of concept learning and theories of memory. Journal of Experimental Psychology: Learning, Memory, and Cognition, 15, 275-281.

Jacoby, L. L., & Brooks, L. R. (1984). Nonanalytic cognition: Memory, perception, and concept learning. In G. H. Bower (Ed.), The psychology of learning and motivation. Vol. 18. New York: Academic Press.

Johnson, D. M. (1972). A systematic introduction to the psychology of thinking. New York: Harper & Row.

Levine, M. (1966). Hypothesis behavior by humans during discrimination learning. Journal of Experimental Psychology, 71, 331-338.

McClelland, J. L., Rumelhart, D. E., & Hinton, G. E. (1986). The appeal of parallel distributed processing. In D. E. Rumelhart & J. L. McClelland (Eds.), Parallel distributed processing: Explorations in the microstructure of cognition. Vol. 1. Cambridge, MA: MIT Press/Bradford Books.

Medin, D. L., & Shaffer, M.M. (1987). Context theory of classification learning. Psychological Review, 85, 207-238.

Medin, D. L. & Smith, E. E. (1981). Strategies and classification learning. Journal of Experimental Psychology; Human Learning and Memory, 7, 241-253.

Miller, L. A. (1971). Hypothesis analysis of conjunctive concept learning situations. Psychological Review, 78, 262-271.

Murdock, B. B. (1982). A theory for the storage and retrieval of item and associative information. Psychological Review, 89, 609-626.

Murdock, B. B. (1989). Learning in a distributed memory model. In C. Izawa (Ed.), Current issues in cognitive processes: The Tulane Flowerree symposium on cognition. Hillsdale, N. J.: Erlbaum.

Nahinsky, I. D. (1970). A hypothesis sampling model for conjunctive concept identification. Journal of Mathematical Psychology, 7, 293-316.

Nahinsky, I. D., & Morgan, S. (1983). Episodic components of concept representation indicated in concept acquisition. Psychological Reports, 52, 931-960.

Nahinsky, I. D., & Oeschger, D. E. (1975). The influence of specific stimulus information on the concept learning process. Journal of Experimental Psychology: Human Learning and Memory, 1, 660-670.

Nahinsky, I. D., Oeschger, D. E., & O'Leary, D. (1977). Stimulus memory and contextual cues in the abstraction process. Canadian Journal of Psychology, 31, 102-112.

Nosofsky, R. M. (1986). Attention, similarity, and the identification-categorization relationship. Journal of Experimental Psychology: General, 115, 39-57.

Nosofsky, R. M. (1988). Exemplar-based accounts of relations between classification, recognition, and typicality. Journal of Experimental Psychology: Learning, Memory, and Cognition, 14, 700-708.

Nosofsky, R. M., Shin, H. J., & Clark, S. E. (1989). Rules and exemplars in categorization, identification, and recognition. Journal of Experimental Psychology: Learning Memory, and Cognition, 15, 282-304.

Posner, M. L. (1969). Abstraction and the process of recognition. In G. H. Bower & J. T. Spence (Eds.), The psychology of learning and motivation. Vol. 3. New York: Academic Press.

Posner, M. L., & Keele, S. W. (1968). On the genesis of abstract ideas. Journal of Experimental Psychology, 77, 353-363.

Reed, S. K. (1972). Pattern recognition and categorization. Cognitive Psychology, 3, 382-407.

Restle, F. (1962). The selection of strategies in cue learning. Psychological Review, 69, 329-343.

Rosch, E. (1973). On the internal structure of perceptual an semantic categories. In T. E. Moore (Ed.), Cognitive development and acquisition of language. New York: Academic Press.

Rosch, E. (1975). Cognitive representation of semantic categories. Journal of Experimental Psychology: General, 104, 192-233.

Rosch, E. (1977). Human categorization. In N. Warren (Ed.), Advances in cross-cultural psychology. Vol. 1. London: Academic Press.

Rosch, E., & Mervis, C. B. (1975). Family resemblances: Studies in the internal representation of categories. Cognitive Psychology, 7, 573-605.

Shoben, E. J., Wescourt, K. T., & Smith, E. E. (1978). Sentence verification, sentence recognition, and the semantic-episodic distinction. Journal of Experimental Psychology: Human Learning and Memory, 4, 304-317.

Smith, E. E., Shoben, E. J., & Rips, L. J. (1974). Structure and process in semantic memory: A featural model for semantic decision. Psychological Review, 81, 214-241.

Tulving, E. (1972). Episodic and semantic memory. In E. Tulving & W. Donaldson (Eds.), Organization and Memory. New York: Academic Press.

Wattenmaker, W. D., Dewey, G. I., Murphy, T. D., and Medin, D. L. (1986). Linear separability and concept learning: Context, relational properties, and concept naturalness. Cognitive Psychology, 18, 158-194.

Whittlesea, B. A. W. (1987). Preservation of specific experiences in the representation of general knowledge. Journal of Experimental Psychology: Learning, Memory, and Cognition, 13, 3-17.

Wickens, T. D., & Millward, R. B. (1971). Attribute elimination strategies for concept identification with practiced subjects. Journal of Mathematical Psychology, 8, 453-480.

Commentary

Episodic Components of Concept Learning and Representation, I.D. Nahinsky

MARY E. LASSALINE
EDWARD J. WISNIEWSKI
DOUGLAS L. MEDIN
University of Michigan

In this chapter, Nahinsky suggests that episodic information plays a more central role in conceptual structure than generally has been proposed by category learning models. In the context of category learning, episodic information includes some specific representation of the category item encountered, as well as the context in which it occurred. This context may specify temporal information about the other category members that preceded or followed the particular item during learning.

Nahinsky argues that current models have under-emphasized the role of such episodic influences. In particular, abstraction models (e.g., prototype models) typically construct a summary representation of category members which fails to capture information about any particular item. Exemplar models, while explicitly representing information about individual category members, fail to represent the temporal context in which an item occurs. Furthermore, these models typically have viewed categorization as a somewhat indirect process. In particular, categorization of an item is based on the overall similarity of that item to previously stored items of that category. Nahinsky suggests that episodic information about an item may include an explicit item-category membership association. As a result, categorization of a previously seen item might involve directly accessing the item and this association.

Nahinsky describes a number of studies suggesting that episodic information may include specific item-category associations and that people may categorize old items by directly accessing the item, rather than by computing its similarity to previously stored items (as in some exemplar models). Furthermore, Nahinsky provides evidence that the temporal ordering of items during learning strongly affects their subsequent retrieval. In the last section of the chapter, Nahinsky sketches a processing model of category learning (based on both abstracted and episodic information) which is sensitive to such temporal ordering effects and which represents specific item-category membership associations.

Percepts, Concepts and Categories
B. Burns (Editor)

11

Modeling Category Learning and Use: Representation and Processing

DORRIT BILLMAN
Georgia Institute of Technology

I. INTRODUCTION

As the area of concept learning has matured, a variety of process models have been developed and fit to data. The successes and failures have lead to substantial increases in our understanding. However, as possibilities are developed, it becomes more and more valuable if whole classes of models can be systematically evaluated. This allows identification of the factors responsible for failure or success and can provide a more coherent and systematic view of the theoretical options remaining.

One pervasive contrast has been between theories which assume that abstract representations (such as rules or prototypes) are needed, versus those which assume that classification judgments are based on comparison to the learning instances, without any intermediate, abstract representation involved. Instance comparison models offer parsimony in that they propose that the same representations and processes needed in other "simple" tasks such as similarity judgments and memorization, are also sufficient for the "fancy" task of categorization. They capitalize on the idea that items within a category tend to be more similar to each other than to items in other categories. Similarity models often suggest that categorization is based on a stable, explanatory similarity metric which describes how similar a target instance is to known members of alternative categories.

It would be desirable to test the whole class of models and see when instances are sufficient and when rules or abstract representations are needed. However, generalization from failures of particular models to the whole class of instance or rule models has proved essentially impossible because of mutual mimicry. Reed (1972) was one of the earliest to note the problem of perfect mutual mimicry; he noted that the average-distance similarity model (where generalizations are made based on average distance to old exemplars) and the average prototype model (where an average is updated from examples and distance from this controls classification) specify the identical mapping from learning instances to classification judgment of a target item. Barsalou (1990) has noted that this is inherent in the "rule" versus "instance" contrast; any model of one class can be mimicked by some model in the other. In short, simply restricting representation while leaving claims about processing open is no restriction at all. Alternatively, trying to test claims about processing with no restrictions on representation would also pose difficulties. Both must be jointly constrained.

The theme of this volume is the role of processing and structure in psychological models of concepts and categorization. I use the term "representation" to refer to relatively static forms that preserve information. I take this to be synonymous with one sense of "structure". However, "structure" can also refer to "architecture" of the mind, such as multiple memory stores, or to distinct forms of thought, such as Piagetian constructs. The issue addressed in this chapter is how we can investigate general classes of concept and categorization models, given the tradeoff and indeterminacy between claims about the nature of processing and of representation. This chapter explores two approaches to this problem.

One approach is to move away from the mechanism language of "representation" and "processing" in order to test classes of models. First, instead of detailing a mechanism, the effort is to precisely specify a class of mapping functions that the representation and process--whatever they may be--jointly compute. Here I mean the standard mathematical notion of a function: a structure that maps every element or value in its domain onto a single element or value in its range. In the case of category learning, the function domain is a set of learning instances plus the target items to categorize, and the function range is the set of possible category judgments. Thus, any such function maps from the learning experience plus current example onto a categorization judgment of the current example. Testing a class of models, then, consists of testing a specified class of functions.

Second, one must identify some property inconsistent with the specified class of mapping functions and the property can then be tested. Any test showing the necessity of the critical property is then a demonstration of the inadequacy of the specified class. To illustrate, consider some classic but far removed examples in psychophysics and decision theory. Simple theories of color vision might stipulate that effects of hue are independent of effects of saturation on color judgments; interaction between these variables would be inconsistent and a sufficient basis for rejecting this class of functions. Many simple theories of risk preference, including subjective expected utility theory, stipulate that effects from dollar amount of a possible win are independent of effects from current level of wealth on preference for different gambles; interaction between these variables would be inconsistent and sufficient to reject the class. If evidence for the critical property is found, all models computing the specified class of mapping functions are insufficient and rejected. The focus is on identifying some property inconsistent with a broad class of models, but which proves necessary to account for the data. This would demonstrate the insufficiency of the target class of model; here, tests that disconfirm are informative.

This focus on general classes of functions and *disconfirmation* is complementary to development of mechanism theories. Mechanism theories typically strive to provide one particular model with power sufficient to generate a target class of behavior and experiments are designed to confirm or show good fit to the model. The first section of this chapter defines a general class of models, which are motivated by the informal distinction between simple instance models and rule abstraction models, and reports on its adequacy.

So far, I have been sounding rather negative about development of mechanism models and specification of particular representation and processing claims. Pessimism stemming from the underdetermination of theory by data is a much discussed (c.f. Anderson, 1990) and long-standing concern. Why not return to the behaviorist agenda of simply describing behavior (classification judgments) as a function of the environmental stimuli (training and test examples)? But mechanism models do provide an important function. So far I have been stressing the importance of one type of generalization--generalization across all models that compute a certain type of function. There is a second type of generalization that is also important--generalization across tasks. It is here that mechanism models are very helpful. A mechanism model should provide a way of thinking about a whole set of related tasks. For example, a categorization model should tell us something about the function specifying induction judgments as well as the function specifying the usual classification judgments. Viewing performance on a set of related tasks as generated by the same mechanism model does two things. First, it encourages us to see relations and make predictions we otherwise might not, thus using the model to extend and guide research. Second, as information about multiple tasks is gathered it provides a richer set of constraints or challenges for the process model to meet.

This brings us to the second approach to testing general classes of models considered in the work reported here: investigate a broader set of tasks to generate a richer set of constraints on viable process models. When new but related tasks are considered, models may make contrasting predictions when previously they did not, or some of the models might be directly applicable while some are not. Either possibility provides a basis for preferring one model over another. In Section IV we summarize some data from tasks that require category learning, but that also invite subjects to make use of any abstract knowledge in inference and transfer tasks.

II. ASSESSING CLASSES OF MODELS: STATIONARITY

A. Background Intent in Model Assessment

This section outlines the first approach to testing general classes of models: specifying and testing a principle critical to the whole class. This contrasts with the majority of theoretical work in concept learning. Most work has tested the sufficiency of individual models by showing a good fit to the data or by considering the relative fit of multiple models. Testing a precisely specified, general class of models is more typical in psychophysics and decision theory (Krantz & Tversky, 1971; Miyamoto & Eraker, 1988). There are, however, at least two examples of testing and rejecting a general class of models relevant to concept learning. This involved identifying a key principle incompatible with any member of the class, and testing for this key principle. If found, this rejects the class.

Tversky and Gati (1978; Tversky, 1977) assessed models of similarity judgments and identified several key properties, including *asymmetry*, inconsistent with any form of metric (or "distance") similarity judgment. Because they found similarity judgments were asymmetric, they demonstrated that judgments in their task could not be modeled by any form of metric judgment, regardless of the dimension or distance rule used. This excluded any model with this sort of function, regardless of its other properties. Nonmetric capabilities that allow asymmetry are necessary and any model without this is insufficient.

For a second example, Medin and Schaffer (1978) tested the general principle of independence (here, additivity) of the effect of attribute matches on categorization judgments. They did assess the fit of one specific model but in the process took care to specify and test an alternative, broad class of independent cue models. They showed that categorization judgments could not be modeled by any function in which the effect of each attribute was independent of the values of other attributes (given correctly specified attributes). Attribute independence was rejected and some form of relational coding was found necessary; any model without this is insufficient.

The work reported here is concerned with testing another broad principle about the function specifying classification judgments. Before detailing the principle and class of models tested, one point about the intent of the project should be made explicit. In specifying a class of functions mapping from learning instances plus target to the classification judgement, I am making a claim about general characteristics of the entire behavior. Many times models specify a mechanism for some component but don't make claims about the entire mapping. For example, a model might propose how, given a set of instance representations, these are used to make similarity judgments but leaves open the question of how the instances were selected. Alternatively, a model might specify when rules are formed, but leave open the question of how one rule is selected over or combined with others. Or a model might make claims about representation without specifying use. To get predictions in such cases requires additional bridging assumptions. In contrast, when I talk about different classes of functions, I am talking about differences between one entire mapping function and another, without addressing what aspect of representation or what stage in processing produces the difference.

B. Motivation of the Contrast Tested

The informal contrast between "instance" and "rule" models is primarily one of representational specificity, but there are several correlated properties on which "instance" and "rule" models typically differ. The two types of models tend to capture contrasting intuitions.

1) "Instance" models use a relatively distributed representation, a set of stored instances. "Rule" models are more likely to have information concentrated in few, centralized representations. 2) "Instance" models naturally preserve and use context, or detailed relations among characteristics; "Rule" models naturally preserve regularities across contexts. 3) "Instance" models can rely on stored instances without other, intermediate representations. No special process for category learning need be invoked, but a single and unchanging process of storing instances and making similarity judgments can be used for item memory and for classification. "Rule" models entail construction of some new, intermediate representation or some new decision criterion which changes categorization judgments. They need additional processing, beyond instance memory, for constructing and using these representations or for altering criteria.

Some of the contrasts between "instance" and "rule" models are matters of degree; others produce no empirically distinguishable judgments. *Stationarity* is a property which is intuitively important and testable. The contrast between *Stationary* models and *Dynamic Abstraction* models is derived from the third contrast discussed above. The critical property of Stationary models is that they must have a fixed classification function that does not change over the course of learning. New types of representations or rules cannot be added; changes in parameters, attribute weights, or form of the classification function are not allowed. Of course, *judgments* of a given test item must change systematically as new instances are encountered if any learning is to occur; as the arguments of a function (learning instances plus item to classify) change, its values (the classification judgments) will change. However, the function mapping from the learning instances plus the test item to a classification judgment need not change. This is one appealing, simple, property which restricts the class of functions substantially. This is the class of functions which is specified and tested. I next give examples of Stationary and Dynamic Abstraction models to illustrate the contrast before returning to a more detailed definition.

C. Examples of Stationarity and Dynamic Abstraction

To illustrate the principle of Stationarity, we will describe several pairs of models in the literature, a stationary model (usually labeled an "instance model") and a related, more powerful dynamic abstraction model. Medin and Shaffer's (1978) model as described in their Eq. 1 is a stationary model. It includes salience parameters for attribute weights, but these parameters are fit to data averaged over the entire course of learning and no basis for change is specified. While Medin and Shaffer mention fluctuations in encoding due to changes in hypothesis testing or attention, no systematic procedure for change is described, specified in the equations, or modeled in fitting the data. Noskofsky's models (1984, 1987) are based on the Medin and Shaffer model, but are dynamic abstraction models. They specify that the weighting parameters change over learning, so that the attributes which distinguish between contrasting categories come to be weighed more heavily.

Hayes-Roth and Hayes-Roth (1977) describe a family of related models. One example of a stationary model that they describe stores information about the power set of attribute values and makes classification judgments based on the frequency of all sets in the power set. The best fitting model was a related dynamic abstraction model. It stores information about all sets in the power set, but classifies in accord with the most discriminative member of the power set. Since the subject can only determine which is most discriminative over the course of learning, this is not a stationary model.

Simple prototype models which store average or most frequent attribute values are stationary as are the homologous models which compute and average distances of the test item to prior instances; in either of these types of models, the dimensions, or attributes, may be unequally weighted. To change these into dynamic abstraction models, the weights of the attributes might change over learning, for example, by emphasizing attributes which best distinguish between categories.

In addition, Brooks (1978) and Whittelsea (1987) describe stationary models, as does Hintzman & Ludlam (1980). While MINERVA (Hintzman & Ludlam, 1980) is clearly a stationary model, MINERVA II (Hintzman, 1986) adds a dynamically constructed "second echo"; the effects of the second echo may serve as a dynamically constructed abstraction changing the classification function; the status of this model may be interesting to analyze. Other Dynamic Abstraction models include Anderson, Kline, & Beasley (1979); Holland and Reitman (1978); and Billman & Heit (1988).

D. Description of Stationary Models

In Stationary Models, while arguments of the function change, the function itself does not. The key to viewing learning this way is to define both the training instances and the test instances (and only these) as the arguments of the classification function. As more instances are seen, the argument to the function changes and the output also changes, resulting in learning. However, the function itself does not change as a function of learning. Stationary functions may have parameters, so long as these are fixed over the course of learning; they might be reset for different types of stimuli, for example. Thus, any parameter which can be specified in advance is allowed. These might include "most similar instance", or fixed "salience" weights on different attributes. However, any way in which the function is altered in light of what has been encountered is excluded.

The contrasting class of Dynamic Abstraction Models allows change in the classification function. This might be a change in the way instances are represented, a change in the way intermediate information is represented, or a change in the way represented information is used in making classification judgments. The core idea is that the way information is used changes with learning. Not only do classification decisions change as additional instances are encountered but the decision function itself changes.

A more complete description of Stationary Models and background assumptions is provided in Billman and Richards (1988). I summarize four properties here.

1. Representation Assumptions

As in testing any class of models, there must be some correspondence between the representation intended and manipulated by the experimenter and that perceived by the subject. In addition, the set of input instances and the target instance must be representable by a matrix, where rows in the matrix represent different instances and columns represent different attributes. Cells in the matrix may have missing values for unobserved attributes. The notation for specifying the representation is similar to but more general than the attribute-value matrix presented in Estes (1986); loosely, a frame representation is assumed. The representation of instances might use just individual attributes, store 2-ples and 3-ples, or code relations among attribute values by representing each member of the power set of attribute values. Attributes may have equal or unequal weight, reflecting unequal salience or attention. Any coding scheme is allowed with any weighting of components, so long as this is unchanging. This assumption about instance representation is shared by virtually all models of categorization. It is intended to be very broad and uncontroversial.

2. Similarity and Classification.

Classification of a target instance is a function of the similarity between the target and the learning instances. A target instance is more likely to be classified into one category as its similarity to that category increases and its similarity to contrasting categories decreases. Each matching attribute value increases similarity; mismatching values decrease similarity. The method for combining matches and mismatches is open; matches might be integrated instance-by-instance (e.g. comparing the target to each old instance and then combining information across instances) or attribute-by-attribute (e.g. comparing each attribute-value of the target to other values of the same attribute and then combining across attributes). The similarity might be metric or nonmetric, relational or independent cue. The only restriction on the classification function is that it must be specifiable in advance, rather than dynamically altered over the course of learning. Neither change in the method for representing instances nor in the method for using the representations to produce a classification judgment is allowed.

To illustrate the Stationary Model side of this distinction, the classification rules specified in Reed (1972) based on his Minkowski distance (Eq. 1), the Medin & Shaffer (1978) classification rule for the *context model*, and Hintzman's classification rule for Minerva are all examples. They all specify functions that compute similarity between the test instance and the learning instances to produce a classification judgment that remains fixed across learning. On the Dynamic Abstraction side of this distinction, the Nosofsky extended model (1984, 1987) specifies a similarity function that must change to weigh diagnostic attributes more heavily. The trial by trial criterion change typical of rule models is perhaps most clear in models that specify a trial by trial process. For example, in Anderson, Klein, & Beasley's model (1979) selective information about each learning item is preserved on every trial, either in the form of a new rule or as altered strength of the selected rule. Given a particular learning or test instance, the new rule formed or the old rule selected both change with learning. In Billman & Heit (1988), the probability of sampling particular attributes changes to reflect their history of predictive success. Kruscke's (1990) connectionist model changes attribute weights. Almost all Dynamic Abstraction models change the classification function in a manner that includes (and perhaps can be reduced to) changes in attribute weights.

3. Nature of Sensitivity to Covariation.

Relational or multiplicative similarity metrics allow sensitivity to covariation among attribute-values and can yield judgments sensitive to which attribute values occurred together. A stationary model can be sensitive to co-occurrence as a byproduct of preserving the whole pattern of co-occurrences in the original data--the instances. It can not abstract or identify which attributes are reliably correlated, or shift criteria to give these any special role. Again, the key requirement is that the function is sensitive to the same aspects of input, in the same way, from start to finish.

A clear understanding of this property is critical to understanding the current studies, since sensitivity to 'correlations' is frequently associated with instance models. I distinguish between co-occurrence and correlation. In this paper, I will use *correlation* to refer to a recurring association between values of particular attributes and *co-occurrence* to refer to any combination of attribute values. For a model to be sensitive to correlation, in this stricter sense, the model must have some basis for computing what attributes are associated and the nature of this association. For a model to be sensitive to co-occurrence, it need only preserve and use some information beyond independent attributes (e.g. instances that specify the set of attribute values co-occurring in it). Testing for selective reliance on those co-occurrences which formed reliable associations, i.e., on correlation, is one way of testing whether the function changes with learning.

Our experiments test whether subjects show selective reliance on correlated attributes or whether simple sensitivity to co-occurrence is sufficient.

We know from prior work that subjects are sensitive to co-occurrence among individual attribute values (Medin, Altom Edelsom, & Freko, 1982), but these particular findings are consistent with a stationary model. Consider a quintessential, stationary instance model: one that stores a veridical record of all input. This model will have "passive" access to all the co-occurrence information ever presented; it need not show any selective reliance on correlated attributes. Prior work was not designed to distinguish unbiased sensitivity to all the co-occurrence information from preferential sensitivity to reliable correlations. This distinction between correlation and co-occurrence is critical for the studies reported here and we will return to it in discussing the logic of the design.

4. Stochastic But Stable Variation Allowed

Finally, Stationary models may have a stochastic classification function, with random fluctuations in judgments, so long as the distribution of these fluctuations remains constant. Thus, error or noise in processing which causes judgments to change randomly over time, is still an instance of a stationary model. Only changing bias or change in the sampling distribution would imply a nonstationary function.

Many classification models assume a stochastic classification function, i.e., specifying *probabilities* of selecting or storing something on a particular trial. For these models the same item might be judged differently on different trials, even supposing no learning had happened. Whatever the basis for this variations--low level neural noise or high level fluctuations in hypothesis testing--so long as the variation is stable over time and the random function does not change, it is still a Stationary classification function.

5. Summary

Beyond the general requirements for the existence of a representation matrix and of a classification function, Stationarity is the only restriction on the class. Thus, Stationary Models form a very broad class. Stationarity of the classification function is the only substantive restriction. While arguments (the learning set and the item to classify) change, the function itself is fixed over learning.

E. Importance of Stationarity

Distinguishing between Stationary and Dynamic Abstraction Models is valuable for several reasons. First, it pins down one central contrast between the informal notions of "instance similarity" versus "rule abstraction" models. The core idea of abstraction models is change in the classification function, typically by adding new representations. The related, well-defined class of Dynamic Abstraction models is more general as it allows change in the classification function either from change in the kind of information represented or the way in which this information is used. This side-steps potential ambiguity as to whether it is the way information is being stored or the way it is used which changes; change of either sort puts a model into the Dynamic Abstraction class. In contrast, instance similarity models typically gain explanatory force by assuming that a fixed, identifiable comparison procedure is sufficient. Similarity models emphasize capturing the contextual richness of input, not changing the type of representation or process over the course of learning. "Similarity" is often described as a relatively simple, stable thing, e.g. as akin to confusibility (Medin & Shaffer, p. 209). Most Stationary functions are (or can be mimicked by) instance comparison models.

Second, the contrast marks a profound change in power, and power of a model is very important. Two models that can do anything cannot be distinguished. Increased power aggravates the problem of mutual mimicry. For example, by letting an instance model reweigh attributes, it can perfectly mimic many rule models. If an instance model can shift all its weight to whatever dimension is specified by the "rule", it will classify all red items into the "red" category, regardless of other attributes. One of the core ideas of "the priority of the specific" is to see how far one can get with less powerful processes.

Third, Stationary and Dynamic Abstraction models are specified in terms of properties of the entire judgment process. The model specification maps directly onto testable predictions about the representation/process combination without additional assumptions. This is not true of "instance" and "rule" models and this is what leads to the mutual mimicry problems there. Instance models typically claim the type of representation is constant, while rule models claim it is changing. But processing may be unconstrained. Success of a particular model may not imply anything about form of representation per se, but to the model's ability to change the mapping function itself over time. "Rule" models that fail (e.g. distance from averaged prototype), may fail because they do not specify a changing function, that is, they can be mimicked by a model where clearly no change is involved (sum of average distance from instances). On the other side, "instance" models that succeed may succeed because they do allow change in the categorization function. It would be valuable to know the factors necessary for success.

Finally, should it be the capacity to *change* the classification function that is necessary for a model's success (rather than specifying one fixed, correct method) then it becomes an important research question to specify the way in which the classification function (not just its arguments) changes over time, and the different ways that change can be produced in a mechanism model.

III. EXPERIMENTAL TESTS OF STATIONARITY MODELS

A. Identifying and Controlling a Critical Property

Testing the class of Stationarity Models required testing sensitivity to some criteria or parameter which cannot be specified in advance (as required by Stationary classification functions), but must be derived over the course of learning. We looked for preferential reliance on correlated attributes. Since correlations are specified only over a set of instances, they cannot be identified prior to knowing characteristics of the particular *collection* of instances involved. Thus any preferential reliance on correlated attributes could only be due to abstracting this regularity over learning and changing the classification function as this is discovered. Our tests assessed whether the classification function abstracted and preferentially used information about correlations. As described above, we distinguish between preferential use of *correlation* and nonselective use of the entire pattern of *co-occurrence*.

By manipulating the pattern of co-occurring attributes, we could identify judgments that preferentially preserve reliable correlations identified in input. If subjects are preferentially sensitive to preserving the systematic correlations over preserving any idiosyncratic co-occurrence of attribute values, they must be abstracting structure, not just using a stationary though multiplicative metric.

Since a multiplicative similarity metric is typically described as showing sensitivity to "correlations", we need to be very explicit about how our studies test the adequacy of a stationary but multiplicative metric. In a multiplicative similarity metric, a new instance is more similar to a learning set if it matches a few instances closely, than if it matches many instances a little bit. Matching one instance on four attributes is better than matching two instances on two attributes each. The distribution of attribute matches across instances matters, not simply the total number of matches. Suppose two attributes are correlated in input. In general, an instance which preserved that correlation will be more similar than one which does not, since the correlated instance is likely to have many attribute matches to a few instances rather than a few attribute matches to many instances.

However this does not depend on any abstraction, discovery, or preferential use of the correlated attributes. The metric is sensitive to which values co-occur on the correlated attributes just as it is sensitive to co-occurrences among the values of all attributes.

The experiments summarized below distinguish 1) generalizations which could only be based on abstracting and preferentially using information about correlated attributes from 2) generalizations sensitive to co-occurrence among all attribute values simply as a byproduct of a multiplicative similarity metric correlated.

To do this, we developed a measure, the *similarity profile*, which allowed us to define schematically the number and arrangement of attribute value matches. A schema provides a way of specifying "roles" for how attributes are related to one another independently of assigning a particular attribute to a particular role. In addition, we needed to control for any effects of intrinsic differences in attribute salience, independent of their schematically defined roles. By counterbalancingwhich attributes played which roles, we could separate initial salience or importance of attributes from importance based on discovering their schematically defined predictive role over the course of learning.

An example, illustrated in Figure 11.1 and detailed in Table 11.1, will help explain how this was done. Instances are represented as vectors in which each column represents an attribute. We constructed sets of learning instances which included correlated attributes, for example head type and tail type of pictured animals. Forced choice novel pairs tested whether a subject had abstracted and used the correlation. These pairs were designed so that both test items would be equally similar to the learning set, over any stationary similarity function.

Equating similarity across any stationary similarity function requires matching at a much more detailed level than just the number of times each attribute value in a test instance occurred in the learning set. For example, matching two instances on 2 attributes would not be the same as matching one instance on 3 attributes and one instance on 1 attribute, if the function used a multiplicative combination rule. A similarity profile shows the distribution of attribute matches between one instance (test item) and a set of instances (learning items). It is a vector with as many columns as there are instances in the learning set. The first number in the vector is the number of attribute matches on the

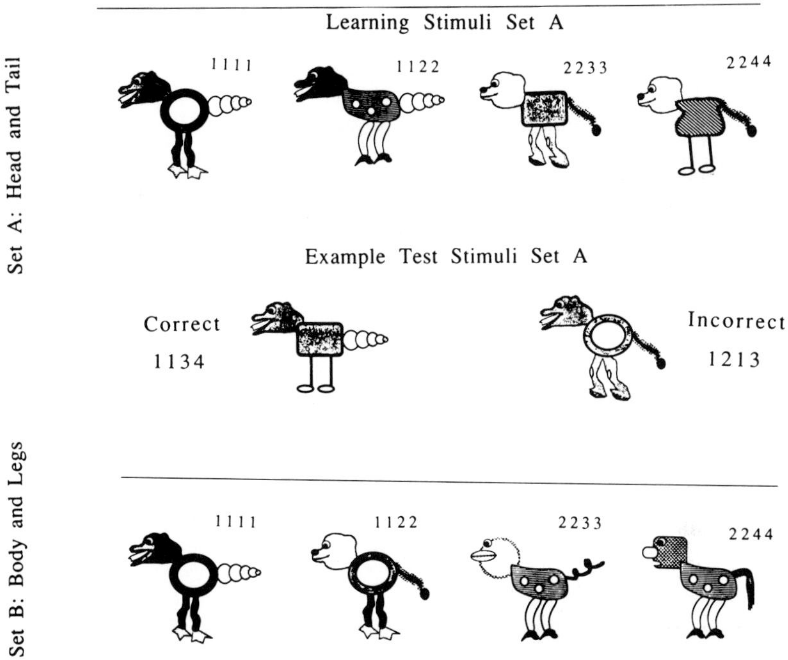

Figure 11.1. Animal stimuli used in Experiment 1. Two sets of four animals were used for different subjects. Head and Tail correlate in Set 1; Body and Legs correlate in Set 2.

most similar learning item; the second is the number of attributes to the next most similar item, and so on. A similarity profile of [332110] means that the target instance matches two old instances on 3 attributes, one on 2 attributes, two on 1 attribute, and one on 0 attributes. (Since our cases all have the same number of attributes we do not here need to represent mismatches explicitly.)

If two test instances have identical similarity profiles to a target set (and differences in initial attribute salience are controlled), their similarity is the same for any stationary comparison function. Figure 11.1 shows 4 learning instances and two test instances, one of which preserves a correlation found in the learning instances and the other of which disrupts it. The four attributes are type of head, body, leg, and tail.

As shown in Table 11.1, Attributes 1 and 2 are correlated in the learning instances. For example, suppose that Head and Tail are the attributes assigned to the first column of the schema shown in Table 11.1. Head 1 always occurs with

Table 11.1. Experiment 1: Example Similarity Profiles for Test Items Preserving & Violating a Correlational Rule.

Similarity Profile Chart for Test Item Preserving Correlation: **1134**

Learning Stimuli Attribute	Simple Representation Attribution Matches				Power Set Presentation Attribute Matches														
	A1	A2	A3	A4	A1	A2	A3	A4	A12	A13	A14	A23	A24	A34	A123	A234	A134	A124	A1234
1111	+	+	-	-	+	+	-	-	+	-	-	-	-	-	-	-	-	-	-
1122	+	+	-	-	+	+	-	-	+	-	-	-	-	-	-	-	-	-	-
2233	-	-	+	-	-	-	+	-	-	-	-	-	-	-	-	-	-	-	-
2244	-	-	-	+	-	-	-	+	-	-	-	-	-	-	-	-	-	-	-

similarity profile of 1134 for simple representation = [2211]
for power set representation = [3311]

Similarity Profile Chart for Test Item Preserving Correlation: **1213**

Learning Stimuli Attribute	Simple Representation Attribution Matches				Power Set Presentation Attribute Matches														
	A1	A2	A3	A4	A1	A2	A3	A4	A12	A13	A14	A23	A24	A34	A123	A234	A134	A124	A1234
1111	+	-	+	-	+	-	+	-	-	+	-	-	-	-	-	-	-	-	-
1122	+	-	-	-	+	-	-	-	-	-	-	-	-	-	-	-	-	-	-
2233	-	+	-	+	-	+	-	+	-	-	-	-	+	-	-	-	-	-	-
2244	-	+	-	-	-	+	-	-	-	-	-	-	-	-	-	-	-	-	-

similarity profile of 1134 for simple representation = [2211]
for power set representation = [3311]

Tail 1 and Head 2 with Tail 2. Test item 1134 preserves this correlation while 1213 disrupts it. However, they have identical similarity profiles. Each matches two old items on two attributes and two old items on one attribute [2211], assuming that the four intended attributes are indeed the way the instances are represented. Now suppose instead that the subject uses an explicitly relational coding wherein a match is scored for each member of the power set of combinations of attribute values. Notice that identical similarity profiles for the simple coding imply an identical similarity for the power set coding.

So far we have equated the number and distribution of attribute matches. Now let us consider how possible effects of initially unequal attribute salience were controlled. Control is important; if subjects simply thought the correlated attributes were intrinsically more important, this would give a basis for favoring the correlated test item, but not because correlation *per se*, or any abstraction of structure, was important. Hence, we randomized or counterbalanced which attributes play the role of A1 and A2; sometimes head and tail correlate, sometimes body and legs. Thus, if one set of attributes or combinations of attribute values is particularly salient, it will weight some judgments in one direction, but an equal number in the opposite direction. Subjects would always prefer the match on these salient attributes but this would sometimes lead to choice consistent with the correlation and sometimes systematic preference for the other member of the forced choice pair.

Let us consider one more complication. In any case where stimulus attributes are manipulated, there must be some correspondence between experimenter-intended and subject-perceived attributes for predictions to be supported. We designed experimental stimuli so that a mismatch of perceived and intended representation would be unlikely to produce a spurious result. First, for a given type of stimuli, we counterbalanced or randomized which attributes correlated; any spurious alternate encoding would be unlikely to have a consistent effect across stimuli versions. Second, across experiments we used different kinds of stimuli as well as different values for the same attributes. Thus, if we consistently get preferences in accord with abstracting a correlation, it is very unlikely to be due to alternative encodings, different from those intended.

We ran a series of experiments all of which presented some test items that preserved a reliable correlation and some that disrupted the correlation. Test items were created in pairs with matched similarity profiles, hence they were equally good category members by any stationary function. If subjects preferentially picked the test item which preserved a reliable correlation from the learning set, they must have abstracted something about the structure of the stimulus set and used this preferentially in generalization. This would be an example of a Dynamic Abstraction function and inconsistent with any function from the Stationarity class. Are Stationary models sufficient, or are the more powerful functions of Dynamic Abstraction models required?

We wanted to assess Stationary models in a task which might be expected to favor a simple, fixed-similarity instance comparison process. Following Brooks (1978) lead, we looked at concept learning in an implicit task, where subjects were unaware of the category to be learned at the time of exposure to instances. Our subjects were told they were in a memorization task and training involved tasks designed to induce memorization of individual items.

B. Experiment 1

1. Method

The first experiment used pictures of animals composed of head, body, tail, and legs, each of which (across subjects) might take on any of four values. For some subjects head and tail were the correlated attributes and for others body and legs correlated. The head-tail set are shown in Figure 11.1 and the schematically specified relations in Table 11.1.

Pictures were displayed on overheads to small groups of subjects. Twenty eight college students participated, for class credit. Subjects were told that they were to memorize the instances. Subjects did the easier, naming task first and then the construction task. In the naming task, subjects were shown the four animals and each one was named. Then subjects attempted to write down the names as each animal was displayed. The set of four animals was displayed 6 times. In the construction task, subjects had six types of each body part on cardboard "puzzle pieces". Subjects were told the name of an animal and asked to construct it from memory, thus recalling it. Subjects corrected their errors and ran through the four animals two or three times. No mention of categories, correlations, or groupings was made during learning.

After learning, subjects were told that all the strange animals they had seen came from another planet, Tumbolia. They were shown pairs of novel test items and told to pick which animal in a pair also came from Tumbolia. In each forced choice pair, one animal preserved the correlation and the other disrupted it. Both items had identical similarity profiles. Thus, there would be no basis for preferring one over the other by any stationary categorization function or similarity metric, be it additive or multiplicative.

There were 12 Old Values test items. These items only used attribute values which had occurred in the original learning animals. These items assessed whether subjects had abstracted the reliable correlational pattern available in the learning instances.

There were 6 Novel Values test items. These tested whether subjects would preserve a correlation between previously correlated attributes even when new values of the correlated attributes were introduced. First, a new animal with *all new* attribute values was presented (schematically, 3355). The correlated attributes are always the first two columns in the schema notation. In particular, the two correlated attributes each assumed a novel, third value; for example, if head and tail had been the correlated attributes, subjects had never seen Head 3 nor Tail 3. This animal was identified as coming from Tumbolia as well. Then subjects saw six more forced choice pairs of novel items. All items disrupted some of the pattern of which attribute values went together in 3355 example. But one item in each pair disrupted the pairings on the previously correlated attributes and the other did not. We were assessing whether subjects had decided that preserving the correlation on the reliably correlated attributes was still a priority, even when those attributes assumed new values and there was little evidence about co-occurrence of those particular values. Again, items in the pair were equally similar to the original four learning animals and to the newly introduced animal. For example both items in the pair [3366/3456] matched the new animal on two attribute values [33xx/3x5x] and had two attributes never seen in any prior animal [xx66/x4x6]. Consistent judgments on this test would suggest that subjects had learned that correlations between attributes, not just combinations of particular values, are important.

2. *Results*

Responses were considered correct if they were in accord with the correlational rule. Scores were significantly nonnormal, highly skewed toward all or nearly all correct. Hence, the usual t-test comparison (though significant for both the Old and Novel Values test) was inappropriate. Instead, we asked if the number of high scores was more than expected by guessing; even if subjects were simply guessing, some high scores would result. We treated each subject as a binomial "success" if that subject individually got more right than expected by chance; ten of twelve items and six of six are the scores needed. Then we figured the probability of the actual number of "success" subjects (successes=x) from the number of subjects (trials=n) using the probability that a subject would get 10/12 (or 6/6) by chance (chance probability of individual scoring above criteria=p). On the Old Values test, 12 of 28 subjects scored 11 or 12 (none scored 10); on the Novel Values test, 14 of 28 subjects scored 6/6. Both proportions are drastically unlikely under a guessing model (p's<<.0001). Learners were roughly evenly distributed across the two stimulus versions. In sum, about half the subjects learned and used the correlational rule in generalizing.

A smaller pilot experiment, with different values for animal parts found 6 of 8 subjects scored 6/6 correct; again this many high scores is unlikely if subjects were indifferent to the two items.

C. Experiments 2, 3, and 4

1. Methods

Three more experiments were conducted which had much the same logic, but differed in stimulus characteristics. These experiments used text descriptions of people, for example, "John comes from Houston, reads Discover, plans to be a pilot, is Russian, goes to Berkeley, and sails." First, this allowed us to generalize over type of stimuli. Second, it allowed us to randomize which conceptual attribute (hometown, hobby, etc.) played the schematically defined role of the correlated attributes, since it was practical to produce many different attribute assignments. This ensures more completely that any effects of attribute salience would sometimes work for and sometimes against the choice defined by the correlation.

Each experiment had a condition where the correlation had no exceptions and a condition with exceptions. In the exception conditions one or two learning instances disrupted the correlation: for example, if reading Discover went with sailing while reading Byte went with tennis, one or two individuals switched this pairing.

Across experiments, the number of instances, attributes, and attribute values varied. The three experiments used five (Experiment 4) or six (Experiments 2 and 3) attributes and used seven (Experiment 4) or 12 (Experiments 2 and 3) instances. Uncorrelated attributes took on 12 (Experiment 2), 6 (Experiment 3), or 4 (Experiment 4) values. As the complexity of stimuli increases, it is increasingly difficult to produce pairs of items with identical similarity profiles but which also differ in whether they preserve or violate a correlation present in the learning set. Where similarity profiles were not identical (some test items in Experiments 2 and 3), the item which preserved the correlation was slightly *less* similar to the learning set than was the uncorrelated item. Experiment 4 had an intermediate level of complexity but again allowed perfect matching of similarity profiles. Similarity, which was matched within pairs of items, varied across test pairs. Details of stimuli construction and schema are described in Billman and Richards (1988).

The learning phase of the experiment was again presented as a memorization task with no mention of categories, correlations, or patterns. Subjects began by studying the printed descriptions of the individuals for 10 minutes. After this, a description of an individual was presented on computer and the subject provided the name. Then the study and name cycle was repeated.

After the learning phase, subjects were told that all the people came from the same organization and shared similar characteristics. Subjects first did a Generalization Task. One description was displayed and subjects judged whether or not the individual belonged. In the following Forced Choice Task, two descriptions were displayed and subjects picked which belonged in the organization. Subjects were finally interviewed about how they made their choices.

2. *Results*

A summary of relevant results are presented in Table 11.2. First, across all conditions and experiments, 27 of 78 subjects were learners (scoring significantly above chance) as measured on the Generalization Test, and 30 of 78 scored above chance on the Forced Choice test. Of the 38 subjects in the No Exception conditions, 19 were learners, compared to 8 of 38 in the Exception conditions. Thus, many subjects again learned and used the correlational rule available in input. The number of learners was halved when exceptions were introduced, a significant decrease in all experiments. However, over a fifth still abstracted and used the correlation in generalizing.

To reject the sufficiency of Stationary models and to determine a need for Dynamic Abstraction, we need only show that subjects use some decision function inconsistent with Stationarity. It is not necessary that subjects only or always use the particular decision function we were testing. Nevertheless, it is interesting that about half the subjects did not reliably generalize in accord with the correlational rule. A few subjects in the exception condition mentioned noticing the correlation but then discounting it when they saw exceptions, but in general there was a good match between reporting the correlational rule and making judgments consistent with it. Thus it seems that many subjects did not notice the pattern.

Recall, however, that the test circumstances were selected to bias the learner toward simply trying to memorize instances and away from looking for rules, patterns, or categories. For Experiments 2-4, the material had many attribute values and was difficult to memorize; by design, furthermore, there was little prior reason why some characteristics should go together over others. While necessary for control, this made difficult the task subjects were trying to achieve in the initial 10 minute study period-- memorization. The other task, of naming the individuals given their description, could be done by remembering one distinctive property of each individual and thus was no incentive to look at multiple properties or relations among properties at all. These experiments, then,

Table 11.2. Experiments 2, 3, and 4: Number of Learners and Mean Percent Correct.

	Number of Learners/ Total Subjects		Mean % Correct	
	General-ization	Forced Choice	General-ization	Forced Choice
Experiment 2A No Except.	6/12*	6/12*	73.7%	72.9%
Experiment 2B Except.	1/12	2/12	50.6%	50.0%
Experiment 3A No Except.	7/15*	7/15*	67.9%	70.4%
Experiment 3B Except.	5/15*	5/15*	59.8%	58.3%
Experiment 4A No Except.	6/12*	7/12*	68.0%	74.0%
Experiment 4B Except.	2/12	3/12	51.0%	57.0%

*Numbers of Learners: value greater than chance, $p \leq .05$, conducted under the binominal guessing model.

show that people can discover and apply a correlational rule even in difficult conditions.

D. Summary of Stationarity Tests

Three points from these results are noteworthy. First, these experiments identified conditions when the predictions of Stationarity classification functions contrasted with Abstraction classification functions. This is difficult, because even the restricted class of Stationarity functions is very powerful. It is difficult to identify a basis for generalization which cannot be produced by some powerful but still stationary function, for example, a multiplicative metric or power set encoding. Our judgments required preferential use of correlations occurring reliably across the set of learning instances; such a classification function could not be specified in advance, but only dynamically abstracted from experience with the set of examples. By matching similarity profiles we did not just equate the number of matching attribute values between the target instance and the learning set (as would be needed for an independent cue similarity metric). We also equated the distribution of these matches across learning instances, thus equating similarity as specified by a multiplicative similarity metric as well (and any other stationary classification function which takes the matrix of instance attribute values as its domain). Our effort here has been at testing, and finding the insufficiency of, the whole class of Stationary Models. This therefore shows the necessity of the more powerful Dynamic Abstraction functions.

Second, the experiments identified one way in which subjects change their classification function over learning: abstraction and preferential reliance on correlational information. These experiments produced additional evidence that learners are sensitive to relations among attribute values (e.g. Medin, Altom, Edelson, & Freko, 1982). Further, consistent correlations are preferentially used over idiosyncratic covariation. Of course, correlations, like the value of any summary statistic, can be preserved simply by preserving the whole instance set. But, in our experiments subjects showed *preferential* use of the reliable correlations. This was required for reliably selecting the particular correlated items in our generalization tests. It is perhaps most clearly illustrated in the Novel Attributes Test of Experiment 1. Performance here suggests information about the importance of the attribute per se, not just relations between values, is abstracted[1].

Third, the probability of discovering and using the correlational pattern in generalizing was dramatically decreased by the introduction of one or two exceptions from a set of seven or twelve examples.

[1]One caveat is in order. The studies show sensitivity to correlations and preferentially to that between A1 and A2 in the schema. A1 and A2 differ from other attributes in that they have fewer values, as well as being reliably associated. The studies do not tell us which factor is responsible for subjects' selective preservation of their correlation. Either basis would constitute selective abstraction.

IV. ASSESSING THE NATURE OF ABSTRACTION VIA MULTIPLE TASKS

A. Looking For Evidence of Abstract Products of Concept Learning in Induction Tasks

The classification studies presented above show that people do dynamically abstract structure during learning and that models with a stationary classification metric are insufficient. Given this evidence for dynamic abstraction in classification, there are two reasons why one might seek evidence from other tasks. First, it proved quite arduous to find a diagnostic test. Both classes of models are quite powerful and identifying a test that does not allow mutual mimicry was difficult. Thus, it would be nice to gain additional evidence that this is a difference that makes a difference. Second, the prior experiments are limited in what they tell us. We know viable models are in the Dynamic Abstraction class and include relational coding. But they leave most of the details of a mechanism model unconstrained. An important motivation for looking at multiple tasks is to generate more detailed constraints on the underlying mechanism. Results from first set of experiments could be produced in many ways: by adding new rules ("Cookie head and rooster tail go together" or "Head and tail are related"), by reweighing parameters (increasing salience of an attribute every time it's value is correctly predicted by with another), or by many more complicated changes in procedure. We do not know much about whether the change comes from a change in representation or change in procedure.

Informally, one of the key contrasts between representation and process is that of stability. "Representation" emphasizes a stable bundle of information that can be accessed on different occasions and for different tasks. "Process" emphasizes information or operations which carry out one particular function and are tailored to a particular task.

In asking about the nature of the information abstracted, then, we can look for evidence that information derived in one task is accessible in another. If information is available from one task to the next or from one problem to the next, this suggests it might best be understood as forming and using new representations. If information is not used beyond the initial problem and task, this suggests a parameter change in a specific procedure or some procedural, task-oriented change. Of course, generality of learning is a broad and fundamental question, independently of what it can tell us about the informal contrast between representation and process.

Thus, the style of investigation here is distinct from and complementary to that of the first section. We are not testing precisely specified classes of models, but gathering suggestive evidence across a widened range of tasks. I will briefly discuss two induction tasks we used to find evidence that people abstract information from one learning task to guide inferences both over new attributes and new categories. This work is largely inspired by Goodman's (1954/1983) comments about the way background knowledge is brought to bear in guiding induction. The first task was done by college students (Experiments 5,6,&7), the second by preschoolers (Experiment 8).

The agenda in these studies is asking how subjects use category knowledge to guide inferences about properties of new instances or new categories, not how subjects use information about the properties of instances to form categories. Looking at how categories are used in induction may provide information about categories complementary to that from classification tasks.

B. Inductions About New Properties

S. Gelman's (Gelman & Markman, 1986; Gelman, 1988) work on induction shows that children (and adults) will generalize a new property known to be true of one entity (a blackbird) to another identified as the same category, "bird", but perceptually very dissimilar (a flamingo) over an entity which is perceptually similar but identified as a different category, "bat". One interpretation of these findings is that subjects conjecture that simply being in the same category (or the "special" natural kind categories) makes two entities likely to share additional, unknown properties. This would give categories quite an abstract role in mediating induction. A second interpretation is that subjects are cued to or reminded of other, known properties of the category (birds sing, fly, live in nests, lay eggs, etc.). When these properties are taken into account, the blackbird and flamingo are simply more similar and generalization is based on similarity. Whatever the interpretation, the findings illustrate that children are not just creatures of appearance; but only the first interpretation would imply that children are using category membership per se, distinct from similarity, to guide induction.

One way around this is to teach subjects new categories where the similarities to known category members can be controlled. In three experiments we taught college students novel categories of alien marine animals. In all experiments categories were determined by correlated attributes. For example, the animals that lived in the coral reefs ate with a toothed tongue and the animals that lived in the tidal zone ate with an oral disk. In some conditions, this was the only basis for grouping animals into two categories. In Experiments 5 and 6 a category label was also provided; "skillasos" might live in coral reefs and ate with a toothed tongue while "basatrops" lived in the tidal zone and ate with an oral disk. Experiment 7 used "unsupervised" learning and no category labels were provided; only the correlation among attributes provided a basis for determining two contrasting categories.

1. Methods

Experiment 5. Forty-eight college student subjects participated, divided among a control, Unlabeled Induction, and Labeled Induction conditions. The experimental groups learned about eight animals divided into two categories. Each description gave the animal's name, specified four attributes (habitat, locomotion, method of eating, and method of defense) and provided a category label ("skillasso" or "basotrop"). Two attributes correlated, and two sets of stimuli were created, one with habitat and method of eating determining the categories and the other with means of reproduction and method of locomotion. Example descriptions follow:

> YORKY is a skillaso, reproduces by internal fertilization, gathers food through an oral disk, moves by means of jet propulsion, and lives on the ocean surface.
>
> ZIPPY is a basotrop, reproduces by ... gathers food with a toothed tongue, moves... , and lives on the ocean floor.

Thus in the first stimulus set all 4 creatures living on the ocean surface also ate with an oral disk, while the four on the ocean floor used a toothed tongue. Subjects were given 5 minutes to study eight descriptions of animals. Subjects then practiced by providing the species that went with a description and name. There was another study period and then subjects provided the description and category that went with the individual animal's name.

After learning about the animals, the experimental subjects did the induction task. The induction task assessed how subjects used category knowledge to generalize novel properties from source to target animals. The properties ranged from biochemical to global, such as having particular enzymes or being disrupted by the earth's tides.

Subjects read descriptions of triplets of novel animals. Each triplet consisted of a first, "source instance" and two "target instances". All three appeared on one page. Each description included values of all four familiar attributes and the category label was provided for all source animals as well. The source instance was described first and a novel property ("disrupted by earth's magnetic field") asserted about it. The two target instances differed from each other in that one was from the same category as the target and one was from the other category. Stimuli were constructed so that both target instances were equally similar to the source instance; each target animal matched the source animal on a different two of its four attributes. Each of the triplets dealt with induction about a different, novel property; none had been introduced in any prior animal description.

Control subjects began with the induction task, with no prior learning about the animals. Unlabeled-Induction subjects had no category labels for the target animal descriptions. Labeled-Induction subjects had category labels for the *target* as well as source animals.

After reading the three descriptions, subjects rated each of the target animals on a seven point scale, estimating how likely they thought that target was to share the novel property true of the source animal. We assessed whether subjects were more likely to generalize to the same-category target than to the contrast-category target. If subjects showed this preference, it could not be due to greater similarity between the source and the preferred target: numbers of attribute matches for each target were equal and the attributes on which similarity was based were counterbalanced across subjects. Performance in the learning condition could be compared to chance and to performance in the control group.

Experiment 6. Experiment 6 used within-subject comparison (35 subjects) between property inductions made before learning and inductions about a distinct but similar set of properties made after learning; category labels were provided during learning along with correlated properties to determine category membership. At induction half of the induction triples included the category label and half the triples dropped it.

Experiment 7. Experiment 7 used the same within-subject approach as Experiment 6, but never provided category labels, either during learning or induction. Thus subjects were not in the position of being taught labels, but then having them "taken away" on induction trials. We focused subjects on the existence of the categories to ensure that subjects had learned the categories.

After study and before proceeding to the induction phase they had to describe the two kinds of animals and what made them distinct.

2. *Results*

In all three experiments, subjects used category information from learning to guide inductions. The difference between likelihood ratings of the same-category and the different-category target item, the Discrimination Score, was the primary measure. Discrimination Scores could range from -6 to +6 with zero being chance. The Discrimination Score was significantly positive where labels were consistently provided in Experiment 5 (Discrimination Score=1.23, $t(15)=4.01$, $p<.001$) and in Experiment 6 (Discrimination Score=1.15, $t(34)=4.22$, $p<.001$). In these two experiments subjects were not above chance when labels were provided at learning but removed at induction. In Experiment 7 where no labels were used at all, subjects again showed sensitivity to category membership in their inductions (Discrimination Score 1.07, $t(29)=3.58$, $p=.001$). In addition, in Experiment 5, subjects who had been taught the categories scored significantly higher than those who had not. In Experiment 6, while there was a main effect of whether items were labeled at induction, the difference in judgments before and after learning was not significant.

In addition, subjects reported in debriefing heavy reliance on prior theories about how different properties should interact. For example, an animal that lives in the deep ocean is unlikely to be affected by anything having to do with the sun, whether or not it is in the same category as an animal that is thus affected. The prevalence of such reasoning was striking in its own right and it provided a powerful alternative for classification to the one investigated. Nevertheless, there was a small but significant preference for generalizing to the same-category target item, above and beyond any influence of prior theory.

These experiments suggest that people will use categories to guide the scope of induction of new properties, much as suggested by Goodman, and that status as same-category-member, per se, influences judgments, independently of the properties that members of the same category share. They also incidentally illustrate the prevalence of importing background theory to reason about novel categories and unfamiliar properties.

Thus, the experiments show that people have and use information in a sufficiently abstract form to guide reasoning about completely novel properties. Further, this reasoning is not just based on the similarity between source and

Category	Attribute	Values
Blocks	*color*	red, blue, green, yellow, black, white
	shape	cube, rectangle, T, window, plane, stair

Category	Attribute	Values
Creatures	*covering*	sponge, feathers, wool, scales, fur, leather
	parts	wings, claws, paws, webbed feet, fins, insect legs
Vehicles	*base*	stilts, wheels, legs, landing pad, rocket, pontoon
	load	spaceman, telescope, umbrella, computer, signal, steering wheel
Containers	*shape*	bowl, plate, tray, pot, pitcher, cup
	material	plastic, clay, metal, paper, glass, wood

Note. Underlined attributes were used as Relevant for the Given Category.

Figure 11.2. Stimuli description for Experiment 8. Four types of objects with two versions each.

target instance; even when number of matching attributes are equated, subjects prefer to generalize to the same category instance.

C. Inductions Over New Categories

Just as people can make sensible conjectures about the scope of novel properties, so can they make conjectures about the basis for novel categories. Again influenced by Goodman, we looked at how knowledge about homogeneity or heterogeneity within one category might affect inferences about what attributes would be consistent in a novel but related category (Macario, Shipley, & Billman, 1990). People may know that a type of category (e.g. animal species) are homogeneous in some properties (vocalization) but not others (name). This could then be used to guide inferences about a novel category, given just a single instance. For example, presented with an aardvark which grunts and is named

"Henry", people might infer that while other aardvarks are likely to grunt, they are not likely to be named "Henry". We asked if children would use information that an attribute was heterogeneous versus homogeneous in a familiar category and pick the homogeneous attribute as the basis for extending a new, sister category (Experiment 8). Experiment 8 had two factors: homogeneous versus heterogeneous objects and presence versus absence of a label for the relevant attribute.

Thirty-two preschool children were introduced to a classification task presented as what different puppets would like. This has proved a useful and comprehensible way to structure a classification task for preschoolers (Waxman & Gelman, 1986).

Figure 11.2 shows the stimulus sets used, each varying in two very salient attributes (vehicles: load and support, containers:material and shape, animals:body covering and leg type, and blocks:color and shape). I will use the block stimuli to describe what children saw and did. Experiment 8 had four conditions: homogeneous vs heterogeneous crossed with presence or absence of a label. In all cases, subjects were shown one puppet with a collection of three objects which the puppet liked; conditions differed in what objects the first puppet was given and what the child was told about them. Then a second puppet was introduced with one liked object. The child was presented with a series of objects and asked whether the second puppet would like it (indirectly, whether it belonged with the second puppet's original object).

Figure 11.3 summarizes the differences in materials and procedures across the four conditions. In the heterogeneous conditions the first puppet was given three things which were homogeneous in one attribute (e.g. color), but varied in the other (e.g. shape). Thus the first puppet might get three blue blocks, one cube, one T-shaped, one 'window'. Since one attribute varied while the other did not, this provided a basis for preferring the attribute which did not vary in generalization. When the second puppet was introduced, he would be given one block with a new value of both attributes, a green rectangle. In these conditions, the child would have the option of recognizing that color was constant but shape varied, so color was the attribute relevant to categorization; then green objects of several shapes should be given to the second puppet, but not rectangles of different colors.

In the homogeneous condition the first puppet was given three objects each of which had the same shape and color. This would give the child no basis

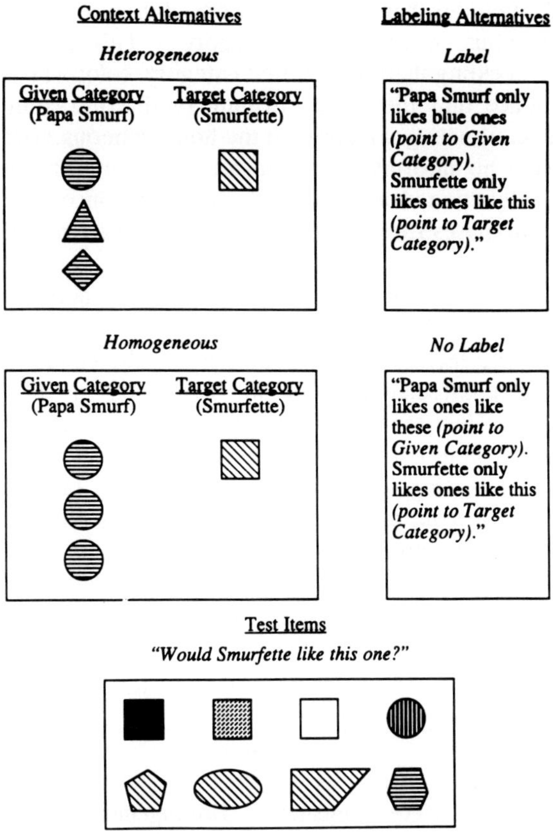

Figure 11.3. Overview of contrasts in stimuli and procedure for Experiment 8 across Heterogeneous versus Homogeneous and Labeled versus Unlabeled Conditions.

for deciding that shape or color was more important for one category and hence might be relevant to a contrasting category.

The labeled conditions differed from the unlabeled conditions in that children were told the puppet "likes blue ones", while in the unlabeled condition the child was told the puppet "likes ones like these". Thus the label pointed out the attribute value as relevant to one category, though it mentioned only the value, not the attribute itself.

In the heterogeneous, labeled condition the child would have information both from the label and from having only one dimension homogeneous and the other heterogeneous in the initial category. This provides two types of information for deciding that, for the first category, color was relevant and shape was not and hence that the same attribute (with contrasting value) might be relevant for the contrasting category. In the homogeneous, no label condition, the child would have no information about an organizing principle for the initial category. This functioned as a control condition, since the child was left to express any initial biases for classification, with no clues from the sister category.

The number of objects that a child sorted in accord with the intended attribute are shown in Table 11.3. Chance performance would be 4 out of 8 choices. When both factors were available to aid the child, the mean score was 6.44. Scores were affected by both the label and the heterogeneity factor. Further, even when just the trials with no label were considered, children used information about heterogeneity of the attribute irrelevant in the contrast category to guide sorting into the novel category. Thus, subjects could do systematic induction from a single instance by using information about which attribute was homogeneous in a sister category.

Table 11.3. Experiment 8: Mean Number of Correct Responses (Out of 8).

	Context		
Labeling	**Heterogeneous**	**Homogeneous**	**Overall**
Label	6.44	6.14	6.29
No Label	5.69	4.86	5.27
Overall	6.06	5.50	5.73

Notes. N = 8 per cell.
Context effect p = .031; Label effect p = .0004; Context X Label p = .2887.

D. Summary

These studies have focused not on learning to classify instances into categories, but on how established category knowledge is used in reasoning about new properties or new categories. They showed that children and adults are willing to use categories in fairly abstract ways across tasks. How does this relate to the studies in the first part of this chapter and what does this suggest about category representation?

Categories are important across a variety of tasks, including classification and induction. In looking for a parsimonious account of the common knowledge in evidence across tasks we can learn something about the nature of category representation. Categories are more than an assemblage of similar instances. No model which treats classification as a function of fixed similarity is sufficient. Subjects dynamically decide on and use selective information about what is relevant to guide classification (Experiments 1-4 and 8). Further, categories are more than a collection of linkages between particular attribute values and a label. Subjects know about the relations among attributes within categories and about the relevance of contrasting values of the same attribute across contrasting categories. Further, category membership per se can play a privileged role in guiding the scope of induction for novel properties and does so even when no category label at all was provided.

V. CONCLUSIONS

The research summarized in this chapter used two complementary approaches to guide theorizing about the representation and processing of categories. The first approach identifies one characteristic critical for any model's success and tests the adequacy of a class of models rather than one- by-one assessment. We identified and rejected the broad and powerful class of Stationarity models by focusing on the entire mapping function from input to classification judgment (Experiments 1-4). Dynamic Abstraction models are required even with types of stimuli and circumstances of learning which might favor simple instance comparisons.

Successful test of a whole class is difficult given the surprising amount of mutual mimicry among models. In particular, power of representation and power of processing can trade off and the same classification judgments can be achieved in multiple ways. Nevertheless, inferences can be made over classes of models and necessary properties can be identified. This required 1) precisely specifying the models at the level of the entire function that maps from the domain of learning-instances-plus-novel-test-item to the range of classification judgment, 2) identifying a property to which one but not the other class is sensitive, and 3) then testing for this property.

This first approach tells us something very certain about the whole classification function but suggests very little about the respective roles of representation and processing. The second approach is less certain but makes more specific suggestions about role of representation and processes within the viable class of models.

Is abstract information about categories the result of stable representations or task-specific and more transient processing invoked anew for each judgement? Results from Experiments 5-8 can most parsimoniously be understood by assuming that quite abstract category representations are formed, representations that organize information so it can be used across tasks. Investigation across a variety of tasks, not just classification was critical here. We found 1) that subjects abstract information about what attributes are important (Exp 1-4), 2) that this information is used in induction as well as generalization tasks (Exp 5-8), 3) that people can apply abstracted information about attribute importance to new categories and new attribute values (Exps 8 & 1), and 4) that category membership (distinct from known shared properties) can be used as a basis for controlling induction (Exp 5-7).

This cross-task evidence has two broad implications. First, it provides converging evidence for Dynamic Abstraction models. These related tasks seem to require information that is relatively stable and decontextualized from the original instances. Dynamic Abstraction models, but not Stationary models, would easily predict the sort of information needed in these induction tasks would be abstracted in classification tasks. Second, the evidence suggests something about the nature of the abstractions and the role of representation versus process change. Similar, abstract information appears to be used in several different tasks: abstraction apparently results from stable, relatively context independent representations, rather than from a fleeting process specific to some task which leaves no enduring product.

A combination of methods helps us make progress in identifying necessary principles required for any viable model and in suggesting plausible characteristics within the viable class. It is possible to test general principles despite indeterminacy, at a more specific level, from representation-process tradeoffs. We can also gather more suggestive evidence about the role of representation versus processing within the class of viable models. Such findings are of interest directly. In addition, they can further constrain implementation of a mechanism model. Finally, they can guide our thinking in articulating other general principles governing category learning and use.

ACKNOWLEDGEMENTS

Thanks to Larry Barsalou, John Palmer, and Jeffery Richards for critique and discussion; and to Jeffery Richards, Mickey Mathews, Evan Heit, Andy Masou, and Peter Koopman for aid in implementing the experiments.

REFERENCES

Anderson, J.R. The Adaptive Character of Thought (1990). Hillsdale, NJ: Erlbaum.

Anderson, J.R., Kline, P.J., & Beasely, C.M. (1979). A general learning theory and its application to schema abstraction. In G. H. Bower (Ed.), The Psychology of Learning and Motivation, Vol. 13. New York: Academic Press.

Barsalou, L.W. (1990). On the indistinguishability of exemplar memory and abstraction in category representation. In T.K. Srull & R.S. Wyer (Eds.) Advances in Social Cognition, Vol. 3. Hillsdale, NJ: Erlbaum.

Billman, D. & Heit, E. (1988). Observational learning from internal feedback: A simulation of an adaptive learning method. Cognitive Science, 12, 58-69.

Billman, D, & Richards, J. (1988). Abstraction of correlational rules in implicit concept learning tasks. Manuscript.

Brooks, L. (1978). Nonanalytic concept formation and memory for instances. In E. Rosch & B.B. Lloyd (Eds.) Cognition and Categorization. Hillsdale, NJ: Erlbaum.

Estes, W.K. (1986). Array models for category learning, Cognitive Psychology, 18, 500-549.

Gelman, S.A. (1988). The development of induction within natural kind and artifact categories. Cognitive Psychology, 20, 65-95.

Gelman, S.A. & Markman, E. (1986). Categories and induction in young children. Cognition, 23, 183-209.

Goodman, N. (1983). Fact, Fiction, and Forecast. 4th edition. Harvard University Press: Cambridge, MA.

Hayes-Roth, B., & Hayes-Roth, F. (1977). Concept learning and the recognition and classification of exemplars. Journal of Verbal Learning and Verbal Behavior, 16, 321-338.

Hintzman, D.L. (1986). Schema abstraction in a multiple-trace memory model, Psychological Review, 93, 411-428.

Hintzman, D.L., & Ludlam, G. (1980). Differential forgetting of prototypes and old instances: Simulation by an exemplar-based classification model. Journal of Memory and Cognition, 8, 378-382.

Holland, J.H., & Reitman, J.S. (1978). Cognitive systems based on adaptive algorithms. In D. A. Waterman, & F. Hayes-Roth, Pattern directed inference systems. New York: Academic Press.

Krantz, D.H., & Tversky, A. (1971). Conjoint-measurement analysis of composition rules in psychology. Psychological Review, 78, 151-169.

Kruscke, J.K. (1990). ALCOVE: A Connectionist Model of Category Learning. Cognitive Science Research Report #19. Indiana University.

Macario, J.F., Shipley, E.F., & Billman, D.O. (in press). Induction from a single instance: Formation of a novel category. Journal of Experimental Child Psychology.

Medin, D.L., Altom, M.W., Edelson, S.M., & Freko, D. (1982). Correlated symptoms and simulated medical classification. Journal of Experimental Psychology: Learning, Memory, and Cognition, 8, 37-50.

Medin, D.L., & Schaffer, M.M. (1978). A context theory of classification learning. Psychological Review, 85, 207-238.

Miyamoto, J.M. & Eraker, S.A. (1988). A multiplicative model of the utility of survival duration and health quality. Journal of Experimental Psychology: General, 117, pp 3-20.

Nosofsky, R.M. (1984). Choice, similarity and the context model of classification. Journal of Experimental Psychology: Learning, Memory, & Cognition, 10, 104-114.

Nosofsky, R.M. (1987). Attention and learning process in the identification and categorization of integral stimuli. Journal of Experimental Psychology: Learning, Memory, & Cognition, 13, 87-108.

Reed, S.K. (1972). Pattern recognition and categorization. Cognitive Psychology, 3, 383-407.

Tversky, A. (1977). Features of Similarity. Psychological Review, 84, 327- 352.

Tversky, A., & Gati, I. (1978). Studies in similarity. In E. Roach & B. B. Lloyd (Eds.), Cognition and Categorization. Hillsdale, NJ: Erlbaum.

Waxman, S.R., & Gelman, R. (1986). Preschoolers' use of superordinate relations in classification and language. Cognitive Development, 1, 139-156.

Whittlesea, B.W.A. (1987). Preservation of specific experiences in the representation of general knowledge. Journal of Experimental Psychology: Learning Memory, Cognition, 13, pp. 3-17.

Commentary

Modelling Category Learning and Use: Representation and Processing, D. Billman

IRWIN D. NAHINSKY
University of Louisville

Billman reviews important distinctions among category learning models. The instance based-rule based distinction is perhaps the most extensively studied distinction. Billman introduced the "stationary abstraction" - "dynamic abstraction" dichotomy as a related means of discriminating between models. Stationary abstraction models include a fixed rule for mapping instances into categories, while dynamic abstraction models provide for a process that can modify the mapping rule over the course of learning in response to evidence. When one considers the large number of conceptual strcutures that may partition a stimulus domain into categories, it is evident that an organism without rule-modifying capabilities would be hard pressed to organize experience.

The author effectively demonstrates the applicability of the dynamic abstraction principle in a series of studies that show sensitivity to correlational structure as a basis for category membership. She found that if category membership depends upon correlation between attribute values on an attribute pair, subjects discover the defining rule and generalize beyond the specific learning values. This occurs despite the fact that other clues to category membership, including overall similarity, are controlled for. The ability to abstract rule-based structures of this type presents a challenge to models that require a fixed and limited process for generating a conceptual structure. Research of this type should stimulate more work on models incorporating rule-testing procedures.

Percepts, Concepts and Categories
B. Burns (Editor)

12

Learning Categories With and Without Trying: Does It Make A Difference?

THOMAS B. WARD
ANGELA H. BECKER
Texas A&M University

I. INTRODUCTION

Humans are remarkably adept category learners, capable of acquiring new groupings for both novel and known objects under a variety of conditions. In some situations (e.g., formal educational settings) individuals make a deliberate attempt to learn the category membership of particular sets of items, and, many a disappointed instructor's beliefs notwithstanding, learning does take place. In other situations, learning about category membership occurs despite the fact that individuals may not deliberately attempt to learn. Put differently, people do learn when they are trying to learn, but they also learn when they are trying to do other things.

Because category learning can occur both intentionally and incidentally, it is important to examine the nature of the learning process under both types of conditions. However, with some notable recent exceptions (e.g., Fried & Holyoak, 1984; Kemler Nelson, 1984; Lewicki, 1986) most research concerned with category acquisition has focussed on intentional learning conditions. Typically, the research participant in an intentional learning study is given a set of objects, either successively or all at once, with the explicit task of learning the "correct" category assignments of the objects. Feedback is given until some criterion is reached or until some fixed number of trials occurs.

Such intentional learning tasks differ in many respects from a large number of day-to-day learning experiences in which people appear to acquire vast amounts of category knowledge simply as a result of interacting with objects. Given the pervasiveness of incidental category acquisition in everyday settings, there may be an entire domain of human learning about which little is known. This lack of research attention to incidental category learning may not be a serious problem if the principles that govern learning are exactly the same for incidental and intentional types of learning. In effect, if incidental and intentional learning are highly similar then we may have already incidentally learned about incidental learning by studying intentional learning. However, there are hints even in the small literature available that there may be important differences in the nature of category learning across incidental and intentional conditions (e.g, Kemler Nelson, 1984; J. Smith, 1989). Thus, the time is long overdue to begin to intentionally learn about incidental learning. The present chapter focusses on the extent to which there are differences in the nature of what people learn and in the types of processes they use in acquiring categories intentionally and incidentally.

The plan of the chapter is, after a few preliminary comments, to consider the ways in which incidental category learning has been hypothesized to differ from intentional learning and to consider the strength of support for those proposed differences. In the first major section, particular attention is given to the suggestion that incidental learning may result in a more holistic mode of processing with the consequence that acquired category structures differ from those that are acquired under intentional learning conditions (e.g., Kemler Nelson, 1984). In different subsections within that first major section we also examine the idea that a holistic mode of processing may be well-suited to learning about real-world categories, and we consider what it means to claim that incidental and intentional learning are different and what types of evidence would support a fundamental difference. Because the term "holistic" has been used in different ways, we also consider different senses of that term and the evidence in support of "holistic" processing in each of those senses. As will be seen, the conceptual and operational definitions of holistic that one chooses are directly linked to the question of what it means for incidental and intentional learning to differ in fundamental ways.

In a second major section we explore the more general idea that what individuals learn about the structure of a category is more determined by the attentional demands of the task than whether or not the individual intends to learn. We consider the extent to which it is possible to predict from the attentional demands of an incidental task exactly which attributes will predominate in subsequent category decision-making. This second section is concerned primarily with attention to single attributes. In the third major section, we explore the extent to which attribute correlations are abstracted during the course of incidental learning. This section also explores the link between the concepts of analytic and holistic processing and the tendency for category decision-making to be dominated by correlated attributes. It also expands on ideas introduced in the first major section that a) category learning, either intentional or incidental, does not occur in a vacuum but rather in the context of particular activated knowledge structures, and b) these knowledge structures can direct attention to particular attributes regardless of the learner's intent to acquire categories.

II. SOME PRELIMINARY COMMENTS

Psychologists are prone to think of humans as being fundamentally adaptive organisms that are constantly about the business of adapting to their environments. With this view, one might legitimately ask whether there is ever a time when we are entirely without the *intent to learn* in interacting with novel materials in novel settings. In response we can say that people can have different goals in mind in interacting with materials. Assuming that multiple goals may be operative at a given point in time, we might say that the goal of learning is higher or lower in some goal hierarchy, that it is closer or farther from consciousness, that it is somewhere on the continuum from deliberate to implicit.

Whether or not humans are ever entirely without the general intent to learn *something*, we may still create experimental situations (and there presumably exist natural situations) in which the more specific intent to learn the *category membership* of a set of items is greatly reduced. This can be accomplished with cover tasks that lead the learner to believe that the primary goal of the task is something other than learning category membership or that do not provide any information that the items are members of categories at all. The present chapter is concerned with these kinds of situations.

As the focus of this book is on structure and process, some preliminary comments on these issues are also relevant. Throughout this chapter, there are at least three different ways in which the term "structure" could be used. One is to refer to the set of attributes and attribute correlations that the experimenter (or nature) has manipulated to generate groupings to be learned. This can be thought of as the *external* or *attribute* structure. This first sense of the term could also include aspects of the material to be learned such as integrality of the component dimensions, assuming that integrality is a stimulus construct (see e.g., Garner, 1974).

The other two uses of "structure" are concerned with the individual's internal representation of categories. One is to refer to the set of items that the learner considers to be members of the category in question (i.e., category extension). The other is the form in which category information is represented internally (e.g., rules, exemplars, prototypes, feature frequencies). We later give a more detailed distinction between these last two types. Our belief is that understanding of differences between incidental and intentional category learning depends on a clear separation of these different senses of the term "structure".

With these three senses of the term in mind, structure-process interaction might reasonably be thought of as sets of interactions linking external structures, learning and retrieval processes, and internal structures. Specific learning processes interact with available external structures with the consequence that particular internal structures are generated. With the problem framed in this way we can ask whether or not processes differ across intentional and incidental conditions in such a way that, on exposure to the same external structure, different internal structures result.

A. Attention to Attributes in Intentional and Incidental Learning

There is intuitive appeal to the idea that, when individuals deliberately attempt to learn categories, they bring the full weight of their cognitive resources to bear on the task. They engage in sophisticated strategies, analyze the stimuli into their component features and test hypotheses. This active mode of processing contrasts sharply with intuitions about individuals who interact with objects without the intent to learn about their category membership. These latter individuals are thought of as much less likely to use strategies, to analyze stimuli or to test hypotheses.

It is this intuitive distinction between intentional and incidental learning that Kemler Nelson (1984) captured in a more formal way. She proposed that intentional learning conditions are likely to encourage more sophisticated modes of processing than incidental conditions. As a *processing consequence*, there should be more analytic processing in the former and more holistic processing in the latter types of situations. As a *structural consequence*, component attributes should be more prominent in internal category structures resulting from intentional learning and overall similarity should predominate in categories learned incidentally (see Brooks, 1978; J. Smith, 1989; for related ideas).

If holistic modes of processing are elicited in incidental learning situations, it raises the intriguing possibility that there may be a match between a mode of processing elicited in everyday encounters with objects and the attribute structure of the categories to which those objects belong. As considered in detail below, such a match could facilitate real-world category learning. In coming to understand this potential facilitation, it is useful to examine the "primitive" nature of holistic processing and the presumed attribute structure of naturally occurring categories.

B. Holistic Processing as Primitive

There is a sense in which holistic modes of processing are thought of as easier, more primitive, or less developed than analytic modes; holistic processing has been referred to as a fallback mode (J. Smith, 1989). The evidence for this general view comes largely from studies in which individuals are shown small sets of objects and asked to classify in a group the objects that belong together. These restricted classification tasks typically allow a contrast between groupings that result from selectively attending more to one dimension than to another and groupings based on overall (unweighted) similarity (see however, L. Smith, 1989). The former types of grouping are referred to as analytic and the latter as holistic.

Results from restricted classification tasks reveal a preponderance of holistic groupings by young children (Shepp, Burns, & McDonough, 1980; L. Smith & Kemler, 1977; Ward, 1980) and by retarded and learning disabled children (Burns 1986; Kemler, 1982; Ward, 1986). In addition, impulsiveness, field-dependence, and lower scores on certain IQ subtests (e.g., Block Design) are associated with higher levels of holistic responding than are reflectiveness, field-independence and higher subtest scores among nonretarded individuals (J. Smith & Kemler Nelson, 1988; Ward, 1983; Ward, Foley & Cole, 1986; Ward, et al. 1989). Finally, when adults are required to respond rapidly or to perform tasks with a cognitive load they tend to respond more holistically (J. Smith & Kemler Nelson, 1984; Ward, 1983; Ward, et al., 1986). All of these results are consistent with the idea that a holistic mode of processing is in some sense less "advanced" than an analytic mode (see however criticisms of this position by Cook & Odom, 1988 and Wilkening & Lange, 1990). As a result it is reasonable to predict that holistic modes may be more likely in incidental learning tasks that do not evoke high-level strategic behavior.

C. Structure of Real-World Categories

Because holistic processing might be viewed as more primitive than analytic processing, a natural assumption would be that it would lead to poorer category learning performance than analytic processing. However, the success of a given mode of processing may depend on the attributes of the category to be learned. The detrimental effect of holistic processing might only be observed with categories that possess a single defining attribute that the learner must discover in order to classify accurately. If category membership information is distributed across many component attributes, then a holistic approach might be less detrimental and could even provide the learner with an advantage.

Since the pioneering work of Rosch and her colleagues (e.g., Rosch & Mervis, 1975) there has emerged a general consensus that the attribute structure of most of the concrete object categories that people encounter in everyday settings is best described as obeying family-resemblance principles. Rather than sharing single, obvious defining attributes, members of naturally occurring categories are thought of as sharing characteristic attributes that are true of some, but not all members. Different members are thought to possess different clusters of those characteristic attributes.

One result of possessing a set of characteristic attributes could be that the members of any given category would be highly similar to one another in an overall way, even though there is no one attribute that could be used to reliably classify all members. This type of category structure can reasonably be thought of as being learnable by way of a holistic approach. Indeed, it could even be claimed that a holistic approach would be better than an analytic one, since the most extreme version of the latter would entail a search for one and only one defining attribute on which to base category decisions; with no defining attributes, such an extreme form of analytic processing is bound to fail. (Less extreme versions of analytic processing described below are less prone to failure with family-resemblance categories.)

The above analysis (no pun intended) suggests a happy coincidence of process and structure that could conceivably facilitate real-world category acquisition. In day-to-day encounters with objects, without the explicit intent to learn, sophisticated analytic modes of processing may not be engaged. Hence rather than selectively attending to the component attributes of the objects, we may process them more holistically. If we are then asked whether some new object is a member of the same category, we will compare it to the stored holistic representation without selectively attending to any of the individual component attributes in the process. Because the kinds of object categories that we encounter in such settings may have family-resemblance structures, such an approach would lead to highly accurate generalization to new instances of the categories in question. Relying on any single attribute to determine category membership might lead to errors because that attribute may not be present in some category members.

D. Evidence for a Structure-Process Link in Incidental Learning

One way to begin to assess the link between structure and process in incidental and intentional learning is to attempt to simulate in the laboratory various aspects of family-resemblance category structures and incidental learning conditions. A particularly good example of this approach is the work of Kemler Nelson (1984). She presented college students with cartoon faces that were divided into categories according to family-resemblance principles. The faces varied along attributes such as type of hair (straight, wavy, curly), nose (tall-thin, medium, short-fat), ears (large, medium, small) and mustache (handle-bar, medium, clipped).

An example of a category structure used by Kemler Nelson (1984, Experiment 2) is shown in Table 12.1. Notice that the members of category A all have straight hair and that they take on a characteristic but not defining value for the other attributes. In contrast, the members of B all have curly hair and characteristic values on the other attributes that are opposite to those of category A. Assuming a roughly equal salience of the four attributes, the members of each category are more similar to each other overall than they are to members of the opposing category. Thus, the person could learn category membership by an analytic approach of focussing on the defining attribute of hair-type or by a holistic approach of focussing on overall similarity.

Kemler Nelson had some subjects learn these categories intentionally and other subjects learn them incidentally. On any given trial, the intentional learners saw a single face, predicted whether it was a doctor or fireman and received feedback. In contrast, on any given trial, the incidental learners saw pairs of faces from the same category and were asked to pick the face that best fit their idea of a fireman (or doctor depending on which category was shown on the particular trial). Information about category membership was available (in the form of uniforms) to the subjects in the incidental condition, but they were not told to try to learn the category membership of the presented faces.

Analytic and holistic approaches were differentiated by way of performance on a transfer task. Examples of critical test items for differentiating analytic and holistic responding are shown in Table 12.2. Note that T3 has the hair of category B but the values that were characteristic of category A for the other three attributes. T4 is exactly the opposite. The logic is that an analytic individual who has isolated the defining attribute will assign T3 to B and T4 to A, whereas, a holistic processor would be expected to assign T3 to A and T4 to B. Using a criterion of 16 consistent responses out of the 20 presentations of the critical test items, Kemler Nelson classified 22 intentional subjects as criterial attribute (analytic) learners and 22 as family-resemblance (holistic) learners. In contrast, only 7 incidental learners were identified as analytic processors, whereas 29 were identified as holistic. The results appear to provide compelling evidence for the idea that incidental learning conditions are more likely than intentional ones to result in holistic modes of processing.

Note that, as described to this point, the distinction between analytic and holistic processing has been based solely on the idea that category decisions will involve selective attribute weighting in the former but not the latter. This way of considering analytic and holistic processing allows a relatively straightforward way of providing evidence *for analytic processing*; if selective weighting occurs with materials composed of roughly equally salient attributes, then processing is analytic. However, this approach is less adequate in providing evidence against analytic processing and for the existence of holistic processing as a qualitatively distinct mode; holistic processing is inferred from the *absence* of clear evidence for analysis (e.g. selective weighting). The problem is that a highly analytic approach of assessing matches on each attribute and using a majority rule would produce the pattern of performance described above as indicating "holistic" processing. Thus, in the next subsection, we consider a more qualitative distinction between analytic and holistic processing and relate that distinction to the more central question of what it means to argue that incidental and intentional learning are fundamentally different.

E. What Does It Mean To Be Different?

A further consideration of Kemler Nelson's results allows us to begin answering the question of the extent to which incidental and intentional learning are different. In considering this question, it is useful to consider what we mean by "different". By the *strong version* of the term, one could mean that fundamentally different learning principles are involved. That is, the fundamental processes of storage and retrieval or the internal representations on which those processes operate might be different. So for example, if category information is stored exclusively in terms of rule-based abstractions in the case of intentional learning and exclusively in terms of separate whole exemplars in the incidental case, and if category decision-making is based on satisfying a rule versus similarity to exemplars then one could argue that intentional and incidental learning operate according to fundamentally different principles.

There is also a *weaker sense* in which one might claim that intentional and incidental learning differ. This weaker case would be satisfied if different patterns of classifying were observed under intentional and incidental conditions, but those patterns were interpretable by appeal to the same underlying learning principles. Illustrating this point can also help in differentiating different senses of the term "structure".

Consider again Kemler Nelson's results. Incidental learners tended to classify T3 as a member of A and intentional learners tended to classify it as a member of B. Clearly then, different patterns of classifying are observed in intentional and incidental conditions. Thus, if the term "structure" is taken to mean the extension of the category (i.e., the exact items that an individual considers to be members of the category), it could be argued that the category structures for the two types of learners are different. On the other hand, if "structure" refers to the underlying form of the information used in making category decisions then a different interpretation is possible. For example, both intentional and incidental learners might have represented the categories in the form of exemplars and both might have made category decisions by way of similarity to known exemplars. The different patterns of classifying T3 and T4 would then be attributed to the fact that the intentional learners assigned more weight to feature 1 than to the other features in computing the similarity of test items to known exemplars whereas the incidental learners assigned equal weight to the attributes. By this interpretation, the observed difference may be a matter of breadth of attention rather than of differing structural principles. Because we favor this latter view of structure we will reserve the term "structure" to refer to whether a category representation is rule-based, exemplar-based or some other form.

We will refer to differences in exact classifications which do not necessarily imply different underlying structures as being differences in "category extension". By this interpretation, structure differences would indicate the strong version and extension differences the weak version of a difference between intentional and incidental learning.

It is clear from Kemler Nelson's (1984, Experiment 2) results that different category extensions can result from intentional and incidental learning conditions. It is not certain whether the results also support the stronger case that the underlying structures are different.

A related issue is whether Kemler Nelson's results support a version of holistic processing that is qualitatively different from analytic processing or simply an analytic but nonselective form of processing in incidental learning. A version of holistic processing that is qualitatively different from analytic processing implies no access to the underlying component attributes whereas a quantitative version does not (see e.g., Kemler Nelson, 1989). The analytic, nonselective mode would imply explicit representation of component attributes but equal weighting of those attributes in computations of similarity.

Kemler Nelson's (1984) third experiment was designed to provide evidence regarding the qualitative versus quantitative difference views described above. She used categories that were linearly separable (LS) and those that were not (NLS). The rationale for using these attribute structures was that independent cue models (that assume independent assessment and additive combination of matches and mismatches along component attributes) generally predict an advantage for LS categories. Interactive cue models that assume multiplicative computations of similarity (e.g., Medin & Schaffer's, 1978 context model) can predict an NLS advantage if those multiplicative computations lead to high similarity of some individual exemplars to one another (as opposed to moderate similarity of all members to one another).

Kemler Nelson (1984; 1989) reasoned that evidence favoring multiplicative models such as the context model would be consistent with holistic processing (see J. Smith, 1989 for a similar argument). Presumably, this is based on the idea that multiplicative computations mimic the action of component attributes on one another; attributes are thought to be evaluated "in the context of" other attributes of the same objects (as well the situation in which the object occurs). In a more formal sense, Kemler Nelson (1989) has drawn attention to the link between multiplicative computations of similarity in categorization models and Euclidean similarity metrics, the latter being appropriate to judgments of stimuli composed of integral dimensions. Because integral dimensions are assumed to be processed holistically, in the strong sense of the term, categorization behavior that can be described with mathematical formulae appropriate for integral materials could be taken as an indicator of holistic representation and processing. By this reasoning, Kemler Nelson's (1984) demonstration that NLS categories are learned more readily than LS categories in incidental conditions could indicate a kind of holistic processing in the sense that component attributes of the objects may not be assessed independently of one another.

If an NLS advantage could result *only* from the qualitatively different version of holistic processing, Kemler Nelson's (1984, Experiment 3) findings may also provide evidence for the strong sense of a difference between incidental and intentional learning. Given the importance of this claim, it is essential to carefully scrutinize its support. One source of concern is that it is not clear that the type of interactive cue model that Medin and Schaffer had in mind is really consistent with holistic processing in the strong sense as described by Kemler Nelson (1989); the model does *explicitly* represent separate component attributes in a way that seems inconsistent with the spirit of holistic processing. On the other hand, hue, saturation and brightness may well be separate component attributes that must be represented in computations of similarity between color stimuli despite the fact that subjects process them holistically in the strong sense.

A more important concern is that the patterns of performance that the context model predicts by way of multiplicative computations of similarity can be mimicked by additive combinations of independent cues in computations of similarity in models that have been described since Kemler Nelson first presented her results (e.g., Hintzman, 1986). Hintzman's MINERVA2 accomplishes this by computing similarity as an additive combination and specifying that activation of traces is a nonlinear (e.g., cubic) function of similarity. If the formal mathematical principles thought to underlie simulations of holistic processing are unnecessary to account for performance, then the holistic processing argument is undermined.

Finally and most importantly, the context model was originally designed to account for *intentional* category learning, and it has had a fair amount of success in accounting for intentional category learning phenomena. Thus, even if multiplicative similarity did imply holistic processing, we would have to conclude that holistic processing also occurs in intentional learning tasks. This returns us to the question of strong and weak versions of a difference between intentional and incidental learning. If both types of learning can be represented with the same assumptions about underlying exemplar representations and "holistic" modes of determining similarity, then the results described above support the weak rather than the strong version of differences based on intent. It is important to note, however, that referring to those differences as "weak" is in no way a statement that they are unimportant. Any demonstrable differences between intentional and incidental learning are important, and Kemler Nelson's results clearly reveal differences.

Before leaving this issue of modelling, it should be noted that there are many aspects of both incidental and intentional category learning that the context model, as originally stated, has difficulty accounting for (e.g., sequence effects). In the third major section, we consider some of these phenomena.

F. The Reliability of Incidental Versus Intentional Learning Differences and the Value of Analytic Processing

Putting aside for the moment the question about weak versus strong versions of a difference between intentional and incidental learning, it is worth considering how stable and reliable any types of differences might be. It is also worth considering whether there could be reasons other than selective versus either holistic or nonselective processing for the observed performance differences. Ward and Scott (1987) and Ward (1989) for example, claimed that analytic modes of processing and selective weighting of component attributes could have operated in both the intentional and incidental conditions to produce the pattern of Kemler Nelson's (Experiment 2) results. They noted that an analytic processor who had focussed selectively on any of the three characteristic attributes would have assigned T3 to A and T4 to B, exactly as predicted for "holistic" processors. Such individuals would be referred to as attribute-plus-exception learners because they could classify two-thirds of the original learning items on the basis of information from any of the individual characteristic attributes, but would have to use additional information to classify the items that did not have the characteristic value of the attribute (i.e., the exceptions). Ward and Scott (1987) and Ward (1989) considered several models for these attribute-plus-exception learners; the favored one was an exemplar model in which one characteristic attribute was heavily weighted but the remaining attributes all figured into computations of similarity of a test item to individual known exemplars.

Before examining Ward and Scott's results, it is useful to consider why analytic processing in this form of selective weighting of component attributes of exemplars might be expected in naturally-occurring incidental learning situations and how it could facilitate learning to recognize members of real-world object categories. The idea is that the external structures of many real-world categories possess several partially informative attributes so that selective attention to any of those attributes could facilitate identifying category members. In addition, some attributes (e.g., shape) may be much more informative than others. The attributes that are particularly informative may vary with the broader domain of which the new items are a subcategory (e.g., natural versus artifact or animal versus plant) and of which the learner has some knowledge. Thus, acquisition of categories and attention to attributes is not random as if it were occurring in a vacuum. Rather, the person's knowledge can facilitate attention to highly informative attributes.

The context in which learning occurs presumably activates particular knowledge structures. Among other things, those knowledge structures are assumed to contain information about attributes that are useful for differentiating members of differing categories within the domain organized by that knowledge structure. When new objects are encountered, the processor selectively attends to their status on the relevant attributes. This selective processing would not require deliberate strategic processing by the learner, but rather is driven by the currently activated knowledge structures.

To illustrate the ideas described in the previous paragraphs, consider the learner who is walking through a zoo. Presumably, that behavior will activate something like an "animal" knowledge structure. Among other things, that structure may contain information that shape and texture are crucial determinants of membership in various animal categories. Thus, on encountering new species, the individual may store a trace that, with some relatively high probability, contains information about the shape and texture of the new creature. No intent would be required for the encoding to be biased in terms of the activated attributes.

The ideas described in the previous paragraphs are part of the *attribute availability hypothesis* which was originally developed to help to account for young children's relative success in learning real-world categories (Ward, 1990b; Ward, Vela, & Hass, 1990). It is described here to illustrate the point that analytic processing (i.e., selective processing of component attributes) can be expected from very young children and from adult incidental learners despite the potential limited involvement of sophisticated strategic processing in each case. It is also described to illustrate that a holistic mode of processing is not the only type that might facilitate learning of family-resemblance categories. Analytic processing, with selective attention guided by knowledge structures could also be helpful.

Ward and Scott (1987) noted that responses to the full set of test items shown in Table 12.2 should be used to separate true holistic responders from attribute-plus-exception learners. They went on to conduct a study very similar to that reported by Kemler Nelson (1984, Experiment 2), and using the full set of test items, identified 19 and 6 analytic and holistic learners, respectively, in the intentional condition and 13 and 6 analytic and holistic learners, respectively, in the incidental condition. The incidental learners were less likely than the intentional ones to focus on the defining attribute and more likely to focus on one of the characteristic attributes, but they were no less likely to show selective attribute weighting and no more likely to show holistic patterns. Using only T3 and T4 as test items would have led to the erroneous conclusion that the incidental learners were more likely to show holistic (or nonselective) modes of processing. The Ward and Scott results provide no support for an analytic versus holistic distinction between intentional and incidental learning. Rather, they indicate analytic modes of processing can predominate in incidental as well as intentional learning situations.

Although a reanalysis of the Kemler Nelson data along the lines suggested by Ward and Scott did reveal some shifting away from holistic responding in the incidental conditions, the incidental learners from that study were still significantly more likely than the intentional ones to exhibit holistic modes of transfer task performance (Kemler Nelson, 1988). Thus there is a discrepancy between results from otherwise fairly similar procedures, and there has been some debate about the reasons for that discrepancy (Kemler Nelson, 1988; Ward, 1988). There is a variety of subtle methodological differences that may be responsible.

Other data from Kemler Nelson's laboratory (e.g., Kemler Nelson, 1984, Experiment 1) are consistent with the link between incidental learning and holistic processing of cartoon faces. In contrast, other data from our laboratory (Ward, 1989) are consistent with analytic processing by incidental learners with quite similar materials. One is tempted to conclude that Swarthmore undergraduates are simply better at following instructions than Texas A&M undergraduates and thus more likely to show differences between conditions. On the other hand, Ward and Scott's intentional learners were also more likely than Kemler Nelson's to show analytic patterns. Texas A&M boosters would undoubtedly be happy if they could claim that their students were simply more analytic overall than Swarthmore students.

To avoid appeals to intercollegiate chauvinism, it is useful to consider what other evidence exists on the issue of analytic versus holistic modes of processing in intentional and incidental category learning. Data reported by J. Smith (1989) appear to tip the balance in favor of Kemler Nelson's hypothesis. He used four-element CVCV letter strings rather than faces as stimuli. As an example, the members of Category I were BUNO, KUNO, BYNO, and BUNA, whereas the members of Category II were KYPA, BYPA, KUPA, and KYPO. Note that the members of Category I all have N in the third position and that the members of Category II all have P in that position. Note also that the other three positions take on characteristic values, but that there is one noncharacteristic value for each of those positions.

Some subjects were instructed to learn the groupings and others performed a speed-reading cover (incidental learning) task. Using test items of the form KYNA and BUPO Smith found that individuals instructed to learn the categories tended to generalize by way of the criterial attribute and those who performed the speed-reading task were more likely to generalize in terms of family-resemblance. The result follows directly from Kemler Nelson's hypothesis.

Whether a strong or weak version of holistic processing in incidental learning provides the best interpretation of Smith's results is not clear. In the absence of compelling evidence to the contrary it may be more prudent to suggest that there may be quantitative differences in the breadth of attention observed in intentional and incidental learning conditions rather than a qualitative difference of analytic versus holistic processing. When viewed in this way, one could state that "all other things being equal" intentional tasks will lead to a narrow focus of attention on a more limited set of attributes than will incidental tasks.

The obvious problem raised by the statements in the previous paragraph is that all things are generally not equal. There may be no such thing as an incidental (or intentional) task that is neutral with respect to the breadth of attention that is fostered. Consider again, Smith's incidental task. Speed-reading certainly seems to be the sort of task that would encourage learners to treat CVCVs as word-like stimuli and thus process them as "holistic" units. Thus the effects obtained could be due to the fact that reading letter-strings as words leads people to treat them as holistic units rather than as a collection of separate letters. By this reasoning it is the specific processing demands of the task, and not the learner's intent, that determine what is learned. This issue is explored more fully in the next section.

III. TASK DEMANDS, STIMULUS EFFECTS AND TRANSFER-APPROPRIATE PROCESSING

It appears that incidental conditions do not lead inevitably to modes of processing that are qualitatively different from those that occur in intentional learning. Some experiments reveal striking differences; others do not. In addition, even when differences are found it is not always certain that they are due to differences in the learners' intentions, per se. The pattern of findings leads us to conclude that "intent" may not be the most important factor in producing changes in processes and acquired structures. Rather, other task differences may be more important. These can include the attributes that are made salient by the particular processing demands of the task, the nature of the materials, and the link between the processing required in the learning task and the type of response required in the transfer task. Therefore, rather than focussing exclusively on the differences between incidental and intentional learning tasks, it may be fruitful to examine the kinds of factors that might produce differences in the outcome of otherwise similar incidental learning conditions.

A. Predicting Which Attributes Are Selected

Data reported by Ward and Scott (1987) and Ward (1989) clearly show that selective use of attributes does occur in incidental learning situations. However, it is not clear whether that selectivity reflects a random choice of any available attribute or something more systematic. Therefore it is useful to ask how predictable selective weighting of specific attributes might be in incidental learning.

The idea explored in this section is that the specific processing demands of the incidental learning task can affect a) the nature of the category information that is stored and b) the nature of subsequent category decisions that are based on that information. The ideas represent an extension of models concerned with intentional learning to the incidental learning domain. Specifically, it is widely acknowledged that selective weighting of attributes during intentional learning will affect a) the probability that those attributes are stored veridically and b) the extent to which those attributes influence subsequent category membership decisions (Estes, 1986; Medin & Schaffer, 1978; Nosofsky, 1987). Put more simply, people should be most likely to store what they attend to. As a working hypothesis we suggest that selective weighting of component features can and does also occur during the performance of incidental learning tasks, and that this selective weighting influences which features get stored and used in making category membership decisions. Thus, we argue that intent is not required to produce a selective weighting of stimulus attributes or predictable selective attribute learning. What the individual does with presented information is more important than what the individual intends.

As means of examining this issue further, it may be useful to consider the ideas within the framework of a specific model. The specific framework chosen to illustrate the points is an episodic trace model of category learning (see, e.g., Estes, 1986; Hintzman, 1986). Previous models of this type have been concerned with intentional learning, but we suggest that the same principles apply to incidental learning situations. One basic assumption of such a model is that each experience with a category exemplar (whether in an intentional or incidental learning task) constitutes an episode and results in a trace. The trace of each episode can be characterized as a set of feature values. Most importantly, not all features of any given exemplar have an equal probability of being stored veridically. Rather, those features which are given more attention or processing resources during the presentation of the exemplar have a higher probability of being stored veridically. In an intentional learning task, this greater amount of processing resources might occur because the learner is testing a hypothesis regarding that particular feature. In an incidental learning task, the learner is not aware of the need to store information about which items are members of which categories and thus is unlikely to test hypotheses about category membership. Nevertheless, as described earlier, there may be aspects of the incidental task which draw his or her attention selectively to particular features. As a result, those features would have a higher probability of being stored on that particular presentation. Thus, across trials more traces will contain an accurate record of that feature and it will come to have more of an effect than the others on category decisions regarding novel items.

It would be relatively easy to bias a person's attention to a particular feature in an incidental learning task. For example, the person could be shown faces from one of two categories and be asked to compare those faces in terms of their hair without being told to try to learn category membership. The individual would most likely follow instructions and attend primarily to hair, and subsequent category decisions would probably reflect the use of that attribute. However, such a transparent manipulation might result in the desired pattern of performance simply because the subject determined what the experiment was about and *intentionally* produced the pattern of responses that he or she thought was expected. Thus, we sought a more subtle manipulation to bias the subject's attention, in which the demand characteristics would be less obvious. Ideally the task should be one in which there is no *explicit* direction to the participant to attend to particular attributes.

An extrapolation of principles discovered by Gati and Tversky (1984) in their work on similarity judgments suggested a possible task. Their work indicates that in making similarity judgments between objects, participants weight both the common and distinctive features of those objects. More importantly, there are properties of the stimuli and task that appear to influence whether common or distinctive features will be weighted more heavily. Specifically, distinctive features are weighted more heavily in comparing pictorial materials whereas common features are weighted more heavily in comparing verbal descriptions of those materials (Gati & Tversky, 1984). Thus, a similarity judgment task may provide an incidental learning procedure by which it is possible to bias an individual's attention toward or away from particular attributes by manipulating the nature of the material to be judged.

The bias toward the distinctive features of pictures and toward the common features of verbal descriptions is highly reliable, having been found for nearly all of the sets of materials that have been examined (see e.g., Gati & Tversky, 1984), including cartoon faces and verbal descriptions of those same faces. In the first study described in this section (Ward, Hass, & Vela, 1990) we used cartoon pictures and verbal descriptions of faces that varied along the features of head covering and mustache. Those features served as either common or distinctive features depending on the condition in which a subject participated. More specifically, the faces and descriptions varied along the two informative features of type of head covering and type of mustache. In addition, the third feature of glasses (none, red, purple, grey, green, blue and yellow) was varied orthogonally to category membership so that greater variability could be included in the set of faces to be learned. Four different types of category pairings were created: pictorial with head covering as the distinctive feature, pictorial with mustache as the distinctive feature, verbal with head covering and verbal with mustache as the distinctive feature. The categories were structured so that when one feature served as a distinctive feature, the other served as a common feature. Type of category pairing served as a between subjects variable.

The conditions will be referred to in terms of which feature was distinctive. For the *pictorial/head covering* contrast, all of the members of one category (e.g., Smiths) had head coverings (hats or berets) whereas none of the

Figure 12.1. Sample materials from the **pictorial/head covering** condition. Note that the hat is a distinctive feature (i.e., present in one member of the pair and not the other), and that the mustache is a common feature that takes on a different value (long versus short) in each member of the pair.

members of the opposing category (e.g., Joneses) had head coverings. Thus, head covering served as a distinctive feature. In addition, all members of one category had short mustaches whereas all members of the other category had long mustaches. Thus, mustache served as a common feature, and the exact value (long versus short) differed across the categories. An example pairing is shown in Figure 12.1.

For the *pictorial/mustache* contrast, all of the members of one category had mustaches (long or short) whereas none of the members of the opposing category had mustaches. Thus, mustache served as the distinctive feature. In addition, all of the members of one category had berets and all of the members of the other category had hats. Thus, head covering served as a common feature, and the exact value (hat versus beret) allowed a separation of members of the opposing categories.

The structures for the *verbal/head covering* and *verbal/mustache* conditions were the same as described for the pictorial conditions. The only difference was that the stimuli consisted of verbal descriptions rather than cartoon drawings of the faces.

On each trial, subjects saw a pair of faces (descriptions) and rated their similarity. The members of the pair were always from the opposing categories (i.e., one was Smith and one was a Jones) and they appeared side-by-side on a computer screen. Subjects rated the similarity of each pair on a scale from one to nine by pressing one of the number keys on the keyboard. The pairs remained on the screen until the subject responded.

For any given subject, the Smiths always appeared on one side of the screen and the Joneses on the other. The label "Smith" always appeared under each Smith and the label "Jones" always appeared under each Jones. Subjects were asked to settle a family argument about the similarity of the two families. The were told that they could tell who was a Smith and who was a Jones by way of the labels, but they were not asked to try to learn which faces were members of which category.

For any given subject there were eight different Smiths and eight different Joneses created by variations in the exact color and pattern of the hat or beret and the exact color, shape and bushiness of the mustache. Additional variability was provided by varying the color of glasses orthogonally to category membership. The verbal descriptions contained information about the color and patterning of the head coverings, mustaches and glasses.

The Smiths and Joneses were randomly paired to create 32 unique pairs with the restriction that each individual face appear four times across the 32 pairs. Subjects judged similarity for all 32 pairs and then participated in a surprise test in which they had to indicate the correct category membership for the 16 original learning faces and 16 test items. The test items were created by re-pairing the type of head covering and type of mustache. So for example, if the Smiths in the similarity judgment task had head coverings and short mustaches and the Joneses had no head coverings and long mustaches, test items would have head coverings and long mustaches or no head coverings with short mustaches. In this particular example, calling the former type of test item a Smith would be considered a response to the distinctive attribute (head covering) whereas calling it a Jones would be considered a response to the common attribute (long mustache).

The mean number of responses consistent with the use of the distinctive feature (out of a maximum possible of 16) was 10.5, 11.7, 6.3, and 9.8 for the verbal/mustache, pictorial/mustache, verbal/head covering, and pictorial/head covering conditions respectively. Of most interest is the fact that the mean number of distinctive feature responses was significantly higher for subjects in the pictorial conditions than for those in the verbal conditions, $F\ (1, 86) = 4.90, p < .05$. Although the difference appears to be greater for conditions in which the hat was the distinctive feature, there was no interaction of feature with material type, $F\ (1, 86) = 1.12, p > .25$.

Table 12.1. Formal structure of a category learning problem used in the Kemler Nelson (1984) and Ward and Scott (1987) studies. The faces denoted are ones presented during either incidental or intentional learning.

Category A					Category B				
Face	Hair	Must.	Ear	Nose	Face	Hair	Must.	Ear	Nose
1.	1	1	1	2	4.	3	3	3	2
2.	1	1	2	1	5.	3	3	2	3
3.	1	2	1	1	6.	3	2	3	3

Note. The values 1, 2, and 3 refer respectively to straight, wavy, and curly for the attribute of hair, to clipped, medium-length, and handlebar for mustache, to small, medium and large for ears, and to tall-thin, medium, and short-fat for nose. Four such problems were created so that each of the four attributes could serve equally often as the critical attribute.

The results are what would be expected from the presumed weighting of features during the incidental learning task. That is, distinctive features should be weighted more heavily for pictures and common features more heavily for verbal descriptions. A reasonable interpretation is that, as a result of the differential weighting of the common and distinctive features during the incidental similarity judgment task, subjects' category representations were more or less dominated by one or the other of those types of features. Thus, selective weighting of features can have predictable consequences for incidental as well as intentional learning.

In a second study we also examined incidental category learning with pictorial and verbal materials. In this study we wanted to have a direct comparison to materials used in our previous work on incidental category learning (Ward & Scott, 1987). Thus, we used cartoon pictures of faces that varied along the attributes of hair, nose, ears, and mustache as one set of materials, and verbal descriptions of those faces as a second set of materials. (Data from the pictorial conditions are briefly described in Ward, 1989). As in the Kemler Nelson (1984) and Ward and Scott (1987) studies, the faces (descriptions) were divided into categories that could be learned either by way of a criterial attribute, one of the other characteristic but noncriterial attributes, or overall similarity. A formal representation of one set of such categories is shown in Table 12.1, but it is im-

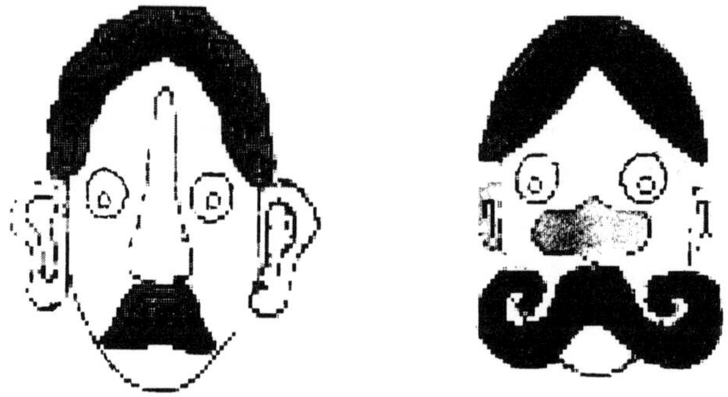

Figure 12.2. Examples of cartoon faces differing on all four component attributes.

portant to note that, across subjects, each attribute served equally often as the criterial attribute.

Examples of cartoon faces used in the pictorial conditions are shown in Figure 12.2. The materials for the word conditions were verbal descriptions of the faces used in the picture conditions. The labels used were "curly", "wavy" and "straight" for hair, "short-clipped", "medium length" and "handlebar" for mustache, "large", "medium" and "small" for ears, and "tall-thin", "intermediate" and "short-wide" for nose. The description for any given face appeared in two rows of two descriptors each.

Information about category membership was made available to the participants in that the faces (descriptions) always appeared atop fire fighter or doctor "uniforms". Participants were told that they could tell who was a fire fighter and who was a doctor on the basis of the uniforms, but they were not instructed to try to learn which faces were members of which category. Rather, they saw two faces (descriptions) on any given trial and were asked to rate the similarity of the faces. For subjects in the Within-Category condition, the members of the pairs being judged on any given trial were always from the same category (e.g., two faces from A or two faces from B). For subjects in the Cross-Category condition the members of the pairs were always from different categories (one face from A and one from B). Notice the impact of these different comparison conditions on the status of the criterial attribute (see Table 12.1). In the Within-Category condition the criterial attribute (in this case, hair) is always the same across the members of any given pair being judged. In the Cross-Category condition the criterial attribute is always different across any given pair being judged.

Table 12.2. Transfer items for use with learning items in Table 12.1.

Face	Hair	Must.	Ears	Nose
T1	1	1	1	1 (Category A Prototype)
T2	3	3	3	3 (Category B Prototype)
T3	3	1	1	1
T4	1	3	3	3
T5	1	3	1	1
T6	3	1	3	3
T7	1	1	3	1
T8	3	3	1	3
T9	1	1	1	3
T10	3	3	3	1

After making a series of 48 similarity judgments, participants were shown the faces or descriptions one at time and were asked to report the category membership of each face or description. In addition to the original faces which were used to determine if participants had learned their category membership, test faces were presented to determine which features were weighted heavily in the participants' decisions (Table 12.2).

It is important to note that, for this new study, the concept of *difference* versus *commonality* is slightly different than in Gati and Tversky's work. In their work (as in Ward, Hass & Vela, 1990), the concern was with *distinctive* features which are present in one member of a pair being judged but not the other, and *common* features which are present in both members of the pair being judged. In the present study, *differences* refer to different levels of the same type of attribute whereas *commonalities* refer to a match in terms of the level of that attribute. However, a pilot study revealed that commonalities, as used in the present sense, are weighted more heavily in judging verbal descriptions whereas differences are weighted more heavily in judging pictures.

The predictions for this study were as follows. After making similarity judgments in the Cross-Category condition with pictorial materials, the subjects' category decisions should be biased toward the criterial attribute. In contrast, after making similarity judgments in the Within-Category condition with pictorial materials there should not be as strong a bias toward the criterial attribute. This is because a) differences should be weighted more heavily than commonalities in judging the similarity of pictures, b) the criterial attribute always represents a difference between the pairs in the Cross-Category condition and always represents a commonality in the Within-Category condition, and c) attributes that receive more weight during exposure to category members (whether intentionally or incidentally) should become more important in the representation of that category. The exact opposite predictions hold for the verbal descriptions because commonalities should be more important in judgments of verbal materials. Three hundred and eighty-four college student participants were, as nearly as possible, evenly distributed across the cells in the design.

This second experiment also examined the impact of both similarity and dissimilarity judgments on subsequent category membership decisions. The major reason for this manipulation was that Gati and Tversky (1984) have observed that commonalities are weighted more heavily in similarity judgments whereas differences are weighted more heavily in dissimilarity judgments. Thus, all other things being equal, commonalities should be more important in category decisions following similarity judgments whereas differences should be more important following dissimilarity judgments. However, the type of judgment had no impact on performance and will not be discussed further.

Participants who correctly classified the original learning items on 10 of the 12 transfer task presentations of those items were identified as learners. Of the total number of participants, 182 met this criterion. Performance on the test items (see Table 12.2) was used to determine for these learners whether they had learned by way of a single attribute, holistic overall similarity or some other rule.

Of the 182 learners, 95 could be clearly identified as responding on the basis of one or another of the component attributes and 7 were identified as holistic. The remainder showed ambiguous patterns, most of which tended to approximate single attribute rules. Of the clear single attribute learners 40 used the criterial attribute for their decisions and 55 used one of the other three noncriterial attributes.

Table 12.3. Frequencies of criterial attribute learners in the four major cells of the design defined by the two levels of Material and two levels of Contrast Type. Frequencies are shown separately for the incidental and intentional learning conditions.

	Contrast x Material Condition			
Learning Type	Within/ Picture	Cross/ Picture	Within/ Word	Cross/ Word
Incidental	7	18	11	4
Intentional	12	6	11	10

Note. The row labelled Incidental is for the Similarity and Dissimilarity conditions combined because there was no difference between those conditions.

Of more interest than the overall distribution of criterial and noncriterial attribute responders is the distribution of criterial attribute responders across the different conditions. Table 12.3 shows the number of individuals whose responses appeared to be determined by the criterial attribute separately for each of the four conditions defined by the crossing of Contrast Type and Material. As can be seen, the greatest numbers of criterial attribute learners occurred in the Cross-Category/Picture and Within-Category/Word conditions. An analysis of the frequencies in Table 12.3 indicated that the distribution is significantly different from what would be expected if Contrast Type and Material independently influenced criterial attribute responding, X^2 (1) = 7.78, p < .05. Put differently, Contrast type and Material interacted significantly in determining the frequencies of criterial attribute learners. The results are as predicted on the basis of the features which should be weighted heavily in making similarity judgments in those conditions. When the criterial attribute always represented a difference between pictures in a similarity judgment task (Cross-Category condition) it was used more frequently in subsequent category decisions regarding those pictures. When the criterial attribute always represented a commonality across verbal descriptions (Within-Category condition) it was used more frequently in subsequent category decisions regarding those verbal descriptions. The results indicate that, when individuals judge the similarity of objects, they learn the category membership of those objects, and the attributes that are weighted heavily in those similarity judgments come to play a prominent role in the newly formed category representations of the objects whose similarity was judged.

It is important to note that in the studies just described, the similarity judgment task was an incidental learning task. Participants were not asked to try to learn the category membership of the presented faces. Rather, they were simply asked to make similarity judgments. An important implication of the results, then, is that the deliberate intent to learn category membership appears to be unnecessary to produce selective weighting and storage of stimulus attributes.

In our view, what the learner does with presented information is more important than what he or she intends. This conclusion is certainly not novel within the general area of cognitive psychology. For example, in the verbal learning literature, the differential impact of different incidental orienting tasks on the duration and strength of memory traces has been well established (e.g., Craik & Tulving, 1975). However, within that literature the studies typically involve the retention of lists composed of items with which the participants are already highly familiar. In the present studies, the concern was with how individuals learned new categories and with the attributes that came to predominate in those new category representations. Thus, the principle that actions are more important than intentions extends to the domain of acquiring new categorical information as well as to the domain of storing previously acquired items.

B. Some Important Cautions

There are a number of reasons for being extremely cautious in interpreting the results reported in the present section, particularly those from the second study. For example, in the second study slightly less than half of the participants performed well enough in classifying the original items to be considered learners. In addition, because the predictions for that study had to be framed in terms of criterial attribute responding, the most important data come from those learners who relied heavily on the criterial attribute. As a result, the conclusions rest on a relatively small proportion of the total number of participants tested. As a partial defense, it should be noted that the data from the first study come from a much higher proportion of the subjects.

As a more general concern, the procedures used in these studies are admittedly artificial. It is unlikely that individuals learning natural categories would experience a series of pairs from one or two categories and repeatedly judge the similarity of the members of those pairs. Nevertheless, one prime means of developing new categories or extending old ones may well be the comparison of the similarity of new objects to one another or to internal representations of exist-

Figure 12.3. Examples of Identi-Kit faces differing on all four component attributes.

ing categories. (If this is not the case, then the emphasis on similarity in many models of category learning may be misplaced.) The present results confirm that category learning can occur as a result of making similarity judgments and that different attributes can become more or less important in the representation of the newly formed categories depending on whether or not they figure prominently in computations of similarity. Thus similarity judgments may be a means of forming new categories and extending old ones and they may play an important role in determining which of the many possible partitionings of a set of objects will occur.

The most important caution, however, is that we have recently completed a study in which heavy weighting of a specific attribute during an incidental learning task was unrelated to what people learned about category membership (Ward & Becker, in preparation). In that study we used faces constructed from Identi-Kit materials that varied in hair (curly versus straight), nose (large versus small), mustache (long versus short) and eyes (round versus narrow). Sample items are shown in Figure 12.3.

We had two different incidental tasks: subjects judged either the similarity or the relative honesty of Smiths and Joneses. The intent was to evoke the subjects' knowledge structures regarding the concept of honesty, and we anticipated that they would attend more to the eyes when making honesty judgments (see e.g., Berry & McArthur, 1985).

Data from a control task indicated that the attributes were equally influential in judgments of similarity but that the eyes were much more salient than any of the other attributes in judgments of honesty. Despite the enormous differences in the weighting of the eyes in the two incidental tasks, subsequent categorization tasks did not reveal a significantly greater number of eye responders in the "honesty" condition (3 eye responders versus 11 other single-attribute responders) than in the similarity judgment condition (2 versus 14).

We do not yet have an explanation for this discrepancy. Some possibilities that should be examined in future research are that a) there is no link between the weighting of an attribute in an incidental learning task and how much information is stored about that attribute, b) there is a nonlinear relation between weighting and storage, and c) there are different patterns of attribute weighting in the incidental learning task and the subsequent category decision task and those differences mask our ability to observe a relation between weighting and storage that occurs during the incidental task.

The first of these explanations would indicate a striking difference between intentional and incidental learning. It is normally assumed that what the person attends to determines what he or she learns. But, by this explanation, there is a dissociation between attention and learning in incidental category learning.

The last explanation suggests an important methodological consideration for future studies. If the processing demands of the transfer task are significantly different from those of the incidental learning procedure, then effects of the incidental task may be obscured. One manifestation of this could be that requiring a deliberate categorization decision in a transfer task might lead the individual to process more analytically than during the learning task with the result that incidental learning appears more analytic than it really is. This last problem indicates why it is important to demonstrate specific links between performance in the incidental task and responding during the transfer task (see e.g., Ward, 1988).

Finally, it should be noted that with the Identi-Kit faces we have found for the first time in our laboratory more holistic responding in incidental than intentional learning tasks. Specifically, with these more realistic faces, the majority of incidental learners show holistic patterns regardless of whether the incidental orienting task is similarity judgment (43 holistic to 18 single attribute responders) or ratings of the honesty of the presented faces (38 holistic to 13 single attribute responders). In contrast, the intentional learners showed a bias toward single-attribute patterns (23 holistic to 58 single attribute responders).

The fact that we have never found large numbers of holistic responders in incidental tasks with cartoon faces indicates that there may be important stimulus parameters that determine whether or not differences between incidental and intentional learning will be observed (see Ward, 1989). Being more realistic in appearance, the Identi-Kit faces present more compelling configurations and the individual features are much less distinct from the "whole". For the cartoon faces, the individual features appear much more separable from one another. Thus it may be that there is a subtle difference between the modes of processing evoked in some incidental tasks and some intentional tasks that can result in more responding to overall appearance in the former. When the features are too readily separable from the whole objects, this subtle difference is not observed because only a small amount of analysis is required to respond to the components, and that amount of analysis can occur regardless of learning intent. Differences may be more likely to be observed with moderately separable materials. As an aside, the same may hold true for observing developmental differences in analytic and holistic responding (see Ward, 1989).

It should be noted that Kemler Nelson (1984) has obtained holistic responding in incidental tasks with her cartoon faces. Perhaps her faces represented stronger configurations than those used in our laboratory. One bit of evidence in support of this claim is that even Kemler Nelson's intentional learning subjects show more holistic responding than do our intentional learning subjects. What all of this means is that future studies will have to more carefully control the exact nature of the stimuli used if a true picture of incidental versus intentional learning differences is to be observed.

To summarize the results in this section, judging the similarity of objects is one of the ways of incidentally learning category membership and category structures. The extension of new categories formed in this way appears to be flexible. The exact form that the new category extensions take is sometimes, though not always related in predictable ways to interactions between the type of material judged and the types of contrast made. In the present studies, the materials were pictures and written descriptions, but there are other ways (e.g., auditory) of presenting category relevant information. An important goal for future research is to come to an understanding of how different modes of presentation and judgment interact to determine what is learned both intentionally and incidentally.

IV. ABSTRACTING FEATURE CORRELATIONS

The focus of the first sections of the present chapter has been on selective processing of individual attributes. Clearly however, category representations are much richer than could be revealed by a consideration of attributes in isolation. Thus in considering potential differences between intentional and incidental learning it is useful to examine the role of feature correlations in category structures and decision-making.

In relating research on feature correlations to the analytic versus holistic processing distinction, it should be noted that analytic processing is not limited to selective attention to single attributes. Any evidence that attention is being devoted selectively to particular attributes or attribute combinations out of a larger set of otherwise equally perceptually salient attributes would also signal analytic processing.

Work reported by Billman and Richards (1989) on the abstraction of feature correlations is relevant to the question of analytic versus holistic processing in incidental learning, although their focus was on a slightly different aspect than that addressed in the studies described above. Billman and Richards were interested in testing a class of *fixed similarity* models against a class of *abstraction* models in terms of their ability to account for incidental learning of category information. According to Billman and Richards, the models that they identified as fixed similarity models (e.g., Brooks, 1978; Hintzman's MINERVA2, 1986; Medin & Schaffer's context model, 1978) do not allow the function for computing similarity to change systematically during category learning. Such models certainly seem in the spirit of the type of holistic processing described by Kemler Nelson. That is, an individual operating in a primitive, nonstrategic holistic mode would not be expected to engage any sort of processing mechanism that would systematically alter the weights associated with the component attributes of the objects being judged. In contrast, the class of *abstraction* models described by Billman and Richards allow for systematic changes in similarity functions (e.g., altered weightings of component attributes).

In principle, the type of selective weighting in incidental learning conditions that was described in the previous section could well be represented in the type of model that Billman and Richards (1990) have referred to as fixed-similarity. That is, a given incidental task could encourage selective weighting of a given attribute from the start. No readjustment of the pattern of weighting need occur during the course of performing the incidental learning task. Thus analytic category decisions at transfer would be based on the fact that the weighted attributes were better represented in memory and the fact that the same metric would be applied in the category decision task. Thus, the results presented so far do not challenge a *fixed similarity* interpretation of incidental category learning. Billman and Richards' results, however, do.

Billman and Richards report four studies in which subjects learned a set of objects without being told that they were members of a category. The sets all involved correlated attributes. For example, in one condition of their first experiment subjects studied a set of novel animals in which the type of head and type of tail were perfectly correlated. After learning that set, subjects were told that the studied creatures all came from the same planet. They then saw more novel animals that either preserved or violated the correlation and had to indicate which ones were also from that planet. The crucial aspect of these test items was that the former were no more similar to the original exemplars than were the latter whether similarity was computed additively or multiplicatively. As a consequence, fixed similarity models have no way to predict an advantage for either type of test item.

The results were quite straightforward. Regardless of whether the materials were pictures of novel animals or descriptions of people (Experiments 2-4), subjects chose items that preserved the correlations reliably more often than items that violated the correlation. In addition, many subjects reported a conscious awareness of the correlation. Finally, when subjects were shown completely new variants that preserved the correlation but with different specific values than those that had been used in the original items, their generalizations based on these new exemplars were also consistent with preserving the correlation. Thus, they had not just learned that one particular head was associated with one particular tail. Rather, they had learned the more general idea that type of head and type of tail were correlated. One interpretation of these findings is that subjects had abstracted a rule about the correlation during the incidental learning trials. At the very least, it appears that their attention shifted to the correlation across the learning trials even though they were learning about the category incidentally via memorization of specific exemplars.

On the surface, Billman and Richards' findings seem more consistent with the idea that analytic processing occurs during incidental learning, and thus the balance may have tipped closer to the neutral point on the incidental/holistic hypothesis. The abstraction of a rule certainly seems to qualify as involving analysis. It is important to note however that the studies did not include an intentional category learning task as a comparison. Thus, although the findings indicate that a considerable amount of abstraction does occur during incidental category learning, it is not clear whether more abstraction would occur under intentional learning conditions.

Before leaving the issue of correlated attributes, it is important to note that instance models of the fixed similarity type do allow for the preservation of information about attribute correlations. Indeed Medin, Altom, Edelson, & Freko (1982) provided support for the context model by showing that subjects would prefer stimuli that preserved correlations over those that violated correlations. However, the context model achieves its ability to predict such a result indirectly, as a result of its use of a multiplicative similarity function. High similarity to one known exemplar is thus more influential than moderate similarity to two known exemplars. In general, test items that preserve a correlation that was true across learning exemplars, should be more similar to those exemplars by way of the multiplicative computation. They should thus be more likely to retrieve those exemplars than test items that do not preserve the correlation. The context model does not achieve the prediction through a shifting of selective attention to the attributes that are correlated during the course of experience with those exemplars.

The members of each pair of test items used by Medin et al. (1982) were equally similar via additive similarity computations; they had equal numbers of characteristic attributes. However, those items that preserved the correlation were more similar to known exemplars via multiplicative computations than were items that violated the correlation. Hence, its assumption of instance-based multiplicative computation of similarity allows the context model to account for the observed pattern of results. However, because Billman and Richards' test items were equally similar by any stationary similarity metric based on some function of attribute values (including a multiplicative one), the appeal to multiplicative computation of similarity does not provide an adequate mechanism for explaining their results. That is, Billman and Richards' pairs of test items not only matched old exemplars on the same total number of attributes, they also matched in the sense of being highly, moderately or weakly similar to the same number of known exemplars.

Because fixed similarity models provide no mechanism for *systematically adjusting* the weighting of (attention to) particular attributes or attribute correlations of exemplars, they can not account for the selective reliance on consistently correlated attributes that Billman and Richard's subjects showed. Variants on instance models (e.g., Nosofsky, 1984; 1987) do allow for systematic shifts in weighting of attributes, but the alterations in weighting are generally based on feedback, which would not be present in incidental learning conditions.

This raises an important question about how attribute correlations become selectively encoded in incidental learning. Before examining that question, however, it is useful to consider other aspects of relational information that have yet to be examined in incidental learning.

A. Relational Information

The types of feature co-occurrences that have been examined in the preceding paragraphs can be thought of as one form of relational information. The form would be referred to as empirically-derived or data-driven in the sense that they derive from repeated exposures to stimuli that maintain the correlation and in the sense that subjects would not be expected to have preexisting knowledge that would unite those features independently of the repeated exposures. The absence of effects due to knowledge is typically insured by carefully counterbalancing the features that are correlated and observing the same effects regardless of the particular pairings. Knowledge could still come to play a role in these types of correlations in the sense that it could be abstracted from the observed correlation if it is brought to consciousness. That is, once having empirically observed a correlation between attributes, the person might generate an idea, the form of which would be constrained by other knowledge, about why those attributes are empirically correlated. But, the involvement of knowledge would be after the fact.

Relational information is potentially much richer than this type of empirically-based feature co-occurrence and may, under some circumstances, be knowledge-driven. People do have knowledge that they bring to bear on the task, and that knowledge can result in selective use of a relation between otherwise separate features. Many researchers have pointed to the importance of knowledge or theories in producing higher-order relational information (e.g., Murphy & Medin, 1985; Wattenmaker, Dewey, Murphy, & Medin, 1986).

It is quite apparent that knowledge structures influence the extent to which independent cues or relational cues are extracted and used in intentional category learning (Wattenmaker, et al., 1986). The nature of what is learned appears to depend on an interaction between the attribute structure of the materials to be learned and the particular knowledge structures that are brought to bear on the learning task. There is, however, little evidence on the extent to which knowledge structures interact with stimulus materials to determine what is learned in incidental learning tasks. Consistent with our position that the same fundamental processing principles operate in intentional and incidental learning situations, one would expect that similar effects would be observed regardless of the learner's intent.

The ideas in the last paragraph should not be taken to mean that goals will not influence the way in which comparisons are made. Certainly, different goals might activate different knowledge structures with the result that different features or relations become more salient. However, the basic operating principles that a) knowledge structures are activated and b) they influence what is encoded about presented exemplars would be the same regardless of the goals of the learner (e.g., learn categories versus store individual instances).

B. Mechanisms for Selectively Weighting Relational Information

Although feedback would not provide a mechanism for adjusting the weights of component attributes in incidental learning, other implicit mechanisms might. For example, Medin and Florian (1990) have recently interpreted order effects in category learning in terms of exemplar-guided encoding. According to this idea, exemplars are not encoded in strict isolation. Rather, encoding of a given exemplar may be "guided" or in some way influenced by immediately preceding exemplars. Although Medin and Florian did not describe specific mechanisms that would produce the effect, in a relatively simple way, if stimuli 1011 and 1000 occur in succession, they are more likely to be in short-term store at the same time than if they are separated by intervening exemplars. Their presence in short-term memory together could draw attention to the attributes that they share, even if the individual is not trying to learn what makes them members of the same category. Hence, the shared attributes (10**) could be selectively stored for those exemplars.

Billman and Richards' subjects may well have had many experiences of exemplars with the same head and tail occurring in succession. Because the remaining attributes of that particular pair of exemplars would be different, these specific values of the correlated attributes would be favored. Thus, exemplar-guided encoding could help to explain subjects' responses to test items that either preserved or violated the correlation of a specific head and tail. The fact that Billman and Richards' subjects also generalized the correlation to new specific values of heads and tails that were not even present in the initial learning task is more difficult to account for with this interpretation. It seems to represent a more generalized attention to the correlation rather than attention to and storage of the specific values of attributes that would be in short-term store together during a learning task.

It is important to note that exemplar-guided encoding effects need not be thought of as sophisticated hypothesis testing in service of category learning; Medin and Florian note that order effects are similar whether subjects are attempting to learn category membership or not. Thus, the same types of processes could have occurred in Billman and Richard's studies. This serves to highlight the point that analytic processing, as indexed by selective weighting of component attributes, should not be confused with conscious, deliberate, strategic behavior (see also, Ward, 1990b for a similar point).

The mechanisms by which common or distinctive features or relations are noticed may be more straightforward for intentional than incidental learning. For example, Medin and Florian (1990) have noted that if a new item reminds a learner of a known exemplar (i.e., retrieves the exemplar), the learner might predict that the new item is a member of the same category. If feedback indicates that the prediction was correct, the learner may then selectively attend to the commonalities of the known and new item. If feedback indicates an error, the learner may attend to the differences.

In some incidental tasks, the subject does not have the goal of learning about category membership and may not even know that the items being compared are members of the same category. Thus, it could be argued that there is no compelling reason to attempt to determine the ways in which the objects are alike or different. Two possible resolutions can be suggested. The first is that individuals could store the exemplars during incidental learning without focussing on relational information. On hearing that the exemplars are members of a category during the transfer phase, the person could then retrieve and compare known exemplars. At that point, knowledge structures might result in "latent" relational information being noticed and abstracted; such relations would be knowledge-driven. Alternately, even without relevant knowledge-structures, because objects with shared correlated features may be generally more similar to each than other objects, they may be more likely to be retrieved together which would then allow the possibility of noticing shared attributes. Such relations may be viewed as empirically-driven. Note however, that these forms of "post-acquisition" attention to correlated attributes could occur only if the exemplar representations contained information about the attributes to begin with.

According to the "post-acquisition" view described in the previous paragraph, exemplar repetitions, remindings and theories would play little role in extracting relational information during incidental storage of exemplars. Billman and Richard's data, however, are inconsistent with this suggestion in that their subjects appeared to become aware of the attribute correlations during the incidental learning task itself. In addition, the fact that order effects are operative in incidental conditions (see Medin & Florian, 1990) suggests that extraction of correlated features occurs during the learning task itself.

As an alternate resolution, it could be suggested that human beings possess what might be termed a "generic comparison" procedure and that we automatically compare items that are present together in working memory whether or not our goal is to find out why they are members of the same category. In interacting with a set of novel materials (or known materials whose categorical relation to one another is not yet known) there may be an automatic comparison of the ways that they are the same and different. Put differently, people may not generally be cognitively idle even when not explicitly instructed to learn (particularly not while participating in an experiment). Being the prodigious categorizers that we are, we may nearly always be implicitly engaged in the process of determining the ways in which objects available in working memory are the same (categorizing) or different (differentiating). What features or relations are considered in those comparisons would be determined by a) aspects of salience that are independent of knowledge (e.g., intensity, focal versus peripheral color), b) contextual factors which could include recently encountered as well as currently presented objects, and c) properties determined by the particular knowledge structures that are activated at the moment. Such comparisons could afford the opportunity to abstract information about shared features and relations.

In the case of incidental learning, subjects do not predict category membership and there is no feedback, but there is sometimes category information available to the subjects (e.g., labels; uniforms) and they are asked to compare objects from the same or different categories (e.g., Kemler Nelson, 1984; Ward, 1989; Ward & Scott, 1987). Perhaps the mere presence of category information is enough to draw attention to commonalities if objects are from the same category and differences if objects are from different categories. This selective drawing of attention could allow for selective storage of single attributes and attribute correlations during the incidental task itself.

There are incidental category learning tasks in which the category information is not available (e.g., Billman & Richards, 1990). Once again, however, because of our overwhelming tendency to categorize, we may not be able to resist forming categories even if they take the relatively simplistic form of "objects presented during this experiment". In this situation the bias would presumably be to seek commonalities and thus they would be selectively favored.

V. CONCLUSIONS

To return to the sense of the question posed in the title of this chapter, is the acquisition of category information fundamentally different when the learner intends to acquire that information as opposed to when the learner has some other intent? The most obvious answer is that "it depends". In our view, considering on what "it depends" is a more fruitful approach than attempting to provide a "yes" or "no" answer to the question.

There are certainly several instances in which incidental learners exhibit more holistic-looking patterns of generalization on transfer tasks than intentional learners (Kemler Nelson, 1984; J. Smith, 1989; Ward & Becker, in preparation). On the other hand, there are several demonstrations of highly analytic performance by subjects operating without the intent to learn (Ward, 1989; Ward & Scott, 1987) or even the knowledge that a category exists to be learned (Billman & Richards, 1990). Clearly then, whether or not category extensions will differ following intentional and incidental learning depends on the exact details of the incidental and intentional tasks, the nature of the material to be learned and interactions between the two. One could conceivably conduct a large number of incidental category learning studies and determine whether the bulk of findings favored holistic processing, but such an approach is likely to fail for the basic reason that any pattern of results will always be based on the exact processing demands of the incidental task; amassing a large number of "holistic" findings would be most likely attributable to a selection of tasks that favor holistic processing; amassing a large number of "analytic" findings would occur from an equally clever selection of tasks that bias attention to particular attributes.

A more important aspect of the question of differences between intentional and incidental learning concerns what one means by a "difference", and a consideration of this issue returns us to the topic of structure and process raised earlier in the chapter. We distinguished between weak and strong senses of the term "difference" and related that distinction to the ideas of structure and process. A weak version of "difference" would be supported if different patterns of category extension were observed, but those patterns could be accounted for by the same underlying category structures (e.g., exemplars) and the same basic processes. A strong version would imply fundamentally different representation formats (e.g., exemplars versus rule-based abstractions) with consequent differences in processes (e.g., exemplar retrieval via similarity versus matching of input to rules). We reserved the term "structure" to refer to those internal representation formats. Thus, a consideration of the meaning of structure and process has helped in conceptualizing the possible nature of differences between incidental and intentional category learning.

If all that is meant is that a slightly different pattern of category extension will occur, then clearly differences between intentional and incidental conditions are sometimes observed, although it is still not clear whether those differences are based on intent per se or on the specific processing demands of the tasks that have been examined to date. If, however, what is meant is that the fundamental structural and processing principles by which category information is stored and retrieved differ across intentional and incidental conditions, then the case for a difference is not particularly strong. That is, even if we were to take the differences that have been observed to be based on intent rather than on specific processing demands, they could still be interpreted simply in terms of a difference in breadth of attention rather than a difference of rule-based abstractions versus exemplar storage. As noted earlier, appeals to multiplicative computations of similarity are neither compelling evidence for holistic processing, nor are they even capable of explaining some of the correlated attribute use that has been observed (e.g., Billman & Richards, 1990).

Exemplar models have developed considerably since Medin and Schaffer (1978) presented the context model. They now include mechanisms for abstraction of feature correlations by way of exemplar-guided encoding. Some of these changes were prompted by observations of phenomena in intentional category learning (e.g., sequence effects). However, such phenomena are also beginning to be found in incidental learning situations (Medin & Florian, 1990). Thus the advances in exemplar models also make them better able to explain incidental category learning.

So, we are left with a dilemma. Intuition tells us that people should learn differently when they try to learn than when they try to do something else. On the other hand, the data supporting this idea are not overwhelming. Two resolutions exist; our intuitions are incorrect or future research must provide a more compelling case for the supposed differences. Carefully considering what we mean when we say that incidental and intentional learning differ will aid in this latter quest.

ACKNOWLEGEMENTS

The research reported in this paper was supported by grant BNS-8608916 from the National Science Foundation and grant MH-43356 from the National Institute of Mental Health. The authors thank Dorrit Billman, Barbara Burns, and Doug Medin for their helpful comments on earlier drafts of this chapter. Address correspondence to Thomas B. Ward, Department of Psychology, Texas A&M University, College Station, Texas 77843.

REFERENCES

Berry, D. S., & McArthur, L. Z. (1985). Some components and consequences of a babyface. Journal of Personality and Social Psychology, 48, 312-323.

Billman, D., & Richards, J. (1990) Specifying and testing a general class of concept learning models: Fixed-similarity versus Selective Abstraction. Unpublished manuscript.

Brooks, L. (1978). Nonanalytic concept formation and memory for instances. In E. Rosch, & B. Lloyd (Eds.), Cognition and categorization (pp. 169-211). Hillsdale, NJ: Erlbaum.

Burns, B. (1986). Relationship of perceived stimulus structure and intelligence: Further tests of a separability hypothesis. American Journal of Mental Deficiency, 91, 196-200.

Cook, G. L., & Odom, R. D. (1988). Perceptual sensitivity to dimensional and similarity relations in free and rule-based classification. Journal of Experimental Child Psychology, 48, 319-338.

Craik, F. I. M., & Tulving, E. (1975). Depth of processing and the retention of words in episodic memory. Journal of Experimental Psychology, 104, 268-294.

Estes, W. K. (1986). Array models for category learning. Cognitive Psychology, 18, 500-549.

Fried, L. S., & Holyoak, K. J. (1984). Induction of category distributions: A framework for classification learning. Journal of Experimental Psychology: Learning, Memory, and Cognition, 10, 234-257.

Garner, W. R. (1974). The processing of information and structure. Potomac, Md.: Erlbaum.

Gati, I., & Tversky, A. (1984). Weighting common and distinctive features in perceptual and conceptual judgments. Cognitive Psychology, 16, 341-370.

Hintzman, D. L. (1986). "Schema abstraction" in a multiple-trace memory model. Psychological Review, 93, 411-428.

Kemler, D. G. (1982). Classifying in young and retarded children: The primacy of overall similarity relations. Child Development, 53, 768-779.

Kemler Nelson, D.G. (1984). The effect of intention on what concepts are acquired. Journal of Verbal Learning and Verbal Behavior, 23, 734-759.

Kemler Nelson, D. G. (1988). When category learning is holistic: A reply to Ward and Scott. Memory & Cognition, 16, 79-84.

Kemler Nelson, D. G. (1989). The nature and occurrence of holistic processing. In B. Shepp & S. Ballesteros (Eds.), Object perception: Structure and process (pp. 357-386). Hillsdale, NJ: Erlbaum.

Lewicki, P. (1986). Processing information about covariations that can not be articulated. Journal of Experimental Psychology: Learning, Memory, and Cognition, 12, 135-146.

Medin, D. L., Altom, M. W., Edelson, S. M., & Freko, D. (1982). Correlated symptoms and simulated medical classification. Journal of Experimental Psychology: Learning, Memory, and Cognition, 8, 37-50.

Medin, D. L., & Florian, J. E. (1990). Abstraction and selective coding in exemplar-based models of categorization. Unpublished manuscript.

Medin, D. L., & Schaffer, M. M. (1978). Context theory of classification. Psychological Review, 85, 207-238.

Murphy, G. L., & Medin, D. L. (1985). The role of theories in conceptual coherence. Psychological Review, 92, 289-316.

Nosofsky, R. M. (1984). Choice, similarity, and the context theory of classification. Journal of Experimental Psychology: Learning, Memory, and Cognition, 10, 104-114.

Nosofsky, R. M. (1987). Attention and learning processes in the identification and categorization of integral stimuli. Journal of Experimental Psychology: Learning, Memory, and Cognition, 13, 87-108.

Rosch, E., & Mervis, C.B. (1975). Family resemblances: Studies in the internal structure of categories. Cognitive Psychology, 7, 573-605.

Shepp, B. E., Burns, B., & McDonough, D. (1980). The relation of stimulus structure to perceptual and cognitive development: Further tests of a separability hypothesis. In F. Wilkening, J. Becker, & T. Trabasso (Eds.), Information integration by children (pp. 113-145). Hillsdale, NJ: Erlbaum.

Smith, J. D. (1989). Analytic and holistic processes in categorization. In B. E. Shepp, & S. Ballesteros (Eds.), Object perception: Structure and Process (pp. 297-323). Hillsdale, NJ: Erlbaum.

Smith, J. D., & Kemler Nelson, D. G. (1984). Overall similarity in adults' classifications: The child in all of us. Journal of Experimental Psychology: General, 113, 137-159.

Smith L. B. (1989). A model of perceptual classification in children and adults. Psychological Review, 96, 125-144.

Smith, L.B., & Kemler, D.G. (1977). Developmental trends in free classification: Evidence for a new conceptualization of perceptual development. Journal of Experimental Child Psychology, 24, 279-298.

Ward, T. B. (1980). Separable and integral responding by children and adults to the dimensions of length and density. Child Development, 51, 676-684.

Ward, T. B. (1983). Response tempo and separable-integral responding: Evidence for an integral-to-separable processing sequence in visual perception. Journal of Experimental Psychology: Human Perception and Performance, 9, 103-112.

Ward, T. B. (1986). Classifying by learning disabled and nondisabled children: Use of overall similarity versus dimensional relations. Bulletin of the Psychonomic Society, 24, 131-134.

Ward, T. B. (1988). When is category learning holistic? A reply to Kemler Nelson. Memory & Cognition, 16, 85-89.

Ward, T. B. (1989). Analytic and holistic modes of category learning. In B. E. Shepp & S. Ballesteros (Eds.), Object perception: Structure and process (pp. 387-419). Hillsdale, N.J.: Erlbaum.

Ward, T. B. (1990a). Further comments on the attribute availability hypothesis of children's category learning. Child Development, 61, 611-613.

Ward, T. B. (1990b). The role of labels in directing children's attention. In J. T. Enns (Ed.), The development of attention: Research and Theory. Amsterdam: Elsevier Science Publishers.

Ward, T. B., & Becker, A. H. (in preparation). Attribute variability, intent to learn and selective attention in category learning.

Ward, T. B., Foley, C. M., & Cole, J. (1986). Classifying multidimensional stimuli: Stimulus, task, and observer factors. Journal of Experimental Psychology: Human Perception and Performance, 12, 211-225.

Ward, T. B., Hass, S. D., & Vela, E. (1990). Selective weighting of attributes in incidental category learning, Unpublished manuscript.

Ward, T.B., & Scott, J. (1987). Analytic and holistic modes of learning family-resemblance concepts. Memory and Cognition, 15, 42-54.

Ward, T. B., Stagner, B. H., Scott, J. G., Marcus-Mendoza, S. T., & Turner, D. (1989). Classification behavior and measures of intelligence: Dimensional identity versus overall similarity. Perception & Psychophysics, 45, 71-76

Ward, T. B., Vela, E., & Hass, S. D. (1990). Children and adults learn family-resemblance categories analytically. Child Development, 61, 593-605.

Wattenmaker, W. D., Dewey, G. I., Murphy, T. D., & Medin, D. L. (1986). Linear separability and concept learning: Context, relational properties, and concept naturalness. Cognitive Psychology, 18, 158-194.

Wilkening, F., & Lange, K. (1990). When is children's perception holistic: Goals and styles in processing multidimensional stimuli. In T. Globerson & T. Zelniker (Eds.), Cognitive style and cognitive development. Norwood, NJ: Ablex.

Commentary

Learning Categories With and Without Trying: Does It Make a Difference, T. B. Ward & A. H. Becker

TARA C. CALLAGHAN
St. Francis Xavier University

In this chapter Ward and Becker ask the question of whether differences exist in the nature of category learning under incidental and intentional learning conditions. A search was made for two basic types of differences: A difference in what is learned (i.e., structure of the category) and a difference in the way in which the category was learned (i.e., the process of category learning). The authors point out that whereas most studies of category learning have focused on intentional learning, it is important to clarify the nature of incidental learning, since most real-world category learning is accomplished without specific instructions to learn.

Kemler Nelson's (1984) claim that incidental learning results in more wholistic than analytic processing, and hence categories based on overall similarity structure rather than critical features, was first assessed. Ward and Becker make the strong argument that Kemler Nelson's findings are more consistent with the view that in incidental learning the observer spreads attention over all attributes (breadth of attention) than the idea that overall similarity judgments are made (wholistic processing). The authors draw on the results of their own research to support this claim. Noting the differences between stimulus and task factors in the two sets of experiments Ward and Becker nonetheless argue that what is learned about category structure is determined more by attentional demands than intention to learn. In a final section the authors note the strong trend in category learning situations--incidental or intentional--for learners to abstract correlated attributes and for the possibility of prior knowledge to guide attribute selection and weighting. They make the interesting conclusion that one reason why very little difference is found between incidental as compared to intentional learning is that we as humans may be inherently comparing creatures and whether instructed to or not we simply jump in and compare when interacting with a novel set of materials.

I am left with one basic question after reading this very interesting and thorough study of the issues of structure and process in incidental learning. By allowing for the 'abstraction of correlated attributes' and the 'equal weighting of all attributes' to fall under the concept of analytic is one not arguing for the elimination of the concept of wholistic processing altogether? Is there no way that this is to be saved as a quality of processing that is unique? From the arguments presented in this chapter it appears that Ward and Becker would be happy to exchange 'wholistic processing' with 'wide breadth of attention.' Similar arguments have been made vis a vis the dichotomy between preattentive (another kind of incidental learning) and attentive processing by myself and others (Callaghan, Lasaga & Garner, 1986; Treisman and Gormican, 1988). There is however, one way that preattentive and wholistic processing appear to differ substantially from attentive and analytical processing and that is in the obligatory nature of the former processes. In preattentive tasks one cannot help but attend to all the features simultaneously, in attentive tasks one may choose the feature to be singled out, or the combination of features. It is as though the breadth of focus is fixed in the former case and changeable in the latter. In wholistic processing the weighting of all features is fixed (perhaps equally, though this may not be necessary--perhaps there are alternative, but fixed, patterns of weighting for different objects or pairs of objects), and in analytic processing one may change the weighting as the task demands. Although my thinking on these matters has led me to believe that the obligatory difference is a significant one, those from other perspectives may conclude (to borrow a phrase)--It depends.

Callaghan, T.C., Lasaga, M.I. & Garner, W.R. (1986). Visual texture segregation based on orientation and hue. Perception & Psychophysics, 39, 32-38.

Kemler Nelson, D. (1984). The effect of intention on what concepts are acquired. Journal of Verbal Learning and Verbal Behaviour, 23, 734-759.

Treisman, A.M. & Gormican, S. (1988). Feature analysis in early vision: Evidence from search asymmetries. Psychological Review, 95, 15-48.

Percepts, Concepts and Categories
B. Burns (Editor)

13

Not Just Any Category: The Representation of the Self in Memory

JUDITH F. KROLL
FRANCINE M. DEUTSCH
Mount Holyoke College

I. INTRODUCTION

If there is a single principle that characterizes research in cognitive psychology in the past three decades, it is the notion that knowledge is organized and that the resulting organization has profound consequences for the way in which we conceptualize and categorize our experiences (Bower, 1970; Mandler, 1967; Medin, 1989; Rosch, 1975b). A great deal of cognitive research has therefore focused on perception, comprehension and memory for object categories. More recently, social psychologists have argued that the general cognitive principles that appear to describe object categories may apply to social categories as well (Cantor & Mischel, 1979; Fiske & Taylor, 1984; Kihlstrom, Cantor, Albright, Chew, Klein, & Niedenthal, 1988). A central question in applying the cognitive framework to the social domain is whether knowledge about social categories, and the self, in particular, can be conceptualized in the same terms as knowledge for object categories such as fruits and vegetables or animals and tools (Greenwald, 1981; Greenwald & Pratkanis, 1984; Holyoak and Gordon, 1984; Lingle, Altom, & Medin, 1984).

In this chapter we first consider evidence for the notion that people have category-like representations for the self. We then review a program of research we have pursued on this topic. Finally, we consider whether self-knowledge is represented and accessed in the same way as object knowledge, and whether differences between self and object representations, particularly in the superiority for the self that is often observed in comprehension and memory tasks, can be attributed to structural differences in the underlying mental architecture or to process differences in the way in which those structures are used. One possibility is that memory representations for the self are organized according to structural principles that differ from those that describe ordinary mental categories. Alternatively, the self may share the basic features of ordinary object categories, but differ from other categories in the process of evaluation and revision. To anticipate our conclusions, we will argue that knowledge structures for the self do share features of ordinary object categories, but also differ in some important ways, especially in the degree to which they undergo modification across the lifespan.

A. Category Representations For the Self: Self Schemas.

Markus (1977) introduced the term "self-schema" to refer to cognitive structures that organize and summarize information about the self in particular domains.[1] Markus hypothesized that individuals who possessed self-schemas would have more rapid and efficient memory access for relevant characteristics than individuals who were aschematic for a given dimension. For example, if an individual possesses a self-schema for the domain of intelligence, they are likely to define themselves along this dimension, to efficiently process and remember

[1] For the purposes of our discussion we will not make fine-grain distinctions between categories and schemas.

information relevant to intelligence, and to judge others on this dimension. In an important series of studies (e.g., Markus 1977; 1983; Markus, Crane, Bernstein, and Siladi, 1982; Markus & Nurius, 1986; Markus & Wurf, 1987), Markus and her colleagues demonstrated that individuals who are schematic for a given dimension, as measured by extremely high self-descriptive ratings, are faster to endorse self-descriptive traits and have better recall for schema-related information.

Related to the claim that the self is represented categorically, is the proposal that the self-schema renders the individual an "expert" for schematic domains (Markus, Smith, & Moreland, 1985). Such expertise carries with it the cognitive costs and benefits that are normally observed when acquiring skill in a particular domain (Chase & Simon, 1973; Chi, Feltovich, & Glaser, 1981; de Groot, 1965). Access to and memory for domain-specific knowledge becomes fast and efficient, but reliance on high-level cues for categorized information may also lead to errors of overinclusion whenever the category cues are present.

One hypothesis about self-schema effects is that they reflect the operation of a highly elaborated memory representation for the self, one that is perhaps even more distinctive than memory for ordinary cognitive categories. Support for the elaborated nature of self-knowledge has been provided by a series of studies on the self-reference effect (Bower & Gilligan, 1979; Keenan & Baillet, 1980; Rogers, Kupier, & Kirker, 1977). In an initial study, Rogers et al. (1977) examined the self-reference effect by extending the depth-of-processing paradigm introduced by Craik and Tulving (1975). In the standard depth-of-processing paradigm, subjects are asked to judge a set of words on the basis of their structural properties (e.g., whether they appear in upper or lower case), their phonemic properties (e.g., whether they rhyme with a given word), or their semantic properties (e.g., whether they are synonyms of a given word or fit into an incomplete sentence to render it meaningful). These judgments form the encoding phase of the experiment; subjects are then given a later test of memory. The typical result in the depth-of-processing paradigm is that words that are more deeply encoded are better remembered.[2] Rogers et al. (1977) added a self-judgment task to the standard paradigm, in which subjects were required to judge a set of trait words for self-descriptiveness. The comparison of the self-judgment encoding with the standard encoding tasks showed that memory was superior for

[2]Although the depth-of-processing framework has come under attack for a variety of reasons that we will not address in more detail here (e.g., Baddeley, 1978), the initial interpretation of these data was that the more meaningful the encoding task, the more elaborated the resulting memory traces.

words encoded relative to the self. If the explanation of depth-of-processing effects is that they reflect the degree of elaboration in memory, then finding even better memory for self judgments than for semantic judgments suggests that the self is represented by an even more highly elaborated memory structure than meaningful objects.

Another source of converging evidence to support the interpretation of the self as a specially elaborated cognitive structure comes from studies of the self as a cognitive reference point (Holyoak & Gordon, 1983; Srull & Gaelick, 1983). Previous research on object categories has shown that some objects are more representative of the categories to which they belong than others (Mervis & Rosch, 1981; Rosch, 1973; 1975b; Smith, Shoben, & Rips, 1974) and that these prototypical instances can serve as cognitive reference points in the process of comparing concepts (e.g., Holyoak & Mah, 1982; Rosch, 1975a). In one experiment, Holyoak and Gordon asked subjects to compare themselves to a friend on a variety of social and physical dimensions. The critical empirical result in studies of cognitive reference points is an asymmetry in the degree of rated similarity as a function of whether the hypothesized reference point is the referent or the subject in the question. An atypical instance (e.g., a mango) is rated as more similar to a typical instance (e.g., an apple) than vice versa. Holyoak and Gordon found precisely this asymmetry in ratings of the friend and the self: the friend was seen as more similar to the self than vice versa, indicating that the self functions as a cognitive reference point. This conclusion was qualified, however, by the results of a second experiment in which the self was compared to a variety of social stereotypes. Holyoak and Gordon reported that whether the self acted as a reference point was a function of the degree of knowledge the person had about the stereotype. The self was a reference point only for less familiar stereotypes, suggesting that the self may have a status equal to, but not greater than, other highly organized familiar object categories.

A number of important implications can be derived from the research on self-schemas and self-reference. At a specific level, these studies suggest that self-schemas possess many of the same cognitive properties that characterize object categories and that under some conditions the self-schema may function as a superior cognitive structure. Because this issue has been a central focus of recent debate, we will examine it in more detail at a later point in the chapter and consider whether the unique features of the self-schema are best attributed to their structural properties or to processes that are enabled by those structures. At a general level, the research we have reviewed shows the utility of adopting a cognitive approach to problems that had previously been considered exclusively within the domain of social or personality psychology. The cognitive approach to the study of self-knowledge has the particular benefits of addressing the dynamic as well as the descriptive aspects of self-knowledge, and of focusing on the content in addition to the evaluation of the self.

II. MEASURING THE SELF-SCHEMA: THE SPONTANEOUS TRAIT GENERATION (STG) TASK

The starting point for our research on this topic was a concern with the way in which self-schemas have typically been identified. In the standard procedure (e.g., Markus, 1977) a large number of subjects rate themselves on a large number of personality dimensions. Dimensions are considered schematic for a given subject if she or he assigns extreme ratings to those traits and also rates those dimensions high in importance. This general approach has been criticized on the grounds that trait ratings of self-descriptiveness are reactive in that subjects must respond to dimensions that the experimenter provides (e.g., Deutsch, Kroll, Weible, Letourneau, & Goss, 1988; Kihlstrom & Cantor, 1984; McGuire & McGuire, 1982). Thus, the dimensions subjects endorse may not be central to their personalities. We have argued elsewhere (Deutsch et al., 1988) that this may not be an important problem if the goal of the research is to compare schematic and aschematic people on a particular dimension. However, if the research goal is to identify the most salient or central self-schemas for a given individual, then the reactive nature of the rating procedure may make it an insensitive method. A second criticism of the rating procedure is that it confounds schematicity and extremity. Individuals may be likely to possess self-schemas for domains in which they assign themselves extreme ratings. However, the assignment of ratings that are not extreme does not necessarily imply than an individual is not schematic for that dimension.

The method that we have developed to study the self-schema uses a simple procedure that increases the probability of identifying the central aspects of an individual's self representation and does not rely on the extremity of trait ratings. We call the procedure the *spontaneous trait generation* (STG) task because it involves asking subjects to list traits that they believe most closely reflect their personality (Deutsch et al., 1988). Subjects are free to provide any trait words they like, whether they would assign extreme ratings to a given word or not. Once we identify this list of words as representing a portion of the self-schema, we can then ask whether the words in the self-generated list possess any special information processing characteristics.

In our work we have used the STG task to address three types of research questions concerning the representation of self-knowledge. First, we compared the processing of self and other traits for generated and rated traits to see whether the generation method provided an advantage over the typical procedure. Second, we examined the way in which self-schemas identified in this way changed over time. Finally, we examined the relations between self-schemas identified by the STG procedure for pairs of friends.

A. Processing Self and Other Traits

In the first experiment, we asked undergraduate women to generate self-descriptive traits that were central to their personalities (Deutsch et al., 1988). The trait generation task was always performed in a first session that was separated by at least one week from the subsequent parts of the experiment. Subjects later rated a large set of trait words on self-descriptiveness. The set of traits included words they had generated in the first session and traits generated by other subjects. In a final session, they performed a "me" or "not me" reaction time task for the entire set of trait words. In the speeded reaction time task the trait words were presented one at a time and subjects had to decide whether each word was self descriptive. The time to make the decision and whether subjects responded "yes" or "no" was recorded. Following the reaction time procedure subjects were given an incidental recall task.

The goal of the initial experiment was to compare subjects' responses for their own traits with their responses to a randomly chosen other's traits. The procedure we used to generate the self-other comparison is illustrated in Figure 13.1. For each subject, another subject was randomly designated as the "other." In addition to calculating mean ratings, reaction time, and percent recall for self-generated traits, we also calculated these values for a given subject on the random other's generated traits. Thus, in the example shown in Figure 13.1, we computed measures for Subject A's responses to her own traits and Subject A's responses to Subject B's traits, where Subject B was the designated "other" for Subject A. One advantage of this procedure is that it maintains any biases inherent in the subject's own self-generated list. For example, if people tend to generate positive traits in describing themselves, that bias should be equally true, across the sample, for both the self and the other conditions. It is essential to note that subjects were never asked explicitly to judge anyone other than themselves; their responses on another subject's traits were responses they made *about themselves* on a set of words generated by another person.

Step 1: **Subjects generate self-descriptive traits**

Subject A's Traits	Subject B's Traits
1. intelligent	1. determined
2. quiet	2. energetic
3. sensitive	3. honest
4. lazy	4. intelligent
5. reasonable	5. happy
6. friendly	6. athletic
7. dependable	7. artistic
8. emotional	8. intense
9. practical	9. musical
10. caring	10. funny

Step 2: **Random subjects are paired such that Subject A's responses are computed for her own traits and for Subject B's traits**

Self — Subject A's Responses: Subject A's Traits

Other — Subject A's Responses: Subject B's Traits

	Self: Trait	RT	Recalled	Other: Trait	RT	Recalled
	Intelligent[a]	672	Yes	determined	1422	Yes
	quiet	529	No	energetic	1072	No
	sensitive	646	No	honest	921	No
	lazy	646	Yes	**Intelligent**	672	No
	reasonable	563	Yes	happy	2785	Yes
	friendly	371	Yes	athletic	732	No
	dependable	475	No	artistic	941	Yes
	emotional	391	Yes	intense	1072	No
	practical	419	Yes	musical	972	Yes
	caring	417	No	funny	1135	No
Mean Responses:		513 ms	60%		1172 ms	40%

a. Analyses were performed without overlapping traits.

Figure 13.1. Illustration of the procedure used to create self-other pairs.

The results on all of the measures supported the self-schema hypothesis: subjects consistently rated themselves higher, responded more rapidly, and remembered more of the self-generated traits than the other-generated traits (Deutsch et al., 1988, Experiment 1), suggesting that the STG task was reflecting access to a self-schema.

To see whether the extremity of self-descriptive ratings could account for the pattern of results, we performed a second analysis on these data in which another sample of traits were drawn on the basis of the ratings that subjects had assigned to them during the second phase of the experiment. Recall that subjects had been asked to rate the self descriptiveness of a large group of traits that included those they had generated as well as those generated by the other subjects. It was thus possible that when presented with a trait word that had not been generated previously, a subject might nonetheless rate it as extremely self descriptive. The question we asked in this analysis was whether there would be any processing differences for generated and non-generated words that were assigned the same ratings. In the matched analysis, traits were explicitly chosen to match the ratings the subject had previously assigned to the original "self" and "other" trait words. For example, although we haven't shown ratings in the illustration in Figure 13.1, imagine that Subject A had assigned a rating of "10" (very self-descriptive) to the trait "friendly." In the matched analysis we identified all the non-self generated traits that were assigned a rating of "10" and randomly sampled one of them to be matched to "friendly." We did this for each of the ten traits, resulting in a new set of matched traits that had been assigned ratings identical to the self. We then repeated the procedure for the other-generated traits. If the extremity of self-descriptive ratings can account for the self-schema effect, then a similar pattern of self-other differences on the reaction time and recall tasks would be expected when the traits were matched on the ratings. The results are shown in Figure 13.2 for the original and matched trait words. Self-generated traits were judged more rapidly than a set of matched trait words, suggesting that the generation procedure accessed a knowledge representation that conferred greater processing advantages than mere extremity on the rating task.[3]

[3]A puzzling aspect of these data was that reaction times for the other-generated traits were also faster than those for the matched traits. If these two conditions represent traits that were not generated by the self and were equally self descriptive, then there should have been no principled reason to expect a difference between the two conditions. However, a difference between these two conditions is that the initial set of "other" traits were traits that an individual subject actually generated, whereas the matched "other" traits were randomly drawn from the larger pool of traits on the basis of the assigned ratings.

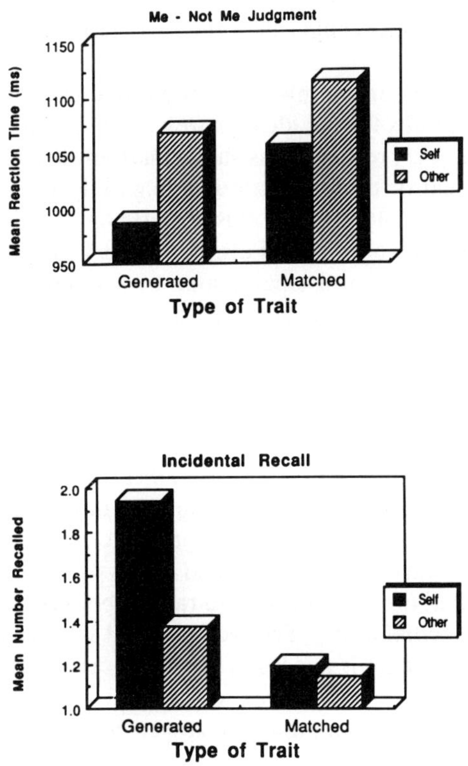

Figure 13.2. Mean reaction times (in milleseconds) and number of traits recalled for self-generated and matched traits (adapted from Deutsch et al., 1988, Experiment 1).

A similar pattern of results was obtained in the incidental recall task, suggesting that there is an advantage for self-generated traits in memory.

Overall, the results of this experiment suggested that the STG procedure was an effective method for identifying self-schemas. Self-generated traits were consistently processed more rapidly, rated higher, and remembered better than other-generated traits. Moreover, the self-generated traits reflected a greater processing advantage over and above traits that were assigned similarly high

It is possible that traits actually generated by an individual possess some additional properties, such as being higher frequency or more desirable (Deutsch et al., 1988).

ratings, suggesting that the generation task may provide a more sensitive means for operationalizing the self-schema than the use of extreme ratings.[4]

A second experiment in this series replicated this pattern of results and showed processing benefits for the ideal as well as the actual self (Deutsch et al., 1988, Experiment 2). Past research has shown that self-knowledge often includes information about ideal or desired attributes (e.g., Higgins, Klein, & Strauman, 1985; Markus & Nurius, 1986). In this experiment, subjects were asked to generate ten self-descriptive traits and also ten traits which described their ideal self. Traits from both types of lists generated by a group of subjects were then pooled and presented in a rating task and a self-judgment reaction time task similar to those described above. The critical results are shown in Table 13.1 where ratings and reaction times are given for self-generated traits, self-ideal-generated traits, and other ideal traits (i.e., ideal traits generated by another subject). Self-generated traits were rated as more highly self-descriptive than ideal-generated traits, but there was not a significant difference in reaction time for the self and ideal-generated traits. These data thus suggest that individuals' self-conceptions include self-schemas that incorporate ideal as well as actual traits. Note that the comparison with the other-ideal allows us to rule out the alternative interpretation that ideal traits are processed rapidly because they tend to be more highly desirable than actual traits. The fact the ideal self was so highly accessible suggests that the self-schema is not just a cognitive category that, by virtue of its organization, facilitates information processing, but also a potentially dynamic mental structure that directs attention to change. We considered this possibility in the next experiment.

[4]An alternative interpretation of our results is that the STG task is effective not because it reflects more central dimensions of the self than rating tasks do, but because it requires the cognitive act of generation which, in and of itself, is known to confer processing advantages in later tests of memory (e.g., Graf, 1980; Graf & Schacter, 1985; McDaniel, Waddill, & Riegler, 1988; Slamecka & Graf, 1978). In another series of experiments (Deutsch, Kroll, & Basile, 1989; Kroll, Deutsch, & Basile, 1991) we examined the contribution of generation effects to results obtained with the STG task. If the results we have described reflect the process of generation per se rather than access to a well-organized knowledge structure for the self, then similar effects would be expected when subjects generate information about people other than themselves. These experiments involved comparing performance on self-generated traits to a variety of other-generated conditions. The results showed that there were generation effects in both reaction time and recall measures, but that information generated *about the self* produced a processing advantage above and beyond generation effects that were produced for stereotypes, or by other generation tasks (e.g., completion or transposition).

Table 13.1. Mean ratings and reaction times as a function of trait category and time of testing.[a]

	Type of Trait		
	Self	Ideal	Other Ideal
Ratings[b]	8.1	7.4	7.3
Reaction Times[c]	948	981	1022

a. Adapted from Deutsch et al. (1988), Experiment 2.

b. Ratings are based on 10-point bipolar scales.

c. RT is reaction time in milliseconds.

B. Self-Schema Changes Over Time

To see whether the aspect of self-knowledge that we measured using the STG task changes over time, we conducted a longitudinal study on a group of first year college students (Deutsch et al., 1988, Experiment 3). Because entrance into college is an important life transition (Astin, 1977; Feldman & Newcomb, 1973) we expected that the changes that students undergo during this period might be reflected in changed self-schemas. Markus (1983; Markus & Nurius, 1986) has proposed that people have "possible selves" in which they entertain aspirations and expectations for the future as well as a representation of what they were like in the past. One advantage of the longitudinal design is that it allowed us to directly identify traits that were added and deleted from an individual's self-schema.

The procedure for this study was similar to the one described above. The first phase of the study occurred early in the first semester. Ninety-five first year students were first asked to generate ten self-descriptive traits. Subjects later completed a trait rating task based on the corpus of pooled traits generated by the entire set of subjects, and in a final session performed the "me-not me" reaction time task. The same procedure was repeated in the second phase of the study during the second semester. The only change was that following the administration of the reaction time task in the second phase, subjects were asked to perform an incidental recall task.

Table 13.2. Mean ratings, reaction times, and percent recall as a function of trait category and time of testing.[a]

Type of trait	Time one		Time two		
	Rating[b]	RT[c]	Rating	RT	% Recall
Self Traits					
Stable[d]	8.64	988	8.58	869	40
Distinctive one (D1)[e]	8.06	1035	7.80	955	20
Distinctive two (D2)[f]	8.00	1059	8.14	914	31
Random Other	6.81	1068	6.99	961	

a. Adapted from Deutsch et al. (1988), Experiment 3.
b. Ratings are based on 10-point bipolar scales.
c. RT is reaction time in milliseconds.
d. Stable traits are traits that subjects generated at both times one and two.
e. Distinctive one traits were generated at time one but not time two.
f. Distinctive two traits were generated at time two but not time one.

To examine the process of self-schema change, we first identified the set of traits that remained constant and the set that changed from the first trait generation procedure to the second. We identified approximately 27% of the traits as identical or "stable" traits at the two times of testing. Of the unique traits, those that subjects generated at the first test only, we called "distinctive one" (D1) traits. Traits generated at the second test only, we called "distinctive two" (D2) traits. The data are shown in Table 13.2, where mean ratings, reaction times, and percent recall are given as a function of trait category and time of testing. If individuals possess a constant self-schema that does not change over time, then the STG task might simply reflect a randomly sampling of personality-specific traits and there would be no reason to have expected differences between the two types of distinctive traits. The pattern of data shown in Table 13.2 clearly allowed us to reject that hypothesis. Subjects were fastest to judge those traits that were part of their current but not past self-schema. They were also consistently faster to process stable traits than distinctive traits. The ratings showed a similar pattern of results. In both measures, however, there was an interesting asymmetry between the data at the two times of testing. At time one, the difference between the distinctive traits was not significant for either measure. That is, at time one, subjects did not appear to differentiate between traits that were included on their self-generated list (but were to be omitted at the second time of testing) and traits that later appeared as new traits at time two. However, at time two, there was an advantage in processing the new compared to the old traits. The recall data at time two provided additional support for this pattern: stable traits were more likely to be recalled than distinctive traits, but new traits (D2) were better recalled than old traits (D1).

Deutsch et al. (1988) interpreted this pattern of results as evidence for increasing differentiation in the self-schemas of first-year students as the school year progressed. The data suggest that the self-schema undergoes a process of revision that is weighted more heavily towards the future self than by the past self. The results also suggest that the STG task provides a sensitive method for examining the factors that influence self-schema changes. We will examine this issue more closely when we consider the relations between the self-schemas of friends.

C. Similarity in the Self Schemas of Friends

If the STG procedure provides a sensitive measure of the self-schema, as we have suggested it does, then it can be used as a tool to study the effects of relationships on self-knowledge. Our assumption in extending the self-schema work in this direction, was that friendship is a vehicle for self-change and that friends use each other (in a positive sense) as a source of information about aspects of their personalities that might develop (Markus & Nurius, 1986). We have shown already that the self-schema as measured by the STG task appears to be sensitive to an individual's conception of their ideal self (Deutsch et al., 1988, Experiment 2) and to changes within an individual over time (Deutsch et al., 1988, Experiment 3). If our assumptions about using the STG task to measure self-schemas are correct, then we expected that friends would be more likely to generate similar traits about themselves than random pairs, and that even for nonoverlapping traits, there would be some special processing consequences for a friend's traits. If certain constructs are highly accessible to both individuals of a friendship pair (Duck, 1977), then we might expect them to describe themselves in similar terms. Similarity between friends would also be expected to result in higher self-ratings, faster "me-not me" judgments, and better recall for both self and friend-generated traits compared to traits generated by a random other. This study tested these predictions (Goss, Kroll, & Deutsch, 1984).

Thirty-one pairs of close friends participated in the study. One member of each pair was recruited by the experimenter and was then asked to recruit a close friend to participate with her in the study. The procedure was identical to that described by Deutsch et al. (1988), Experiment 1.[5] Subjects first generated seven self-descriptive traits. Two weeks later they were given the trait rating task, and at least two days following the rating task they returned to the laboratory to perform the "me-not me" reaction time task. Following the reaction time phase of the experiment, subjects were given an incidental recall task in which they were asked to list as many of the traits words as they could remember from the experiment.

It is important to remember that subjects were never asked to describe or rate their friend. They responded only with respect to themselves and understood the request to bring a friend to participate in the experiment in terms of our need for additional subjects. We also requested that they not discuss the study with their friend or anyone else. The empirical question was whether individuals would process traits generated by their friends, in a different manner than traits generated by random others, although they did not explicitly know which traits their friend generated.

Each subject was paired with their friend and with a random other. We first computed the overlap between generated traits. The data are shown in Table 13.3. As predicted, trait overlap was higher between friends than between random pairs. Although the result was only marginally reliable in this experiment, it has since been replicated in other studies (Deutsch & Mackesy, 1985; Deutsch, Sullivan, Sage, & Basile, 1991).

The results for the rating, reaction time, and recall tasks are also shown in Table 13.3. On each of these measures, the predictions were supported by the data. Subjects rated their own traits more highly than their friend's traits and their friend's traits more higher than a random other's traits. Likewise, reaction times and recall for the friend's traits fell between those for the self and the other, suggesting that a friend's traits have greater status than those of a random other.

[5]The subjects in this study were a subset of those who participated in Deutsch et al. (1988), Experiment 1. The data have been analyzed here to examine the friendship hypothesis. In addition, in this study subjects were asked to generate seven self-descriptive traits in the STG task. That value was somewhat arbitrary, and in later studies we consistently asked subjects to generate ten traits. The larger number of traits produced more reliable results.

Table 13.3. Mean overlap, ratings, reaction times, and percent recall as a function of source of trait.[a]

	Source of Trait		
	Self	Friend	Other
Overlap [b]		1.16	.77
Ratings [c]	8.71	7.39	7.02
Reaction Times [d]	980	1051	1095
Percent Recall	37	25	20

a. From Goss et al. (1984).
b. Mean number of overlap out of seven generated traits.
c. Ratings are based on 10-point bipolar scales.
d. Reaction time in milliseconds.

The method that we have described for assessing similarity between friends contrasts with previous research comparing self-ratings on a large corpus of traits. To see whether the similarity we obtained was revealed because of the identification of the self-generated traits, we replicated the procedure used for assessing similarity in previous studies by computing overall correlations between friends ratings on the entire corpus of traits (Bailey, Finney, & Helm, 1975; Fiedler, Warrington, & Blaisdell, 1952; Miller, Campbell, Twedt, & O'Connell, 1966; Thompson & Nishimura, 1952). The resulting correlations of self ratings were not higher between pairs of friends than between random pairs. The absence of self-concept similarity when overall correlations are examined replicates previous results. It is only when the central dimensions of the self-concept were identified, as we have claimed is the case using the STG task, did the similarity between friends emerge in their self-ratings.[6]

An obvious question in considering these data, is whether friends were simply more similar from the start, or whether they became more similar over time. If the self is represented as a dynamic schema (Markus & Wurf, 1987), then

[6]Likewise, in a later study of similarity between friends (Deutsch & Mackesy, 1985, Study 1), a measure of similarity was obtained by summing the difference between self-ratings assigned by pairs of subjects on 40 traits. On this measure friends were no more similar than nonfriends, suggesting again that it was only on dimensions identified by the STG task that similarity between friends was evident.

friendship, in which one is continually seeking validation and elaboration of one's own interpretations and structuring of experience (Duck, 1979), may provide a means for modifying self-knowledge. Friends may allow us to explore possible selves (Markus & Nurius, 1986) and to affirm current self-conceptions.

Although changes in the self-schema that occur as friendship develops may be best studied in the context of longitudinal research, it was possible to indirectly examine the alternative interpretations for self-schema similarity among friends observed in the present study. If friends are influenced by each other's self-schemas, that influence may be most apparent in domains in which their self-conceptions differ. For example, if friends actively reconsider their self-conceptions along those dimensions in which they differ, we would expect the process of change to be reflected by the way they respond to self-related information in those domains. In a final analysis, we examined the effect of friendship on the processing of traits that friends did *not* share.

In previous studies, reaction times for the "me - not me" judgment have been analyzed as a function of self-descriptive ratings (Kuiper, 1981; Rogers, 1981). In this analysis, average reaction times to traits in the speeded judgment task are plotted as a function of the self-descriptive rating that subjects previously assigned those traits. The function that emerges has an inverted-U shape. Individuals are fastest to judge traits that they rated as highly descriptive and also fast to judge (and presumably reject) traits that they rated as highly undescriptive. Reaction times for moderately descriptive traits are judged more slowly, suggesting that they may reflect the contribution of an extended decision stage in processing.[7]

We first examined individual's reaction times to trait words as a function of the ratings previously assigned to those traits, to see whether we could replicate the inverted-U shaped pattern. These data are shown in Figure 13.3. The function labeled "additional" includes all of those traits that the subject and her friend did *not* generate. The data for these traits clearly conformed to the predicted inverted-U shaped pattern reaction times were fastest for traits previously rated as highly self-descriptive or not at all self-descriptive, but were slower for intermediate rating categories. The critical comparison for present

[7]The inverted-U shaped function resembles evidence for other two-stage decision models in the cognitive literature in which initial decisions are made rapidly for information above or below criterial thresholds and slowly for information that falls between (e.g., Atkinson & Juola, 1974; Smith, Shoben, & Rips, 1974).

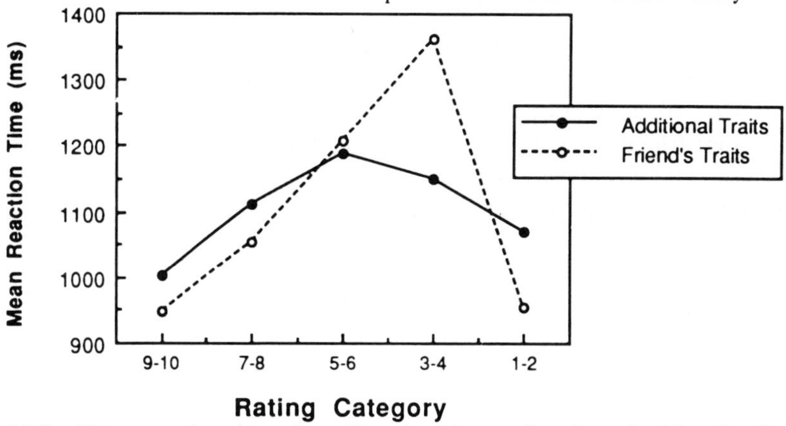

Figure 13.3. Mean reaction times (in milleseconds) as a function of subjects' ratings of additional and friends' traits. The rating scale ranged from highly self-descriptive (10) to not at all descriptive (1). Values displayed in the figure are averaged across adjacent rating categories (from Goss et al., 1984).

purposes was with traits generated by the friend. To evaluate the effect of friendship on this inverted-U shaped function, we plotted a separate function for those traits generated by the friend. These are the data labeled in the figure as "friend's traits." These are again an individual's reaction times as a function of *her* own rating scores but on the friend's generated traits. The striking result was that the function for the "friend's traits" was also an inverted-U shaped pattern, but it deviated from the "additional" traits in one portion of the curve: those traits rated as *not* very self-descriptive of the individual, but generated by the friend, produced extremely long reaction times compared to other conditions in the experiment.

If the initial similarity between two individuals was the sole factor determining whether or not they became friends, then pairs of friends would be expected to differ on specific traits, but their responses to traits that were true of their friends, but not self-descriptive, should have been no different than their responses to traits generated by other individuals in this category. The marked difference between responses to friend's traits and other traits in Figure 13.3 may thus be one indication that an individual's self-schema changes as a consequence of friendship rather than simply being a causal determiner of it. Additional evidence for this hypothesis was provided by an analysis of the percentage of "no" responses for the category in which the "additional" and "friend" functions differed. Whereas the percentage of "no" responses in the "4/3" rating category was 67% for the "additional" traits, it was only 39% for the "friends" traits. Subjects thus appeared to be accepting a subset of their friends' traits as true of themselves despite the fact that they initially rated these traits as not particularly self-descriptive. Their reluctance to reject these traits as self-descriptive when they are included in a friend's self-schema may reflect the process of change. These results suggest that friendship may function as a medium for revising the structure and content of self-knowledge. We consider possible mechanisms for the process of self-schema revision in the final discussion.

D. Summary of Findings With the STG Task

The experiments we have reported used the spontaneous trait generation task as a way of identifying self schemas to reflect central aspects of individual's personalities. At a general level, we replicated the pattern of self-others differences reported by Markus and others in that self-generated traits were judged more rapidly and remembered better than other-generated traits. The results go beyond the previous research, however, in demonstrating that self-generated traits produce information processing advantages over traits that are rated equivalently but not generated. In other research (Deutsch, Kroll, & Basile, 1989; Kroll, Deutsch, & Basile, 1991) we have shown that it is not generation per se that is responsible for the self advantage in speed of processing and in memory, because traits that are generated about others or about stereotypes produce generation effects but do not yield the same performance advantages as self-generated traits. Finally, we demonstrated that self schemas can change over time and can be influenced by interactions with friends. Taken as a whole, our research using the STG method shows that when central aspects of an individual's self-schema are identified, there are indeed cognitive advantages associated with the processing of self-referent information in measures of both initial comprehension and later recall.

III. IS THE SELF A SPECIAL CATEGORY?

A number of recent papers have been critical of the claim that self-schema effects are special in either a qualitative or quantitative sense (Bower & Gilligan, 1979; Ferguson, Rule, & Carlson, 1983; Greenwald & Banaji, 1989; Klein, & Kihlstrom, 1986; McDaniel, Lapsley, & Milstead, 1987). The dominant theme in these critiques is the idea that the self does represent a powerful, highly differentiated, and well organized memory structure but that it is not alone in doing so; any comparable domain that is conceptually organized to the same degree will produce similar effects. Some of these critics argue that the reason that previous studies have so consistently shown processing advantages associated with the self is that they have failed to compare self processing with a set of comparable controls. For example, Klein and Kihlstrom (1986) showed that the advantage of self-referent judgments in the standard depth-of-processing paradigm (Craik and Tulving, 1975; Rogers et al., 1977) can be eliminated if the semantic task used for comparison encourages a similar degree of cognitive organization. When they manipulated the degree of organization that a semantic or self-referent task encouraged, they found that it was organization rather than the source of the encoding that accounted for their results.

A related issue that is problematic for the interpretation of some past self-schema research is that self-schema effects are sometimes confounded with the effects of social desirability. Because subjects will tend to describe themselves in relatively favorable terms, it is not clear whether processing benefits that appear for the self are attributable to the nature of memory organization for the self or to the fact that desirable traits may be easier to process than other information. Ferguson et al. (1983) addressed this possibility in a depth-of-processing study in which subjects were asked to decide whether positive and negative traits described themselves, a well-liked, disliked, or neutral other person. In another condition of the study, a different group of subjects were asked to decide whether the same set of words were desirable, imageable, meaningful, or familiar. All subjects were then given tests of recall and recognition memory. The results provided only partial support for the self-schema hypothesis because although memory for self-rated traits was quite good, memory for traits that had been rated for desirability was just as good. Ferguson et al. (1983) concluded that this paradigm was not sufficiently sensitive to study the organization of self-knowledge, although they also acknowledged the possibility that evaluative judgments are always made with respect to the self. Good memory for traits judged for desirability may therefore be another manifestation of the self-schema.[8]

Using a different approach, McDaniel et al. (1987) also came to the conclusion that the self is not a privileged memory system. These authors used the release from proactive inhibition (PI) paradigm to examine the categorical properties of self-knowledge. In this paradigm, a series of to-be-remembered items that are all sampled from the same dimension are presented. Proactive

[8]Note that the potential confounding of self-traits with trait desirability is not a problem in research using the STG task because the critical comparison is always with traits that have been generated by another person. Because each individual's self-generated traits serve as other-generated traits for another subject, within experiments the average desirability of trait words were the same in both conditions.

inhibition refers to the fact that there is typically a decline in memory performance across trials. On a subsequent critical trial new items are presented from either the same or a different dimension. If the dimension is represented in memory as an organized category, then there is typically a recovery of the memory function, or "release" from PI (Wickens, 1970). McDaniel et al. (1987) reasoned that if the self functions as a distinctive category, then it should be possible to obtain release from PI in an experiment in which self-descriptive traits were contrasted with traits that described a well-known person or neither the self nor a well-known other. The results were mixed because release from PI was obtained for self-rated traits in one experiment, but the trait categories used in that experiment were confounded with desirability or likableness. When this confounding was removed in a second experiment, no release from PI was observed. McDaniel et al. (1987) concluded that self-schematic encoding in memory is less automatic than has been generally believed.

How should we interpret these critical findings? It is possible that the self-schema results are not as ubiquitous as the earlier findings and our own research seemed to indicate. However, the three studies we have just reviewed can themselves be criticized on the same grounds that we initially criticized Markus' method for identifying self schemas: they all use measures of self-reference that are essentially reactive and rely on the extremity of self-ratings to identify self schemas. The failure to observe a distinctive pattern of results for the self in these studies may thus be attributable to a lack of centrality in operationalizing the self-schema. There are also wide variations among studies in how the "other" or "non-self" conditions are defined. For example, in the McDaniel et al. (1987) study, both extremely high and extremely low self-descriptive ratings were used to identify the self and non-self categories. However, it is possible that traits that are *extremely* nondescriptive are also represented as part of the self compared to traits that are more neutral. It is therefore unclear whether the failure to obtain distinctive results for the self is attributable to the manner in which self categories were defined, to the sensitivity of the measure, or to the reason the authors suggest, namely that memory for the self is not specially represented.

One source of complexity in considering the literature on this topic is that two different, but related, phenomena have been reported. Research on self-reference and self-schema involves access to knowledge structures that contain information about the self. In these experiments subjects are asked to decide how well a set of trait words describe them. However, research on self-generation effects, such as the work recently reported by Greenwald and Banaji (1989), concerns the effectiveness of the process of self-generation per se, but not necessarily the effectiveness of eliciting information about the content of self-knowledge. In the Greenwald and Banaji (1989) study, individuals were asked to generate a list of friend's names and then to construct sentences by pairing the names with a specified target noun (e.g., If the name was *Barbara* and the target noun was *cafe*, the subject might produce the sentence, "On her last vacation Barbara drank cappucino at the cafe"). In some conditions the names were generated by the subject but in other conditions they were provided by the experimenter. After creating sentences, subjects were given a distractor task. The critical phase of the experiment was a test of free recall for the target nouns. The main result was consistently higher levels of recall for target nouns paired with self-generated names. However, Greenwald and Banaji (1989) argued that the superiority for self-generated material in this study did not require the conclusion that the self is an extraordinary memory system. Because the superiority of nouns associated to self-generated names could be modeled quite well by associative memory models that were intended to account for data from standard memory experiments, they proposed that the self is simply a familiar and highly organized knowledge structure in memory that confers all of the usual mnemonic benefits that other effective memory cues have also been observed to provide. However, because the information that subjects generated in this study was not explicitly about the self, it is not clear that this evidence bears on the central claim that the self is a superior memory structure.

Unlike the generation task performed in the Greenwald and Banaji (1989) study, the STG procedure requires subjects to generate information specifically about themselves. We have argued that our research using the STG task demonstrated that the self may indeed have special category representation in memory when measures are taken to ensure that the defined self-schema includes central aspects of the self. Thus, some of the apparent controversies in the literature on the self as a special category may be understood in methodological terms. If past studies have relied too heavily on measures of the self-schema that are not necessarily central, then past results may not accurately reflect the special properties that the self-schema is hypothesized to possess.

However, the question is not merely one of methodology; there are a number of distinct theoretical alternatives for conceptualizing how the representation of the self might differ from other categories (see Greenwald and Banaji, 1989, for a related discussion). One possibility is that the difference is quantitative rather than qualitative. The self-schema may possess the same functional architecture as other mental categories but the connections between the underlying units within the structure may be better differentiated or more strongly linked with other concepts. This quantitative proposal is essentially a proposal that the difference between the self and other categories is a matter of degree. Whether or not a particular domain is represented categorically may initially be a function of how much knowledge an individual possesses within that domain and whether that knowledge has some integrated organization. For example, higher-level hierarchically-based structures may replace attribute lists once a category has been formed (e.g., Lingle et al., 1984). However, after a category structure exists within a particular domain, differences between different types of categories may be a function of degree of knowledge and elaboration. If the self-schema is continually used to evaluate incoming information about the self and others, it may not only acquire the features of an expert knowledge structure, it may become the ultimate expert memory system.

An alternative proposal is that the self differs from other categories in terms of its basic structure. For example, Cantor and Mischel (1979) argued that compared to object categories, social categories are less likely to be disjoint, and are therefore less likely than object categories to be represented as discrete hierarchies. The issue of how categories are structured in memory (e.g., Smith & Medin, 1981) and whether there are category structures or schema representations at all (e.g., Rumelhart, Smolensky, McClelland, & Hinton, 1988) is a topic that has received a great deal of attention in the cognitive literature of the past decade and a detailed consideration of these issues is beyond the scope of this chapter. Moreover, because there is still disagreement in the cognitive literature about the merits of alternative conceptualizations of category structure (e.g., Medin, 1989), it is not even clear that a comparison of object categories with self, on the basis of available evidence, would lead to any clear resolution of this issue. What we can do, however, is to briefly consider the evidence that we do have that bears on category or schema structure.

Typically a set of empirical results must co-occur for the inference to be drawn that a particular domain is represented categorically or schematically (Alba & Hasher, 1983). Of these results, the most important are that information consistent with a schema is better recalled (Bartlett, 1932), more quickly recognized (Rosch & Mervis, 1975), and more likely to be falsely recognized (Bransford, Barclay, & Franks, 1972) than is information that is not consistent with that schema. All of these critical results have also been reported for self-schemas in the research we have already reviewed. If we take this evidence as bearing on category or schema structure, we are led to the conclusion that the evidence is more consistent with the notion that the self and other categories share the same underlying structure but differ in quantitative ways that may have important processing implications. (It is possible, of course, that the wrong assumptions have been made about these "critical" results and that these empirical results do not sufficiently constrain models of memory structure to determine what the structures are, let alone whether they differ.) We believe, however, that there is not compelling evidence to reject the hypothesis that the memory representation for the self is structurally similar to memory representations for object categories.[9]

Where do these considerations leave us in evaluating the self-schema in terms of structure and process? We would like to speculate that the self-schema shares the fundamental architecture of general object categories. Whether that architecture ultimately turns out to be a hierarchy, a network, or a set of connections among basic units in a distributed representation, is not something we can determine at present. However, the evidence from our own research using the STG paradigm suggests that the self does have special properties. We think that on structural grounds, the self may differ quantitatively from other categories and that these quantitative differences may have important implications for processes that are enabled by the goodness of the architecture. In particular, we would like to propose that over the course of an individual's life and in different situations the self-schema must be able to change much more dramatically than analogous representations for object categories.[10] In the next section, we consider possible mechanisms for self-schema change.

IV. CHANGE AND DEVELOPMENT IN SELF-SCHEMAS OVER THE LIFESPAN

We have argued that self-schemas have many of the same information-processing properties as other schemas. Just as with object schemas, social information that fits into an existing self-schema is better remembered, and

[9]Lingle et al. (1984) suggest that mixed category representations, including aspects of both probabilistic and exemplar models, may best account for available categorization data. They speculate that one way that social and object categories may differ is in terms of the relative weighting of each component structure. It is not clear, however, how this proposal might be tested.

[10]Connectionist models may ultimately be an attractive option from the perspective of self-schemas because they do not require a fixed memory representation; activation among units at any given time gives rise to a pattern that possesses schema-like properties. Because we will argue that the uniqueness of the self-schema is related to its dynamic properties, a functional architecture that facilitates change over time is desirable.

information that doesn't quite fit may be distorted in order to assimilate it to that self-schema. Although these information-processing characteristics of schemas account for the maintenance and consistency of self-knowledge over time, it is important to consider why, despite its schematic representation, change and development in self-knowledge do occur across the lifespan. Some recent research comparing the cognitive performance of young and elderly adults has shown that schematic processing is age-invariant, suggesting that although older adults may perform less well in some types of memory tasks, they are not impaired in using schemas (Arbuckle, Vanderleck, Harsany, & Lapidus, 1990) and semantic categories (Light & Burke, 1988) to comprehend and remember information. It makes it reasonable from a cognitive perspective, then, to assume that schematic processing occurs throughout the life span. The issues on which we wish to focus are why self-schemas appear to be *more* amenable to change than other kinds of schemas, and how these changes might occur.

The self-schemas that have been studied most extensively are traits. Each trait schema can be considered an umbrella under which a set of behaviors can be organized. So, for example, if a person considers herself generous, there will be a particular set of acts in which she engages that will express her generosity and will lead her to evaluate herself as generous. The major difference between this kind of schema and a nonsocial schema is the greater degree of variability in the specific information that is included within a given schema, both variability across individuals and variability across groups. Relative to nonsocial information which may exist independent of social context, the meaning of social information is strongly affected by the social context. Thus, when nonsocial information is abstracted into schematic representation by different individuals, the resulting schemas are more similar than schemas abstracted in the social domain. For example, a schematic representation for a car will include information pertaining to its serving as a vehicle and having four wheels, no matter who is holding the schema (excluding children, of course), whereas a schema for generosity may vary widely depending upon both the individual and the culture in which she is embedded. Presenting an admired household object to a guest as a gift may be the essence of generosity within one culture, yet may be viewed in another, as not only irrelevant to generosity, but as grossly inappropriate behavior. Certainly schemas within the nonsocial domain also vary between individuals, and nonsocial categories are influenced by contextual factors, but the nature of this variability is more related to expertise and experience in a particular domain than to a qualitatively different organization of information (see Holyoak & Gordon, 1984, and Lingle et al., 1984, for discussions of category variability among social and nonsocial categories). Self-schemas are also different from other kinds of schemas because of their centrality in thought. Previous research suggests that the self is the central cognitive structure that serves as a referent for processing all kinds of information (Kuiper, 1981; Kuiper & Rogers, 1979). The sheer frequency of invoking self-schemas, and their integration in all aspects of social information-processing expose them to more challenges. Given the variability described above, these challenges predispose them to change.

Aside from the cognitive characteristics that distinguish self-schemas from other types of schemas, unlike more stable aspects of reality, the self itself changes. Most current theories of human development assume that development continues throughout the lifespan (e.g., Erikson, 1980; Levinson, 1978; Ruble, 1987; Stewart, 1982). Biological and social changes create new selves along the way (Markus & Wurf, 1987). If self-schemas are to reflect the self in any way they must change as well. (This is not to argue that because of their conservative properties changes in self-schemas may lag far behind changes in the nature of the actual self. Consider the 70 year old person who is surprised to see an old person peering back from the mirror.)

Much of the previous work on change and development in self-schemas has focused on the role of discrepant information in inducing change. Fiske and Taylor (1984) review three major theories of how change in social schemas is elicited. The focus in each is on the role of exposure to discrepant information. The models are: (1) the bookkeeping model, (2) the conversion model, and (3) the subtyping model. In the bookkeeping model, each experience that is discrepant with a self-schema changes that schema incrementally, such that aggregated discrepancies lead to substantial change. In the conversion model, the degree of discrepancy of a given instance is more relevant to the likelihood of change. One dramatic inconsistency can fundamentally alter the schema. In the subtyping model, discrepant information leads to the development of subcategories within the overall schema. Thus, change implies greater differentiation of existing schemas rather than abandonment of them. Although the empirical work on these models does not directly address their application to self-schemas, each could be applied to development and change in self-knowledge. For example, imagine a person whose self-schema includes the trait "organized." According to the bookkeeping model, this self-schema would change most readily as a result of a series of behaviors that revealed some disorganization such as losing a wallet, or forgetting appointments. The conversion model implies that change occurs because of an extremely discrepant act, such as forgetting to turn in a tenure file. Finally, in the subtyping model, the self-schema of organized would not be abandoned, but instead elaborated; she might think of herself as organized but forgetful in certain contexts.

There are additional ways in which self-schemas may change that are not the result of grappling with discrepant information. Attention to new or previously unnoticed information can lead to the adoption of new self-schemas. Because each person has an enormous amount of information available about the self, one function that schematic-processing serves is to reduce the amount of information to which the perceiver must attend. Self-schemas focus the perceivers attention on selected domains of their behavior. This implies that many behaviors are ignored, that the particular organization influences how the behaviors that are noticed are interpreted, and that the importance of emerging behaviors in changing self-knowledge is contingent on their being noticed and integrated into new self-schemas.

We propose that there are two primary ways in which new self-schemas are adopted: (1) reorganization of information about the self into new trait categories; and (2) increased salience of existing trait categories. Each of these can occur without changes in behavior, but simply through cognitive restructuring. These processes may also occur as concomitants of behavioral change.

A. A Model of Change in Self-Schemas

We noted earlier that the self-schemas of friends are similar. We propose that the ongoing conversation between people in important relationships (e.g., friendship, marriage) provides a mechanism by which new self-schemas may be adopted. In particular, conversations about people play a key role. Because the empirical work that we have done has been limited to friendship, the discussion here will focus on that relationship. However, the same theoretical analysis could be applied to any relationship in which conversation about people plays a central role. How can talk between friends produce change in self-schemas? Previous research has shown that dimensions central to individuals self-conceptions are used for describing others (Shrauger & Patterson, 1974) and for processing information about others (Markus & Smith, 1981). Thus, in conversations between friends, each friend presumably uses his or her own self-schema dimensions or constructs to describe others, allowing the friend to become aware of those habitually used. Subsequently each might begin to use the friend's dimensions to describe other people and ultimately to describe the self. Imagine two first year roommates getting to know each other. Both are excellent students and enjoy the challenge of their new courses. At the outset, however, one views herself as creative and the other does not. During their ongoing conversation the "creative" roommate refers to other students as creative or not, comments on creative activity, and perhaps even labels some of her new friend's actions as creative. Subsequently, the friend comes to see her own behavior in a new light. Writing a paper may now be viewed as a creative act, rather than a studious one. The sum of this kind of redefinition leads to the reorganization of self-knowledge into a new self-schema. The example described entails reorganization of behaviors from the category of studious to the category of creative. However, the adoption of a self-schema for creative could have also occurred because the conversations between friends simply increased the salience of the dimension of creative, and behaviors that had previously been ignored were now attended to. In this case, a new self-schema would be adopted without substantial reorganization of information.

Several implications follow from the notion that self-schemas change as a result of the ongoing conversations between friends about other people: (1) the self-conceptions of friends should be more similar than those of nonfriends; (2) it is specifically talk in friendship that is associated with the development of similarity, as opposed to more affective dimensions of friendship; and (3) conversations about others provide a means of exposure to new interpersonal dimensions, and the impetus for their subsequent adoption as self-schemas. Studies that we have conducted provide evidence for each of these implications, supporting this model of change. First, several studies have confirmed that the self-schemas of friends are more similar than those of nonfriends (Deutsch & Mackesy, 1985; Deutsch, Sullivan, Sage, & Basile, 1991; Goss, Kroll, & Deutsch, 1984). As we noted earlier, our findings contrast with earlier correlational studies that failed to find similarity in the self-concepts of friends. We proposed that these initial failures to uncover similarity were probably due to limitations in the methods employed. When the STG method was used, similarity between friends was revealed. As we have already noted, its advantage over previous methods of assessing self-concept is that it is nonreactive; it taps the salient dimensions used by different people to construct their own self-knowledge.

Second, a study was conducted to assess the relation between various dimensions of friendship and similarity between friends (Deutsch et al., 1991). In this study, fifty-eight pairs of first year roommates completed the STG task and a questionnaire about their perceptions of college life. Embedded in the questionnaire were six questions pertaining to their relationship with their roommate, including questions about their liking for their roommate and the amount of time they spent talking. As in the earlier study described, similarity was assessed by the number of overlapping traits friends generated in the STG task. In the critical analysis, trait overlap was regressed on the measures of talking and liking. As hypothesized by the model, the amount of time spent talking together was a more important predictor of similarity than was liking.

Consistent with these findings, our model specifies that it is specifically conversation between friends that allows for the exposure to and ultimate adoption of new self-schemas dimensions.[11]

Third, we previously showed that friends' self-schema dimensions take on special importance to each other (Goss et al., 1984). According to the proposed model these dimensions are adopted for both describing others and for describing the self. Two studies were conducted to examine this aspect of the model.

In the first study (Deutsch & Mackesy, 1985, Study 1) sixteen pairs of female friends separately read character sketches and wrote their impressions of a target person. These sketches included descriptions of behaviors characteristic of 20 preselected personality traits (e.g., creative, concerned, opinionated). Half of the subjects completed the STG task before reading the sketch and writing their impressions, the other half completed the STG task afterward. The key result was that when subjects described the target person, they used a greater proportion of their friend's traits than of a random other's traits. They used each other's self-schemas even though the friends had been separated and had never discussed this particular person. This result provides additional evidence that friends' self-schemas take on special importance to one another.

Finally, as an analogue to the model we have described in friendship, we examined the process that occurs when two unacquainted people discuss a target person (Deutsch & Mackesy, 1985, Study 2). In this study 40 female undergraduates were randomly assigned to pairs. In the first part of the study they carried out all activities independently. Subjects completed the STG task, read a depiction of a target person (TP1), and wrote their impressions. Then they individually read descriptions of a second target person (TP2), discussed their impressions of the second target person with their partner, and *independently* wrote descriptions. Finally, subjects again completed the STG task, and rated various aspects of their impressions of their partner. All of the predictions based on the proposed model of change were confirmed. Subjects used more of their partners' self-schema dimensions in describing TP2 than in describing TP1. Moreover, the number of partners' self-schema dimensions that were included in subjects' self-descriptions increased from the first to the second time subjects

[11]We are aware that these correlational results do not preclude the alternative explanation that friends who already possess more similar self-schemas spend more time in conversation than do less similar friends. A longitudinal study is planned to disentangle these competing explanations.

completed the STG task. Finally, trait overlap between partners increased after the discussion of the target person. Thus, presumably subjects used their own self-schema dimensions to discuss a target person, exposing their partner to those dimensions which the partner then used to describe the target person and herself. This process results in the emerging similarity between partners. Although we expect that these are relatively superficial changes, repeated exposure to friends' ways of conceiving others would result in more lasting changes in self-conception.

Taken together, these results support the claim that conversation provides a means for acquiring new self-schema dimensions for organizing information about the self. Although this presumably occurs within friendship, it is the exposure to new ways of organizing information about behavior that is key and any context in which it occurs can theoretically facilitate the adoption of new self-schemas. Hence, the study in which previously unacquainted strangers begin to use each others self-schemas after a relatively brief period of exposure to their respective ways of thinking illustrates that the key component in the process is a cognitive one.[12]

Likewise, we are arguing that exposure to particular trait dimensions occurs in conversation and ultimately leads to a reorganization of information about the self. Perhaps the simplest and most common way that exposure occurs is through conversation. However, one could also imagine exposure occurring to new or seldom used trait dimensions through other routes, such as listening to a lecture, a sermon, or even reading a book. In future research we hope to explore more fully the conditions under which mere exposure to trait dimensions can lead to their adoption as self-schemas.

The changes we have been describing thus far entail the adoption of new self-schemas in the absence of any behavioral change. However, we are not arguing that behavioral change is irrelevant. As we noted earlier, there is certainly ample evidence that self-knowledge undergoes significant transformation during

[12]This argument is consistent with findings from priming studies in which behavior is perceived differently depending on dimensions with which the perceiver has previously been primed. For example, subjects evaluated a target named Donald, who "shot rapids, drove in a demolition derby, and planned to learn skydiving." Their evaluations were more favorable after they had been exposed to the trait dimension, adventurous, than when they had been exposed to the trait dimension, reckless, notwithstanding that this exposure occurred in an ostensibly unrelated context. Presumably, exposure per se led to the reorganization of the behavioral information (Higgins, Rholes, & Jones, 1977).

life transitions such as entry into school, menarche, becoming a parent and retirement (e.g., Ruble, 1987). Each of these life transitions entails substantial behavioral change with concomitant change in self-conception. Our work underlines the importance of considering cognitive processes even when new behaviors are the impetus for change in self-view. Changes in self-conception occur through either challenges to old self-schemas or adoption of new self-schemas.

Let us consider a woman undergoing a major life event--the transition to motherhood. This woman finds herself experiencing a multitude of new thoughts and feelings, and engaging in a variety of new behaviors. Imagine that previously her self-schema entailed the trait, carefree. Subsequent to her child's birth, hourly she is reminded of her new responsibilities and the constraints on her life that accompany them. These challenges may ultimately lead her to revise the notion of herself as carefree. Based on the theoretical models presented earlier, the challenges may work incrementally or suddenly and dramatically to finally convince her that carefree no longer describes her, or her notion of carefree may become more differentiated, allowing her to retain the self-schema of carefree as applied to a particular set of circumstances.

New self-schemas are adopted during life transitions in order to organize the influx of new behavioral information about the self. However, it is important to note that cognitive restructuring mediates this process. It is not performing the behaviors per se that leads to a change in self-knowledge, but that change is mediated by the organization of those behaviors into new self-schemas. For example, most new mothers are careful to prevent their newborn babies from falling, yet only some come to define themselves as protective. Consistent with this argument, recent work has shown that information-seeking is related to the development of a positive self-view during the transition to motherhood (Deutsch, Ruble, Fleming, Brooks-Gunn & Stangor, 1988). Information-seeking may be another means of acquiring exposure to new constructs, and thus, a route to self-change. New mothers who have read and discussed mothering may have some readily available constructs under which to organize and understand their newly acquired behaviors. This research illustrates the interactions between cognitive and behavioral processes that we believe allow for the development of new self-conceptions across the lifespan.

V. SUMMARY AND CONCLUSIONS

In this chapter we considered evidence that suggests that individuals possess a memory schema for knowledge about the self. Beyond the demonstration of self-schemas, however, is a debate about whether knowledge for social categories such as the self is organized and represented in the same way as knowledge for object categories. A review of our own research on spontaneously generated self-schemas showed that when central aspects of the self-schema were identified, there appeared to be processing advantages in comprehension and memory tasks for the self compared to other categories, and also evidence that self-schemas undergo modification across the lifespan. Although the available empirical findings do not allow a strict determination of whether self and object categories differ in their mental structure, we believe that they do point towards a model of processing differences between self and object categories that may be related to the degree of expertise associated with self-knowledge. We described a model of self-schema change that focuses on the way the self-schema changes in response to changes in the cognitive information available in different social contexts. In future work we hope to further examine the processes by which new dimensions are incorporated into self-conception as a means to examine the interplay between the dynamics of change and the structure of mental categories.

ACKNOWLEDGEMENTS

This chapter was written while the first author was a visiting fellow at the Netherlands Institute for Advanced Study. The support and resources of that institution are gratefully acknowledged. The first author also acknowledges support of grant MH44246 from the National Institute of Mental Health. We would like to thank the many students who have collaborated on this project over the years, especially Nicoletta Basile, Rebecca Goss, Lynn Letourneau, and Adrienne Weible.

REFERENCES

Alba, J.W., & Hasher, L. (1983). Is memory schematic? Psychological Bulletin, 93, 203-231.

Arbuckle, T.Y., Vanderleck, V.F., Harsany, M., & Lapidus, S. (1990). Adult age differences in memory in relation to availability and accessibility of knowledge-based schemas. Journal of Experimental Psychology: Learning, Memory, and Cognition, 16, 305-315.

Astin, A.W. (1977). Four critical years: Effects of college on beliefs, attitudes, and knowledge. San Francisco: Jossey-Bass.

Atkinson, R.C., & Juola, J.F. (1974). Search and decision processes in recognition memory. In D.H. Krantz et al. (Eds.), Contemporary developments in mathematical psychology, Volume 1. San Francisco: W.H. Freeman.

Baddeley, A.D. (1978). The trouble with levels: A reexamination of Craik and Lockhart's framework for memory research. Psychological Review, 85, 139-152.

Bailey, R.L., Finney, P., & Helm, B. (1975). Self-concept support and friendship duration. Journal of Social Psychology, 96, 237-243.

Bartlett, F.C. (1932). Remembering: A study in experimental and social psychology. London: Cambridge University Press.

Bower, G.H. (1970). Organizational factors in memory. Cognitive Psychology, 1, 18-46.

Bower, G.H., & Gilligan, S.G. (1979). Remembering information related to one's self. Journal of Research in Personality, 13, 420-432.

Bransford, J.D., Barclay, J.R., & Franks, J.J. (1972). Sentence memory: A constructive vs. interpretive approach. Cognitive Psychology, 3, 193-209.

Cantor, N., & Mischel, W. (1979). Prototypes in person perception. In L. Berkowitz (Ed.), Advances in experimental social psychology. Volume 12. New York: Academic Press.

Chase, W.G., & Simon, H.A. (1973). The mind's eye in chess. In W.G. Chase (Ed.), Visual information processing. New York: Academic Press.

Chi, M.T.H., Feltovich, P.J. & Glaser, R. (1981). Categorization and representation of physics problems by experts and novices. Cognitive Science, 5, 121-152.

Craik, F.I.M., & Tulving, E. (1975). Depth of processing and the retention of words in episodic memory. Journal of Experimental Psychology: General, 104, 268-294.

de Groot, A.D. (1965). Thought and choice in chess. The Hague: Mouton.

Deutsch, F.M., Basile, N., Sage, C., & Sullivan, L. (in press). The effects of talking versus liking on similarity between friends. Personality and Social Psychology Bulletin.

Deutsch, F.M., Kroll, J.F., & Basile, N. (1989, June). Thinking about the self: Generation effects in self-schema research. Poster presented at the First Annual Meeting of the American Psychological Society, Alexandria, VA.

Deutsch, F.M., Kroll, J.F., Weible, A.L., Letourneau, L.A., & Goss, R.L. (1988). Spontaneous trait generation: A new method for identifying self-schemas. Journal of Personality, 56, 327-354.

Deutsch, F.M., & Mackesy, M.E. (1985). Friendship and the development of self-schemas: The effects of talking about others. Personality and Social Psychology Bulletin, 11, 399-408.

Deutsch, F.M., Ruble, D.N., Fleming, A., Brooks-Gunn, J., & Stangor, C.S. (1988). Information-seeking and maternal self-definition during the transition to motherhood. Journal of Personality and Social Psychology, 55 (1), 420-431.

Duck, S.W. (1977). Inquiry, hypothesis and the quest for validation: Personal construct systems in the development of acquaintance. In S.E. Duck (Ed.), Theory and practice in interpersonal attraction. London: Academic Press.

Duck, S.W. (1979). The personal and the interpersonal in construct theory: Social and individual aspects of relationships. In P. Stringer & D. Bannister (Eds.), Constructs of sociality and individuality. London: Academic Press.

Erikson, E.H. (1980). Identity and the life cycle. New York: W.W. Norton.

Feldman, K.A., & Newcomb, T.M. (1973). The impact of college on students. San Francisco: Jossey-Bass.

Ferguson, T.J., Rule, B.G., & Carlson, D. (1983). Memory for personally relevant information. Journal of Personality and Social Psychology, 44, 251-261.

Fiedler, F.E., Warrington, W.G., & Blaisdell, F.J. (1952). Unconscious attitudes as correlates of sociometric choice in a social group. Journal of Abnormal and Social Psychology, 47, 790-796.

Fiske, S.T., & Taylor, S.E. (1984). Social cognition. New York: Random House.

Goss, R., Kroll, J.F., & Deutsch, F.M. (1984, August). Shared selves: The influence of friendship on self-schema. Poster presented at the meeting of the American Psychological Association, Toronto, Canada.

Graf, P. (1980). Two consequences of generation: Increased inter and intra word organization of sentences. Journal of Verbal Learning and Verbal Behavior, 19, 316-327.

Graf, P., & Schacter, D.L. (1985). Implicit and explicit memory for new associations in normal and amnesic subjects. Journal of Experimental Psychology: Learning, Memory, and Cognition, 11, 501-518.

Greenwald, A.G. (1981). Self and memory. In G.H. Bower (Ed.), Psychology of learning and motivation, Volume 15. New York: Academic Press.

Greenwald, A.G., & Banaji, M.R. (1989). The self as a memory system: Powerful, but ordinary. Journal of Personality and Social Psychology, 57, 41-54.

Greenwald, A.G., & Pratkanis, A.R. (1984). The self. In R.S. Wyer & T.K. Srull (Eds.), Handbook of social cognition, Volume 3. Hillsdale, NJ: Erlbaum.

Higgins, E.T., Klein, R., & Strauman, T. (1985). Self-concept discrepancy theory: A psychological model for distinguishing among different aspects of depression and anxiety. Social Cognition, 3, 51-76.

Higgins, E.T., Rholes, W.S., & Jones, C.R. (1977). Category accessibility and impression formation. Journal of Experimental Social Psychology, 13, 141-154.

Holyoak, K.J., & Gordon, P.C. (1983). Social reference points. Journal of Personality and Social Psychology, 44, 881-887.

Holyoak, K.J., & Gordon, P.C. (1984). Information processing and social cognition. In R.S. Wyer & T.K. Srull (Eds.), Handbook of social cognition, Volume 1. Hillsdale, NJ: Erlbaum.

Holyoak, K.J., & Mah, W.A. (1981). Cognitive reference points in judgments of symbolic magnitude. Cognitive Psychology, 14, 328-352.

Keenan, J.M., & Baillet, S.D. (1980). Memory for personally and socially significant events. In R.S. Nickerson (Ed.), Attention and performance, Volume 8. Hillsdale, NJ: Erlbaum.

Kihlstrom, J.F., & Cantor, N. (1984). Mental representations of the self. In L. Berkowitz (Ed.), Advances in experimental social psychology, Volume 17. New York: Academic Press.

Kihlstrom, J.F., Cantor, N., Albright, J.S., Chew, B.R., Klein, S.B., & Niedenthal, P.M. (1988). Information processing and the study of the self. In L. Berkowitz (Ed.), Advances in Experimental Social Psychology, Vol. 21. New York: Academic Press.

Klein, S.B., & Kihlstrom, J.F. (1986). Elaboration, organization, and the self-reference effect in memory. Journal of Experimental Psychology: General, 115, 26-38.

Kroll, J.F., Deutsch, F.M., & Basile, N. (1991). Thinking about the self: Generation effects or self-schemas? Unpublished manuscript, Mount Holyoke College, South Hadley, MA.

Kuiper, N.A. (1981). Convergent evidence for the self as a prototype: The "inverted-U RT effect" for self and other judgments. Personality and Social Psychology Bulletin, 7, 438-443.

Kuiper, N.A., & Rogers, T.B. (1979). Encoding of personal information: Self-other differences. Journal of Personality and Social Psychology, 37, 499-514.

Levinson, D.J. (1978). The seasons of a man's life. New York: Knopf.

Light, L.L., & Burke, D.M. (1988). Patterns of language and memory in old age. In L.L. Light & D.M. Burke (Eds.), Language, memory, and aging. New York: Cambridge University Press.

Lingle, J.H., Altom, M.W., & Medin, D.L. (1984). Of cabbages and kings: Assessing the extendibility of natural object concept models to social things. In R.S. Wyer & T.K. Srull (Eds.), Handbook of social cognition. Volume 1. Hillsdale, NJ: Erlbaum.

Mandler, G. (1967). Organization and memory. In K.W. Spence & J.T. Spence (Eds.), The psychology of learning and motivation (Vol 1). New York: Academic Press.

Markus, H. (1977). Self-schemata and processing information about the self. Journal of Personality and Social Psychology, 35, 63-78.

Markus, H. (1983). Self-knowledge: An expanded view. Journal of Personality, 51, 543-565.

Markus, H., Crane, M., Bernstein, S., & Siladi, M. (1982). Self-schemas and gender. Journal of Personality and Social Psychology, 42, 38-50.

Markus, H., & Nurius, P. (1986). Possible selves. American Psychologist, 41, 954-969.

Markus, H., & Smith, J. (1981). The influence of self-schemata on the perception of others. In N. Cantor & J.F. Kihlstrom (Eds.), Personality, cognition, and social interaction (pp. 233-262). Hillsdale, NJ: Lawrence Erlbaum.

Markus, H., Smith, J., & Moreland, R.L. (1985). Role of the self-concept in the perception of others. Journal of Personality and Social Psychology, 49, 494-512.

Markus, H., & Wurf, E. (1987). The dynamic self-concept: A social psychological perspective. Annual Review of Psychology, 38, 299-337.

McDaniel, M.A., Lapsley, D.K., & Milstead, M. (1987). Testing the generality and automaticity of self-reference encoding with release from proactive interference. Journal of Experimental Social Psychology, 23, 269-284.

McDaniel, M.A., Waddill, P.J., & Riegler, G. (1988, November). Further support for a multifactor theory of the generation effect. Paper presented at the Twenty-ninth Annual Meeting of the Psychonomic Society, Chicago, IL.

McGuire, W.J., & McGuire, C.V. (1982). Significant others in self-space: Sex differences and developmental trends in the social self. In J. Suls (Ed.), Psychological perspectives on the self, Volume 1. Hillsdale, NJ: Erlbaum.

Medin, D.L. (1989). Concepts and conceptual structure. American Psychologist, 44, 1469-1481.

Mervis, C.B., & Rosch, E. (1981). Categorization of natural objects. In M.R. Rosenzweig & L.W. Porter (Eds.), Annual Review of Psychology, 32, 89-115.

Miller, N., Campbell, D.J., Twedt, H., & O'Connell, E.J. (1966). Similarity, contrast, and complementarity in friendship choice. Journal of Personality and Social Psychology, 32, 205-213.

Rogers, T.B. (1981). A model of the self as an aspect of the human information processing syste. In N. Cantor & J.F. Kihlstrom (Eds.), Personality, cognition and social interaction. Hillsdale, NJ: Erlbaum.

Rogers, T.B., Kuiper, N.A., & Kirker, W.S. (1977). Self-reference and the encoding of personal information. Journal of Personality and Social Psychology, 35, 677-688.

Rosch, E. (1973). On the internal structure of perceptual and semantic categories. In T.E. Moore (Ed.), Cognitive development and the acquisition of language. New York: Academic Press.

Rosch, E. (1975a). Cognitive reference points. Cognitive Psychology, 7, 532-547.

Rosch, E. (1975b). Cognitive representations of semantic categories. Journal of Experimental Psychology: General, 104, 192-233.

Rosch, E., & Mervis, C.B. (1975). Family resemblances: Studies in the internal structure of categories. Cognitive Psychology, 7, 573-605.

Ruble, D.N. (1987). The acquisition of self-knowledge: A self-socialization perspective. In Eisenberg (Ed.), Contemporary topics in developmental psychology (pp. 243-270). New York: Wiley.

Rumelhart, D.E., Smolensky, P., McClelland, J.L., & Hinton, G.E. (1988). Schemata and sequential thought processes in PDP models. In J.L. McClelland, D.E. Rumelhart, & the PDP Research Group (Eds.), Parallel distributed processing: Explorations in the microstructure of cognition, Volume 2. Cambridge, MA: MIT Press.

Shrauger, J.S., & Patterson, M.B. (1974). Self-evaluation and the selection of dimensions for evaluating others. Journal of Personality, 42, 569-585.

Slamecka, N.J., & Graf, P. (1978). The generation effect: Delineation of a phenomenon. Journal of Experimental Psychology: Human Learning and Memory, 4, 592-604.

Smith, E.E., & Medin, D.L. (1981). Categories and concepts. Cambridge, MA: Harvard University Press.

Smith, E.E., Shoben, E.J., & Rips, L.J. (1974). Comparison processes in semantic memory. Psychological Review, 81, 214-241.

Srull, T.K., & Gaelick, L. (1983). General principles and individual differences in the self as a habitual reference point: An examination of self-other judgments of similarity. Social Cognition, 2, 108-121.

Stewart, A.J. (1982). The course of individual adaptation to life changes. Journal of Personality and Social Psychology, 42, 1100-1113.

Thompson, W.R., & Nishimura, R. (1952). Some determinants of friendship. Journal of Personality, 20, 305-314.

Wickens, D. (1970). Encoding categories of words: An empirical approach to meaning. Psychological Review, 77, 1-15.

Commentary

Not Just Any Category: The Representation of the Self in Memory, J. F. Kroll & F. M. Deutsch

THOMAS B. WARD
ANGELA H. BECKER
Texas A&M University

Kroll and Deutsch provide an interesting perspective on the self as a special kind of category. They make the important observation that, rather than differing from simple object categories in structural properties, the self is special primarily in the extent to which it undergoes modification over the lifespan.

Although their point is well-taken, my suspicion is that object categories also undergo significant change. To use Kroll and Deutsch's example of the new mother (or in my case father), it is certainly true that the self undergoes change as a result of the experience, but personal observation tells me that many other categories also undergo change. Baby clothes now possess the property of "cuteness" rather than "silliness." Cars are now evaluated in terms of the ease of getting a baby in and out of a car seat. Hundreds of other categories have all undergone comparable significant changes. Thus, for any given new experience, object categories may change as much as the self. Perhaps the self undergoes greater modification throughout the lifespan because it is involved whenever change occurs whereas baby clothes, for example, are only affected by this very specific change.

In addition to the important difference identified by Kroll and Deutsch, it may be that there is also a structural difference in the sense that the self is at once more specific and more general than simple object categories. Because it refers to a single individual, the self is more specific than the category of cars, for example. Although there may be possible, ideal and future selves, these are nevertheless *interpretations* of a single continuous entity. At the same time, the self is much richer and more general in the sense that so much more information is perceived as relevant to the self than to typical object categories. The self may be thought of as a broader knowledge structure or theory which helps to organize categories. Although one can have a theory of cars, the cars themselves are entities that are organized by the theory. In contrast, the self itself may *be* the organizing theory.

PART D. Higher-Order Representation and Processing

Percepts, Concepts and Categories
B. Burns (Editor)

14

Perceptual Representations of Choice Alternatives

ALBERT F. SMITH
State University of New York at Binghamton

I. INTRODUCTION

In the beginning is a stimulus set. At least, a stimulus set is at the beginning of most experiments designed to investigate issues in the psychology of choice. These sentences, crafted to parallel the opening sentences of Garner's (1978a) paper on aspects of a stimulus, differ in one crucial way from those of Garner: Whereas he asserted that a *stimulus* is at the beginning of every experimental procedure designed to investigate human information processing, I have asserted that there is a stimulus *set*. (Garner, of course, would endorse the significance of the stimulus set; much of his work emphasized the importance of an explicit or implied stimulus set in perception of the single stimulus, e.g.,

Garner, 1962, 1974.) The distinction between the stimulus and the stimulus set is critical here because the behavior observed in experiments on choice and preference appears to depend crucially on perceived relations between stimuli--relations that arise from the comembership of stimuli in a stimulus set. This chapter will summarize evidence concerning how variation of experimental stimulus situations influences choice behavior. That choice is context-dependent is a point that has been made by others (see, e.g., Tversky & Sattath, 1979); the theme of this chapter is not only that stimulus context induces a structure on a set of alternatives, but also that the stimulus attributes that serve as the basis of choice may be quite different in different contexts.

Luce (1959), in presenting his axiomatic theory of choice behavior, noted that a consideration that, although outside the scope of his theory, was essential to its application, was the definition of the *set* of choice alternatives. Luce assumed the existence of a well-defined set of choice alternatives. However, he observed that "how these sets come to be defined for organisms, [and] how they may or may not change with experience, are questions that have received but little illumination..." (pp. 3-4) and noted that the lack of answers to such questions posed the principal limitation to the applicability of theories of choice. Luce concluded his monograph by stating that "until a theory is created that describes how organisms form categories out of the raw material of sensation..., we can hardly feel secure that the empirical identification we make of alternatives in this or any other choice theory are appropriate" (p. 131).

This chapter reviews evidence for the notion that the perceived structural composition of a stimulus-- specifically, a choice alternative--depends on the set of stimuli within which it is presented. Choices are assumed to depend on perceived properties of available and, sometimes, inferred alternatives. Since one of the objectives of theories of choice is to relate choice behavior across different subsets of a master set of alternatives, the application of such theories would, at least, be substantially more complicated if the perceived structure of the stimuli varies over subsets.

The chapter is organized as follows: First, the types of data for which theories of choice seek to account are summarized. Since an important consideration in the development of theories of choice has been the influence of similarity between alternatives on choice behavior, similarity effects are reviewed; this leads to a discussion of the nature of stimulus attributes and the types of perceived similarity relations to which they give rise. Finally, examples of the variability of stimulus properties over contexts are described: Such variability may preclude simple relationships between choices over contexts defined by different stimulus sets.

II. PROBABILISTIC THEORIES OF CHOICE

One of the goals of choice theories is to characterize systematic relations among choices from different sets of alternatives, and particularly to relate choices from different subsets of a master set of alternatives. For example, if X, Y, and Z are the members of such a master set (e.g., vegetables on a menu; modes of transportation between two cities; styles of dining-room chairs), a choice theory might relate choices from the pairwise offerings of {X,Y}, {X,Z}, and {Y,Z} to choices from the entire set, {X,Y,Z}, or might relate choices from any two pairs to the choice from the third pair. Probabilistic theories of choice suppose that individuals' preferences exhibit some random fluctuation, and use the probability of choosing one alternative over another as an index of the strength of the preference for that alternative over the other, or as a measure of the perceived relative value of the alternatives.

Luce (1959) proposed a theory that characterized choices as based solely on the values of the alternatives from among which the choices were made; the theory assumed that these values were inherent in the alternatives. Among the most important consequences of this formulation was that choices consistent with the model satisfy the constant ratio rule: The ratio of the choice probabilities for two alternatives is independent of and constant over other alternatives that might be available. For example, if the odds of selecting X over Y from a set that included just X and Y were 2 to 1, then these should be the odds as well for selecting X over Y from a set of X, Y, and Z, or from a set of X, Y, Z, and W, regardless of the choice probabilities of the other alternatives. (This is called independence from irrelevant alternatives.) The constant ratio rule implies that two alternatives that are treated equivalently in any choice context should be treated equivalently in every context. Luce (1959) noted that this formulation could be expected to hold only when the alternatives under consideration were sufficiently homogeneous so that they were evaluated consistently on some utility scale. Subsequent writers have argued that such consistency cannot generally be expected (e.g., Tversky, 1972), but have agreed that some kind of homogeneity among alternatives should result in choices that satisfy the constant ratio rule.

Two specific consequences of similarity between alternatives cast doubt on the possibility of a general formulation of choice according to some invariant value inherent in the alternatives (see Tversky & Sattath, 1979, pp. 548-549).

These effects--*interchangeability* and *commensurability*--are particular violations of the constant ratio rule that may be illustrated by considering choices from among subsets of the three alternatives X_1, X_2, and Y, in which X_1 and X_2 are similar to each other and different from Y. For the following discussion, let P(A,B) denote the probability of choosing A from the pair of A and B, and let P(A; A,B,C) denote the probability of choosing A from among A, B, and C.

Suppose that the similarity of X_1 and X_2 makes them interchangeable with respect to Y in pairwise choices, that is, suppose that $P(Y,X_1)$ is approximately equal to $P(Y,X_2)$. Due to the similarity of X_1 and X_2, it is quite possible that, in a choice from among all three alternatives, the probability of choosing Y will not be reduced by the availability of both X_1 and X_2, so that the observed choice probabilities would be such that $P(Y; X_1,X_2,Y) = P(Y,X_1) = P(Y,X_2)$. Assuming a non-zero choice probability for each alternative from the full set, this equality would imply a failure of the constant ratio rule, according to which $P(Y,X_1)/P(X_1,Y)$ should equal $P(Y; X_1,X_2,Y)/P(X_1; X_1,X_2,Y)$ and $P(Y,X_2)/P(X_2,Y)$ should equal $P(Y; X_1,X_2,Y)/P(X_2; X_1,X_2,Y)$ (see Tversky, 1972).

Commensurability is relevant to choices between X1 and X2: By virtue of their similarity, X_1 and X_2 should be easy to compare, and so one may dominate the other. In selecting a dessert, a perfect piece of fruit and a slightly flawed piece of fruit may be chosen with equal probability in pairwise choices with a piece of cake, but from the pair that includes the two pieces of fruit, the perfect piece will almost always be chosen over the flawed piece. If choices are based solely on a consistent value inherent in each alternative, then interchangeability of X_1 and X_2 with respect to Y would imply indifference between X_1 and X_2 and would preclude dominance of X_1 over X_2 (see Tversky, 1972). In this example of the selection of a dessert, the contribution of the flaw to the evaluation of the flawed piece of fruit is *contingent* on the available subset of alternatives. The flaw is treated as irrelevant relative to the cake, but determines the choice relative to the unflawed piece of fruit.

Whenever the contribution of some stimulus property to the evaluation of an alternative is contingent on the available subset of alternatives, choices based on that evaluation are likely to violate the constant ratio rule. Luce's axiom, as embodied in the constant ratio rule, asserts that choices are independent of perceived subset-dependent relations among the members of a stimulus set. For choices to be consistent with the axiom, the existence of the stimulus properties on which choices are based must precede logically the existence of a stimulus set, and the values of those stimulus properties and their combinations must also exist prior to the existence of a stimulus set. If the functional properties of stimuli depend on the particular available stimulus set, then the applicability of the constant ratio rule is seriously threatened.

III. ATTRIBUTES OF CHOICE ALTERNATIVES AND THEIR IDENTIFICATION

To assert that choices depend on perceived relations between members of a stimulus set is equivalent to asserting that choices depend on stimulus properties that vary across subsets of stimuli.

In this section, some principal results concerning the perceptual organization of stimulus sets are reviewed. These come primarily from experimental studies in which subjects judge the similarity or dissimilarity of pairs of objects. Although these studies do not concern choice explicitly, the importance of interstimulus relations (i.e., similarity and dissimilarity) to choice necessitates their consideration.

Studies of perceived stimulus structure may be classified according to whether they use stimulus sets that are constructed systematically or that are collected haphazardly. Systematically constructed stimulus sets (e.g., color chips selected so that the set will be an orthogonal combination of values of hue and chroma; rectangles that are an orthogonal combination of levels of area and shape) may be used to assess whether the subjects' judgments are consistent with an hypothesized model of stimulus representation (e.g., Gati & Tversky, 1982; Tversky & Gati, 1982). A representative experimental question is whether subjects' judgments reveal the perceived attributes of the stimulus set to be isomorphic to the manipulated attributes (e.g., Burns, Shepp, McDonough, & Wiener-Ehrlich, 1978). The principal concern of studies in which stimulus sets are constructed haphazardly (e.g., by sampling members of a semantic category such as fruits or animals) is more exploratory--to identify the stimulus attributes that contribute to judgments of interstimulus relations. (Tversky & Hutchinson, 1986, presented an extensive catalog of studies that have employed each of these types of stimulus set.) These types of stimulus set almost certainly differ in the number of attributes on which their member stimuli vary: In systematically constructed stimulus sets, the stimuli are constrained to vary in the limited number of ways manipulated during the construction of the set, whereas in sets sampled randomly (or even with constraints) from a larger stimulus population, n stimuli may vary in as many as ${}_nC_2$ -1 ways--each stimulus may differ in a unique way from every other stimulus. Both the functional nature of the attributes and the structure of the set, as recovered through scaling procedures, may differ in the two kinds of stimulus set.

The remainder of this section will review a distinction between two types of stimulus attribute--dimensions and features--that has proven heuristically useful in research on human information processing, and will discuss two classes of models of perceived interstimulus relations--geometric models and set-theoretic models.

A. Stimulus Attributes

Garner (1978a) distinguished between *dimensions* and *features* as component attributes of perceptual stimuli and, therefore, of choice alternatives. These are two distinct types of variation among members of a stimulus set. Garner (1978b) showed that the distinction was meaningfully reflected in the performance of subjects in information-processing tasks. Other investigators have shown differences in the similarity relations among members of stimulus sets generated from these types of attribute (e.g., Gati & Tversky, 1982; Tversky & Krantz, 1969).

In stimulus sets defined according to dimensions, each member of the stimulus set is characterized by some level of each dimension (Garner, 1978a). The levels of any dimension may be unordered (nominal) or ordered. For example, the primary filling of a sandwich (e.g., tuna salad, turkey, Swiss cheese) is a level of an unordered dimension; every sandwich has such a filling. The size of a shirt is a level of an ordered dimension. A set of sandwiches might be defined on dimensions of type of bread and primary filling; a set of shirts might be defined on dimensions of size and color. Dimensions are properties of a *stimulus set* and the levels of the dimensions are properties of *stimuli.*

A feature is a binary attribute, the two levels of which are "feature-absent" and the feature itself (Garner, 1978a). This gives the term "feature" a dual meaning: As a two-valued attribute, it is a property of the stimulus set, and as the positive value of that attribute, it is a property of stimuli. Examples of features are eyeglasses on a face, garlic on a pizza, or pickle on a sandwich. In Garner's sense, the definition of feature is restrictive in that only one type of glasses might occur, and garlic and pickle would occur in only one quantity.

For Garner (1978a), the crux of the distinction between dimensions and features is what a *single stimulus* implies about the stimulus set of which it is a

Table 14.1. Featurally and Dimensionally Defined Stimulus Sets with Common Members.

(a) Set defined on two features

		Mushrooms: Absent	Mushrooms: Present
Garlic	Absent	Plain Pizza	**Mushroom Pizza**
	Present	**Garlic Pizza**	Mushroom and garlic pizza

(b) Set defined on one dimension

Onion Pizza
Mushroom Pizza
Garlic Pizza
Spinach Pizza

Note. The pizzas that occur in both sets appear in **bold** type.

member. Garner suggested that the full set of dimensions (of a stimulus set defined on dimensions) is knowable from any member, but that the full set of features (of a stimulus set defined on features) is knowable only from the complete stimulus--the stimulus composed of the "present" level of every feature. It is unlikely that such an absolute distinction can be made, because only from a set of stimuli, and not from a single stimulus, can attributes of stimulus variation, whether they be dimensions or features, be knowable. Dimensions and features, as attributes of stimulus sets, describe variation over stimulus sets and are deduced from stimulus sets. A single stimulus does not vary: Alternatives may be inferred from a single stimulus, but the inference of the alternatives is inductive.

Consider, for example, the two sets of pizzas shown in Table 14.1. The set shown in the upper panel was constructed by crossing the two levels--present and absent--of two features--garlic and mushrooms--and adding them to a common stem of crust, tomato sauce, and cheese to yield plain pizza, mushroom pizza, garlic pizza, and garlic-and-mushroom pizza. The members of the set shown in the lower panel may be conceptualized as representing four levels--onion, garlic, mushroom, and spinach--of an unordered dimension. No single garlic pizza or mushroom pizza reveals the attribute structure of the set from which it came; the garlic and mushroom pizzas in the two sets are the same. The stimulus attributes of the pizzas common to the two sets are the same, but the attributes of the stimulus sets differ. Some exposure to or experience with a stimulus set, rather than with just a single stimulus, is required for the mental representation of a stimulus set to reflect its true attribute structure.

The similarity relations within stimulus sets defined on features are distinctly different from those within sets defined on dimensions. Given stimuli composed of levels of separable dimensions, the perceived dissimilarities between pairs of stimuli with equivalent differences are often equal, whereas this is typically not true for stimuli generated from features. Consider first stimuli defined on a pair of dimensions, with a and b as levels of one dimension and h and k as levels of the other, so that a single stimulus is denoted by a pair of letters indicating its levels on the two dimensions and the four possible stimuli are ah, bh, ak, and bk. Physically, the difference between stimuli ah and bh is equivalent to that between ak and bk--within each pair, the level of the second dimension is constant, and the difference between the stimuli on the first dimension is between a and b. Tversky and Krantz (1969) found, using a set of schematic faces generated from dimensions, that subjects judged the difference between ah and bh to be equal to that between ak and bk.

Consider next a stimulus set generated by adding features to a stem that is common to all members of the set; for example, the stem might be a circle within which facial features are placed. If S denotes the stem, and h and k denote the positive values of two features, the four stimuli that can be generated are S, Sh, Sk, and Shk. Physically, the difference between stimuli Sk and S is the same as that between Shk and Sh--the stimuli in each pair differ only in whether k is present. However, Gati and Tversky (1982) found, using schematic faces constructed by adding features to a stem, that subjects judge the dissimilarity between Shk and Sh to be significantly less than that between Sk and S. For stimuli generated from features, the contribution of each feature added to a stem decreases as the salience, importance, or perceptual weight of the stem increases: In the pair Shk and Sh, h is functionally part of the stem, so the role of k in Shk is in general less than its role in Sk. As a consequence, the dissimilarity of Shk and Sh is judged to be less than that of Sk and S.

Tversky and Gati (1982; see also Gati & Tversky, 1982) proposed a comprehensive set-theoretic approach to stimulus attributes in which each level of a stimulus attribute is conceived as consisting of a set of elements. They distinguished between two classes of attributes, or features. Nested features are ordered by amount and characterized by an inclusion relation, so that each level of any attribute consists of the elements that constitute the adjacent lower level, and more: Because the elements that constitute the lower level are common to the two levels, adjacent levels are distinguished by the elements that are contained in the higher level and lacked by the lower level. Chained features are characterized by an overlapping relation: Although adjacent levels of an attribute share some elements, each level also has unique, or distinctive, elements.

Whether either the dimension-feature distinction or the nested-features-chained-features distinction is significant for choice behavior is an open problem: That perceived similarity relations between stimuli depend on the attribute structures of stimulus sets indicates the possible importance of these distinctions for choice. The different patterns of perceived similarities for stimuli from sets of objects defined on dimensions and on features might result in different preference relations for alternatives generated from these two types of attribute. For example, the perceived value of a feature may depend on the nature or value of a stem, whereas the perceived value of a level of a dimension may be invariant over the members of a set of alternatives.

The importance of these distinctions likely depends, at least in part, on the extent to which a presented set of items implies alternatives, and on the knowledge or experience of the individual making a choice. Consider once again the four pizzas in the lower panel of Table 14.1: An individual with some knowledge of pizzas and how they are made might, if asked to make a selection from this set, treat the four levels on the dimension on which these pizzas vary as features, and might infer a choice set of 16 pizzas that ranged from plain to a deluxe pizza with all four toppings.

The attributes of a stimulus set are a conceptual abstraction that is deduced from the set. The nature of the variation among a set of stimuli is a conceptual consequence, rather than a physically-real structural foundation, of the stimuli themselves: The stimulus attributes are primary, and those of the stimulus set are secondary.

B. Structural Models of Perceived Interstimulus Relations

Two principal approaches--geometric and set-theoretic--have been used to model perceived proximity relations between stimuli (i.e., perceived similarities and dissimilarities). Geometric models represent the members of a stimulus set as points in a multidimensional space so as to maximize the correspondence between interpoint distances and judged interstimulus dissimilarities. The dimensions of the space may correspond to perceived dimensions of the stimulus set. The set-theoretic approach treats each stimulus as a collection of features and models perceived interstimulus relations as functions of common and distinctive features (see, especially, Tversky, 1977). The appropriateness of any model depends on the degree to which particular properties of the representation correspond to properties of the modeled data. The fundamental ideas of each approach are summarized in this section, and their relevance to choice is described later.

Geometric models. Distances between points in a multidimensional Euclidean space satisfy minimality, symmetry, and the triangle inequality: The distance between each point and itself is zero; interpoint distances are indifferent to which point serves as origin of the measurement; and the direct distance between two points is less than or equal to the sum of the distances between each of them and any other point (see Kruskal & Wish, 1978; Shepard, 1972; Tversky & Gati, 1982). In addition, in a low dimensional space, the number of points that can share a nearest neighbor is quite small (see Tversky & Hutchinson, 1986). Because certain of these characteristics either correspond to or might be expected of interstimulus proximity judgments, the geometric model has been used to model interstimulus relationships.

Set-theoretic models. For many stimulus sets, perceived interstimulus relations may be modeled more effectively as functions of the common and distinctive features of the stimuli than as interpoint distances in a multidimensional space (Gati & Tversky, 1982; Sattath & Tversky, 1977; Tversky, 1977; Tversky & Gati, 1982). The dissimilarity between two stimuli, X and Y, is represented in the contrast model (Tversky, 1977) as

$$d(X,Y) = (a * f(x\text{-}y)) + (b * f(y\text{-}x)) - (t * f(x \text{ INT } y)),$$

where d(X,Y) is the dissimilarity of X and Y, x-y are the distinctive features of X (i.e., the features of X that are not features of Y), y-x are the distinctive features of Y, x INT y are the features common to X and Y, f is a measure of feature importance or salience, and a, b, and t are weights on the measured subsets of features.

The contrast model accounts for those aspects of proximity judgments captured by geometric models, and accounts, in addition, for certain observed properties of judgments of interstimulus relationships that are not accommodated straightforwardly by geometric models (although see Krumhansl, 1978): First,

a) **Dissimilarities**

	A	B	C
A	---	3	4
B	3	---	5
C	4	5	---

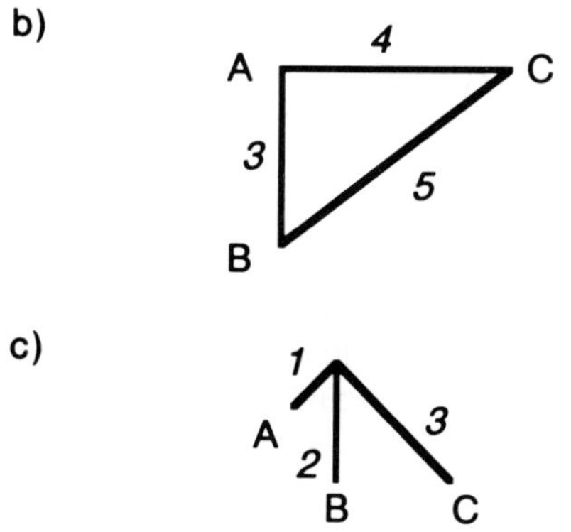

Figure 14.1. (a) A symmetric matrix of dissimilarities among three stimuli, A, B, and C. (b) A representation of the dissimilarities as distances in a two-dimensional space. (c) A representation of the dissimilarities as distances in a tree; the distance between two stimuli is the sum of intervening branch lengths.

observed self-dissimilarities are not equal for all stimuli. The contrast model accommodates this finding by letting f(x INT x) differ from f(y INT y): The self-similarity of a complex stimulus may differ from that of a simple one. Second, judged dissimilarities between two items are not necessarily symmetric; this is modeled by letting a and b, the weights applied to the two sets of distinctive features, differ. Finally, similarity judgments and dissimilarity judgments for the same object set are not always related perfectly. Dissimilarity judgments may be based on an assessment of distinctive features and similarity judgments on an analysis of common features. To the extent that the common and distinctive features of a stimulus pair are isolable, the type of interstimulus judgment that is required of subjects may lead them to apply a set of weights (in the contrast model) appropriate to that type of judgment (see Payne, 1982; Sattath & Tversky,

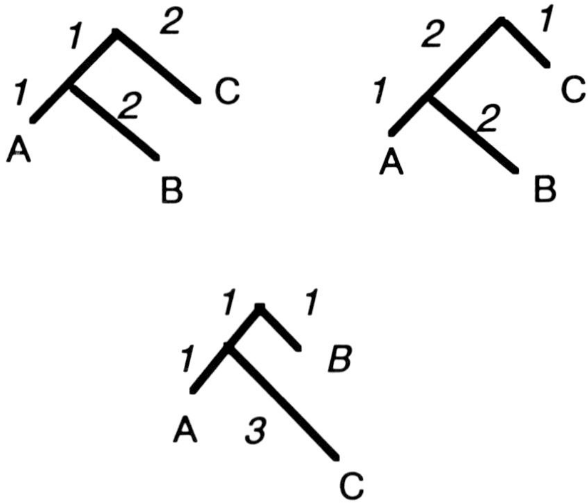

Figure 14.2. Three variants of the tree shown in panel c of Figure 14.1.

1987; Tversky, 1977). Because choice is the outcome of disaggregation process, the goal of which is to exclude some alternatives in favor of others, it is plausible that the features relevant to choice would be the distinctive features rather than the common ones.

Representations of perceived interstimulus relations. Geometric models of stimulus structure are displayed in geometric spaces of low dimensionality, whereas set-theoretic models of interstimulus relations are displayed as trees. In such tree configurations, stimuli are represented as terminals of branches, the lengths of which may be adjusted so that the lengths of paths between terminals correspond maximally to perceived dissimilarities between the items (see Carroll, 1976; Corter & Tversky, 1985; Sattath & Tversky, 1978). In such representations, the branches may be interpreted as features or as feature groups and the length of any branch may be interpreted as the importance of the feature or features represented by that branch.

Figure 14.1 shows a matrix of dissimilarities among three stimuli, A, B, and C, and depicts these dissimilarities both as distances in a two-dimensional space and in a tree. These representations are not unique. Any pair of orthogonal axes could be laid across the representation of the distances in two-dimensional space (or, analogously, the triangle defined by the three points could be rotated to any position in two-dimensional space). The tree configuration could be "bent" at any point to define nodes or to define common features (see Tversky & Hutchinson, 1986; Tversky & Sattath, 1979). Figure 14.2 illustrates some variants of the tree displayed in Figure 14.1.

Several comparative studies indicate that psychological dissimilarities are modeled best by a representation that corresponds to the structure from which those distances originated. Pruzansky, Carroll, and Tversky (1982) sampled distances from planes and from trees and showed that interpoint distances from planes were modeled better by a multidimensional scaling procedure than by a tree-fitting algorithm, but that distances between points in a tree were represented better in trees than in multidimensional spaces. Pruzansky et al. suggested two statistical properties of interstimulus proximity data as aids to decisions about the appropriateness of models of interstimulus relations. Tversky and Hutchinson (1986) showed that measures of centrality and reciprocity, based on nearest neighbor relations implied by proximity data and represented in models, could help diagnose the appropriateness of a structural representation of a stimulus set. The measure of centrality is an index of the distribution of the nearest neighbor relationship over the members of a stimulus set: The centrality of a set of items is low when nearest neighbor status (e.g., being judged most similar to another stimulus) is distributed over members of the set, and is high when nearest neighbor status is concentrated in one item (e.g., one item is judged most similar to each of many other items). Reciprocity indexes the symmetry of the nearest-neighbor relation: A set's reciprocity is highest when the nearest neighbor relation is symmetric, so that each stimulus is the nearest neighbor of its nearest neighbor, and low when a single stimulus is the nearest neighbor of all of the others. Tversky and Hutchinson (1986) emphasized that because it is geometrically impossible for a set of points in a low-dimensional space to have high centrality and low reciprocity, stimulus sets for which proximity data are characterized by such values cannot be represented appropriately in low-dimensional spaces.

Pruzansky et al. (1982) and Tversky and Hutchinson (1986) used both multidimensional scaling and tree-fitting procedures to analyze many sets of interstimulus relations data, and arrived at the consistent conclusion that judged interstimulus relations between members of factorially generated sets are fit better by geometric models than by tree models, whereas relations between randomly selected conceptual stimuli tend to be represented better as trees. Tversky and Hutchinson observed that conceptual stimuli may generally have more attributes than do perceptual stimuli, and that high-dimensional configurations may be better approximated by a tree than by a low-dimensional geometric solution. Stimuli that are generated by crossing levels on a small number of factors vary in just that small number of ways, whereas naturally selected or conceptual stimuli may vary

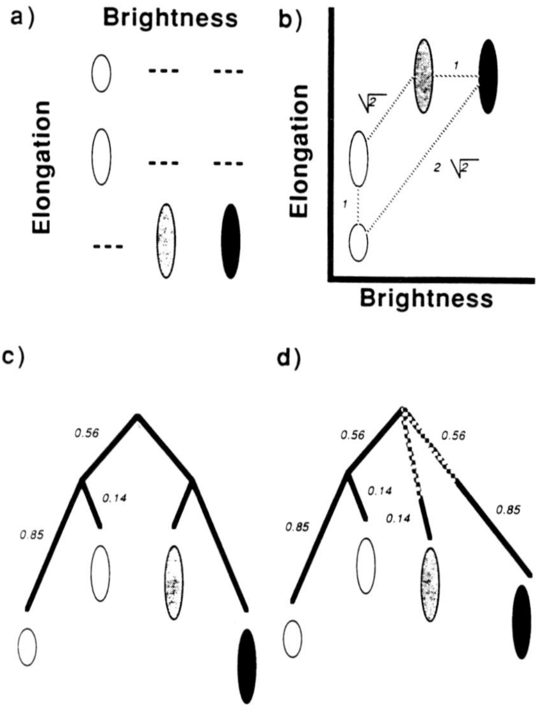

Figure 14.3. (a) Four ellipses constructed by combining levels of two ordered dimensions--brightness and elongation. (b) A two-dimensional spatial representation of perceived dissimilarities (indicated by the numbers) among the four ellipses; it is assumed that the judged differences betweeen adjacent levels of each dimension is 1 and that steps are additive. (c) A tree representation of the perceived dissimilarities among the ellipses in which the major branches represent levels on different attributes. (d) A tree representation of the dissimilarities in which each major branch represents a level of brightness; the lengths of the checkerboarded segments are disregarded in measuring the distance between the terminal nodes at the ends of branches that include the broken segments.

in many ways. Only two stimuli are required to define a feature, whereas many more would seem to be required to support a set of more than two or three dimensions.

Consider, for example, the set of four geometric forms shown in Figure 14.3a. These vary on two ordered dimensions, and it would be reasonable to infer five additional items to complete this set (i.e., the hyphens could be replaced with realizable objects). The perceived relations between these four items could be represented in a two-dimensional space (as in Figure 14.3b) or as a tree in which the major branches represent white forms and elongated ellipses (as in Figure 14.3c). Note that, in such a tree, the major branches represent levels of two different stimulus dimensions. The tree in Figure 14.3c should be contrasted with that in Figure 14.3d, which contains branches that represent levels of brightness, and the marked segments are not counted when measuring the distance in the representation between the elongated items (see Corter & Tversky, 1986, who proposed this representation for items that share features but that are located on different branches of a tree).

IV. MODELS OF PREFERENCE

If preferences depend on perceived similarities, preference judgments might exhibit relationships that are analogous to those expected of similarity judgments. Corresponding to minimality, symmetry, and conformity to the triangle inequality that might characterize dissimilarity judgments for a set of objects are analogous potential properties of pairwise preference relations: An individual should be indifferent between identical copies of some choice alternative; the preference for A over B should be in some sense complementary to the preference for B over A; and transitivity might be expected for preferences over a set of alternatives (see Coombs, 1983; Luce, 1959; Tversky, 1969, 1972). These expectations are captured algebraically by the logarithm of the ratio of complementary pairwise preferences (e.g., $\ln P[A,B]/P[B,A]$); $\ln P(A,B)/P(B,A)$ is denoted here as $\ln R(A,B)$. If $P(A,B) = P(B,A) = 0.5$, then $\ln R(A,B) = \ln R(B,A) = 0$ (i.e., indifference between A and B); in general, $\ln R(A,B) = -\ln R(B,A)$ (complementarity); and if both $\ln R(A,B)$ and $\ln R(B,C)$ are greater than zero, then $\ln R(A,C)$ is greater than zero (transitivity). Although these seemingly reasonable relationships--particularly transitivity--do not characterize expressed choices for alternatives in some sets, it is of interest to summarize how these properties, when they hold, may be illustrated.

Geometric models. Figure 14.4a shows a simple representation of the mean judged values of three items labeled A, B, and C. On average, choices from among these linearly ordered items should be transitive--on average, A should be preferred to B, B preferred to C, and A preferred to C. If judged value is correlated perfectly--at least ordinally--with a single stimulus property (or with a set of stimulus properties that are perfectly correlated among themselves), then the scale shown in Figure 14.4a could be relabeled as that property.

Coombs (1950; see also Coombs, 1983) proposed that individuals and stimuli could be modeled jointly on a scale that represented such a stimulus

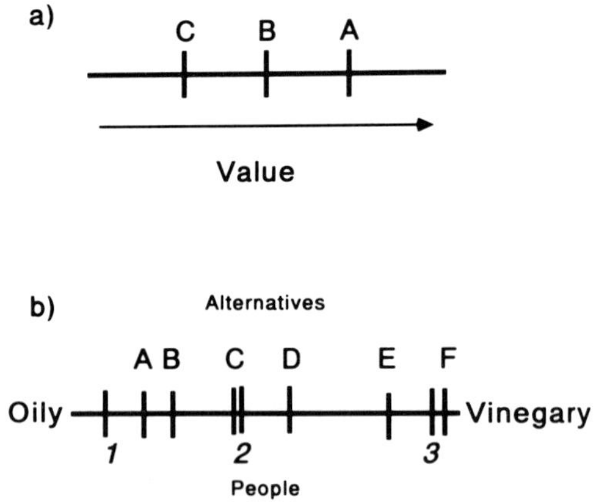

1 's preference ranking: A,B,C,D,E,F
2 's preference ranking: C,D,B,E,A,F
3 's preference ranking: F,E,D,C,B,A

Figure 14.4. (a) Three objects on a scale of perceived value. (b) Locations of six alternative salad dressings (A-F) and three people (1-3) on a scale that ranges from oily to vinegary; the preference ranking of each individual corresponds to the distances between the scale location of that individual and the scale locations of the alternatives.

attribute (i.e., the psychological correlate of a physical property or an intercorrelated group of properties). Stimuli would be located on the scale according to their levels of that attribute, individuals would be located at the levels of the attribute that they most valued, and the value of a stimulus to an individual would be a decreasing function of the distance between the scale positions of the stimulus and the individual. Such a representation captures the preferences of many individuals for sets of alternatives for which there is consensus among people in their perceptions of the properties of the set, but disagreement in the evaluation of those properties.

Consider, for example, a set of salad dressings made of complementary parts of oil and vinegar. People might agree in their perceptions of the set of dressings as varying on a single scale from oily to vinegary, but vary in their preferences for the dressings. One person might value most pure oil, value least pure vinegar, and, in general, value the dressings decreasingly as the proportion of oil decreases. Another might favor almost pure vinegar over all the others, and yet another may prefer an equal mix to all the others. The preference evaluations

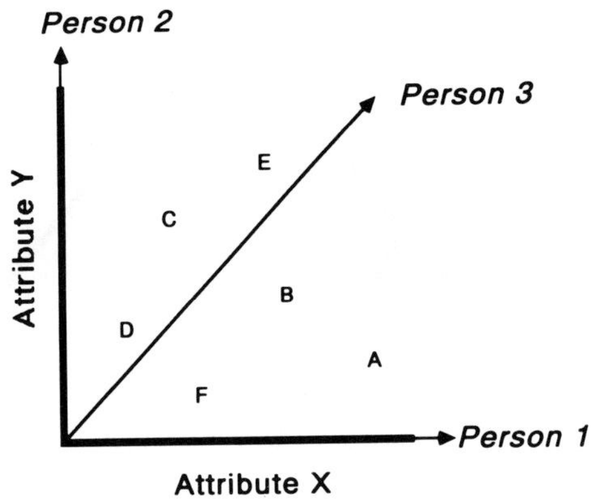

1 's preference ranking: A,B,E,F,C,D
2 ' s preference ranking: E,C,B,D,A,F
3 's preference ranking: E,A,B,C,F,D

Figure 14.5. Six alternatives (A-F) that vary in two dimensions. Value functions for three people are shown; their preference rankings correspond to the projections onto the value functions of the points that represent the alternatives.

of all of these individuals may be captured in a joint representation of people and stimuli (see Figure 14.4b).

This scheme has been generalized to characterize relative evaluative judgments of stimuli whose dissimilarities are represented in more than one dimension. In Carroll's (1980) vector model, points that represent stimuli are arranged in a multidimensional space, and the value function of each subject is represented as a vector through the space; the projection of each point on a subject's vector represents the value of the corresponding stimulus to the subject (see Carroll, 1972; 1980, pp. 244-245). Each individual's value function crosses a set of parallel indifference curves that are determined by his or her personal exchange rate for the attributes. Thus, for a set of interstimulus relations based on judgments of their physical properties (i.e., independent of or prior to expression of preferences), many different evaluative orderings of the stimuli are possible. The value functions of different individuals can be interpreted as differing in the relative contribution of each dimension to perceived value. Figure 14.5 shows a set of points that represents a set of alternatives in a two-dimensional space and vectors that represent the value functions of three individuals. For Person 1, the value of an alternative is associated exclusively with

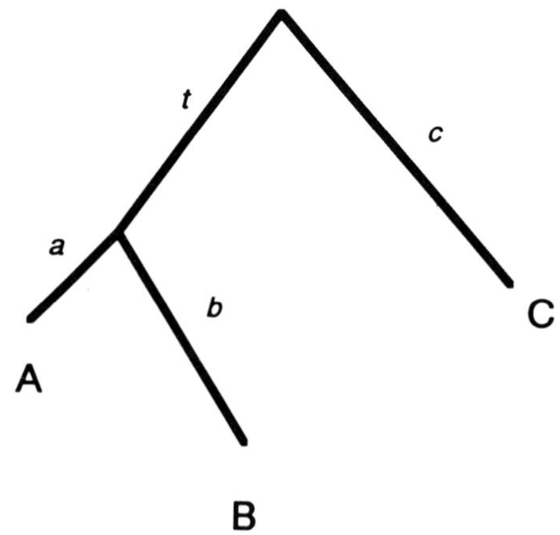

Figure 14.6. The feature structure of a set of three stimuli, A, B, and C. Each branch represents a feature or a group of features. Features a, b, and c are unique to A, B and C, respectively; feature t is common to A and B.

its level on Attribute X, for Person 2, value derives solely from the levels of Attribute Y, and for Person 3, each attribute makes a non-negative contribution to value.

Geometric models of preference represent stimulus alternatives as points in a space the dimensions of which correspond to properties of the stimulus set. Individuals are represented either as vectors or as points in the space, and the value of an alternative to an individual is determined either by the distance of stimulus points to that representing the individual or by the projection of points on the individuals' value vector. In any case, a value scale is determined, and the predictions of choice models can be assessed. Whether the available set of choice alternatives influences the value function, and if so, how, are open questions: The distribution over available alternatives of levels of the attributes of the stimulus set may influence how people weight those attributes in evaluating alternatives.

Set-theoretic models. Tversky and Sattath (1979) proposed a model of choice behavior that treated choice alternatives as sets of features. A set of choice alternatives and their features are modeled as a rooted tree in which the terminal nodes represent alternatives and the branches represent features. Figure 14.6 shows such a tree: The length of each branch represents the value of the represented feature; the value of an alternative is represented by the sum of the lengths of the branches between the root and the terminal node for that alternative. Properties that are common to a subset of stimuli are represented by superordinate branches. In Figure 14.6, features a, b, and c are unique to A, B, and C, respectively; feature t is common to A and B and so is shown as a branch superordinate to the branches that represent the unique features of A and B. Such a tree can be constructed from any set of binary choice that satisfies certain restrictions.

Tversky and Sattath showed that the preference tree model accommodates the interchangeability and commensurability effects of similarity on choice, and that several novel predictions of the model are satisfied empirically. They suggested a choice rule that might be called *independence from irrelevant properties*: This rule asserts that the choice between two alternatives X and Y is based solely on features that distinguish the alternatives (i.e., the two sets of distinctive features), with any features common to the alternatives being disregarded. In particular, according to the preference tree model,

$$P(X,Y) = v(x\text{-}y)/[v(x\text{-}y) + v(y\text{-}x)], \text{ and}$$
$$P(Y,X) = v(y\text{-}x)/[v(x\text{-}y) + v(y\text{-}x)], \text{ so}$$
$$R(X,Y) = P(X,Y)/P(Y,X) = v(x\text{-}y)/v(y\text{-}x),$$

where, for example, v(x-y) is the value of the distinctive features of X. These equations make explicit the earlier suggestion that the stimulus features relevant to choice are distinctive features rather than common ones. The ratio of choice probabilities is the ratio of values of distinctive qualities of compared alternatives. The principle of independence from irrelevant properties accounts for the commensurability effect of similarity by treating as irrelevant to the choice between the similar alternatives the values of features common to those alternatives that would moderate choices relative to a dissimilar alternative.

V. THE INSTABILITY OF STIMULUS PROPERTIES

The literature of psychology is replete with examples of the contingency of perceived stimulus set structure on presentation conditions. Although not all of these examples are explicitly relevant to choice, demonstrations that the perceived attributes of a stimulus--or of a stimulus set--depend systematically on presentation context raise concerns about choice models' assumptions that attributes are fixed. Such variability may limit the applicability of choice models that assume cross-contextual stability of stimulus attributes.

These examples may be categorized roughly into two classes. The first consists of demonstrations that the relative weight accorded to particular stimulus attributes depends systematically on stimulus context. The second is a specialization of the first--not only are the weights applied to stimulus attributes in making judgments about the stimuli changed, but new attributes "emerge" as the functional basis of responding to the stimuli. (The case of emergent attributes could actually be considered reweighting with the weights of one or more attributes being zero in some contexts.)

A. Contextual Contingency of the Salience of Stimulus Attributes

Many investigators have shown that stimulus classification depends on the particular context within which a stimulus is classified. These examples confirm that cognitive processing of alternatives depends critically on the perception of features and dimensions as aspects of stimulus *sets*: Although the properties of a stimulus may be constant over contexts, it is by virtue of the membership of that stimulus in a set that attributes of stimulus sets are induced, and it is these attributes that determine how the individual stimulus is perceived, classified, or evaluated.

Tversky (1977), who called a specific version of this idea the diagnosticity principle, demonstrated the dependence of the featural (or dimensional) structure of a stimulus set on its members as follows: Matched quadruples of stimuli--either country names or schematic faces--were constructed. Each pair of quadruples contained a common triple of items, Z, X, and Y, and a fourth item--either X' or Y'; the matched quadruples may be denoted {Z,X,X',Y} and {Z,X,Y',Y}. Stimulus items were chosen so that the judged similarity of X to Z was roughly equal to that of Y to Z, and X' and Y' were selected so that X' was more similar to X than to Y and Y' was more similar to Y than to X. A quadruple was presented to subjects who were asked either to classify its members into two groups of two each or to choose which of X, Y, and either X' or Y' was most similar to Z. In the classification task, the modal partitions confirmed the similarity relations according to which the sets were constructed: Most subjects divided the {Z,X,X',Y} quadruple into {Z,Y} and {X,X'} and the {Z,X,Y',Y} quadruple into {Z,X} and {Y',Y}. In the similarity task, when asked to choose which of X, X', and Y was most similar to Z, more subjects selected Y than X; however, when the alternatives were X, Y', and Y, more subjects selected X than Y. Thus, the substitution of Y' for X' changed the stimulus with which Z was classified and changed the perceived similarity relations among the stimuli. Tversky argued that the substitution of Y' for X' changes the structural organization of the original stimulus set. McGuire and McGuire (1981) termed a similar idea the distinctiveness postulate. They showed that the first adjective reported by individuals asked to describe themselves depended on the context in which they were asked to carry out this task, or on the context in which they were asked to imagine themselves. For example, a tall, male, contact-lens-wearing basketball player might describe himself as tall relative to his classmates, as a contact-lens wearer relative to his teammates, as a basketball player relative to other athletes, and as male relative to his family.

Lockhead (1989), too, found that the perceived organization of entire sets of stimuli depends on the stimulus context. Lockhead asked subjects to define category boundaries for stimuli that varied in one dimension. He found that increasing the variability of the presented stimuli on that dimension led subjects to specify wider categories with more variable boundaries. Lockhead concluded from a series of studies that whether particular items will be grouped together depends not just on the attribute values or features of the stimuli, but on the relations of stimuli with each other.

L. B. Smith (1988) described experiments that suggest that the similarity relations perceived among a set of objects depend on the particular objects that are presented. In particular, her data indicate that the criteria that subjects use to classify sets of stimuli depend on the composition of the presented set of stimuli. Subjects were asked to partition three different four-membered sets of items that varied on two attributes (color and size); checks were run to ascertain that subjects did not perceive division of the sets into two groups of two as a task requirement. In one stimulus set that consisted of four different stimuli, the opportunity to classify by shared level on one dimension--stimuli that shared a level on that dimension differed substantially on the other--was pitted against the opportunity to classify by overall similarity--similarity on each dimension without identity on either. Each of the two other stimulus sets consisted of two replications of each of two objects: In one of these, the two objects were relatively similar overall, whereas in the other, the two objects were quite different. Smith found that older children classified the set consisting of four different objects by overall similarity, but classified each set consisting of two of each of two similar objects into two groups of two identical objects. In other words, stimuli that were grouped together in one set were classified into separate groups in the other.

B. Emergent Attributes

In some situations, the type of attribute that serves as the basis of classification or judgment appears to depend on the stimulus context.

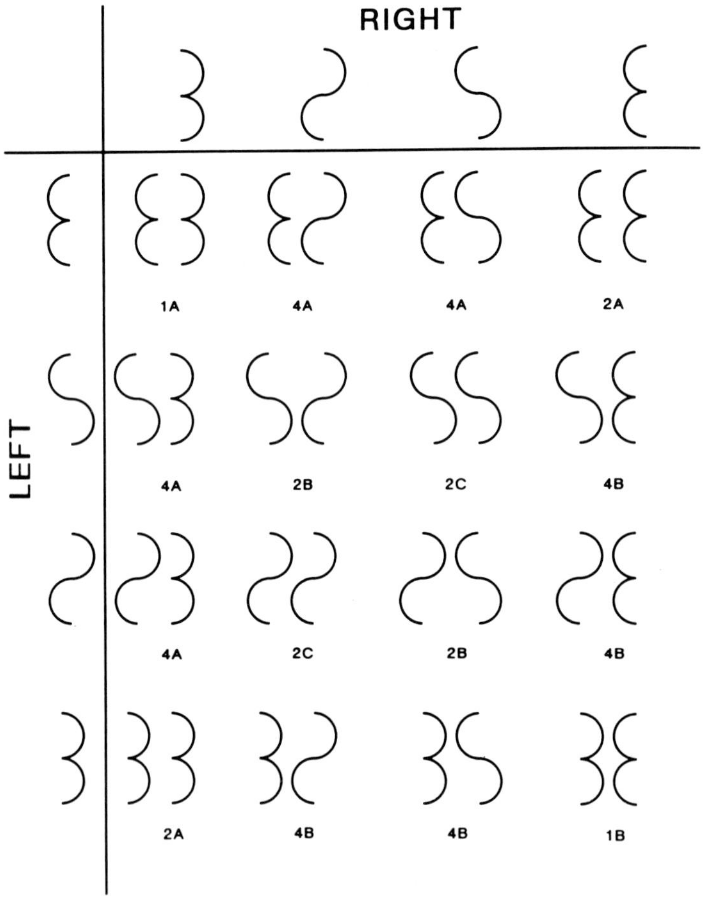

Figure 14.7. A set of 16 stimulus patterns generated by combining factorially four levels on each of two dimensions--the left and right positions of the pattern. The set may also be organized into groups of configurally identical patterns: Patterns with the same label (e.g., 4A) are configurally identical by rotation and/or reflection.

Johnson (1986; see also Johnson & Fornell, 1987) argued that as objects are considered in different contexts, attributes are redefined. Johnson's particular interest was in determining what attributes serve as the basis for choices from sets of objects that share no obvious properties. For example, what stimulus properties are relevant to a choice between a washing machine and a bicycle? Whereas the basis on which two bicycles might be compared is relatively obvious, the basis of comparison of what Johnson called "noncomparable" objects is not.

Johnson showed that as the number of physical attributes shared by to-be-compared alternatives becomes smaller, subjects reconceptualize the alternatives in terms of attributes that are more abstract: The alternatives are redefined from collections of specifiable features into values on abstract dimensions.

A. F. Smith (1984) showed that the basis of classification of visual patterns that were characterized by both dimensional and configural properties depended on the subset presented to subjects. The set of patterns that was used is shown in Figure 14.7. Subjects who were asked to partition the entire set of 16 patterns into groups tended to classify on the basis of component properties, but subjects who were given triples of patterns that were selected to pit component-based classifications against configural classifications tended to classify by configurations. The subset of items presented--for classification or for choice--determine the type of attribute on which responses are based.

VI. THE ROLE OF CONSTANT ATTRIBUTES IN CHOICE

Most of the examples that were described in the preceding section were studies of stimulus categorization. This section is concerned explicitly with choice. It reviews some predictions of the preference tree model, designed to accommodate similarity effects on choice, and summarizes several studies of choice that have yielded results that are discrepant with the predictions of the preference tree model.

A. The Preference Tree Model and Constrained Choice

Tversky and Sattath (1979) proposed that the set of alternatives from which a choice is made is functionally several subsets of alternatives, each of which includes one or more items. Subsets may be organized by similarity relations or by external constraints on the choice process that do not correspond to perceived similarity relations. Consider, for example, three alternatives, X_1, X_2, and Y, with X_1 and X_2 similar to each other and dissimilar to Y. If the decision-maker perceives his or her choice as between the X's and Y, with selection of a particular X if the first decision is to choose an X, the choice set may be represented as $\{X_1,X_2\}\{Y\}$. Alternatively, if circumstances force a decision between one of the X's and a pair consisting of the other of the X's and Y, with subsequent selection from the pair if the pair is first chosen, the choice set could be represented as $\{X_i,Y\}\{X_j\}$, with i,j = 1,2, and i not equal to j.

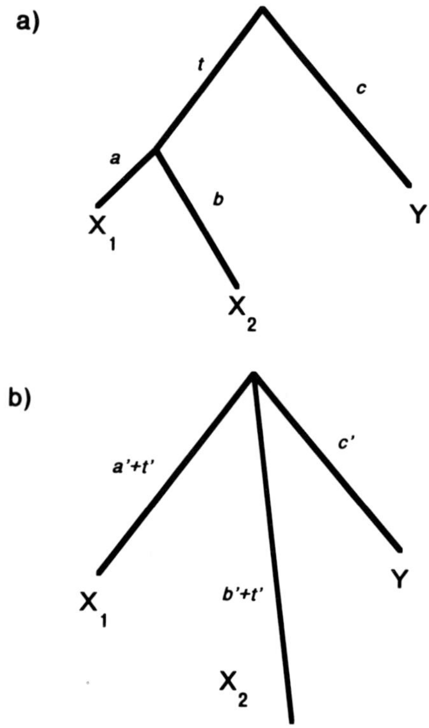

Figure 14.8. (a) Feature structure of a set of three alternatives, two of which are similar to each other (X_1 and X_2) and dissimilar from the third (Y). (b) Hypothesized functional feature structure of the set of alternatives when a similarity-inconsistent constraint is imposed on the choice process. It is assumed that feature weights are rescaled so that the sum of feature weights is the same as in panel a.

According to the preference tree model, the stimulus attributes that are relevant to choices between subsets (i.e., between the pair and the singleton in the examples above) are not necessarily relevant to choices within subsets. For example, consider the featural composition of the alternatives shown in Figure 14.8a: Features a and t of X_1 are both relevant to a choice between X_1 and Y, but because t is common to X_1 and X_2, only feature a is relevant to a choice between X_1 and X_2. This section addresses the conditions under which features are treated as irrelevant and those under which, to predict choices, it may be useful to assume that a new feature has emerged.

Tversky and Sattath (1979) reported two experiments that tested the applicability of the preference tree model to choices made under different constraints on sets of alternatives. They constructed triples of stimulus alternatives (gambles defined by probability of winning and amount to win, and college applicants defined by SAT scores and grade point averages); each triple consisted of a pair of similar items and an odd item, with the similar items defined so that one dominated the other (i.e., X_1 and X_2 were similar to each other and dissimilar to Y, and X_2 dominated X_1). Stimulus triples were presented to subjects as a pair and a singleton, and each subject was to indicate whether he or she preferred the singleton or an alternative from the pair; if the latter, a choice was made from the pair after all of the triples had been presented. The presented partition of each triple was either compatible with the similarity structure of the set (i.e., $\{X_1,X_2\}\{Y\}$) or included the pair of the dominated similar alternative and the odd alternative (i.e., $\{X_1,Y\}\{X_2\}$); these may be called the similarity-consistent and similarity-inconsistent constraints, respectively.

To predict the differences between the choice probabilities for each alternative under the similarity-consistent and similarity-inconsistent constraints, Tversky and Sattath assumed only that the probability of selecting a subset of alternatives is proportional to the value of its constituent features and that the probability of selecting an alternative from a previously selected pair is independent of the probability of selecting the pair; these will be referred to as the proportionality and independence assumptions. Tversky and Sattath predicted that, relative to choice probabilities under the similarity-consistent constraint, the choice probability under $\{X_1,Y\}\{X_2\}$ for the dominated similar alternative (X_1) would be higher, for the dissimilar alternative (Y) would be lower, and for the dominating similar alternative (X_2) would be the same.

The empirical results were only partially in accord with these predictions. Although Tversky and Sattath observed, in each of their experiments, an increase in the probability of choosing the dominated similar alternative (X_1), and a decrease in the probability of choosing the odd alternative (Y), they also found a consistent, significant decrease in the probability of choosing the dominating similar alternative (X_2)--the singleton under the similarity-inconsistent constraint.

The observed results are consistent with predictions generated by applying the proportionality and independence assumptions to the modified feature structure for the similarity-inconsistent constraint that is shown in Figure 14.8b. The feature t, common to X_1 and X_2, which appears only once in Figure 14.8a, is assumed to occur separately for each stimulus under the similarity-inconsistent constraint. (All features are rescaled so that the sum of the weights under each

Table 14.2. Differences Between Choice Probabilities for Alternatives Schematized in Figure 14.8 Under Similarity-Consistent and Similarity-Inconsistent Constraints for Various Sets of Feature Values.

Feature Values						
a	b	c	t	$d(X_1)$	$d(X_2)$	$d(Y)$
.1	.2	.4	.3	.108	−.015	−.092
.2	.4	.3	.1	.039	−.012	−.027
.0001	.0999	.5	.4	.285	−.142	−.143
.01	.09	.4	.5	.280	−.147	−.133
.15	.25	.2	.4	.093	−.036	−.057

Note. Each set of feature values sums to 1. The difference values ($d[X_1]$, $d[X_2]$, $d[Y]$) are differences between the probability of selecting the specified stimulus under the similarity-inconsistent constraint $\{X_1,Y\}$ $\{X_2\}$ and the probability of selecting it under the constraint $\{X_1,X_2\}$ $\{Y\}$. Calculation of these probabilities was based only on the proportionality and independence assumptions (see text). X_1 is the less-preferred similar alternative; X_2 is the more-preferred similar alternative; Y is the dissimilar alternative. a, b, and c are the unique features of X_1, X_2, and Y, respectively; t is the feature common to X_1 and X_2.

constraint is 1.) Table 14.2 shows, for various values of the four features under the similarity-consistent constraint, the differences between choice probabilities for each of the three alternatives under the two constraints. For each of the sets of feature weights shown, the difference for each of these alternatives is consistent with the observations of Tversky and Sattath. The essential effect of the modified feature structure shown in Figure 14.8b is to elevate the relative weight of t in the set of features.

B. Further Experiments on Constrained Choice

Kahn, Moore, and Glazer (1987) pursued further the effects on choice probabilities of constraints on choice sets. Like Tversky and Sattath, Kahn et al.

Table 14.3. Differences Between Choice Probabilities Under Similarity-Inconsistent and Similarity-Consistent Constraints--Comparison of Simulated Values and Observations of Kahn et al. (1987).

Similarity-Inconsistent Constraint	$d(X_1)$	$d(X_2)$	$d(Y)$
$\{X_1,Y\}\{X_2\}$			
Simulation with d	.12	-.23	.11
Kahn et al.	.06	-.17	.12
$\{X_1\}\{X_2,Y\}$			
Simulation with d	-.02	-.04	.06
Kahn et al.	-.01	-.05	.06

Note. The basic feature weights for the model are a = 0.05, b = 0.15, c = 0.5, and t = 0.3; a, b, and c are the unique features of X_1, X_2 and Y, respectively; t is the feature common to X_1 and X_2. These weights are rescaled to allow for d, the pair comembership feature to be 0.5 under each constraint. X_1 is the less-preferred similar alternative; X_2 is the more preferred similar alternative; Y is the dissimilar alternative. The difference values are differences in the probability of selecting the alternative under the specified similarity-inconsistent constraint and the probability of selecting it under the similarity-consistent constraint $\{X_1,X_2\}\{Y\}$.

obtained results that were discrepant from the predictions of the preference tree model. Relative to choice probabilities under the similarity-consistent constraint, Kahn et al. found, under the $\{X_1,Y\}\{X_2\}$ constraint, that the choice probability of the dominated similar alternative (X_1) was higher and that the choice probability of the dominating similar alternative (X_2) was lower. However, in contrast to both the results of Tversky and Sattath and to the predictions of the preference tree model, Kahn et al. found a higher choice probability for the dissimilar alternative (Y). Kahn et al. also presented stimuli under the similarity-inconsistent constraint in which the dominating similar alternative and the dissimilar alternative are paired and the less-preferred similar alternative is the singleton (i.e., $\{X_2,Y\}\{X_1\}$), and found, contrary to the predictions of the preference tree model, that choice probability for the dominating similar alternative was lower and for the dissimilar alternative was higher than under the similarity-consistent constraint.

Kahn et al. hypothesized various effects on choice probabilities that, in some instances, work in the same direction and that, in other instances, cancel each other. The differences in choice probability between each similarity-inconsistent constraint and the similarity-consistent constraint may be characterized adequately by supposing that mere comembership in the pair results in the induction of a feature common to the alternatives in the pair--simply by being in a pair, two stimuli may be treated as sharing a feature that influences choice probabilities in the same way as any of the other stimulus features. (Such a feature could be interpreted as a bias to choose the pair, an explanation considered by Kahn et al.) Table 14.3 summarizes, for the indicated values of the four stimulus features shown in Figure 14.8a and a specified value of d, the emergent pair-comembership feature, the differences between choice probabilities for each alternative under each similarity-inconsistent constraint and the similarity-consistent constraint. Table 14.3 shows also the empirical results of Kahn et al. (1987, Table 3). Hypothesizing the pair-comembership feature yields a set of simulated results based on the proportionality and independence assumptions that is similar in both sign and magnitude to those of Kahn et al. The notion that pair-comembership leads to the induction of an effective feature is consistent with another finding of Kahn et al., that the choice probability for *each* alternative is lower when it is the singleton in a presentation set of a pair and a singleton than when it is one of three singletons presented with no constraints whatsoever.

Still required, of course, is an explanation of the differences between the results of Tversky and Sattath and those of Kahn et al. Why does pair-comembership appear to account for the results of Kahn et al., but seem not to be required to describe the results of Tversky and Sattath? The differences between the experimental procedures used in the two studies may be responsible. Further studies of several different types are required to elucidate these effects.

C. Distribution of Alternatives and Violations of Regularity

The preceding section addressed the preference tree model's predictions about situations in which constraints are imposed on the choice process that are either compatible or inconsistent with perceived similarity relations among the available alternatives. Another prediction of the preference tree model--a prediction shared by constant ratio models--is that choices satisfy regularity.

Regularity is satisfied if expanding a set of alternatives does not increase the choice probability of the members of the original set.

Consider, for example, a menu that offers a choice between a bagel and a croissant. Neither the preference tree model nor a constant ratio model is equipped to predict that the probability of selecting either of these items would be increased by adding some third item (e.g., an English muffin) to the menu. According to the constant ratio model, if the English muffin is never chosen from the augmented set, then the probabilities of selecting the bagel and the croissant should be the same in the augmented set as in the original set. If the English muffin is ever selected, then both the bagel and the croissant should be selected with lower probability in the augmented set than in the original set. In the former case, the constant ratio model treats the presence of the English muffin as utterly irrelevant to the choice between the bagel and the croissant; in the latter case, for the ratio of choice probabilities of bagel and croissant to be equal in the original and augmented sets, the augmenting alternative would have to reduce the choice probabilities of the core alternatives by equal proportions.

According to the preference tree model, the probabilities of selecting bagel and croissant in the augmented set would depend on the similarity of these alternatives to each other, and on the new similarity structure induced by the addition of the English muffin. In fact, the preference tree model predicts that any selection of the augmenting alternative (i.e., the English muffin) would be at the expense of the more similar core alternative. For example, if the English muffin is perceived as highly similar to the bagel and quite different from the croissant, the preference tree model would predict that the choice probability for the croissant in the augmented set would be equal to its choice probability in the core set and that the choice probability of the bagel in the augmented set would be reduced by the probability of selecting the English muffin.

Huber, Payne, and Puto (1982) proposed that if a set of choice alternatives were expanded by adding an item that was similar to but dominated by an item in the core set, systematic violations of both regularity and the similarity-based predictions of the preference tree model would result. Huber et al. demonstrated this effect with several stimulus sets whose members were descriptions, on two attributes, of consumer products (e.g., cars described in terms of ride quality and gas mileage). Augmenting alternatives were almost never selected, but nevertheless modified the relative choice probabilities of alternatives in the core set. Whereas a never-selected alternative would be treated by the constant ratio model as irrelevant and dropped from consideration, Huber et al. demonstrated that the never-selected item is a relevant choice alternative because it establishes a context that modifies the selection probabilities for items in the core set.

Huber and Puto (1983) studied the effects on choice of augmenting items that, although not dominated by the core alternatives, were relatively inferior in that they offered a poorer tradeoff on the described attributes than did the items in the core set. For example, for cars, the sacrifice in ride quality for each additional mile per gallon was greater in switching from members of the core set to the augmenting alternative than in switching between core alternatives. Consistent with the results of Huber et al., expanding the set resulted in increased choice probability for the core item that was more similar to the augmenting item.

Huber et al. and Huber and Puto entertained various accounts for this "attraction" effect. For example, Huber and Puto considered the possibility that the effect was due to a symmetric devaluation of the augmenting item and the core item from which it was more different, resulting in increased choice of the core item to which the augmenting item was more similar. The basic notion is that an individual who starts with either the augmenting item or the core item from which it is most dissimilar would find that the most attractive switch would be to the core item most similar to the augmenting item.

Addition to a core set of an item similar to one of the core alternatives may modify the angular projection of a value vector through the alternative space, that is, the relative weighting of the attributes of the alternatives may depend on the particular set of alternatives available. Huber et al. considered weight shifting as an explanation, but did not pursue the idea. However, Huber and Puto suggested what might be an appropriate way to discriminate between this account and their relative attribute comparison hypothesis. They argued that if the increase in choice probabilities for a core item in the neighborhood of the augmenting item is due to symmetric devaluation, it should be observed only when the augmenting items are relatively inferior to the items in the core set. However, if expanding the core set results in a consistent change over the alternative space in the weighting of the attributes of the alternatives, this should not depend on the value relationship of the augmenting item to the core items. In other words, this hypothesis predicts that choice probability will increase for the similar core item regardless of whether the augmenting item is relatively superior to or inferior to that core item (as long as it did not dominate the core item).

This section reiterates how important an understanding of the attribute structure of choice alternatives is for describing and predicting choices. Characterizing the functional structure of alternatives and of sets of alternatives is as important for understanding choice as the development of models that describe choice processes or models that describe how properties of alternatives are utilized in choice (see also Garner, 1976).

VII. FINAL REMARK

This chapter has discussed some aspects of how the perceived structure of choice sets may influence choices. Choice is essentially a discrimination procedure, the object of which is to exclude some members of a set of items in favor of others. That choice depends on similarity relations among members of a set of alternatives has been recognized for as long as theorists have worked on formal models of choice. Similarity relations among members of sets depend on the particular composition of the sets, and choices, too, depend on the particular composition of sets.

The perceived structure of a set of stimuli depends on the properties of the members of that set. Stimulus items--or choice alternatives--have inherent properties, but how such cognitive operations as perception, classification, and evaluation weight the mental representations of these properties depends on the context established by other stimuli. The relevant contextual stimuli are sometimes present and perceptually available; in other situations, they are available in memory; in still others, they are implicit.

There is a cyclic character to the perception of the structure of a set of things: Properties inhere in objects, but the perception of such stimulus properties is moderated by the available (or, sometimes, implicit) stimulus context. On the basis of this context, the attributes of the stimulus set are deduced; this conceptual structure of the set determines how the attributes of the set are weighted during cognitive processing of the representations properties of the members of the stimulus set. Although the structure of any stimulus may be constant over contexts, the processing of that structure depends on some relevant stimulus context.

Because of the structure of alternatives and of sets of alternatives is important for choice, a distinction between two different types of attributes of stimulus sets--dimensions and features--was reviewed. Stimulus members of sets defined on these different types of attribute are characterized by different interstimulus similarity relations, and the type of attributes on which a stimulus set is defined may influence choices from that set. In particular, the perceived value of a feature may vary more over alternatives within a set than the perceived value of a level of a dimension.

Choices based on subset-dependent evaluations of choice alternatives are likely to violate the constant ratio rule. For choices to be consistent with this rule, the stimulus properties relevant to choices must be constant over contexts. Much evidence indicates, however, that the functional properties of stimuli depend on the particular available stimulus set, so that the constant ratio rule of choice is generally inapplicable.

Several predictions of an alternative choice model, the preference tree model (Tversky & Sattath, 1979), were reviewed. This model was developed to account for interitem similarity effects in choice data. A number of studies that reported systematic violations of the predictions of the preference tree model were summarized; changes in the structure of alternatives perceived by subjects were hypothesized to account for the observed effects. Understanding how contexts guide and constrain the perception and organization of sets of choice alternatives is an essential complement to developing models of choice processes that treat stimulus attributes as given and constant over contexts.

REFERENCES

Burns, B., Shepp, B. E., McDonough, D., & Wiener-Ehrlich, W. K. (1978). The relation between stimulus analyzability and perceived dimensional structure. The Psychology of Learning and Motivation, 12, 77-115.

Carroll, J. D. (1972). Individual differences and multidimensional scaling. In R. N. Shepard, A. K. Romney, & S. Nerlove (Eds.), Multidimensional scaling: Theory and applications in the behavioral sciences: Vol 1. Theory (pp. 105-155). New York: Seminar Press.

Carroll, J. D. (1976). Spatial, non-spatial, and hybrid models for scaling. Psychometrika, 41, 439-463.

Carroll, J. D. (1980). Models and methods for multidimensional analysis of preferential choice (or other dominance) data. In E. D. Lanterman & H. Feger (Eds.), Similarity and choice: Papers in honour of Clyde Coombs (pp. 234-289). Bern: Hans Huber Publishers.

Coombs, C. (1950). Psychological scaling without a unit of measurement. Psychological Review, 57, 145-158.

Coombs, C. (1983). Psychology and mathematics: An essay on theory. Ann Arbor: The University of Michigan Press.

Corter, J. E., & Tversky, A. (1986). Extended similarity trees. Psychometrika, 51, 429-451.

Garner, W. R. (1962). Uncertainty and structure as psychological concepts. New York: Wiley.

Garner, W. R. (1974). The processing of information and structure. Hillsdale, NJ: Erlbaum.

Garner, W. R. (1976). Interaction of stimulus dimensions in concept and choice processes. Cognitive Psychology, 8, 98-123.

Garner, W. R. (1978a). Aspects of a stimulus: Features, dimensions, and configurations. In E. Rosch & B. B. Lloyd (Eds.), Cognition and categorization (pp. 99-133). Hillsdale, NJ: Erlbaum.

Garner, W. R. (1978b). Selective attention to attributes and to stimuli. Journal of Experimental Psychology: General, 107, 287-308.

Gati, I., & Tversky, A. (1982). Representations of qualitative and quantitative dimensions. Journal of Experimental Psychology: Human Perception and Performance, 8, 325-340.

Huber, J., Payne, J. W., & Puto, C. (1982). Adding asymmetrically dominated alternatives: Violations of regularity and the similarity hypothesis. Journal of Consumer Research, 9, 90-98.

Huber, J., & Puto, C. (1983). Market boundaries and product choice: Illustrating attraction and substitution effects. Journal of Consumer Research, 10, 31-44.

Johnson, M. D. (1984). Consumer choice strategies for comparing noncomparable alternatives. Journal of Consumer Research, 11, 741-753.

Johnson, M. D., & Fornell, C. (1987). The nature and methodological implications of the cognitive representation of products. Journal of Consumer Research, 14, 214-228.

Kahn, B., Moore, W. L., & Glazer, R. (1987). Experiments in constrained choice. Journal of Consumer Research, 14, 96-113.

Krumhansl, C. (1978). Concerning the applicability of geometric models to similarity data: The interrelationship between similarity and spatial density. Psychological Review, 85, 445-463.

Kruskal, J. B., & Wish, M. (1978). Multidimensional scaling. Beverly Hills, CA: Sage.

Lockhead, G. R. (1989). Category bounds and stimulus variability. In B. E. Shepp & S. Ballesteros (Eds.), Object perception: Structure and process (pp. 267-295). Hillsdale, NJ: Erlbaum.

Luce, R. D. (1959). Individual choice behavior: A theoretical analysis. New York: Wiley.

McGuire, W. J., & McGuire, C. V. (1981). The spontaneous self-concept as affected by personal distinctiveness. In M. D. Lynch, A. A. Norem-Hebeisen, & K. J. Gergen (Eds.), Self-concept: Advances in theory and research (pp. 147-171). Cambridge, MA: Ballinger.

Payne, J. W. (1982). Contingent decision behavior. Psychological Bulletin, 92, 382-402.

Pruzansky, S., Tversky, A., & Carroll, J. D. (1982). Spatial versus tree representations of proximity data. Psychometrika, 47, 3-24.

Sattath, S., & Tversky, A. (1977). Additive similarity trees. Psychometrika, 42, 319-345.

Sattath, S., & Tversky, A. (1987). On the relation between common and distinctive feature models. Psychological Review, 94, 16-22.

Shepard, R. N. (1972). Introduction to Volume I. In R. N. Shepard, A. K. Romney, & S. Nerlove (Eds.), Multidimensional scaling: Theory and applications in the behavioral sciences: Vol 1. Theory (pp. 1-20). New York: Seminar Press.

Smith, A. F. (1984). Configurations and dimensions in pattern perception. Unpublished manuscript, Yale University, New Haven, CT.

Smith, L. B. (1989). A model of perceptual classification in children and adults. Psychological Review, 96, 125-144.

Tversky, A. (1969). The intransitivity of preferences. Psychological Review, 76, 31-48.

Tversky, A. (1972). Elimination by aspects: A theory of choice. Psychological Review, 79, 281-299.

Tversky, A. (1977). Features of similarity. Psychological Review, 84, 327-352.

Tversky, A., & Gati, I. (1978). Similarity, separability, and the triangle inequality. Psychological Review, 89, 123-154.

Tversky, A., & Hutchinson, J. W. (1986). Nearest neighbor analysis of psychological spaces. Psychological Review, 93, 3-22.

Tversky, A., & Krantz, D. H. (1969). Similarity of schematic faces: A test of interdimensional additivity. Perception & Psychophysics, 5, 125-128.

Tversky, A., & Sattath, S. (1979). Preference trees. Psychological Review, 86, 542-573.

Commentary

Perceptual Representations of Choice Alternatives, A. F. Smith

STEPHEN E. EDGELL
University of Louisville

Since Restle (1961) first proposed that the choice process involved aspects of the alternatives and proposed the first set-theoretic model of choice, many choice models have assumed that components of the stimuli, which are the alternatives, are the perceptual elements for the person making the choice (e.g., Edgell & Geisler, 1980; Tversky, 1972a, 1972b). Restle's idea that common aspects are not considered, explained that the choice between $101 and $100 was really a choice between $1 and nothing. This idea has continued strong in the theoretical work in the area. Unlike much of the work done in this area which has focused on mathematical investigations of the choice model as a function of mapping a utility parameter space to choice probability space, Smith reminds us that the psychology of why the parameter space takes on the values it does in a specific condition is more important. Of course a choice model must be postulated to allow one to investigate the parameter values, but given such, one can investigate how variations in the stimuli, the stimuli set, or the presentation of either affects the parameter values.

Smith's chapter reviews much work that has been done on the issue of the perceived structure of a set of stimuli and ties this work into the field of choice models. It is important to note that the processes are not fully understood and are complex. For example, not only can other stimuli interact in determining the perception of a stimuli, but even stimuli not present, but inferred by the subject can affect the perception of a presented stimuli. Although Smith's chapter raises more questions than it answers, it accurately identifies the direction for research in the area of choice.

Edgell, S.E., & Geisler, W.S. (1980). A set-theoretic random utility model of choice behavior. Journal of Mathematical Psychology, 21, 265-278.

Restle, F. (1961). Psychology of Judgment and Choice. New York: Wiley.

Tversky, A. (1972a). Elimination by aspects. Psychological Review, 79, 281-299.

Tversky, A. (1972b). Choice by elimination. Journal of Mathematical Psychology, 9, 341-367.

Percepts, Concepts and Categories
B. Burns (Editor)

15

The Effects of Representation on the Processing of Probabilistic Information

STEPHEN E. EDGELL
RANDY D. BRIGHT
PAK C. NG
THOMAS K. NOONAN
LAURA A. FORD
University of Louisville

I. INTRODUCTION

In a decision making task the judge (or subject if a laboratory setting is involved) must process information from the environment to reach a decision. Often this decision is a judgment about the state of a criterion variable or event in the environment. The information stimuli (or cues) are often only probabilistically related to the criterion. It is the task of the decision maker to discover and utilize this probabilistic relationship to make judgments with as high an accuracy as possible. A laboratory paradigm for studying these processes of learning and utilizing the knowledge learned is called multiple cue probability learning.

A multiple cue probability learning study involves exposing subjects to a set of stimuli which the subject must learn to utilize to predict a distal event. The subject is presented with one value from each stimulus dimension on each trial and must produce as a response his or her judgment as to the most likely value of the event. The subject is given some form of feedback in order to facilitate learning the environmental relationships between the stimuli and the event. The feedback is usually the actual event for that trial, but can be other types of information. Over trials the subject learns to utilize the information. After sufficient numbers of trials the subject's performance asymptotes. While the learning function can be of interest also, it is usually this asymptotic performance that is analyzed.

A variation of the multiple cue probability learning paradigm is the case where the cue values and the event values are nonmetric. Nonmetric cues or events are ones which are not numerically valued such as the type of cloud formations, the location of the tumor, which drug to give the patient, etc. It is obviously the case in the real world that there is no shortage of examples where decisions are nonmetric. The nonmetric multiple cue probability learning paradigm provides a simpler environment which facilitates studying many aspects of how subjects process probabilistic information. An analysis technique which uses the standard general linear model from a factorial analysis of variance can be applied, in dichotomous nonmetric situations, to the environment to obtain validities for the various cue dimensions and configurations and can be applied to the subject's responses to obtain the subject's weightings of the cue dimensions and configurations. (For details including derivations and examples see Edgell, 1978 and 1980.) This further facilitates studying subjects' performance in such tasks. For a review of earlier work using the nonmetric multiple cue probability learning (NMPL) paradigm and related paradigms the reader is referred to Castellan, 1977.

The paradigm of multiple cue probability learning has been used to explore many important questions in the area of decision making. One of these is the effect of irrelevant information (Castellan, 1973). A number of researchers have used this paradigm for investigating the effect of relevant configural information[1] and the ability of subjects to utilize it (Edgell, 1978, 1980, 1983; Edgell & Castellan, 1973; Estes, 1972; Mellers, 1980; Stockburger & Erickson, 1974). The paradigm has also been used for investigating the effects of different types of feedback (Bjorkman, 1971, 1972; Castellan, 1974; Castellan & Swain, 1977; Hammond, 1971; Hammond & Boyle, 1971). Other studies investigated the effects of changes in the environment (Dudycha, Dumoff, & Dudycha, 1973; Edgell & Morrissey, 1987; Lindberg & Brehmer, 1976a, 1976b; Peterson, Hammond, & Summers, 1965; Ruffner & Muchinsky, 1978; Summers, 1969). The majority of the studies cited here were of the nonmetric type.

Recently, the form in which the information is presented to the decision maker has been a research topic of much interest (Freedy & Johnson, 1982; Gopher & Kimchi, 1989; Wickens & Kramer, 1985). The development of computer graphics capability has greatly expanded the possibilities of different types of information displays. Research in the area has focused mainly on the effects of presenting the information dimensions in a wholistic, single figure, graphical display or in a separated type of display such as one might commonly find in a business or medical decision setting. Following the terminology of Garner (1976) the wholistic, single figure type of display will be referred to as a unitary display and the separated type of display will be referred to as a separable display. (See Edgell and Morrissey (1992) for a full discussion of this terminology.) It is interesting to note that most of the nonmetric multiple cue studies have used unitary stimuli while most of the metric multiple cue studies have used the separable displays. Goldsmith and Schvaneveldt (1984) ran a series of metric multiple cue probability studies investigating the effect of unitary versus separable displays. Edgell and Morrissey (1992) ran a series of studies investigating this question in nonmetric multiple cue probability learning environments.

[1]The term "configural information" as used in decision making refers to relevant information that is in a pattern of the stimuli dimensions and is not in any dimension by itself. See Meehl (1954) for a good definition and examples of "configural information". One should note that Garner and his colleagues use the term "configural" in a different sense. Their use indicates a physical structure of the stimuli with certain properties. The use in this paper and in the area of decision making refers to the relationships between the stimuli dimensions and the environmental criterion or event. If that relationship is nonadditive, then the environment is said to contain configural information. A subject or judge can be using the information nonadditively in making his or her decisions. Such a judge is said to be employing a configural rule.

The Edgell and Morrissey (1992) study found that, on the average, the unitary display facilitated utilization of configural information versus the separable display provided the environment was sufficiently complex. No difference was found, on the average, for the utilization of relevant dimensional information. However, it was found that some dimensions were much more utilized by the subjects than were others. In the first experiment the stimuli in the unitary condition consisted of a geometric figure (a square or a triangle) which was made up from lines (either vertical or horizontal). Thus shape and line direction were the two dichotomous cue dimensions. The subjects showed a significantly higher asymptotic utilization for the line direction dimension than for the shape dimension even though both dimensions had the same environmental validity. This difference in utilization of stimuli with equal validities can only be accounted for by proposing that the different stimuli dimensions had different salience due to some physical properties of the stimuli other than the validity of the stimuli in the task.

This latter finding was replicated and extended in Edgell and Morrissey's second experiment which had three dichotomous cue dimensions. The same dimensions of shape (square or triangle) and line direction (vertical or horizontal) were used along with a new dimension of number of figures (one or two). However, in this experiment only one of the dimensions was relevant. Which dimension was relevant varied across groups of subjects. Each dimension, when relevant, was utilized to a different degree by subjects. Line direction (again) was used more strongly than shape which was utilized more strongly than number of patterns. This was even more surprising than the findings in their first experiment. In their first experiment both cue dimensions were relevant and thus may have competed for attention from the subject. However, in their second experiment only one dimension was relevant for each group. Hence there could not be competition for attention among relevant dimensions. It should be noted that in their second experiment there was relevant pattern or configural information in addition to the relevant dimensional information. The pattern that was relevant was the pattern consisting of the other two dimensions that were not relevant by themselves. This was true for each condition. It can be argued that some patterns distracted more attention from the relevant cue than other patterns, and it was found that the different patterns were utilized at different levels. Even the separable conditions, where the cues were one of two possible symbols (different for each cue dimension) under the heading cue and the number (1, 2, or 3), showed different levels of utilization for the different dimensions when each was relevant.

This result of different salience levels for different cue dimensions was unexpected and had not been predicted by previous work. In fact, in a study similar to the first experiment of Edgell and Morrissey (1992) except that the dimensions were color (red or green) and letter (X or O), no difference was found in the level of utilization (Edgell, 1978). It should be recalled that the differences found in the Edgell and Morrissey studies were at asymptote, which was measured after over 300 learning trials. This implies that the salience effects are long lasting and do not vanish with practice. This finding is quite important to the research area. It implies that the form in which the information is represented to the subject can have a large effect on the cognitive processing of that information by the subject. It further implies that this effect is not only found when the differences between the stimuli representations are large (such as between separable and unitary stimuli), but is also found for smaller differences (such as different representations of level within the stimuli dimension).

II. EXPERIMENT 1. SALIENCE AND UNITARY STIMULI

The Edgell and Morrissey (1992) experiments established the existence of salience effects for different stimuli dimensions in NMPL. That is, they showed that different representations of the levels of a stimulus dimension caused variance in the level of utilization of the stimulus dimension by subjects. However, because those experiments were designed to investigate a different question, the environments used there were complex, containing both dimensional and configural information. The other relevant information, either dimensional or configural, may have interacted with or competed for attention with the relevant dimension. Hence, it was necessary to replicate the basic finding of salience effects in a more elementary environment. Making only one dimension relevant would be best for establishing its salience and resulting level of utilization by subjects. There would be no other relevant information to distract subjects from using the relevant dimension as there was in the Edgell and Morrissey (1992) studies. Their study had used a moderate level of validity. It was proposed to investigate if the salience effects they found would generalize to a higher level of validity. It would be expected that salience effects would be larger with moderate levels of validity because the task is harder. It is more difficult to discriminate a relevant dimension from an irrelevant one when the validity is moderate. As the validity increases, the salience effect might decrease and should vanish as the validity increases to the maximum, where perfect prediction would be possible. A level of validity significantly above the moderate level used by Edgell and

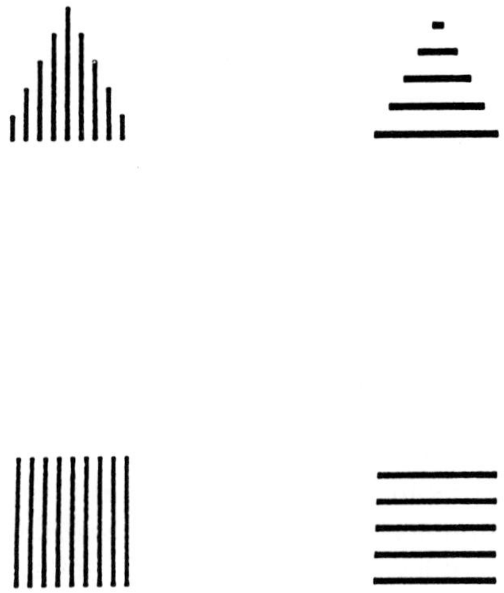

Figure 15.1. The four possible stimuli patterns for Experiment 1.

Morrissey, but well below maximum, is called for. Of course the stimuli should be kept the same as in the Edgell and Morrissey (1992) study.

The Edgell and Morrissey (1992) studies involved an NMPL task with 400 trials. There were two binary valued stimuli dimensions, line direction (vertical or horizontal) and shape (square or triangle). The four possible stimuli patterns are given in Figure 15.1. These same stimuli were used here. As discussed above, two factors were manipulated. The first was which of the two dimensions (line direction or shape) was relevant, and the second was the level of validity of the relevant dimension. The same moderate level of validity (.2) as in the Edgell and Morrissey (1992) study and a higher level (.3) were used. This higher level of validity is one that subjects should utilize very strongly, but it is not so high that the task becomes one which allows almost perfect prediction. It should provide a strong contrast with the more moderate .2 validity level.

Validity is used here in the sense described by Edgell (1978, 1980). The values range from 0, which indicates the information is irrelevant, to .5, which indicates the information is such as to allow perfect prediction. These validities result from a linear model representation of the environmental conditional probabilities of the event given each possible cue pattern. A so-called 70/30 cue with conditional probabilities (of one of the events given each of the cue values) of .7 and .3, would have a validity of .2. An 80/20 cue would have a validity of .3, and so on. (When cue and event base rates are not .5, the situation is more complicated, but that is not of concern here.) This linear model representation was developed to provide an additive validity measure in tasks where the configural information is relevant beyond the additive validity of the dimensional information. It is used here for continuity with much past research which focused on a variety of issues having to do with processing.

A. Method

Subjects. The subjects were 114 students who were fulfilling a course requirement in introductory psychology. The two validity levels were run at different times and thus subjects were not randomly assigned between them. However, for reasons explained below, this is not a problem here, and, moreover, the subjects should be comparable. The subjects were randomly assigned to conditions within a validity level, giving 25 subjects in each group for the two .2 validity groups and 32 per group for the two .3 validity groups.

Apparatus. The apparatus consisted of three computer CRT terminals. They were screened from view of each other so that up to three subjects could be run simultaneously and independently. The stimuli consisted of a square ($C_{1,1}$) or a triangle ($C_{1,2}$) which was made up of vertical ($C_{2,1}$) or horizontal ($C_{2,2}$) lines. ($C_{i,j}$ is the j-th value of cue dimension i.) These four possible stimuli patterns are given in Figure 15.1. Although that figure gives all four possible patterns, it should be kept in mind that on each trial only one of the possible patterns was presented to the subject. Two keys on the terminal were used as response buttons. One had a green piece of tape on it, and the other, a red piece. The subjects' predictions (red or green) and the correct event (red or green) were displayed on the terminal below the stimuli by a short phrase. The experiment was run and the terminals controlled by computer. This is the same apparatus as used in the Edgell and Morrissey (1992) study.

Procedure. At the beginning of the experiment subjects were read instructions in which the task was likened to a medical decision making task. They were told that the cues were abstract to prevent them from using previous learning and biases. They were also told what the cues they would be seeing were, and how to make responses. All subjects were run for 400 trials. Either dimension 1 (line direction) or dimension 2 (shape) was relevant with a validity of .2 or .3 depending upon the condition the subject was in. The event base rate was irrelevant (i.e., the two events were equally likely). Cue base rates were equal, and the cue dimensions were independent of each other (i.e., the four possible patterns were equally likely on each trial). The sequence of the stimuli was generated by the method of Castellan (1970) and Castellan and Brown (1970). The trials were randomized in blocks of 80. The experiment was self-paced. At the beginning of each trial the stimuli were presented on the terminal. After the subject made a response, it was displayed on the terminal, and 0.3 seconds later, the correct event was displayed. The cues remained displayed during this feedback stage. After 1.5 seconds the terminal screen was erased, and, after an intertrial interval of 0.5 seconds, a new trial began. This is the same procedure as used by Edgell and Morrissey (1992).

B. Results and Discussion

The proportions of one of the responses to each cue pattern were calculated for the last 80 trials for each subject. These response proportions were then transformed to the linear model terms as detailed in Edgell (1978, 1980). The application of the linear model representation to the response proportions to obtain utilization weights is exactly analogous to the application of the linear model representation to environmental probabilities to obtain validities as discussed above. The utilization weights have the same range and interpretation as the validities. That is, a value of 0 implies the subject is responding independently of the particular information or is ignoring it. A value of .5 indicates that the subject's responding is totally dependent upon the information. This gave a measure of the asymptotic utilization of the relevant dimension. (Learning curves over trial blocks had reached asymptote.)

The purpose of the experiment was to replicate the findings of Edgell and Morrissey (1992) and investigate if the same results held for a higher validity level. For this reason, simple effects analyses, comparing one condition (line direction relevant) within each validity level with the other condition (shape relevant) within the same validity level, are appropriate rather than the usual factorial analysis of variance. This accounts for the fact that the nonrandom assignment of subjects to validity conditions has no consequences. It should be noted that it is well-known that there would be a main effect for validity level. Further, there is bound to be an interaction across the full range of validity levels (0 to .5) if there is any effect for which dimension is relevant at either of the validity levels tested in this experiment. The reason for this is that, at a validity of 0, the slopes would have to be flat because the dimensions are both irrelevant and neither could not be utilized. At a validity of .5 the slopes would also have to be flat. This is because, at a validity of .5, the event would be perfectly predictable from the one relevant cue dimension. Hence, the task would be a simple problem solving one that all subjects would be expected to solve and perfectly use the relevant dimension regardless of which dimension it was.

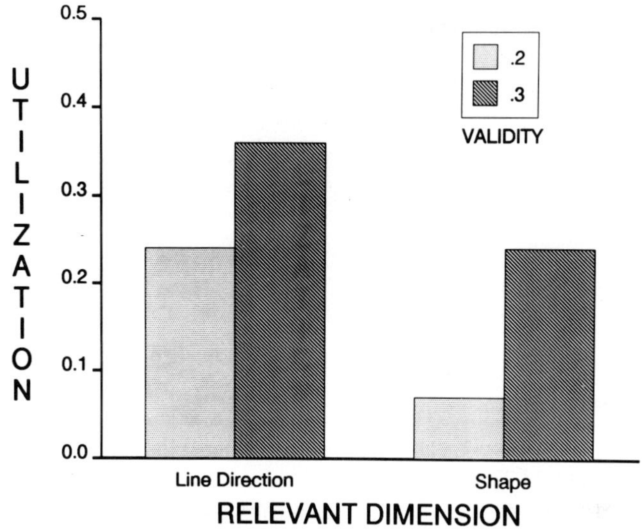

Figure 15.2. A bar graph of the mean asymptotic utilization by which dimension was relevant and the validity level for Experiment 1.

Means across subjects within each condition are given in Figure 15.2. The mean asymptotic utilization in the .2 validity conditions are shown by the lighter bars in Figure 15. 2. As can be seen there, the utilization of the shape dimension was much less when it was relevant than was the utilization of the line direction dimension when it was relevant ($F(1,48)=24.05$, $p<.001$). The mean asymptotic utilization in the .3 validity conditions are given in Figure 15.2 by the darker bars. Again, as can be seen there, the subjects showed considerably less utilization when shape was relevant than when line direction was the relevant dimension ($F(1,62)=9.53$, $p=.003$). It should be noted that the error terms for each of these analyses were calculated only from the two groups involved in the particular analysis. Although this lessens the power of the test, it makes the tests robust to any possible violations of the assumption of equal variance. The present findings replicate the Edgell and Morrissey (1992) findings, but in a simpler environment that had only one dimension relevant. Hence, the salience effect cannot be explained by a competition for attention by other relevant dimensions or patterns. Further the present results show that this finding is not restricted to the moderate .2 level of validity, but rather that it generalizes with no appreciable difference, to a higher level of validity.

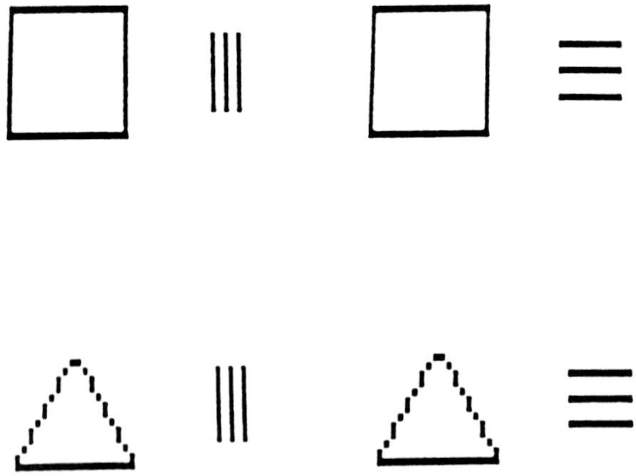

Figure 15.3. The four possible stimuli patterns for one of the two separable conditions (condition 2) in Experiment 2.

III. EXPERIMENT 2. SALIENCE AND SEPARABLE STIMULI

The last study showed that subjects utilize line direction more highly than shape when each in turn is the relevant dimension. The stimuli are unitary in the sense that they are both contained in one figure. (See Edgell and Morrissey (1992) for a discussion of using the term "unitary" for this type of stimuli. It should be noted that this use of "unitary" as opposed to "separable" is the same as defined by Garner, 1976, as was discussed above.) Edgell and Morrissey (1992) found, in a two dimension task similar to the one in Experiment 1, that if the stimuli were separable (that is, the dimensions are physically separated) rather than unitary there was no difference in utilization. Another study of Edgell and Morrissey (1992) used a much more complex environment, which is difficult for subjects, involving three dimensions and a relevant subpattern, and found unequal salience for separated stimuli. However, this may be due to the more complex environment there. The separable stimuli, in the first experiment of Edgell and Morrissey (1992), were much different than those of shapes made from lines. The value of each dimension there was given by a single letter. Thus the obvious question that needed answering was, did the salience effect found in Experiment 1 depend upon the unitary nature of the stimuli. That is, is there something special about having the shape made up of the lines which are the other dimension thereby causing the lines to have a higher salience.

To test this an environment was designed in which the stimuli were still shape (square or triangle) and line direction (vertical or horizontal), but in which they were separated rather than part of the same figure. The square or triangle was presented as an outline drawing and the lines (vertical or horizontal) were presented as several lines next to the figure. The four possible sets of stimuli can be seen in Figure 15.3. In that figure are shown all four possible sets of stimuli. On each trial only one of the possible four was presented to the subject. A single set of stimuli consists of the shape and the lines to the right of the shape. Since which dimension (the shape or the lines) was presented first (on the left) of the CRT could have an influence, it was necessary to run both shape first and lines first. The unitary stimuli of Experiment 1 were also run for comparison. Thus the design of the experiment is a 2 by 3 factorial. There are 3 levels of type of display (unitary, separable shape-lines, and separable lines-shape) and 2 levels of which dimension is relevant (shape or line direction).

A. Method

Subjects. The subjects were 156 students who were fulfilling a course requirement in introductory psychology. They were randomly assigned to the six conditions giving from 25 to 27 subjects per condition.

Apparatus. The apparatus was the same as in Experiment 1 except for the stimuli. The stimuli varied across the three conditions. In the first condition the stimuli consisted of a square ($C_{1,1}$) or a triangle ($C_{1,2}$) which was made up of vertical ($C_{2,1}$) or horizontal ($C_{2,2}$) lines. ($C_{i,j}$ is the j-th value of cue dimension i.) These are the same stimuli as used in Experiment 1. In the second condition the stimuli consisted of the outline of a square ($C_{1,1}$) or a triangle ($C_{1,2}$) with a set of vertical ($C_{2,1}$) or horizontal ($C_{2,2}$) lines to the right of the figure. The stimuli for the third condition were the same as for the second condition except that the set of lines were to the left of the shape. The stimuli from the first condition are shown in Figure 15.1. The stimuli from the second condition are shown in Figure 15.3. The stimuli in the third condition are the same as those shown in Figure 15.3 except that the lines are on the left of the square or triangle.

Procedure. The procedure was the same as in Experiment 1 except that the validity of the relevant dimension (either dimension one or two depending upon the condition) was .2.

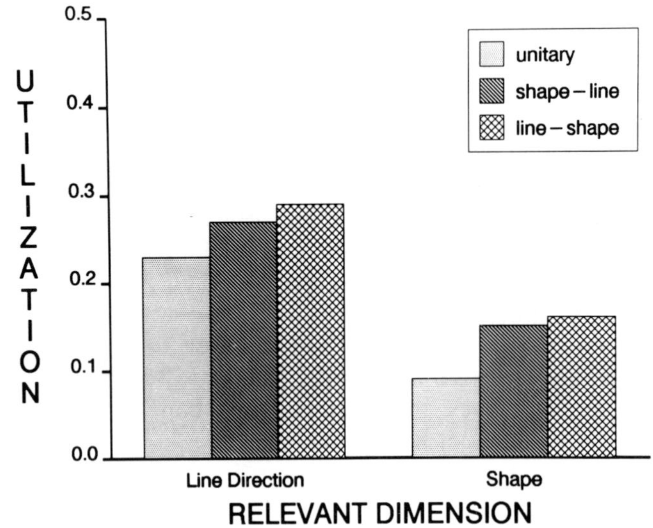

Figure 15.4. A bar graph of the mean asymptotic utilization by type of stimuli in Experiment 2.

B. Results and Discussion

The proportions of one of the responses to each cue pattern were calculated for the last 80 trials for each subject. These response proportions were then transformed to the linear model terms as detailed in Edgell (1978, 1980). This gave measures of the asymptotic utilization of the relevant dimension. (Learning curves over trial blocks had reached asymptote.) Means across subjects within conditions are shown in Figure 15.4. In Figure 15.4 the asymptotic mean utilization for the unitary stimuli (condition 1) are shown by the lighter bars on the left of each set of three bars. The asymptotic mean utilization for the conditions with separated stimuli are given by the darker bars for the shape on the left condition (condition 2) and by the cross hatched bars for the shape on the right condition (condition 3). As can be seen in Figure 15.4, subjects in all three stimuli conditions showed much higher utilization of the line direction dimension when it was relevant than of the shape dimension when it was relevant ($F(1,150)=25.98$, $p<.001$). There was not a significant difference between the line-shape orientation and the shape-line orientation conditions for either the condition when line direction was relevant ($F(1,150)=.08$, $p=.78$) or when shape was relevant ($F(1,150)=.07$, $p=.79$). There were also no significant differences between the average of the two separable stimuli conditions and the unitary one for either the condition when line direction was relevant ($F(1,150)=1.73, p=.19$) or when shape was relevant ($F(1,150)=2.58, p=.11$).

These results leave no doubt that the findings of Experiment 1 were due to the stimulus dimensions that were used. Subjects find shape (square and triangle) to be a less salient cue than line direction, at least as it was generated here. Being unitary or separable did not make any difference in this. While it appeared that the separable stimuli were a little more utilized than the unitary, the difference was not significant and the finding would not be of importance here.

IV. EXPERIMENT 3. SALIENCE AND OTHER SEPARABLE STIMULI

The last experiment showed that even with the dimensions separated, line direction was more salient than shape. In Edgell and Morrissey's (1992) first experiment no difference was found for salience in separable stimuli. However, the stimuli there were quite different in that they were single letters or symbols for each value rather than shapes and line directions. The Edgell and Morrissey environment had both dimensions relevant as well as the pattern. This difference in the complexity of the environment could also account for the difference in findings. It was decided to test the effects of salience for separated stimuli of the type used by Edgell and Morrissey but in an environment with only one dimension relevant, as was done in Experiments 1 and 2 here. Just as it was of interest in Experiment 1 to look at different levels of validity, so is it here. Hence an experiment very similar to Experiment 1 was proposed, the only difference being that the stimulus dimensions would be given by different symbols. It should be noted that this representation of the information is closer to the way information is usually presented in real world decision tasks. The design of the experiment was a 2 by 2 factorial with one factor being which of the two cue dimensions was relevant and the other factor being the validity level (.2 or .3).

A. Method

Subjects. The subjects were 99 students who were fulfilling a course requirement in introductory psychology. As in Experiment 1, the two validity levels were run at different times and thus subjects were not randomly assigned between them. However, for the same reasons as explained there this is not a problem here, and also the subjects should be comparable. The subjects were randomly assigned to conditions within a validity level giving 26 subjects in each group for the two .2 validity groups and 23 or 24 per group for the two .3 validity groups.

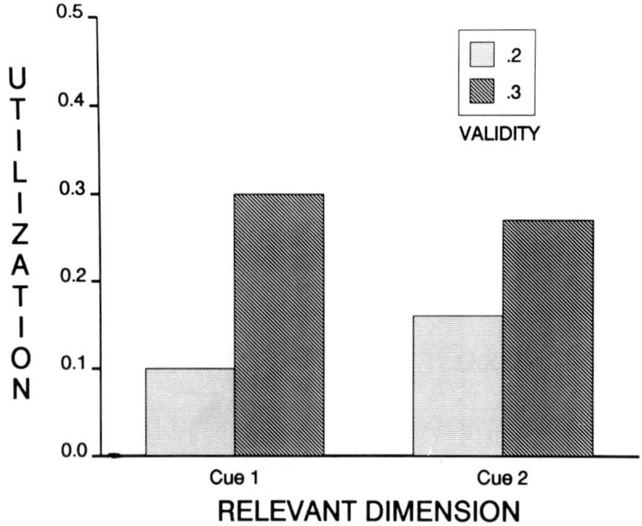

Figure 15.5. A bar graph of the mean asymptotic utilization by which dimension was relevant and the validity for Experiment 3.

Apparatus. The apparatus was the same as in Experiment 1 except for the stimuli. The stimuli consisted of an "X" ($C_{1,1}$) or an "O" ($C_{1,2}$) under the word "CUE 1" and a "+" ($C_{2,1}$) or a "-" ($C_{2,2}$) under the word "CUE 2."

Procedure. The procedure was the same as in Experiment 1.

B. Results and Discussion

The proportions of one of the responses to each cue pattern were calculated for the last 80 trials for each subject. These response proportions were then transformed to the linear model terms as detailed in Edgell (1978, 1980). This gave measures of the asymptotic utilizations of the relevant dimension. The means are shown in Figure 15.5. (Learning curves over trial blocks had reached asymptote.) As was done in Experiment 1 and for the same reasons discussed there, separate analyses were done for each validity level. The asymptotic mean utilizations for the .2 validity conditions are given in Figure 15.5 by the lighter bars. As can be seen the difference was much smaller than was found in Experiment 1 (see Figure 15.2), and the difference was not significant ($F(1,50)=1.34$, $p=.25$). The asymptotic mean utilizations for the .3 validity conditions are given in Figure 15.5 by the darker bars. Again there is only a small difference, and it was not significant ($F(1,45)=.42$, $p=.52$). Looking at Figure 15.5

it can be seen that the direction of the effect in the .2 validity condition was reversed for the .3 validity condition. Because both differences were small and nonsignificant, and because there would be no reasonable explanation for a reversal of which dimension was more salient for different validity levels, this is reasonable evidence that there was no difference. Thus it can be concluded that for separable stimuli of the form used here, there is no difference in salience. The finding in the Edgell and Morrissey (1992) study of equal salience for these same stimuli was therefore not due to the complex nature of the environment in that study.

V. EXPERIMENT 4. SALIENCE AND OTHER UNITARY STIMULI

Experiment 3 showed that for a certain type of separable stimuli, the salience could be equal. However, Experiment 2 showed that for another type of separable stimuli the salience could be quite unequal. Experiment 1 showed that for the type of unitary stimuli used there, the salience was unequal. This raises the obvious question: is it the general rule that unitary stimuli are of unequal salience. It seemed reasonable that there must be unitary stimuli of equal salience. Finding that such exist would not only be of theoretical interest, but would have much practical application in research using the NMPL paradigm. Although it seemed likely that unitary stimuli with equally salient dimensions could be found, it remained to discover if equal salience was more likely or if unequal salience was more likely. It could well be that the particular stimuli used in Experiment 1 were the exception rather than the rule. Line direction may have been more salient because, as can be seen in Figure 15.1, the line direction (vertical or horizontal) was confounded with the spacing between the lines. The vertical lines were half the distance apart as were the horizontal ones. This was due to the type of CRT that they were generated on. This redundant dimension of line spacing may have made the line direction dimension stand out more, and thus be more salient.

Hence it was proposed to run an experiment using unitary stimuli different from that in Experiment 1. It was decided that the dimension of line direction had to be discarded. However, the dimension of shape seemed reasonable to retain. This would also allow testing of whether shape would still be utilized at a low level without the strong salience dimension of line direction. In choosing the particular stimuli, the goal of finding equally salient dimensions was given consideration. It would be useful to have stimuli dimensions where salience might be adjustable so that equality might be achieved through adjustment. This suggested size difference, because the magnitude of the difference is easily adjusted. It was decided to add a third dimension because this would allow the exploration of more dimensions within the same experiment as well as explore this effect within a more complex environment. It was decided to make the third dimension color. Again, the difference between the shades could be adjusted. The colors red and blue were chosen.

The design of the experiment was the same as the previous ones with one dimension relevant per condition. Thus, because there were three dimensions, there would be three conditions here. Given the results from Experiments 1 and 3 of similar findings for different validity levels, it was decided to run only one validity level. A validity level of .25, which is halfway between the two validity levels of Experiments 1 and 3, was used.

A. Method

Subjects. The subjects were 63 students who were fulfilling a course requirement in introductory psychology. Subjects were randomly assigned to the three conditions giving from 20 to 22 per condition.

Apparatus. The apparatus consisted of four color graphics CRT terminals. They were screened from view of each other so that up to four subjects could be run simultaneously and independently. The stimuli consisted of a square ($C_{1,1}$) or a triangle ($C_{1,2}$) which was red ($C_{2,1}$) or blue ($C_{2,2}$) and was large ($C_{3.1}$) or small ($C_{3,2}$). ($C_{i,j}$ is the j-th value of cue dimension i.) Two push buttons on a small chassis located below and in front of the CRT screen were used as response buttons. One was labeled "X" and the other was labeled "O." The subjects' predictions (X or O) and the correct event (X or O) were displayed on the terminal below the stimuli by a short phrase. The experiment was run and the terminals controlled by computer.

Procedure. The procedure is the same as in Experiment 1 except that there were 384 trials randomized in blocks of 64 trials. The validity of the relevant dimension was .25.

B. Results and Discussion

The proportions of one of the responses to each cue pattern were calculated for the last 64 trials for each subject. These response proportions were than transformed to the linear model terms as detailed in Edgell (1978, 1980). This gave measures of the asymptotic utilization of the relevant dimension. (Learning curves over trial blocks had reached asymptote.) The means are shown in Figure 15.6. As can be seen there, the utilizations of the three dimensions were

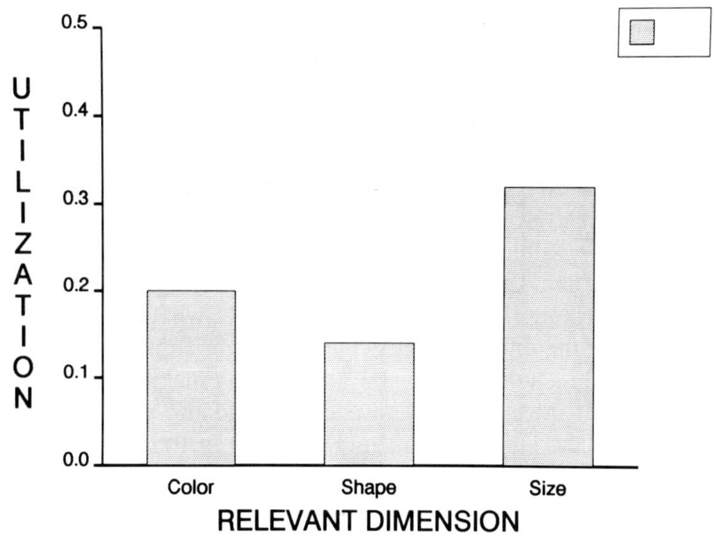

Figure 15.6. A bar graph of the mean asymptotic utilization by which dimension was relevant for Experiment 4.

quite different ($F(2,60)=6.35$, $p=.003$). Again shape (square or triangle) was utilized to a lesser degree than the other two dimensions. Color was utilized a little more than shape, but size was utilized much more than the other two. Recall that in each condition only the one dimension was relevant and thus there was no competition between these different dimensions. Of course all the dimensions were present in all the conditions, but one would expect that after over 300 trials the subject would have discerned which were relevant and which were irrelevant.

The results clearly demonstrate that the differences found in Experiments 1 and 2 are not due to the particular choice of stimuli there. Indeed, given the findings of unequal salience both there and here, this may well be the rule rather than the exception. In other words, different stimuli in general probably all have different saliences. It is interesting that even without the line direction dimension the shape dimension again was found to be of low salience.

VI. EXPERIMENT 5. SALIENCE MANIPULATED

The previous experiment showed that it may well be the case that all stimuli dimensions have different saliences. Much of the research that one would like to run would require dimensions of equal salience. Before beginning the search for such dimensions, it would be useful to answer the question of whether or not the salience can be manipulated. The color and size dimensions were chosen in the last experiment because they can be varied. The relative mix of red and blue in each level of the color dimension could be varied. Also the relative size differential could be varied for the size dimension. Certainly each of these could be made such that discrimination of the different levels was quite difficult or even impossible. At that point it would no doubt be the case that the salience must become 0. In Experiment 4 the size differences and color differences were striking. The interesting question is, what would be the effect of reducing the size of the difference from these levels in Experiment 4, but keeping them well above threshold. It can be argued that making the difference less obvious would lower the salience. However, it could also be argued that making a dimension which the subjects were told was part of the experiment harder to discriminate might call attention to it and thus raise its salience. It was conjectured that a lowering would be more likely.

Given the conjecture of a lowering of salience and the results of Experiment 4 that size was much more utilized than the other two dimensions, it was decided to change the size differential by making it much smaller. Since the change in one dimension, even when not the relevant dimension, could affect the utilization of the other dimensions when they are relevant, it was proposed to run the same three conditions as in Experiment 4, except with the smaller size differential stimuli. The color and shape dimensions were kept the same as in Experiment 4.

A. Method

Subjects. The subjects were 86 students who were fulfilling a course requirement in introductory psychology. Subjects were randomly assigned to the three conditions giving from 27 to 30 per condition.

Apparatus. The apparatus is the same as in Experiment 4 except that the difference in the relative sizes of the stimuli was reduced.

Procedure. The procedure is the same as in Experiment 4.

B. Results and Discussion

The proportions of one of the responses to each cue pattern were calculated for the last 64 trials for each subject. These response proportions were

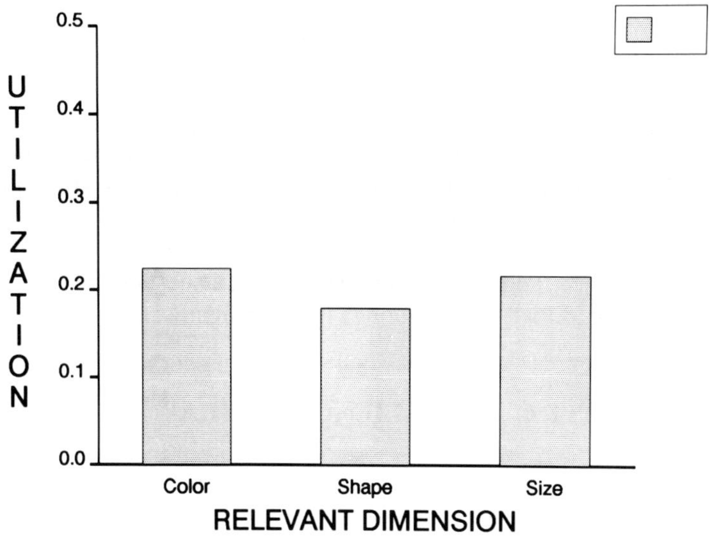

Figure 15.7. A bar graph of the mean asymtotic utilization by which dimension was relevant for Experiment 5.

then transformed to the linear model terms as detailed in Edgell (1978, 1980). This gave measures of the asymptotic utilization of the relevant dimension. (Learning curves over trial blocks had reached asymptote.) The means across subjects within conditions are shown in Figure 15.7. Comparing Figure 15.7 to igure 15.6 from Experiment 4, it can be seen that the utilization of the shape and color dimensions are reasonably equal across the two experiments. However, the utilization of the size dimension is much lower in the present experiment than it was in the last experiment ($F(1,45)=4.37$, $p=.042$). Although subjects were not randomly assigned to conditions between the two experiments, it can be expected that they should be similar. Further the similar findings for the other two dimensions support the conclusion that the difference between the utilization of size in this experiment and in the previous one is not due to differences in the subjects. Rather it must be due to reducing the size differential in the current experiment.

There are two important implications for these results. The first is that varying the apparent salience has the predicted effect on actual salience. That is, lowering the size differential which made the dimension less perceptually striking, lowered the utilization. The other implication is that the saliences of dimensions are independent. Changing the salience of one dimension doesn't affect the salience of the other dimensions. This is further supported by the finding of low salience for the shape (square or triangle) dimension in every experiment where is was used.

Looking further at Figure 15.7, it can be seen that there was relatively little difference in the utilization of the three dimensions. The difference did not reach significance ($F(2,83)=.60$, $p=.551$). However, there is still too large a difference to conclude that equal stimuli have been found. The size and color dimensions are very close, but the shape dimension is still being utilized less than the other two. (Since the goal is to find equality, it is permissible to discuss the difference as being too large even though it is not significant.)

VII. EXPERIMENT 6. THE SALIENCE OF SHAPES

Experiment 5 showed that the salience of the size dimension could be manipulated. In fact the manipulation there produced about equal salience for the color and the size dimensions. However, the shape dimension appears to still be underutilized. Shape is a natural dimension for graphic CRT displays. Given the results of these experiments, that much difference exists in salience for different dimensions, it might well be expected that with different values on the shape dimension the salience of that dimension might be different. Hence it was decided to run an experiment looking at different levels within the shape dimension. Since the goal was to increase the salience, it was thought logical to try shapes that were maximally different. The triangle and the square are both regular polygons differing by only one more or one less side. (Admittedly, these are the two most different polygons which differ by only one side, but salience has been found to be low in several of the previous experiments.) A circle is the limiting case of a polygon with n sides as n goes to infinity, and a triangle is the polygon with the smallest number of sides. Thus it was decided to run one condition (condition 1) with circle and triangle as two levels.

It was certainly possible that the two shapes proposed above would still not give the same saliency as was previously found for color and size (with the reduced differential). Hence, because the goal here was to find three dimensions of equal salience, it was decided to also try possible replacements for the shape dimension. Orientation of the figure is one possible substitution. A triangle (like those used in the previous experiments here) can point upwards, as in the previous studies, or it could point downward. This gives what would seem to be a very different and salient dimension. Thus it was decided to also run a condition (condition 2) with right-side-up and upside-down triangles as the levels of the "shape" stimuli. The figure could also be situated in different locations of

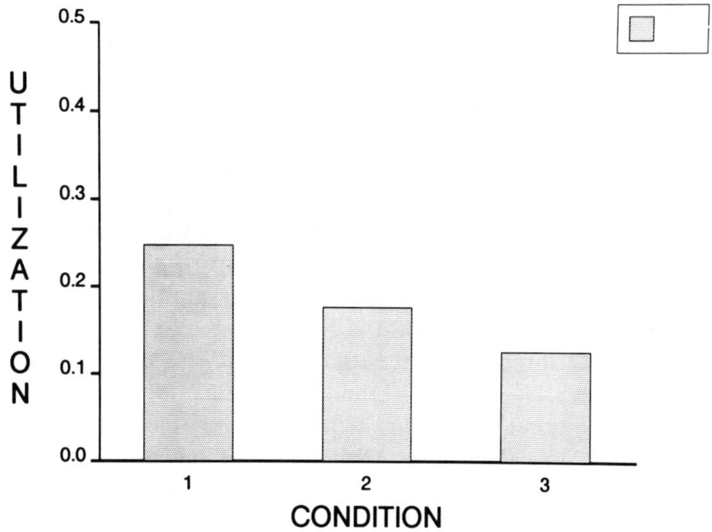

Figure 15.8. A bar graph of the mean asymptotic utilization by which method of representation of the relevant shape dimension was used for Experiment 6. The relevant stimuli were a triangle or a circle for condition 1, a triangle pointing up or down for condition 2, and a square at the top or near the bottom of the screen for condition 3.

the screen. Since the response buttons had a right/left orientation, right versus left orientations for the figure were ruled out. This left up and down as possible levels. It seemed that this should also be a very obvious, and thus a highly salient dimension. It was decided to run a condition (condition 3) with a square up near the top or down in the bottom half (room for feedback was still needed at the bottom of the screen) of the screen as the levels of "shape." In each of the three conditions the shape dimension or substitution for the shape dimension as defined for that condition would be relevant while the other two dimensions, color and size, would be irrelevant and constructed the same as in Experiment 5.

A. Method

Subjects. The subjects were 71 students who were fulfilling a course requirement in introductory psychology. Subjects were randomly assigned to the three conditions giving 23 or 24 per condition.

Apparatus. The apparatus is the same as in Experiment 4 except that the shape stimuli were as described above.

Procedure. The procedure is the same as in Experiment 4.

B. Results and Discussion

The proportions of one of the responses to each cue pattern were calculated for the last 64 trials for each subject. These response proportions were then transformed to the linear model terms as detailed in Edgell (1978, 1980). This gave measures of the asymptotic utilization of the relevant dimension. (Learning curves over trial blocks had reached asymptote.) The means across subjects within conditions are shown in Figure 15.8. As can be seen there, the utilizations of these three different varieties of the shape dimension were quite different ($F(2,68)=3.44$, $p=.038$). Looking at Figure 15.8, the utilization of the square/circle (condition 1) was much higher than the other two. In fact the other two showed quite low utilizations reminiscent of the utilization of the shape dimension in previous experiments. Comparing the utilization of the square/circle in Figure 15.8 to the utilization of the color and size dimensions in Figure 15.7, it can be seen that they are very close to the same. Hence it might be that this version of the shape dimension is the one that will give three equally salient dimensions with color and the modified size dimensions of Experiment 5.

VIII. EXPERIMENT 7. EQUAL SALIENCE FOUND

The last experiment found a shape dimension that seemed to be comparable to the size and color dimensions of Experiment 5. It remained to put these three dimensions together in an experiment with the design of Experiment 4, and thoroughly test this hypothesis. A study not reported here showed that the size dimension still had a little higher salience than the other two dimensions (although it was not significant). Hence, for the present study, the size differential was made a little smaller than in Experiment 5.

A. Method

Subjects. The subjects were 89 students who were fulfilling a course requirement in introductory psychology. Subjects were randomly assigned to the three conditions giving 29 or 30 per condition.

Apparatus. The apparatus is the same as in Experiment 4 except that the shape stimuli were as described above.

Procedure. The procedure is the same as in Experiment 4.

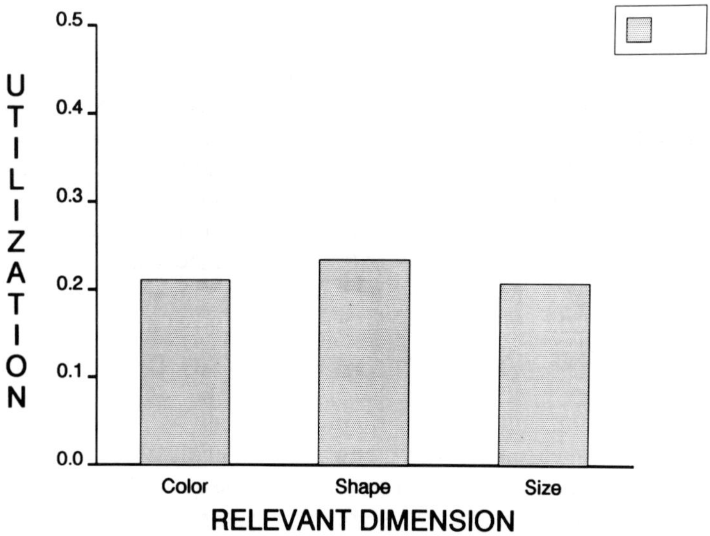

Figure 15.9. A bar graph of the mean asymptotic utilization by which dimension was relevant for Experiment 7.

B. Results and Discussion

Since the purpose of this experiment is to show no difference, very accurate measures are needed. To accomplish this, the proportions should be calculated over as many trials as possible. Inspection of the learning curves over trial blocks showed that learning had reached asymptote before the fourth trial block. Hence the proportions of one of the responses to each cue pattern were calculated for the last 192 trials for each subject. These response proportions were then transformed to the linear model terms as detailed in Edgell (1978, 1980). This gave measures of the asymptotic utilization of the relevant dimension. The means across subjects within conditions are shown in Figure 15.9. As can be seen there, the utilizations of these three different dimensions were very nearly equal ($F(2,89)=.27$, $p=.762$). It can be concluded that these stimuli are a set of stimuli with equally salient dimensions. Since other researchers may want to use these stimuli in research, Appendix 1 gives full details on them. It should be noted that the dimensions being equally salient does not guarantee that the patterns involving pairs of dimensions will also be equally salient. This would need testing before performing any research that required this in addition to equally salient dimensions.

IX. EXPERIMENT 8. SALIENCE EXPLAINED

The several previous experiments have shown that the particular form chosen for the stimuli can have a considerable influence on the subjects' utilization of the information transmitted by the stimuli. It has been shown that one can manipulate this salience by obvious changes in the relevant differences between the levels of the dimension in question. It has further been shown that one can find a set of three binary dimensions of equal, or at least very nearly equal, salience. However the main question remains: why are there differences in salience. As discussed above, it certainly might be expected that any easily discriminable set of stimuli should be equally efficient in giving the subjects the information. This should especially be true after a large number of learning trials. However, the fact is, it is not true.

In checking sequences of stimuli in the lab we often made lists of the stimuli as they appeared on the screen. Occasionally we made errors in these records which we discovered upon rerunning the sequence of stimuli. It was our intuitive impression that our errors were usually on the shape dimension when the stimuli were squares or triangles. Since, in making the records, we wrote the trial down from short term memory after the trial was complete, this led to the hypothesis that perhaps the salience effect was due to a problem with short term memory.

It is easy to see that mistakes in short term memory for the value of a relevant dimension would lead to lower utilization of that dimension. For example, suppose that the shape dimension (square and triangle) is relevant as it was in condition two of Experiment 4. It was a 75/25 cue (validity of .25). That means that on 75 percent of the trials when the stimulus was a square, "X" was the correct event. On the other 25 percent of the trials, "O" was the correct event. For the trials when the triangle was the stimulus, the opposite was the case. That is, on 25 percent of the trials "X" was the correct event, while on the other 75 percent "O" was the correct event. Now if the subject made only 10 percent memory errors in the value of the dimension, then on 90 percent of the trials that the subject recalled that a square was the stimulus it would have in fact been the stimulus, but on 10 percent of the trials when the subject recalled that a square was the stimulus it would have in fact been a triangle. On these mistaken trials the probabilities of the events are the reverse of what they should be. With this assumed 10 percent error rate the probability of an "X" given when the subject recalls square (which is 90 percent correct but 10 percent wrong) is only .7, not .75 as it should be. Thus the cue dimension appears to be a 70/30 one rather than the actual 75/25 one it is. The validity has been reduced from .25 to

.20. It is well-known that the validity of a cue dimension has a strong influence on the level of its utilization by subjects. This can also be clearly seen in the results of Experiments 1 and 3. (See Figures 15.2 and 15.4.)

Thus it was hypothesized that the difference in saliences of the various dimensions could be accounted for by different rates of misremembering of the values of the dimensions which causes perceived validity levels different from actual validity levels for the dimensions. A task was devised to test this. The environment of Experiment 4 was ideal for testing this hypothesis. There were three dimensions each with quite different saliences. If the hypothesis is true, then different levels of errors in memory for the three different dimensions should be observed. Furthermore, these differences should be ordered in magnitude so that the higher the salience the lower should be the number of memory errors. The task must use the same stimuli as in the probability learning task but measure memory instead. It was thought that having the subject do both a memory task and a probability learning task would be so complex as to hinder both, so it was decided to run just a memory task. On each trial one of the eight stimuli patterns from Experiment 4 was presented. Following the presentation of the stimulus pattern the subject was asked to recall the pattern on the trial before the one immediately past. This lag one recall scheme was used in order to produce some errors in memory but not too many. Pilot work showed that the lag one was just about right, and indicated promise for the hypothesis. So it was decided to run this procedure as an experiment.

A. Method

Subjects. The subjects were 29 students fulfilling a course requirement in introductory psychology. This does not include one subject whose data were eliminated from analysis because the subject did not follow instructions and recalled the stimulus just presented rather than the one previous to that.

Apparatus. The apparatus was the same as in Experiment 4 with several exceptions. The stimuli were the same as in Experiment 4, that is, the largest size differential and the square or triangle along with color. As there was no prediction to be made, there was no feedback as in Experiment 4. Responses were made by pressing a number key on the keyboard which was placed where the response buttons were for the previous experiments.

Procedure. At the beginning of the experiment subjects were read instructions in which they were told that the experiment was studying how well people can recall geometric figures. They were also told what the cues they would be seeing were, and what the task involved. They were given examples of the lag one recall that was required using common first names. Any questions that they had about what they were required to do were answered before proceeding. They were told that after the first geometric figure there would be nothing to recall and that they would be instructed on the CRT to just press the "0" key. On the other trials, they were told that they would be given eight choices numbered from one to eight from which to choose the figure that they had seen on the trial previous to the current one. They were informed that the order of the choices would keep changing. Indeed, on each trial the order of the eight possible stimuli patterns was randomly determined. Also the order of the dimensions was randomly determined on each trial. That is, on one trial one of the eight choices would be "small blue triangle" while on another it would be "blue triangle small." This was done to avoid any bias in remembering the dimensions that could be introduced by the nature of the response choice display.

The figure was on the screen for 3.0 seconds. The screen was then cleared for 0.5 seconds. Following this the multiple choice list was presented until the subject made a response by pressing one of the number keys from "1" to "8." Following this the screen was blank for 0.5 seconds after which another trial began. All subjects were run for 128 trials. Each of the eight possible stimuli patterns was equally likely and the order of them was randomly determined in blocks of 64 trials as had been the case in the previous experiments.

B. Results and Discussion

The number of errors for each dimension was scored for each subject. The means across subjects for each dimension are given in Table 15.1. As can be seen in Table 15.1, there are sizable differences in the mean number of errors for each dimension. This was significant using a multivariate repeated measures technique with Wilk's Lambda ($F(2,27)=8.29, p=.002$). Also given in Table 15.1 are the mean utilization scores from Experiment 4 for each dimension when it was the relevant dimension. It is easy to see a pattern there with the more highly utilized dimensions from the probability learning study having the lower error rates in the present memory study. To better test this result a linear equation was fit, using the standard least square error criterion, mapping the error scores to the utilization scores. The resulting predicted utilization scores are shown in Figure 15.10 by the darker bars. For comparison, the utilization scores from Experiment 4 are also shown in this figure by the lighter bars. As can be seen in the figure by comparing the height of the light versus the dark bar in each pair of bars, there is a remarkably good fit. In fact the fit accounts for 99 percent of the variance. Thus not only are the ordering of the saliences of these dimensions accounted for

Table 15.1. Mean memory errors from Experiment 8 and mean utilization scores from Experiment 4 by dimension.

Relevant Dimension	Utilization	Memory Errors
Color	.20	15.6
Shape	.14	16.8
Size	.32	11.9

by the error data but also the relative differences between the saliences are well accounted for by the error data.

These results leave no doubt that the differences in the saliences of the different dimensions are due to memory effects in short term memory. This obviously makes sense because the only way a subject can learn to use the rele-

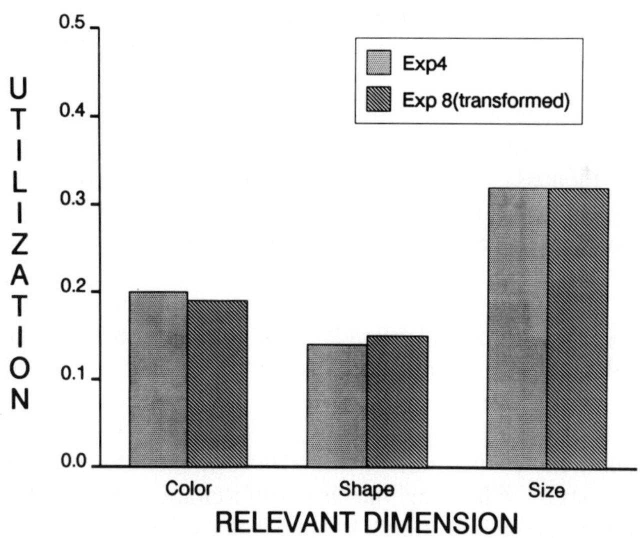

Figure 15.10. A bar graph showing the fit of the utilization predicted by the number of errors from Experiment 8 to the mean asymptotic utilization from Experiment 4 by which dimension was relevant for Experiment 4 or was misremembered for Experiment 8.

vant information is by remembering the past results from feedback and building up an expectancy for the dimensions as was postulated in the Castellan and Edgell (1973) model. If short term memory has some tendency to make errors for a certain dimension, its validity must appear to be less, given it is relevant, than its actual environmental validity. (An irrelevant dimension would still appear to be irrelevant.) When fewer or no errors are made in short term memory for a particular dimension, then the apparent validity of that dimension will be the same as its actual environmental validity. It is well-known, of course, that subjects will show a higher utilization for higher validity. This would be true for apparent as well as actual validity.

X. CONCLUSIONS

The premise that has pervaded the field, namely that a stimulus is a stimulus is a stimulus, is very untrue. One stimulus is not as good as another for presenting the information to the decision maker in tasks such as those used in the studies in this chapter. There are large salience differences between different stimuli. These salience differences translate into quite different levels of utilization of the information by the decision maker, even though the validity of the information is constant. This salience effect is present not only during the initial learning phase of the task, but continues throughout and remains at asymptote. This salience differential is present for many different stimuli dimensions. Different representations of the information, which are contained in a single figure (unitary stimuli) as well as those which are separated, were shown to have different saliences. One study even suggested that separating unitary dimensions, which had different saliences when combined in one figure, did not change the salience difference of them. Whether this will generalize to stimuli other than squares and triangles, and vertical and horizontal lines, is unknown. However, it seems reasonable to at least expect this to hold for many other stimuli and not be unique for squares and triangles, and vertical and horizontal lines. MacGregor and Slovic (1986) found some evidence for salience effects using a face display, however, the comparison is across experiments.

The studies in this chapter demonstrated that salience can be manipulated in rather obvious ways. Changing the size differential and the perceptual difference between the geometric stimuli had the expected effect on the salience. That is, larger perceived differences between the levels of a dimension made that dimension more salient. One can conjecture that this will hold for all stimuli dimensions, but too few examples were studied here to make this a conclusion. It was shown that equally salient dimensions could be found for both separable and unitary stimuli dimensions. This is important for many research projects using the probability learning technique. For example, in studying the effect of relevant dimensional information within and outside of a relevant pattern, it is necessary to have the two dimensions in question (the one inside the pattern and the one not in the pattern) have equal salience.

Several findings from previous studies need to be reevaluated in light of the findings from the studies in this chapter. For example, several authors have found an advantage for a unitary representation over a separable one in information processing tasks while others have found the separable equal or better than the unitary representation (Gopher & Kimchi, 1989). This could be explained by the varying saliences of the different representations of the information in the different studies. It may well be that the salience of the unitary representation was higher for the tasks that found an advantage for it, and the salience for the separable representation was higher for those tasks in which separable did better. As an additional example, consider the results of the third experiment of Edgell and Morrissey (1987). That study involved one dimension relevant from the beginning of learning with the second dimension switching from irrelevant to relevant during learning. It is unknown if the findings of those authors will generalize to other representations of the dimensions or even to reversing the role of the two representations they used by reversing which dimension is relevant from the beginning and which becomes relevant later.

The last study in this chapter showed that the salience effects are correlated with memory errors for the various stimuli representations. That is, the more prone to memory errors the particular representation, the less the salience and thus the less the utilization of that dimension will be. Hence the variance in the utilization can be explained by the salience affecting memory which causes a perceived difference in validity, even though the actual validity was constant across stimuli representations. While it seems reasonable that more noticeable differences in the levels of the dimension might make the dimension more memorable, it is a bit surprising that the levels used here, all of which were very different, would have had the large effects that were found. It remains for memory research to explain the large effects found in these studies.

The findings of these studies have obvious importance for laboratory work involving learning in a probabilistic environment. The question of applications in applied work must also be explored. Many researchers have been studying the use of computer graphics to find unitary stimuli which will aid the processing of several dimensions of complex information in applied settings. (For recent reviews, see Gopher and Kimchi, 1989, and Wickens and Kramer, 1985.) The studies have ranged from laboratory abstractions of real world tasks (e.g., Carswell & Wickens, 1987) to very real world settings such as a nuclear power plant control room (Petersen, Banks, & Gertman, 1981). One of the major differences between studies like those in this chapter and these applied studies is that learning with trial by trial feedback is not the usual in the applied settings. However, subjects will learn as they interact with the system even though the system has been explained to them. The findings of the studies in this chapter will obviously apply with full effect to such learning. Given that memory is the cause for the salience effect, it is hard to conjecture how these salience effects would not also affect performance in a system that was explained to the subjects. Memory is certainly involved in performance in such a system. Perhaps of even more importance, most such systems require the subject to compare the current observed state with earlier remembered states. The memory caused salience effects found in the studies in this chapter will obviously be found to be in full force there.

While laboratory researchers are probably more interested in finding dimensions of equal salience in order to avoid confounding in their studies, applied researchers are probably more interested in finding those stimuli representations with the highest salience. More research is needed to identify such. However, the problem is more complex because most real world tasks involve combining several relevant dimensions. The combining of information from several dimensions can even involve complex configural rules. What is most salient for a single relevant dimension may well not be the most salient when taken in an interaction with other relevant dimensions and relevant patterns of dimensions. While one can tentatively speculate that the salience effect will only be of greater magnitude in more complex information processing environments, only additional research can answer the question.

REFERENCES

Bjorkman, M. (1971). Policy formation as a function of feedforward in a non-metric CPL-task. Umea Psychological Reports, No.49.

Bjorkman, M. (1972). Feedforward and feedback as determiners of knowledge and policy: Notes on a neglected issue. Scandinavian Journal of Psychology, 13, 152-158.

Carswell, C. M., & Wickens, C. D. (1987). Information integration and the object display: An interaction of task demands and display superiority. Ergonomics, 30, 511-527.

Castellan, N. J., Jr. (1970). Determination of joint from marginal distributions in dichotomous systems. Psychometrika, 35, 439-454.

Castellan, N. J., Jr. (1973). Multiple-cue probability learning with irrelevant cues. Organizational Behavior and Human Performance, 9, 16-29.

Castellan, N. J., Jr. (1974). The effect of different types of feedback in multiple-cue probability learning. Organizational Behavior and Human Performance, 11, 44-64.

Castellan, N. J., Jr. (1977). Decision making with multiple probabilistic cues. In N. J. Castellan, Jr., D. B. Pisoni, & G. R. Potts (Eds.), Cognitive Theory (Vol. II, pp. 117-147). New Jersey: L. Erlbaum.

Castellan, N. J., Jr., & Brown, T. V. (1970). Routine for the generation of joint distributions in dichotomous systems. Behavioral Science, 75, 387.

Castellan, N. J., Jr., & Edgell, S. E. (1973). An hypothesis generation model for judgment in nonmetric multiple-cue probability learning. Journal of Mathematical Psychology, 10, 204-222.

Castellan, N. J., Jr., & Swaine, M. (1977). Long-term feedback and differential feedback effects in nonmetric multiple-cue probability learning. Behavioral Science, 22, 116-128.

Dudycha, A. L., Dumoff, M. G., & Dudycha, L. W. (1973). Choice behavior in dynamic environments. Organizational Behavior and Human Performance, 9, 328-338.

Edgell, S. E. (1978). Configural information processing in two-cue nonmetric multiple-cue probability learning. Organizational Behavior and Human Performance, 22, 404-416.

Edgell, S. E. (1980). Higher order configural information processing in nonmetric multiple-cue probability learning. Organizational Behavior and Human Performance, 25, 1-14.

Edgell, S. E. (1983). Delayed exposure to configural information in nonmetric multiple-cue probability learning. Organizational Behavior and Human Performance, 32, 55-65.

Edgell, S. E., & Castellan, N. J., Jr. (1973). Configural effect in multiple-cue probability learning. Journal of Experimental Psychology, 100, 310-314.

Edgell, S. E., & Morrissey, J. M. (1987). Delayed exposure to additional relevant information in nonmetric multiple-cue probability learning. Organizational Behavior and Human Decision Processes, 40, 22-38.

Edgell, S. E., & Morrissey, J. M. (1992). Separable and unitary stimuli in nonmetric multiple-cue probability learning. Organizational Behavior and Human Decision Processes, 51, 118-132.

Estes, W. K. (1972). Elements and patterns in diagnostic discrimination learning. Transactions of the New York Academy of Sciences, 34, 84-85.

Freedy, A., & Johnson, E. M. (1982). Human factor issues in computer management of information for decisionmaking. IEEE Transactions on Systems, Man, and Cybernetics, 12, 437-438.

Garner, W. R. (1976). Interaction of stimulus dimensions in concept and choice processes. Cognitive Psychology, 8, 98-123.

Goldsmith, T. E., & Schvaneveldt, R. W. (1984). Facilitating multiple-cue judgments with integral information displays. In J. C. Thomas & M. L. Schneider (Eds.), Human Factors in Computer Systems (pp. 243-20.). Norwood, NJ: Ablex.

Gopher, D., & Kimchi, R. (1989). Engineering psychology. Annual Review of Psychology, 40, 431-455.

Hammond, K. R. (1971). Computer graphics as an aid to learning. Science, 172, 903-908.
Hammond, K. R., & Boyle, P. J. R. (1971). Quasi-rationality, quarrels and new conceptions of feedback. Bulletin of British Psychological Society, 24, 103-113.
Lindberg, L.-A., & Brehmer, B. (1976a). Retroactive interference in single-cue probability learning. Scandinavian Journal of Psychology, 17, 35-40.
Lindberg, L.-A., & Brehmer, B. (1976b). Transfer in single-cue probability learning. Organizational Behavior and Human Performance, 16, 177-192.
MacGregor, D. & Slovic, P. (1986). Graphic representation of judgmental information. Human Computer Interaction, 2, 179-200.
Meehl, P. E. (1954). Clinical versus statistical prediction. Minneapolis: University of Minnesota Press.
Mellers, B. A. (1980). Configurality in multiple-cue probability learning. American Journal of Psychology, 93, 429-443.
Peterson, C. R., Hammond, K. R., & Summers, D. A. (1965). Multiple probability-learning with shifting weights of cues. American Journal of Psychology, 78, 660-663.
Petersen, R. J., Banks, W. W., & Gertman, D. I. (1981). Performance-based evaluation of graphic displays for nuclear power plant control rooms. Proceedings of the Conference on Human Factors in Computer Systems, Gaithersburg, MO, 182-189.
Ruffner, J. W., & Muchinsky, P. M. (1978). The influence of shifting cue validity distributions and group discussion feedback on multiple cue probability learning. Organizational Behavior and Human Performance, 21, 189-208.
Stockburger, D. W., & Erickson, J. R. (1974). Probabilistic discrimination learning with dimensionalized stimuli. Organizational Behavior and Human Performance, 11, 157-171.
Summers, D. A. (1969). Adaptation to change in multiple probability tasks. American Journal of Psychology, 82, 235-240.
Wickens, C. D., & Kramer, A. (1985). Engineering psychology. Annual Review of Psychology, 36, 307-348.

APPENDIX 1

The stimuli in Experiment 7 were generated on Digital Equipment Corporation VT-240 and VT-340 color graphics CRTs. The triangles were isosceles triangles. The large triangle measured 62 mm high and 57 mm along the base. The small triangle measured 46 mm high and 42 mm along the base. The circles were actually a bit egg shaped with the longer axis in the vertical and the shorter in the horizontal. The large circle measured 50 to 62 mm in diameter while the small circle measured 46 to 52 mm in diameter. The CRT's default colors for blue and red were used. These colors in the HLS (hue/lightness/saturation) specifier system are blue: 0/50/100 and red 120/50/100. Below are given the ReGis graphics commands for generating the stimuli.

Large blue triangle
```
P[400,70]F(V(W(I1))(B)[500,270][300,270](E))
```
Large red triangle
```
P[400,70]F(V(W(I2))(B)[500,270][300,270](E))
```
Large blue circle
```
P[400,170]F(C(W(I1))[+100])
```
Large red circle
```
P[400,170]F(C(W(I2))[+100])
```
Small blue triangle
```
P[400,095]F(V(W(I1))(B)[475,245][325,245](E))
```
Small red triangle
```
P[400,095]F(V(W(I2))(B)[475,245][325,245](E))
```
Small blue circle
```
P[400,170]F(C(W(I1))[+83])
```
Small red circle
```
P[400,170]F(C(W(I2))[+83])
```

Commentary

The Effects of Representation on the Processing of Probabilistic Information, S.E. Edgell, R.D. Bright, P.C. Ng, T.K. Noonan, & L.A. Ford

ALBERT F. SMITH
State University of New York at Binghamton

Edgell et al. investigated subjects' utilization of probabilistic information. On each trial of their experiments, a subject views a multiattribute cuing stimulus, predicts which of two levels of the single varying attribute of a second outcome stimulus will obtain, and observes the outcome event. For each subject, there is an association, albeit imperfect, between one attribute of the set of set of cuing stimuli (the relevant attribute) and the single attribute of the outcome event.

This paradigm may be conceptualized as a set of interleaved probability learning problems (see, e.g., Castellan, 1977; Estes, 1964): Each of the *n* unique cuing stimuli defines a context within which each level of the outcome attribute occurs with specified probability. The most complex description of the task would be as *n* interleaved probability learning problems, but the situation in these experiments can be described more simply as two probability learning problems, each of which is signalled by one level of the relevant cuing attribute. (This is analogous to the reduction of a paired-associates learning problem into a classical concept learning problem with a suitable stimulus set and stimulus-response pairings; see Garner, 1962, Ch. 10.)

If information about the relevant attribute is available to a subject in Edgell et al.'s experimental task, and if the two levels of the relevant attribute are discriminable, the responses of an individual who understands the task should, within a small number of trials, come completely under the control of the relevant attribute (given the levels of relevant-attribute validity used by Edgell et al.). In other words, such an individual should determine which attribute of the cuing set is relevant and should respond according to its level on each trial. To maximize performance, in the sense of maximizing the number of trials on which the outcome event is predicted correctly, the optimal strategy is to predict the likelier outcome event conditional on the level of the relevant attribute of each cuing stimulus (see, e.g., Kroll, 1970).

Edgell et al. found that their subjects did not exhibit such response patterns: Attributes of the set of cuing stimuli, when relevant, differed in the extent to which they were utilized by experimental subjects. Edgell et al. suggest that utilization of the relevant attribute depends on its salience and that the salience of a binary-valued attribute depends on the discriminability of its levels; they demonstrate that reducing the discriminability of the levels of a relevant attribute decreased utilization of that attribute. These ideas about attribute salience and its operationalization are well established within the literature on concept discrimination (see Trabasso & Bower, 1968).

Edgell et al. surmise that at the root of the differential attribute utilization phenomenon is the memoriability of attribute levels--that the levels of less salient attributes are less memorable. Although they present data that purport to show that the memorability of an attribute's levels is related to that attribute's utilization in a probability learning task, their account is not fully satisfying: In the memory experiment, subjects were required to report which cuing stimulus had occurred on the last trial but one. However, what is to be remembered in probability learning is not what stimulus occurred on some previous trial, but rather what outcome occurred given a particular cuing stimulus. In addition, Edgell et al. assessed memory for the earlier cuing stimulus rather than for its three attributes.

A more complete understanding of the differential attribute utilization phenomenon will require a fuller appreciation of why performance in probability learning tasks improves with training. Two rather distinct alternatives are that learning is a consequence of hypothesis testing and confirmation and that learning proceeds essentially involuntarily.

To explore the possibility that learning proceeds by hypothesis testing, one might analyze in more detail responses to individual cuing stimuli. Edgell et al. assert that attribute utilization was measured after subjects had achieved "asymptotic" performance. This may not be sufficiently informative, if what is of interest is the specific pattern of response allocation given each cuing stimulus. Further analysis could reveal what hypotheses are entertained, and it would be of interest to see how many trials would be required for performance, and hence utilization of the relevant cuing attribute, to improve further.

If learning is essentially involuntary, an individual's propensity to predict an outcome, given some cuing stimulus, would be conditioned during exposure to the training sequence (see, e.g., Estes, Campbell, Hatsopoulos, & Hurwitz, 1989, and Nosofsky, Kruschke, & McKinley, 1992, for discussions of possible mechanisms). Differential attribute utilization might be explained by generalization of reinforcement across the levels of the relevant attribute when they are less discriminable. Nevertheless, with sufficient training, utilization of less salient attributes would increase to levels as high as those observed for more salient attributes (see Massaro, Halpern, & Moore, 1968).

A deeper understanding of these issues will likely require examination of performance in task environments more constrained than that used by Edgell et al. Performance might be studied in situations in which the criterion to be achieved by the subject is made more explicit. In addition, performance might be studied in a probabilistic concept acquisition paradigm in which responses would be collected in no-feedback test blocks that would be administered to subjects after they had observed a series of cuing stimuli and their associated outcomes. Finally, comparison of performance among experimental conditions that differ in the nature of the perceived interaction among the attributes used to generate the stimuli might help clarify the phenomenon studied by Edgell et al.

Castellan, N.J. (1977). Decision making with multiple probabilistic cues. In N.J. Castellan, D.B. Pisoni, & G.R. Potts (Eds.), Cognitive theory (Vol. II, pp. 117-147). Hillsdale, NJ: Erlbaum.

Estes, W.K. (1964). Probability learning. In A.W. Melton (Ed.), Categories of human learning (pp. 89-128). New York: Academic Press.

Estes, W.K., Campbell, J.A., Hatsopoulos, N. & Hurwitz, J.B. (1989). Base-rate effects in category learning: A comparison of parallel network and memory storage-retrieval models. Journal of Experimental Psychology: Learning, Memory and Cognition, 15, 556-571.

Garner, W.R. (1962). Uncertainty and structure as psychological concepts. New York, Wiley.

Kroll, N.E.A. (1970). Learning of several simultaneous probability learning problems as a function of overall event probability and prior knowledge. Journal of Experimental Psychology, 83, 209-215.

Massaro, D.W., Halpern, J., & Moore, J.W. (1968). Generalization effects in human discrimination learning with overt cue identification. Journal of Experimental Psychology, 77, 474-482.

Nosofsky, R.M., Kruschke, J.K., & McKinley, S.C. (1992). Combining exemplar-based category representations and connectionist learning rules. Journal of Experimental Psychology: Learning, Memory and Cognition, 18, 211-233.

Trabasso, T. & Bower, G.H. (1968). Attention in learning: Theory and research. New York: Wiley.

Percepts, Concepts and Categories
B. Burns (Editor)

16

Reading Graphs: Interactions of Processing Requirements and Stimulus Structure

C. MELODY CARSWELL
University of Kentucky

I. INTRODUCTION

The "reading" of graphical displays is a skill we often use, as students, researchers, and instructors, and as operators of cars, radios, VCRs, clocks, and a host of other everyday technologies. The line graphs, bivariate point plots, and bar charts sprinkled throughout the present book are examples of graphical displays, as are the dynamic analog displays associated with transportation, consumer goods, and industrial applications. In order to understand how we read such graphs and, in particular, to shed light on why some graphs seem to be so much easier to read than others, researchers have focused heavily on the structural characteristics of these complex visual stimuli. For example, they have tested an assortment of physical dimensions to determine their usefulness as codes for various quantitative variables. In addition, they have looked at the performance effects associated with various combinations and configurations of these dimensions. Research that looks at how such structural differences or formatting variables affect the usability of graphical displays has been called *comparative graphics* (DeSanctis, 1984).

The premise of the present chapter is that emphasis on structural variables alone is insufficient to account for many aspects of graphical communication, even those associated with variations in the efficacy of different graphical formats. One of the most common findings in the field of comparative graphics is that the efficacy of any graphical format is highly task dependent (see reviews by Carswell & Wickens, 1988; DeSanctis, 1984; MacDonald-Ross, 1977). This strongly suggests that we need to take note not only of how our graphs are composed, but also of what our subjects *are doing* with this graphically-presented information. In other words, we must devote attention not only to the structural characteristics of graphical displays, but also to the processing demands of the tasks the graphs subserve. It is essential to describe the ways in which processing demands interact with structural variables to determine the overall performance of the user.

The first part of this chapter is devoted to tracing the history of comparative graphics within the framework of this structure-process, or display-task, distinction. This review is not intended as a compendium of the customary design practices or speculative advice offered in a number of texts on graphing methods (e.g., Brinton, 1914; Everitt, 1978; Schmid & Schmid, 1979; Tufte, 1983, 1990), as good as many of these ideas may prove to be. Instead, the review will be limited to reports of experiments in which display variables, task variables, or both are manipulated by the investigators. The discussion will focus on the gradual refinement of the structural concepts used to distinguish graphical formats, as well as on the growing interest in the processing demands of graphical tasks. This review will be followed by a description of the Proximity Compatibility Hypothesis, one framework which has been developed specifically in response to the need to understand structure-process interactions in the reading of graphs. The validity and utility of this theory will be discussed within the context of experiments directly designed to test its predictions, as well as by attempting to use the theory to summarize the results of a sample of studies in comparative graphics.

II. COMPARATIVE GRAPHICS

Comparative graphics may be defined as the attempt to empirically rank graphical formats in terms of their associated performance efficiency or ease of use. Before describing this search for "bad," "good," and "better" graphs, we need to provide both a limiting definition of graphical displays, as well as a list of characteristics that most graphs share. A critical feature shared by all graphical displays is that each uses variation in one or more physical dimensions as an analog for variation of some directly measured or derived quantity. Using the terminology offered by Kosslyn (1989), the dimension(s) chosen as the information channels for a particular graph will be called the graph's *specifiers.* Typical specifiers include the linear extent of a bar in a bar chart, the horizontal and vertical position of a point in a scattergram, and the angle of a pointer in a meter, dial, or clock. However, when planning a graphical display, the designer must choose not only the specifiers and their configuration, but must also select a *background*, a *framework* (e.g., axes, grids), and alphanumeric or iconic *labels.* Kosslyn (1989) has argued that these are the basic-level constituents of most graphical displays.

A particular combination of this potentially endless variety of specifiers, frameworks, backgrounds, and labels leads to a unique graphical *format.* Formats that are used frequently and consistently tend to be associated with names, some as familiar as "bar charts" and "line graphs," and others somewhat more esoteric, such as "Nightingale petals" and "variable-length whiskers." Given the range of structural variation possible among graphs, the goal of comparative graphics has been to determine which types of variation are critical for performance. Most efforts have focused on formats differing in their choice of specifiers, or on formats that differ in their configuration of specifiers.

A. Early Research

Statisticians and educators performed much of the early research aimed at specifying the characteristics of a good graph. In the 1920s, a debate was waged in the pages of the *Journal of the American Statistical Association* regarding the relative efficacy of two formats that differed both in their framework and specifiers -- pie charts and segmented bar charts. (e.g., Eells, 1926; Croxton, 1927;

Croxton & Stryker, 1927). Even though each of these studies focused on the ability of subjects to estimate proportions, the issue of which format was actually better for this task was not completely resolved, and researchers were still experimentally comparing bars and pies some years later (e.g., Culbertson & Powers, 1959; Sparrow, 1989). Despite their lack of consistent results, these studies represent perhaps the earliest instance of performance data collected on alternative graphical formats.

B. Semantic Congruence and Visual Analogies

At roughly the same time that statisticians were concerned with the relative merits of segmented bars and pies, educational researchers became interested in the semantic compatibility of a graph's specifier with the variable it was chosen to represent. It was reasoned that the more concrete or pictorial the relationship between the specifier and the concept, the easier the graph would be to comprehend and recall. Thus, much of this research was concerned with the relative utility of the pictograph, a format that attempts to reflect some aspect of the variable being represented by using elements of familiar objects as specifiers. For example, in order to represent yearly consumption of coffee by Americans from 1980 to 1985, one might devise a pictograph that utilizes some dimension of a coffee cup, say its height or volume, as the specifier for gallons of coffee. An alternative pictographic method, called the unit pictograph, would vary the number of coffee cups placed end to end to represent the amount of coffee consumed during a given year. Both traditional and unit pictographs may be compared to other graphical methods of presenting this information, such as using the height of a rectangle, a semantically neutral geometric object, to represent coffee quantity in a simple bar chart.

One of the earliest studies of pictographs was carried out by Washburne (1927). Pictographs, along with line graphs, bar charts, and tables of numbers, were used by several thousand jr. high school students as supplemental material for lessons in economic history. Pictographs were found to be relatively poor for enhancing recall of comparative information and specific numeric facts. However, other research soon revealed conflicting results. Wrightstone (1936) found that when compared to bar charts, line graphs, and circle diagrams, students were able to more efficiently locate information in pictographs. They were also more successful at delayed recall when trying to reproduce information from that format. One critical difference between these two studies may well have been the type of pictograph used. Washburne (1927) utilized traditional pictographs (e.g., varying the size of money bags to represent differences in revenue), while

Wrightstone (1936) used unit pictographs. A study by Strickland (1948) provides further evidence that unit pictographs may prove to be especially good formats for use by younger subjects. In her study of elementary school children, Strickland documented the ability of children to answer a variety of questions about information presented in unit pictographs and several other formats including bar charts, line graphs, and pie charts. She found that for first grade students, questions about pictorial formats were answered most accurately. This advantage for the pictograph was not found for older children. In addition, Vernon (1952) found that adult subjects with lower educational levels preferred to use pictographs rather than line graphs.

Most of the research on semantic congruence has used as subjects children (e.g., Washburne, 1927; Wrightstone, 1936; Strickland, 1948) or adults with low levels of education (e.g., Vernon, 1952). Implicit in this choice of subjects is the notion that the mapping of levels of a quantitative variable onto levels of many of the more traditional specifiers (e.g., linear extent, angle, position) is often arbitrary and, thus, must be learned. However, the use of pictographs may provide a shortcut to graphical skills. Returning to our coffee consumption example, the naive subject may have to be taught that the longer the bar or the higher the point, the greater the quantity of coffee that was consumed. However, with the pictograph, the subject can depend on his or her knowledge that larger cups hold more coffee than smaller cups, or in the case of the unit pictograph, that more cups hold more coffee than fewer cups.

As an historical aside, the earliest types of graphical displays were limited to those forms that involved very literal relationships between types of information represented and types of specifiers used (Beniger & Robyn, 1978). Maps, for example, utilized horizontal and vertical extent as specifiers for, on a different scale, horizontal and vertical extent. It was not until the late 18th and early 19th centuries that derivative graphs, graphs that were not limited to literal space-based descriptions, began to come into limited use. A number of historical accounts (e.g., Funkhouser, 1937; Schmid & Schmid, 1979; Wainer & Thissen, 1981) name William Playfair (1759 - 1823), a Scottish political scientist and economist, as the most influential developer and proponent of many of the derivative graphics in use today. However, in introducing his "lineal arithmetic," as he called his graphs, Playfair apparently met with some resistance. Even techniques such as the simple vertical bar chart were not immediately accepted by a public who still thought of graphs as literal depictions of spatial extent. This is nowhere more evident than in Playfair's introduction to the third edition of his *Commercial and Political Atlas* (1801) where he tries to teach his readers to use a bar chart by resorting to the following visual analogy:

> "*Suppose the money received by a man in trade were all in guineas, and that every evening he made a single pile of all the guineas received during the day, each pile would represent a day, and its height would be proportioned to the receipts of that day, so that by this plain operation, time, proportion, and amount, would all be physically combined. Lineal arithmetic then, it may be averred, is nothing more than these piles of guineas represented on paper and on a small scale, in which an inch (suppose) represents the thickness of five millions of guineas, as in geography it does the breadth of a river, or any other extent of country*" (p. 6).

The study of semantic congruence thus reflects important historical trends in the development of graphical displays, and emphasizes the importance of the subject's previous knowledge in the use of graphs. The interest of early investigators in this problem also indicates an acknowledgment that stimulus structure alone could not account for variations in graphical efficacy. Instead, they make an argument for the importance of both stimulus structures and conceptual structures, and further suggest that the match formed by these structures is important to the reading of graphs. Interestingly, in the design of these early experiments, there is also evidence that investigators believed processing variables to be important. Washburne (1926), Wrightstone (1936), and Strickland (1948) all made a point of including different tasks (e.g., recall of specific facts vs. recall of trends) when comparing their various formats. While they provided little in the way of a rationale for including task type as a variable, their data emphasize the importance of doing so. That is, format superiorities frequently did not generalize from one task to another.

C. The Psychophysics of Graphical Specifiers

While the early researchers may have known that processing variables played a role in determining which graphical formats were most easily used, they nonetheless emphasized structural factors. In addition to arguing for the importance of the match between stimulus and conceptual structures, some researchers argued that it was important to determine the resolution with which various specifiers (physical dimensions) could be represented perceptually (e.g., Croxton & Stein, 1932). Graphical efficacy, they argued, must depend on our differential abilities to perceptually encode the various physical dimensions chosen as graphical specifiers. Returning to the pictograph, research showed that these graphs were not universally superior to more abstract forms such as line graphs and bar charts. Traditional pictographs, in particular, resulted in quite poor performance (Washburne, 1927). A psychophysical explanation of these data would suggest that the specifier used to represent quantity in the unit pictograph, (length of a row of objects, or number of objects in the row) may simply be encoded more accurately than the specifiers used in traditional pictographs (i.e., variations in size of irregularly shaped objects). This argument is, essentially, the one used by designers who have promoted the unit pictograph as the preferred method (Brinton, 1914; Neurath, 1944).

Researchers interested in both static statistical and educational graphs, and those interested in dynamic analog displays have frequently focused on our perceptual capabilities to identify and discriminate levels on individual physical dimensions. Cleveland and colleagues (Cleveland, 1985; Cleveland & McGill, 1985; Cleveland & McGill, 1984; Cleveland, Harris & McGill, 1983) have looked at the ease with which subjects can estimate the relative magnitude of values along dimensions commonly used as graphical specifiers. These authors contend that magnitude estimations using different specifiers invoke different 'decoding' processes. These 'elementary perceptual tasks' are assumed to form at least part of the processing demands of many higher-order graphical tasks. Their results suggest that our ability to make exact comparisons is indeed dependent on the physical dimensions chosen by the designer. Cleveland and McGill (1984) provide the following ranking of common graphical specifiers, ordered from the most to least accurately used: (1) positions on common aligned scales, (2) positions on common nonaligned scales, (3) lengths, (4) angles, (5) areas of circles, and (6) areas of irregularly shaped "blobs." Similar data on absolute and relative judgments for various physical dimensions are routinely provided in most elementary expositions on dynamic display design (Hutchingson, 1981; McCormick & Sanders, 1982).

D. Complex Stimulus Structure: Specifier Interactions

The majority of the work on the psychophysics of perceptual specifiers has emphasized the coding of individual quantitative variables as single physical dimensions. Thus, the structural variable of interest has typically been the dimension(s) chosen as specifiers. However, a new approach to describing graphical structure has emerged in the last few decades. Advances in computer graphics have made this a period of innovation in graphical design, with 19th century techniques being resurrected, with traditional statistical formats being animated and used in industry, with industrial formats being used by scientists and statisticians, and with totally new techniques being developed. It is not surprising that several authors have referred to the past two decades as a "graphical renaissance" (Kruskal, 1977; Beniger & Robyn, 1978; Chernoff, 1978; Barnett, 1981). This renaissance has, in turn, produced new challenges for comparative graphics. Researchers have responded to this challenge by emphasizing the structural aspects that characterize each graph as a whole. Thus, rather than treating the specifiers as independent design units, researchers have become interested in the nature of the perceptual interactions that occur among a graph's specifiers.

The *object display* is one multivariate graphical design that, in particular, has led to much speculation about the impact of specifier interactions on user performance. Object displays include any format that utilizes several dimensions of a single perceptual object as its specifiers. Thus, if the height of a rectangle were used to represent one variable and the width of the same rectangle were used to represent a second variable, then the display would qualify as an object display. This formatting technique is usually compared with graphical techniques that use a single dimension of each of several distinct objects to display multiple values. For example, a traditional bar graph is a multi-object display that uses the height of each of several rectangles as its specifiers.

One of the more common object displays is the star or polygon display which uses the distance of several radii from a reference point as the specifiers for several variables. Dynamic polygon displays have found their way into nuclear power plant control rooms where they serve as safety parameter displays (Wood, Wise, & Hanes, 1981). Their use has also been suggested for monitoring the vital signs of hospitalized patients (Siegel, Goldwyn, & Friedman 1971; Cole, 1986), for military command-and-controloperations (Wickens & Scott, 1983), as well as for aviation applications (Wickens & Andre, 1990). A second relatively well-known object display is the Chernoff face display (Chernoff, 1973) which uses variation in the features of a cartoon face to represent up to twenty different variables. Chernoff faces have been used to present multivariate economic data (e.g., Moriarity, 1979) and have been experimentally used for quality control of circuit boards (Hahn, Morgan, & Lorensen, 1983). Perhaps the oldest and most common single-object display, however, is the simple line graph.

Several comparative graphics studies have pitted various object displays against one another in a variety of tasks (Brown, 1985; Mezzich & Worthington, 1978; Wilkinson, 1981). However, the majority of the studies involving object displays have emphasized performance evaluations designed to determine whether single-object displays prove to be superior to multi-object formats (see Wickens, 1986, for review). As summarized by Carswell and Wickens (1987a), much of this research has been driven by theoretical claims that the properties of an object are processed efficiently in parallel (e.g., Duncan, 1984; Kahneman & Treisman, 1984), by the practical appeal of spatially compact displays, and by specific benefits presumed to result from the perceptual interactions of spatially proximal physical dimensions.

The potential performance benefits that might accrue to graphs containing specifiers that interact perceptually, was first discussed by Jacob, Egeth and Bevan (1976). Building on the work of Garner and colleagues (Garner, 1970, 1974; Garner & Felfoldy, 1970), these researchers distinguished between integral and separable displays. Following Garner's description of integral dimensions as two or more physical dimensions that are represented as a single psychological dimension in the observer, Jacob et al. (1976) suggested that some graphical specifiers might be characterized as integral. For example, the designer of a bivariate point display (e.g., a scatter plot) may use the horizontal and vertical position of a single dot to represent the value of two variables. However, upon looking at the graph, the user may not encode these two physical dimensions separately. Instead, two-dimensional position may be coded as a single psychological entity. This situation is contrasted with separable dimensions (and separable displays) in which two or more physical dimensions (or specifiers) remain psychologically distinct. Jacob et al. (1976) argued that the mandatory perceptual integration associated with integral dimensions might prove to be useful as a way of circumventing the demanding logical integrations sometimes required of the information represented by the display. As a demonstration, they compared object displays they considered to be integral (faces and polygons) with relatively separable formats (glyphs and numeric stimuli) for paired associates learning and a subjective clustering task. The results indicated performance advantages for their integral displays. Several studies comparing other presumably integral and separable displays followed, but provided conflicting results regarding the advantages of object displays in general, and integral displays in particular (e.g., Goldsmith & Schvaneveldt, 1984, Goettl, Kramer, & Wickens, 1986; Casey & Wickens, 1986; Carswell & Wickens, 1987a,b).

In an attempt to further determine the circumstances in which object displays might enhance performance, several authors have suggested an alternative to the integral/separable description of perceptual interactions among specifiers. They propose that effective object displays (and effective multi-object displays) contain specifiers that are more aptly described as configural (Barnett & Wickens, 1988; Coury & Purcell, 1988; Sanderson, Flach, Buttigieg, & Casey, 1989). Even Jacob et al. (1976) suggest that their 'integral' displays may actually be 'configural.' Configural dimensions differ from truly integral dimensions in that configural dimensions maintain separate perceptual codes. In addition, and in contrast with totally separable dimensions, properties of the stimulus as a whole may also be coded (Garner, 1974, 1978; Pomerantz, 1981; Pomerantz & Garner, 1973; Pomerantz, Sager, & Stoever, 1977). Some researchers in comparative graphics (e.g., Sanderson et. al, 1988) have argued that these holistic properties, or emergent features, may be an important component of successful graphical design. The emergent feature approach to graphical design suggests that displays will be easily used if an emergent feature is available that serves as a direct cue to a correct response. For example, the formal specifier in a line graph may be the vertical position of a series of dots that are connected by lines. However, the overall shape of the line connecting the dots, its symmetry or overall slope, may quickly provide information essential to the task to be performed by the graph's user.

E. Interactions of Stimulus Structure and Processing Requirements

Whether comparing integral versus separable displays, configural versus nonconfiguraldisplays, single-object versus multi-object displays, or pictorial versus abstract formats, one of the single most consistent findings in comparative graphics is that the efficacy of any format is strongly task-dependent (DeSanctis, 1984; MacDonald-Ross, 1977; Carswell & Wickens, 1988). As previously noted, the importance of processing variables has been recognized by many investigators who have included task type as a factor in their experimental designs. Typical classifications of graphical tasks have included such dichotomies as localization of specific facts versus synthesis of facts (Wrightstone, 1936), or assessment of trends versus finding exact numbers (MacDonald-Ross, 1977). Several authors have recently suggested that if our understanding of graphical comprehension is to proceed, we must develop a theoretical framework for relating stimulus characteristics to the information processing demands of the task for which the graph is designed (MacGregor & Slovic, 1984; Sparrow, 1989; Wainer and Reiser, 1978). This requires the development of a taxonomy of processing requirements, a taxonomy of psychologically- significant structural characteristics, and a method of specifying which structure-process matches will be most compatible.

III. PROXIMITY COMPATIBILITY: MATCHING STRUCTURE TO PROCESS

The *Proximity Compatibility Hypothesis* (PCH) is one attempt to provide a framework for describing how well suited the structural characteristics of a graph are for the processing demands of different graphical tasks. Specifically, the PCH proposes that graphical tasks differ in the amount of *processing proximity* that they require. Additionally, graphical displays differ in terms of the *structural proximity* of the various specifiers used in the construction of the display. The PCH proposes that the level of processing proximity and structural proximity must be carefully matched in order to insure that a given display is a "good" graph -- a graph that allows rapid and accurate performance of the requisite task. Before describing the nature of this structure-process match, the concept of proximity as applied to both graphical tasks and graphical displays will be defined and elaborated.

A. Processing Proximity

High processing proximity characterizes those graphical tasks that require integration of the quantitative information carried by the graph's specifiers. Information integration occurs when there is convergence during processing of inputs from two or more specifiers; the task simply cannot be performed accurately on the basis of a subset of these information channels. A typical example of a task requiring high processing proximity would be the researcher's integration of information regarding a single variable across several points in time in order to predict future values. As an additional example, consider the task of the supervisor of an automated process such as a nuclear reactor who must consider the value of each of several safety parameters in order to decide whether the system is operating normally.

At the other extreme of processing proximity lies the *low processing proximity* task. In such a task, the graph's multiple specifiers are treated, essentially, as input to multiple independent tasks, performed either concurrently or in isolation. Thus both multiple tasks and focusing tasks are considered low in processing proximity demands. A multi-channel task will qualify as a low proximity task if variation in each of the specifiers could be monitored and responded to by a different person, without any interpersonal communication being necessary. An example of a low proximity task might be the aviator's attempt to monitor fuel level while trying to simultaneously maintain the plane's altitude within critical boundaries. An additional example of a low proximity task would be the physician attempting to focus on the more reliable of two test results in order to provide a diagnosis.

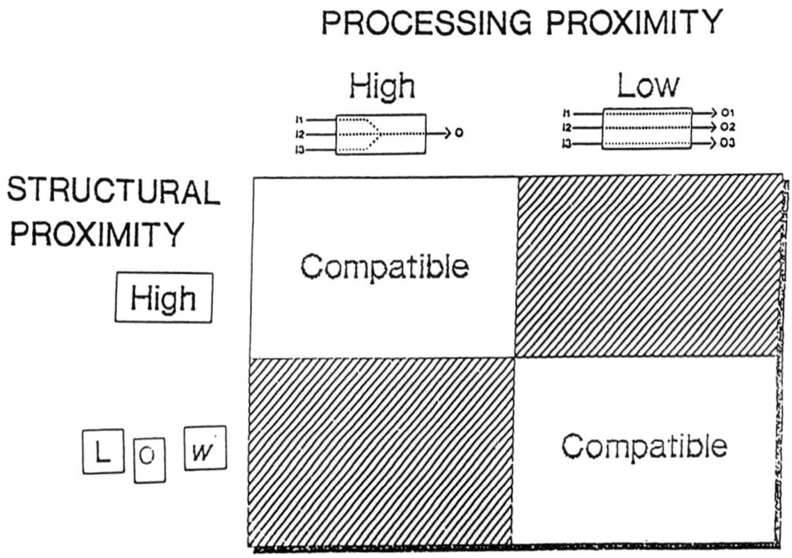

Figure 16.1. Predictions of the Proximity Compatibility Hypothesis (PCH): Highly efficient performance should result when graphs with high structural proximity are used for tasks involving high processing proximity (i.e., information integration). Likewise, graphs with low structural proximity should be used to perform low proximity tasks (i.e., multiple independent tasks, focusing tasks).

B. Structural Proximity

In addition to high and low processing proximity, the PCH also distinguishes between high and low structural (display) proximity. Structural proximity refers to any manipulation of the specifiers that allow these specifiers to group or interact perceptually for the user. Thus, two specifiers that are integral or configural are certainly higher in structural proximity than are two separable specifiers. Further, specifiers that are similar or are spatially proximal are also likely to be high in structural proximity. Thus, object displays would constitute a particularly high level of structural proximity, while multi-object displays such as bar graphs would exemplify low structural proximity.

C. Predicting the Compatibility of Structure-Process Matches

Having described distinctions between both graphical displays (structural proximity) and graphical tasks (processing proximity), the PCH further adds that high proximity tasks will be best performed with high proximity displays. Low proximity tasks, on the other hand, will be best supported by low proximity displays. These predictions are summarized in Figure 16.1, where the two columns represent high and low processing proximity, and the two rows represent high and low structural proximity.

The predictions of efficient processing when a high proximity display is used for high proximity processing (upper left quadrant of Figure 16.1) is supported by theory and research in the areas of perceptual organization and selective attention. In the case that such displays are integral or configural, the perceptual integration of the formal dimensions used by the display designer (i.e., the specifiers) may provide a shortcut to the logical integration that might otherwise be required to complete such tasks. Additional benefits are predicted for such display-task matches by space-based and object-based models of selective attention (Duncan, 1984), which suggest that efficient parallel processing occurs for stimulus dimensions located in close spatial proximity or within the confines of a single object.

However, if displays with high levels of structural proximity are used for tasks requiring low processing proximity, performance may be hindered by the same factors that facilitate performance in high proximity tasks. Integral dimensions used as specifiers may require additional resource-demanding processing in order for the user to ascertain the value of one dimension independent of the other(s). Salient emergent features formed when specifiers configure may distract the user's attention from the relevant "parent" dimensions. The effective parallel processing associated with dimensions that are spatially proximal or part of a single object may lead to response competition, confusions, and intrusions when the dimensions are each used to perform different tasks, or when it is necessary to focus on some subset of the dimensions displayed. These shortcomings associated with high structural proximity may be avoided, according to the PCH, by matching low structural proximity (e.g., multi-object displays) with low processing proximity (lower left quadrant of Figure 16.1). In addition to circumventing problems arising form dimensional interactions and response confusion/conflict, the use of multiple spatially segregated objects for multiple tasks allows for the easy referencing of specifiers to responses by using spatial location. Thus, multi-object displays paired with multiple independent responses provide a stimulus-response ensemble in which spatial congruence can be exploited (see Kornblum, Hasbroucq, & Osman, 1990, for a recent review of issues surrounding such stimulus-response compatibility).

D. Testing the PCH: High Proximity Processing

The predictions of the PCH for tasks requiring high proximity processing, shown in the left half of Figure 16.1, were initially tested in several experiments conducted by Carswell and Wickens (Carswell & Wickens, 1987a,b; Wickens, Kramer, Barnett, Carswell, Fracker, Goettl, & Harwood, 1985) In these studies, a high and low proximity graph were both used to perform a continuous multi-channel integration task. The objective of this task was to detect failures in the input-output dynamics of two simulated "systems." Each system was composed of two continuous, semi-random inputs driving a single continuous output. During "normal" system operation, the arithmetic mean of the two inputs defined the system output. As long as this relationship was maintained by the two systems under their supervision, subjects were not required to make any response. However, if the relationship departed from this prescribed state, the subject was to indicate which of the two systems was failing by pressing one of two keys. This task, therefore, required subjects to perform two simultaneous integration tasks, each of which involved the integration of three values and a "go"-"no go" response.

The displays used by Carswell & Wickens (1987a, Wickens et al., 1986) are shown in Figure 16.2. The bar graph, shown at the left of the figure, was selected to represent a low level of structural proximity. The triangle display, shown at the right of the figure, were chosen as a high proximity alternative. The bar graphs used the height of three separate rectangles (bars) to indicate the value of the two inputs and one output for each system. From left to right, the first three bars represent the inputs and output for System 1, and the last three bars represent the same information for System 2. The four bars with the lower baselines represented system inputs, while the two elevated bars represented the associated system's output.

The triangle display, shown at the right of Figure 16.2, was designed to increase display proximity by combining the three specifiers for each of the systems into a single perceptual object. Thus, the distance of each of the three angles from a reference mark in the center of the triangle's base represented the three values. For example, the distance of the lower left angle from the reference mark represented the value of Input 1, the distance from the lower right angle to the reference point represented the value of Input 2, and the height of the triangle represented the output of the system.

Failure detection performance was measured in terms of percentage of correct detections, number of false alarms, and mean response time for correct failure detections. Data collected from ten subjects who performed the integration task with both high and low proximity displays revealed higher performance efficiency with the triangle displays. Failure detection latencies for the bar graphs ($M = 7.05$ sec) were reliably longer than those for the triangles ($M = 5.99$ seconds) ($F(1,9) = 9.49, p = .01$). Likewise, the bar graph was associated with fewer correct detections ($M = 73\%$) than were the triangles ($M = 90\%$) ($F(1,9) = 36.97, p < .001$). While false alarms were not reliably different for the two displays, the trend was toward more false alarms for the bar graphs ($M = 1.53$/trial, compared to $M = 1.14$/trial for the triangles).

These data suggest that the high proximity triangle displays are, in fact, better suited for performance of the high proximity integration task. This benefit for structural proximity was also found by Carswell and Wickens (1987b) using the same failure detection task with slightly different input-output dynamics for the two systems. In addition, this study used a different arrangement of the bar graphs, in which the four bars representing the inputs for the two systems were aligned vertically and grouped to the left of the display. The outputs were grouped to the right. Still, a benefit was obtained for the more proximal triangle displays.

E. Testing the PCH: Low Proximity Processing

In addition to providing some evidence for the predictions of proximity display benefits for high proximity processing, Carswell and Wickens (1987b) also provided some evidence for the second prediction of the PCH. In an additional experiment, 20 subjects were recruited to perform a low proximity task with the triangles and bar graphs shown in Figure 16.2. Once again, subjects monitored the continuous fluctuations of six variables, but in this task subjects had to respond to each value independently. The subject's goal was to indicate whenever each of the six uncorrelated "output" variables moved above or below a stationary reference point. Six keys (three for each hand) were used to indicate when a reference point crossing had occurred. Subjects performed this task for one-minute trials during which an average of 50 crossings occurred.

The data for this task showed a pattern of display benefits quite different from that found for tasks involving high proximity processing. The mean latency for responses to crossings was now slower for the triangles ($M = 719$ msec) than

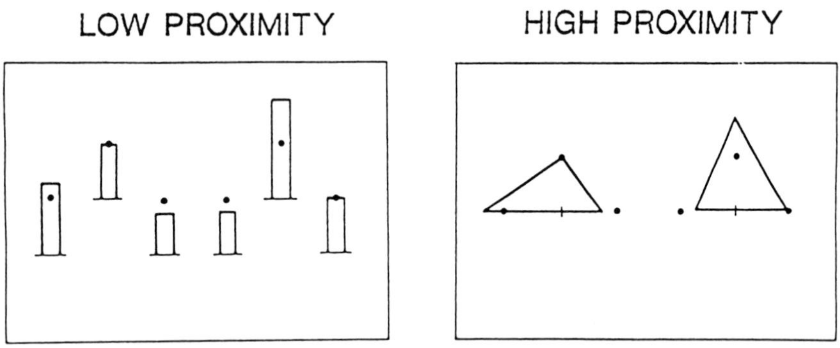

Figure 16.2. Graphical formats compared by Carswell and Wickens (1987a,b) and Wickens et al. (1985). The bar graphs, at left, represent the two inputs and single output of two separate systems. The three leftmost bars represent the two inputs and single output of System 1. The three rightmost bars represent the inputs and output of System 2. For each system, the two bars with the lower baselines represent inputs, while the top bar represents the output value. Dots on the bar graph represent zero-values or reference points for each of the six variables. For the triangle displays, at right, the triangle on the left represents System 1 and the triangle on the right represents System 2. The distance of the two bottom vertices from a line intersecting the base indicates the value of each of the two inputs for the two separate systems. The distance of the top vertex from the base represents the output value for the corresponding system. Once again, zero-values for each of the six variables are referenced by dots. For comparison, the same data set is shown using each of the two formats.

for the bar graphs (M = 675 msec) ($F(1,18) = 9.81, p = .0006$). Results of false alarms showed a trend in the same direction which was, however, only marginally reliable statistically ($F(1,18) = 3.71, p = .07$). The bar graph averaged 6.45 false alarms per trial, while the triangles resulted in an average of 7.60 false alarms per trial. There was no main effect for detection rate, which averaged 79% for both displays.

Overall, the bar graphs may be characterized as leading to quicker and marginally more accurate performance than the triangle displays. These data lend support to the second prediction of the PCH. When faced with a task that requires independent processing of multiple channels, the benefits of display proximity are absent, and an advantage is found for more separable (less proximal) formats.

F. The PCH as a Framework for Comparative Graphics

While the data from the above studies (Carswell & Wickens, 1987a,b; Wickens et al., 1986) do suggest that the PCH may have some merit, one may question whether such data generalize to the wider reaches of comparative graphics. The above studies are limited in that they employ speeded tasks, relevant to industrial tasks like process control and transportation, but not as representative of the less time-stressed memory and problem-solving tasks emphasized in educational studies and research on statistical/scientific visualization. Further, both studies employed tasks that represent relatively extreme endpoints on the processing proximity continuum. An exercise that may provide a stricter test of the utility of the PCH involves actually using the framework to summarize the results of a variety of comparative graphics studies. The contents of Table 16.1 are the end-product of such an exercise.

Table 16.1 represents a review of 42 empirical comparisons of the performance efficacy of different graphical formats. As required by the PCH, each comparison was initially classified according to the level of information processing proximity involved in the experimental task. Initially, the various studies were classified on the basis of whether they required subjects to engage in high, medium, or low proximity processing. Low proximity tasks were mainly those requiring subjects to locate or recall specific numeric values. The one exception was the previously-described study by Carswell and Wickens (1987b) in which subjects performed six independent tasks simultaneously. High proximity tasks, at the other extreme, involved a variety of specific requirements. Several studies involved multicue judgment tasks (Goldsmith & Schvaneveldt, 1984; MacGregor & Slovic, 1986, Goettl et al., 1986). Others involved trend classifications, extrapolations, or comparisons (e.g., Schutz, 1961a; Sparrow, 1989; Wrightstone, 1936; Casali & Gaylin, 1987). Still other tasks were designed to simulate the detection of multiply-determined system states in dynamic, process-control scenarios (e.g., Casey & Wickens, 1986; Carswell & Wickens, 1987a,b; Sanderson et al., 1988). In all cases, it was necessary for the subjects to integrate all of the quantitative information presented in order to make an accurate response. No isolation of particular specifiers was necessary. In contrast, the studies classified as intermediate in processing proximity required both focusing on and integration of specific values. For example, in several of the studies, subjects made simple ordinal judgments of two out of many values shown in the overall graph (e.g., Washburne, 1927; Culbertson & Powers, 1959). Finally, it should be noted that all of the results reviewed did not come from independent experiments. In several cases, task characteristics were used as an experimental variable in a single experiment (e.g., Washburne, 1927; Wrightstone, 1936, Casey

Table 16.1. Summary of studies comparing single and multiple-object displays for performance of tasks requiring high, intermediate, and low processing proximity.

Study	Graphs Compared	Outcome
	Low Proximity Processing	
Casali & Gaylin (1988)	points, lines, 2-d bars, 3-d bars	+
Sparrow (1989)	single multi-line, multiple single-line, stacked bars, bars, pies	+
Barnett & Wickens (1988)	rectangles, bars	0
Carswell & Wickens (1987a)	triangles, bars	0
Schutz (1961a)	single multi-line, multiple single-line	0
Todd & Wickens (1990)	3-d planes, 3-d points, triangles, bar	0
Casey & Wickens (1986)	faces, polygons, bars	-
Carswell & Wickens (1987b)	triangles, bars	-
Peterson et al. (1981)	stars, bars, meters	-
Wainer (1980)	lines, bars	-
Washburne (1927)	lines, bars	-
Wrightstone (1936)	lines, unit pictographs	-
Zhang & Wickens (1990)	rectangles, bars	-
	Intermediate Processing Proximity	
Schutz (1961b)	single multi-line, multiple single-line	+
Sparrow (1989)	Single multi-line, multiple single line, stacked bars, bars, pies	+
Casali & Gaylin (1988)	points, lines, 2-d bars, 3-d bars	0
Culbertson & Powers (1959)	lines, bars	-
Goettl et al. (1986)	triangle, bars	-
Wainer (1980)	lines, bars	-

Table 16.1. (cont.)

Washburne (1927)	lines, bars	-
Todd & Wiskens (1990)	3-d planes, 3-d points, triangles, bars	-
	<u>High Proximity Tasks</u>	
Andre & Wickens (1977)	rectangles, bars	+
Barnett & Wickens (1988)	rectangles, bars	+
Carswell & Wickens (1987a,b)	triangles, bars	+
Goettl et al. (exp 1) (1986)	dots, bars	+
Goldsmith & Schvanaveldt (1984)	rectangles, triangles, bars	+
Jacob et al. (1976)	faces, polygons, glyphs	+
Jones et al. (1990)	polygons, bars	+
MacGregor & Slovic (1984)	faces, polygons, deviation bars, bars	+
Peterson et al. (1981)	stars, bars, meters	+
Schutz (1981a)	lines, bars	+
Sparrow (1989)	single multi-line, multiple single-line, stacked bars, bars, pies	+
Wainer (1980)	lines, bars	+
Washburne (1926)	lines, bars	+
Zhang & Wickens (1990)	rectangles, bars	+
Casali & Gaylin (1987)	lines, points, 2-d bars, 3-d bars	0
Casey & Wickens (1986)	faces, polygons, bars	0
Goettl et al. (exp 2) (1986)	triangles, bars	0
Wickens & Todd (1990)	3-d planes, 3-d points, triangles, bars	0
Wrightstone (1936)	lines, unity pictographs	0
Coury et al. (1989)	polygons, bars	-
Sanderson et al. (1988)	triangles, bars	-

& Wickens, 1986; Barnett & Wickens, 1988). For such studies, comparisons between displays were reviewed separately for each task.

In addition to dictating the classification of experiments based on the processing requirements of the graphical tasks performed by subjects, the PCH also dictates that studies be included that compare graphical formats differing in terms of their structural proximity. For the purposes of the present review, object displays are considered to represent high levels of structural proximity. Multi-object formats, on the other hand, are classified as being low in structural proximity. For the most part, the object displays selected by the various researchers were line graphs or polygons, which were usually compared to some variant of a multi-object bar chart. In a few cases face displays were included as an example of an object display, and unit pictographs were selected as examples of multi-object displays. Traditional pictographs, however, were excluded from this review because of the perceptual difficulties previously noted in the use of their specifiers. In addition, comparisons in which the object display is composed of heterogeneous specifiers (such as combinations of color and form dimensions) were excluded. These heterogeneous object displays will be discussed later.

Table 16.1 provides a directory of both the types of graphical formats compared, and the overall evidence of object display performance advantages ('+', '0'. or '-'), for each of the studies. For all but two of the oldest studies (Washburne, 1927; Wrighstone, 1936) results of statistical analyses were available. For these two studies, graphical advantages were crudely determined by ranking the reported mean performance measures for each format. In studies where more than one multi-object or single-object display was included (MacGregor & Slovic, 1984; Petersen, Banks, & Gertman, 1981; Casey & Wickens, 1986), the graphs with the highest and lowest associated performance efficiencies were used for the classification of display superiorities. Thus, if a study compared four graphs including three object displays and one multi-object format, an object display advantage was coded only if the best graph was an object display *and* the worst graph was the multi-object format. While this summary technique may artificially inflate the reliability of differences between different display conditions within a given study, it remains unbiased in terms of the direction of such differences (i.e., object display advantage vs. disadvantage). Thus, consistencies in the direction of object display effects *across studies* may still be evaluated. One final qualification regarding Table 16.1 arises from at least one comparison (Wickens & Todd, 1990) resulting in contradictory results for different performance measures (speed and accuracy). For this reason, this comparison was coded as showing no overall difference between formats.

The data from Table 16.1 were further summarized to yield an informal, graphical meta-analysis which is presented in Figure 16.3. In this graph, the position of the broken line above the horizontal axis represents the percentage of the reviewed comparisons that revealed performance advantages for the less proximal (multi-object) display. The position of the solid line represents similar information for comparisons that revealed performance advantages for the high proximity (single object) formats. Note that for each of the three task categories, an additional percentage of studies found no statistically reliable differences between graphical formats. The PCH leads to the prediction of an interaction between prevalence of format benefits and type of task performed. Specifically, the incidence of multi-object benefits, represented by the broken line, should decrease from left to right, while the incidence of object display benefits, represented by the solid line, should increase. Figure 16.3 conforms, if somewhat imperfectly. with these predictions.

Perhaps the most interesting aspect of Figure 16.3 (and of Table 16.1) is the identification of studies that clearly provide data contradictory to the predictions of the PCH. For example, two studies comparing graphical formats in a low proximity processing task found performance benefits associated with the use of a high proximity display (Casali & Gaylin, 1988; Sparrow, 1989). Casali and Gaylin (1988) found that for answering questions about specific values, a bar chart proved to be more difficult to use than several other displays including a high proximity line graph. However, their bar chart was configured so that some of the bars were presented behind other bars, in perspective. This three-dimensional approach, as the authors noted, may have made the association of particular specifiers with particular axis labels somewhat difficult. This issue is suggestive of one important modification in the way the PCH is applied. The present formulation of the PCH has emphasized in its description of display proximity only the perceptual relations of different specifiers. Likewise, in describing task classifications, we have only considered the integration or independence of the information represented by the specifiers. Perhaps attention should be paid, as Kosslyn (1989) has suggested, to the perceptual organization (i.e., display proximity, using the terminology of the PCH) of specifiers, frameworks, background, and graphical labels. In addition, we may need to recognize that the requirement to integrate information from framework and labels with information carried by the specifiers is an additional type of high proximity processing, a type that has been largely ignored.

Additional limitations of the PCH are also revealed when considering those high proximity tasks that are better supported by low proximity graphs rather than object displays (Coury, Boulette, & Smith, 1989; Sanderson et al.,

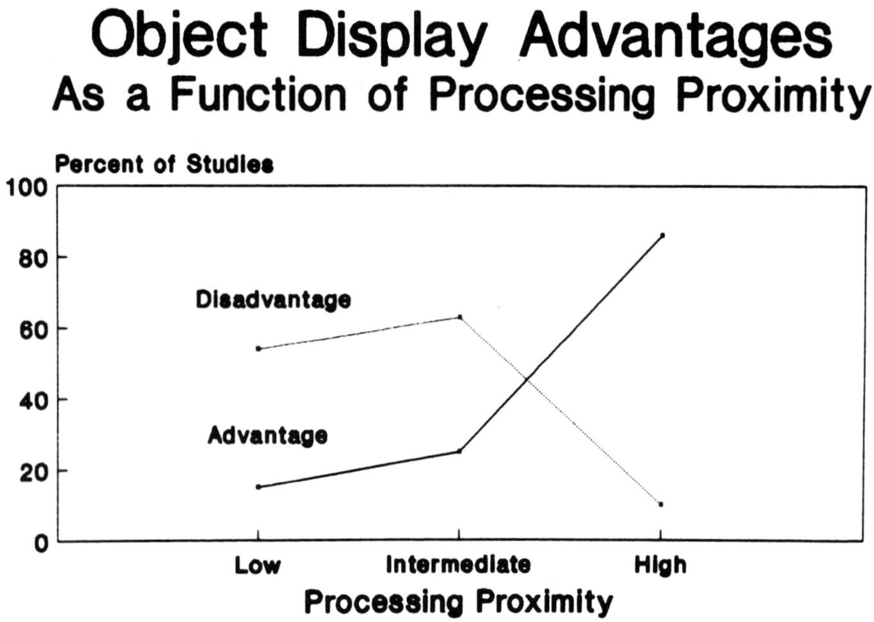

Figure 16.3. The relationship between the proximity of processing demanded by experimental tasks and the proportion of studies showing object display advantages and disadvantages.

1988). In both of these studies, a bar graph display was found to support high proximity processing better than variants of the polygon display. The study of Sanderson et al. (1988) was motivated by concerns regarding the utility of the single-object/multi-object distinction for predicting graphical efficacy. These authors suggested that the key to graphical efficacy was, instead, the presence in a graph of a salient emergent feature that could be used directly as a simple cue for subjects attempting to select an appropriate response. Further, Sanderson et al. (1988) argued, emergent features need not be isolated to object displays. To demonstrate this point, these authors chose as comparison graphs the stimuli of Carswell and Wickens (1987a,b). It will be recalled that these graphs included a bar chart and a triangular object display. The graphs were again used by subjects to perform a failure detection task in a simulated process control scenario, with normal system operation dictated by the relationship of two inputs to a single output. By rearranging the location and ordering of the bar graph's specifiers for the inputs and outputs of the simulated system, Sanderson et al. (1988) were able to show that the bar graph could be used more efficiently than the triangle displays. Importantly, the configuration of bars that proved to be particularly useful was one where the misalignment of the bar tops could be used as an emergent feature indicating abnormal operating conditions for the system. Thus, the objectness of displays per se may not be the only structural factor contributing to their performance efficacy.

G. Refining the Concept of Structural Proximity

Based on the review of the 42 comparisons of single- and multi-object displays described above, it seems that the degree of object integration influences performance in a manner generally consistent with the predictions of the PCH. However, we must question whether object integration is, in fact, the best (or a sufficient) way of conceptualizing display proximity. One problem with the use of object integration as the sole definition of display proximity was encountered when organizing the data for the present review. In several cases, it was necessary to exclude studies from the review because they contained only object displays (Brown, 1985; Mezzich & Worthington, 1978; Wilkinson, 1981), even though these object displays seemed to vary greatly in terms of their subjective unity, and even though there were performance differences in the use of these formats. Additionally, among several of the studies that were included, predictions could not be made about the performance superiority of, for example, a face display versus a polygon (e.g., Casey & Wickens, 1986; MacGregor & Slovic, 1984) or a multi-object bar chart versus a bank of spatially-segregated meters (Petersen et al., 1981). Besides this limitation in precision, the use of the single/multi-object distinction is also subject to the charge by Sanderson et al. (1988) that object integration simply reflects the operation of some other factor, such as the presence of emergent features or integral specifiers.

In order to address the issue of whether object displays are more likely than multi-object formats to contain emergent features (i.e., to contain configural specifiers), Carswell and Wickens (1990) collected performance-based diagnostics of dimensional interactions for pairs of specifiers in thirteen simple (two-element) graphs. A second aim or this study, following from the earlier interest of researchers in so-called integral displays, was to gather information regarding the incidence of integral dimensions in combinations of graphical specifiers.

The thirteen stimulus sets analyzed by Carswell and Wickens (1990) are shown in Figure 16.4. Each stimulus set, which may be considered a rudimentary graphical display, contained four stimuli formed by combining two levels on two physical dimensions. The physical dimensions that served as the specifiers for

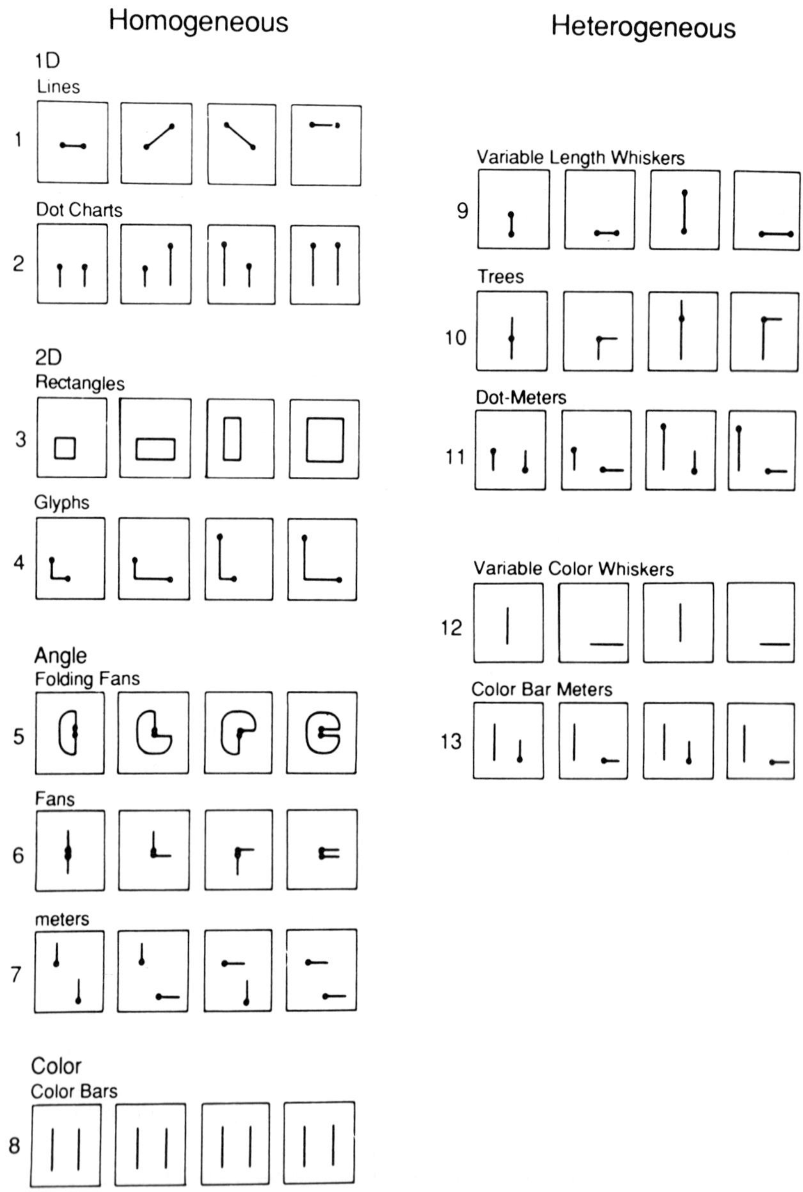

Figure 16.4. Four examples of each of the thirteen graphical displays evaluated for specifier interactions by Carswell and Wickens (1990).

each of the thirteen formats were chosen from the set of dimensions described by Cleveland (1985) as basic graphical elements. These dimensions included linear extent, orientation, and color. Eight of the graphs featured specifiers that were combined into a single perceptual object (Graphs 1, 3, 4, 5, 6, 9, 10, and 12). In the five remaining graphs, the specifiers were assigned to separate objects. In an additional attempt to insure variation in the configurality of the stimulus sets, Graphs 1-8 were composed of repeated pairings of specifiers (homogeneous graphs), while Graphs 9 - 12 were formed by mixing either extent and orientation or color and orientation (heterogenous graphs). This manipulation was based on previous evidence by Garner (1978) that dissimilar dimensions are less likely to configure.

It should also be noted that the graphical stimuli presented in Figure 16.4 represent a variety of analog displays currently used in statistics and industry. Graphs 1 and 2, the lines and the dot charts, are simple variants of the ubiquitous line graphs and bar charts. The rectangles (Graph 3), a simple variant of the polygon-style displays discussed earlier, has been used for both industrial (e.g., Wood et al., 1981) and statistical (e.g., Wilkinson, 1981) purposes. The glyphs (Graph 4), variable-length whiskers, (Graph 9), trees (Graph 10), and variable-color whiskers (Graph 12) are relatively recent innovations in statistical graphics (for a discussion of these techniques see Chambers, Cleveland, Kleiner, & Tukey, 1983; Tufte, 1983; Wainer & Thissen, 1981). The folding fans (Graph 5), fans (Graph 6), and meters (Graph 7) are variants of the dynamic dials and meters commonly found in transportation and industrial settings. Finally, the colorbars (Graph 8), dot-meters (Graph 11), and colorbar-meters (Graph 13) were each chosen to exemplify the combinations of analog displays found in many industrial settings (e.g., a color-coded warning light or vertical moving-pointer display situated next to a meter on a control panel).

As diagnostics for integrality and configurality, Carswell and Wickens (1990) chose the speeded classification methods pioneered by Garner and colleagues (e.g., Garner, 1974, 1976, 1978; Garner & Felfoldy, 1970; Garner & Pomerantz, 1973). To summarize these methods, diagnosis of dimensional interactions is based on the pattern of reaction times obtained when subjects classify a series of stimuli based on several different classification rules. Subjects may be asked, for example, to classify stimuli on the basis of one dimension while the irrelevant dimension always remains at the same value or level from stimulus to stimulus. This *baseline task* is then compared to performance in other classification tasks in which dimensions may be paired orthogonally or redundantly across the stimuli in a series.

Of critical importance for the diagnosis of either integral or configural dimensions is the finding of a decrement for performance of a *filtering task* relative to baseline performance. In the filtering task, the subject may once again classify stimuli based on the value of only one dimension; however, unlike the baseline task, the irrelevant dimension does not remain fixed across trials. If this irrelevant variation results in a performance decrement, then some evidence is provided that the component dimensions interact perceptually to form either integral or configural dimensions. If performance is not disrupted, then the dimensions are likely to be separable. To further discriminate between integral and configural dimensions, two additional tasks must be performed by the subject. In the *condensation task*, subjects again sort a series of stimuli consisting of orthogonally-varying dimensions. However, rather than utilizing variation in only one of the component dimensions, subjects are asked to utilize the value of both dimensions in making their classifications. The efficiency of condensation may be measured by comparing performance in the condensation and filtering tasks. If subjects are able to use both varying dimensions more easily than they can focus on one, then the dimensions may well configure, perhaps because of the presence of a salient emergent feature (Pomerantz & Pristach, 1989). Finally, the integrality of a dimensional pair may be diagnosed by observing performance in a *redundancy task*. In such a task, subjects must classify stimuli based on perfectly redundant pairings of the component dimensions. If performance in this task is more efficient than in baseline conditions, then integral dimensions may be present.

To summarize these diagnostics, integral dimensions are diagnosed if, relative to baseline performance, filtering task performance is slow *and* redundancy task performance is fast. Diagnosis of configural dimensions also depends on finding that, relative to baseline conditions, filtering task performance is slow. However, when compared to filtering task performance, condensation performance must be relatively fast. Finally, separable, noninteractive dimensions are diagnosed when performance in baseline, redundancy, and filtering tasks differs little, and when condensation task performance is relatively slow.

When subjected to these performance-based diagnostics, none of the thirteen graphs appeared to contain combinations of specifiers that clearly fit the pattern for integral dimensions. Instead, the various graphs seemed to fall along a continuum from highly configural to nonconfigural formats. The ranking of graphs in terms of their configurality is presented in Figure 16.5.

This graph is the result of a principal components analysis conducted on summary performance measures derived for each of the thirteen graphs. For each

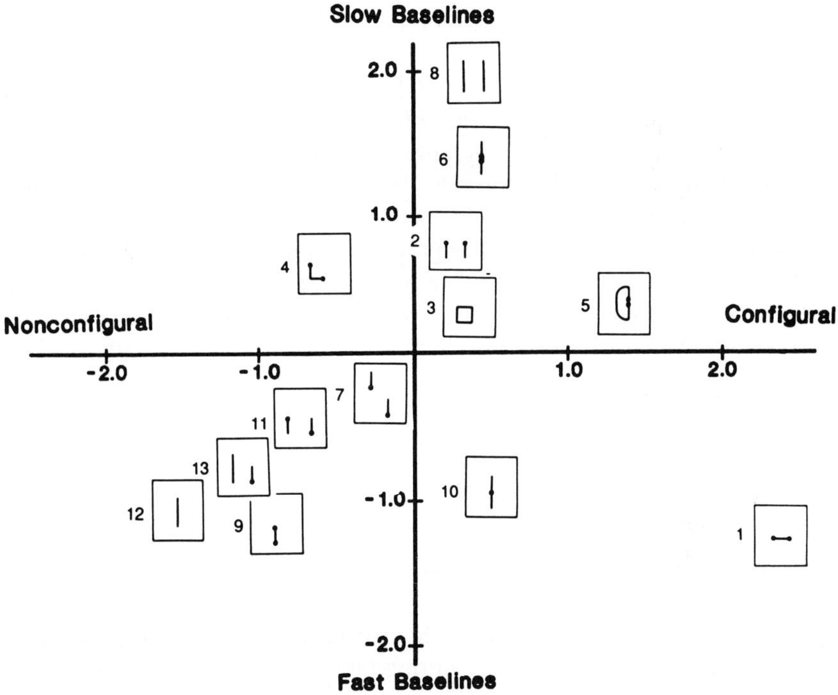

Figure 16.5. Configurality dimension derived from covariation among speeded classification measures taken on the thirteen graphs shown in Figure 16.4. Position along the horizontal axis represents the configurality level of each graph (identified by the numbers used in Figure 16.4). Numbers in bold print represent heterogeneous graphs.

graph, these measures of classification performance were taken from an independent sample of nine subjects. The largest proportion of the variance across the thirteen graphs resulted from the covariation of measures thought to be sensitive to configurality. For example, decrements in filtering performance covaried with the efficiency of condensation performance, but did not covary with the redundancy gain measure necessary for diagnosing integrality. This configurality component is represented by position along the horizontal axis in Figure 16.5. The second principal component, represented by the vertical axis, seemed to be best described as variation attributable to differences in the baseline performances of the thirteen groups of subjects. In summary, the data suggest that homogenous object displays are more configural than homogeneous multi-object displays which, in turn, are more configural than the majority of the heterogeneous displays. The positive relation of homogeneity and configurality was reflected in a point-biserial correlation of $r = .52$ ($p = .06$).

These data support the claim by several researchers in comparative graphics (Sanderson et al., 1988; Barnett & Wickens, 1988; Coury et al., 1989) that configurality (rather than integrality) may prove to be a theoretically useful means of classifying graphical formats. In addition, it appears that the similarity or homogeneity of graphical specifiers may prove to be more important than object integration for indicating when graphs are configural. While homogeneous objects may be somewhat more configural than homogeneous multi-objects, heterogeneous displays seem to show few of the performance patterns associated with configurality, regardless of whether they are single- or multi- object displays. Thus, if the presence of emergent features is central to the benefits found for object displays in the previous studies, heterogeneous object displays should prove to be even more disruptive to performance than the use of homogeneous multi-object displays.

H. Homogeneous and Heterogeneous Object Displays

The PCH proposes that high proximity (single object) displays are particularly well suited for tasks requiring integrative processing. However, the research by Sanderson et al. (1988) suggests that it is actually graphs with emergent features, regardless of whether they are composed of one or more objects, that allow for the most efficient processing. In support of the latter position, Carswell and Wickens (1990) demonstrated that the perceptual interactions associated with a variety of specifier combinations is well described in terms of dimensional configurality. Further, they found that object displays composed of heterogeneous specifiers were considerably less configural (i.e., less likely to produce emergent features) than were homogeneous object displays. The question that presents itself, then, is whether the object display benefits commonly found for homogeneous combinations of specifiers generalizes to heterogeneous object displays as well. Put another way, can object displays without emergent features provide support for high proximity processing?

In order to test the generality of the object display advantage, Carswell (1990) had subjects use both homogeneous and heterogeneous object displays, along with multi-object "control" graphs, to perform several tasks that differed in proximity demands. The graphs used in this series of experiments were identical to those shown in Figure 16.4, allowing direct comparisons of the level of dimensional configurality of the graph with performance in the various tasks. Two

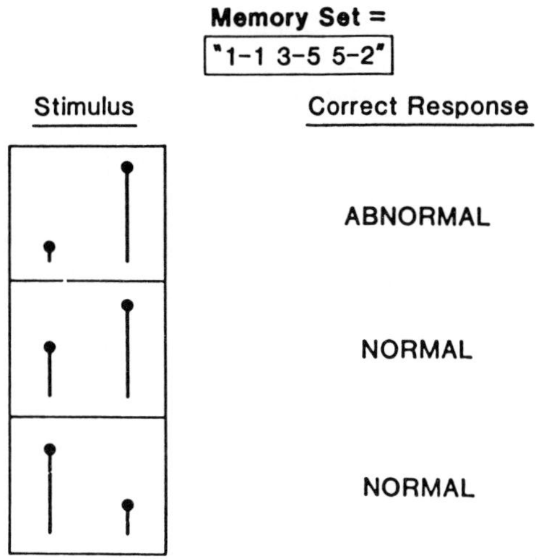

Figure 16.6. A sample comparison task trial using the dot charts (Graph 2, from Figure 16.4). During an actual trial, the six stimuli would appear one after the other, starting with the upper stimulus. Each successive stimulus would appear in the same location on a computer monitor 100 msec after the subject's response to the previous stimulus.

of the tasks required relatively high levels of processing proximity, and therefore present perfect opportunities to compare the strict object-based and configurality-based explanations of graphical efficacy. If an object-based description of display proximity is sufficient to predict performance in high proximity tasks, then subjects should show object display advantages regardless of whether the object display is composed of homogeneous (highly configural) or heterogeneous (nonconfigural) specifiers. However, a configurality-based explanation of display efficacy would propose object display advantages only for homogeneous displays, and would further predict general advantages for homogeneous over heterogeneous displays regardless of whether each display's specifiers are arranged to form one or more objects.

In the first high proximity task studied by Carswell (1990), fifteen subjects performed a simulated 'check-reading' exercise. Check-reading here refers to the task frequently encountered by operators of complex, automated systems in which a quick check must be made of a series of indicators to insure that each is showing values (or combinations of values) that are normal for the ongoing phase of system operation. Thus, subjects were required to proceed through a series of graphs as quickly as possible, indicating whether each was normal or abnormal by pressing the appropriate response key. Each series consisted of six examples of a

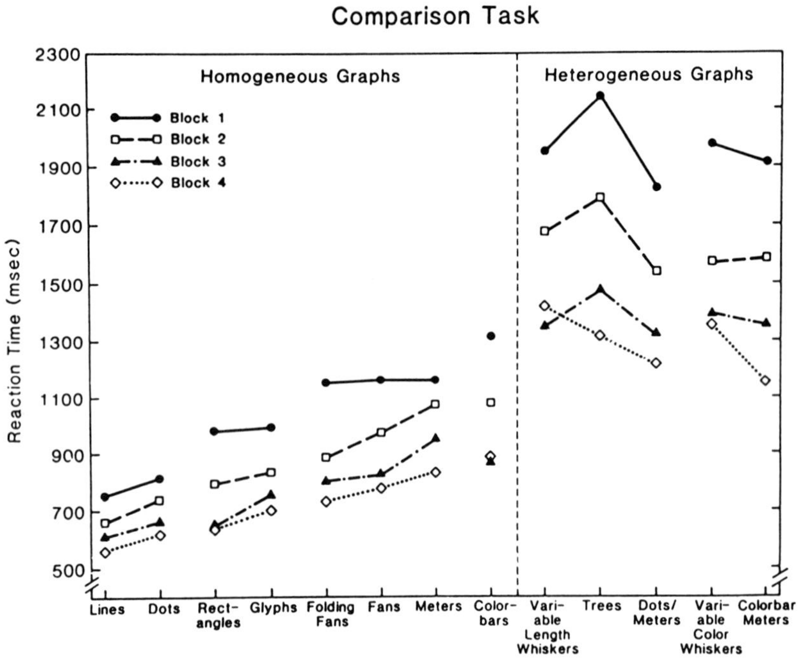

Figure 16.7. Mean reaction times for performing the comparison task. Data are shown for four blocks of trials with each of the thirteen experimental graphs.

single graphical format (i.e., six different sets of values for a given pair of specifiers). The basis for classifying each stimulus as normal or abnormal was a memory set presented to the subject prior to the onset of a trial. This memory set contained information regarding the ordinal relationship that should exist between the two variables presented in each stimulus of the series. Figure 16.6 provides an example of a single trial of this integration task using the dot charts (Graph 2). The memory set, a series of '1's and '2's, indicates whether Variable 1 or Variable 2 should be greater in each stimulus of the series. In this particular example, Variable 2 should be greater than Variable 1 for the first two stimuli, and Variable 1 should be greater for the third stimulus. The pattern then repeats itself. Note that the fifth stimulus in the sample series shows the first variable to be greater in value than the second, thus failing to conform to the memory set, and resulting in a correct 'abnormal' response on the part of the subject.

Each subject performed the check-reading tasks with each of the 13 graphical displays. Prior to collecting data, subjects were trained to a speed-accuracy criterion for 'reading' the values of the various specifiers. Because five levels of each specifier were used to indicate five ordinal values for a variable, subjects had to learn to make five absolute judgments on each of 26 dimensions (13 graphs x 2 specifiers). After initial training, subjects received a full block of check-reading trials. Each block consisted of 10 trials (stimulus series) with each of the thirteen graphical formats for a total of 130 practice trials or 780 responses. Subjects then returned for two additional sessions during which four additional blocks of trials were completed. The order in which the various graphs were used to perform the task was varied randomly for each subject and block of trials.

The mean response times (RTs) for each block of trials and each of the thirteen graphs are illustrated in Figure 16.7. Each line segment in this graph represents a comparison of performance in a single- and multi-object display matched in terms of the dimensions used as specifiers. The left endpoint of each line represents the single-object display and the right endpoint represents its multi-object counterpart. For example, the leftmost set of lines represents relative performance when using lines and dots (Graphs 1 & 2) for each of the four blocks of trials. Positive slopes for any line segment indicate object display benefits. Note that for two sets of graphs (graphs 5, 6, & 7, and Graphs 9, 10, & 11), an intermediate level of object integration was included in these comparisons. This intermediate level of object integration was identified by an additional group of 20 subjects who sorted examples of the various graphical formats into classes based on the number of objects they appeared to contain (Carswell, 1988).

An overall ANOVA (13 graphs X 4 blocks X 15 subjects) on reaction times (RTs) revealed main effects for graphs ($F(12,168) = 53.71, p < .0001$) and blocks ($F(3,42) = 34.57, p < .0001$), as well as a reliable interaction of these two factors ($F(36, 504) = 4.83, p < .0001$). A series of planned comparisons were then conducted to determine whether object integration and/or homogeneity could account for the effect of graphical format. As evident from Figure 16.7, and as predicted by a configurality-based definition of display proximity, the mean RTs for performance with heterogeneous displays were much longer than those for homogeneous (i.e., configural) formats ($F(1,14) = 87.84, p < .0001$). In addition, for each of the three comparisons involving homogeneous single-object versus homogeneous multi-object displays, performance was superior for the format having the higher level of object integration (Graphs 1 vs 2: $F(1,14) = 28.85, p < .0001$); Graphs 3 vs. 4: $F(1,14) = 4.32, p = .052$); Graphs 5 vs. 6 vs 7: $F(2,28) = 4.71$, $p = .017$). However, no such object advantage was found for the two comparisons involving single- and multi-object heterogeneous displays. In one case, there was even an advantage for the multi-object format (Graphs 9 vs. 10 vs. 11: $F(2,28) = 4.14, p = .026$). Because of the overall interaction of blocks and graphical formats, the simple interaction effects of blocks with each of the planned

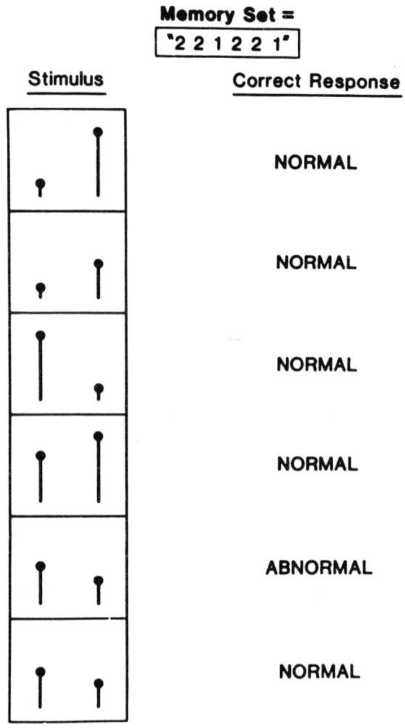

Figure 16.8. A sample conjunction task trial using the dot charts (Graph 2, from Figure 16.4). During an actual trial, the three stimuli would appear one after the other, starting with the upper stimulus. Each successive stimulus would appear in the same location on a computer monitor 100 msec after the subject's response to the previous stimulus.

format comparisons was also assessed. None of the object effects appeared to change across blocks. However, there was some dilution of the homogeneity advantage with practice ($F(3,42) = 15.39, p < .0001$).

These data provide dramatic evidence favoring the configurality-based definition of display proximity. The homogeneity and presumably the configurality of specifiers was clearly related to the efficacy of the graphical formats. Further, the smaller effects found for object integration only favored object displays when they were composed of homogeneous dimensions. These effects for RTs were further supported when a similar analysis was conducted for performance errors (Carswell, 1988). Thus, object integration effects seemed to be clearly subordinate to the effects of homogeneity. However, an additional comparison of these same 13 graphs in a second integration task (Carswell, 1990) revealed that the matter was not entirely settled.

Fifteen additional subjects were recruited to perform a speeded check-reading task, similar in most respects to the ordinal comparison task described above. However, instead of being required to check the relative magnitude of the variables in each display of a series, subjects were required to check for a conjunction of specific values. Figure 16.8 presents an example of a single trial from this conjunction task using, once again, the dot charts. Like the previous comparison task, the subject performing the conjunction task first encountered a six-element memory set. Instead of consisting of '1's and '2's, however, this set contained conjunctions of exact values ranging from 1 to 5. For example, the memory set shown in Figure 16.8 specifies that the first stimulus of the series should contain two variables both equal to 1. However, the first stimulus in the example does not satisfy this requirement, and is correctly classified as abnormal. The memory set further indicates that in the second graph, Variable 1 should be equivalent to 3 and Variable 2 should be equal to 5. Finally, in the third graph of the series, Variable 1 should be equal to 5 and Variable 2 should be equal to 2. Both the second and third graphs in Figure 16.8 match the memory set, and are thus classified as normal. Procedures for the training and experimental sessions were identical to those for the previous task, with the exception that sessions were slightly shorter due to each trial containing a series of three instead of six stimuli.

As illustrated in Figure 16.9, the relative efficacy of the thirteen graphical formats, when used to perform the conjunction task, was quite different from the results obtained for the comparison task. Again, an overall (13 graphs X 4 blocks X 15 subjects) ANOVA on RTs revealed a main effect of graphical format ($F(12,168) = 8.75, p < .0001$). There was also a main effect of block ($F(3,42) = 7.90, p = .0003$), but there was no interaction of graphical format and block. In contrast to the comparison task, performance with the eight homogeneous displays did not differ reliably from that obtained with the five heterogeneous displays. However, when five comparisons were made to test the effects of object integration, three revealed reliable object display advantages (Graphs 5 vs. 6 vs. 7: $F(2,28) = 6.88, p = .0037$; Graphs 9 vs. 10 vs 11: $F(2,28) = 16.67, p < .0001$; Graphs 12 vs. 13: $F(1,14) = 20.14, p = .0005$). Note that the strongest object display benefits were obtained for heterogeneous object displays. Once again, a similar analysis performed on error rates for each condition revealed trends similar to those found for RTs.

In the previous task, the comparison task, homogeneity benefits characterized the performance levels obtained with the different graphical displays. Because homogeneous dimensions are generally more strongly configural than heterogeneous dimensions, these data suggests that a configurality-based

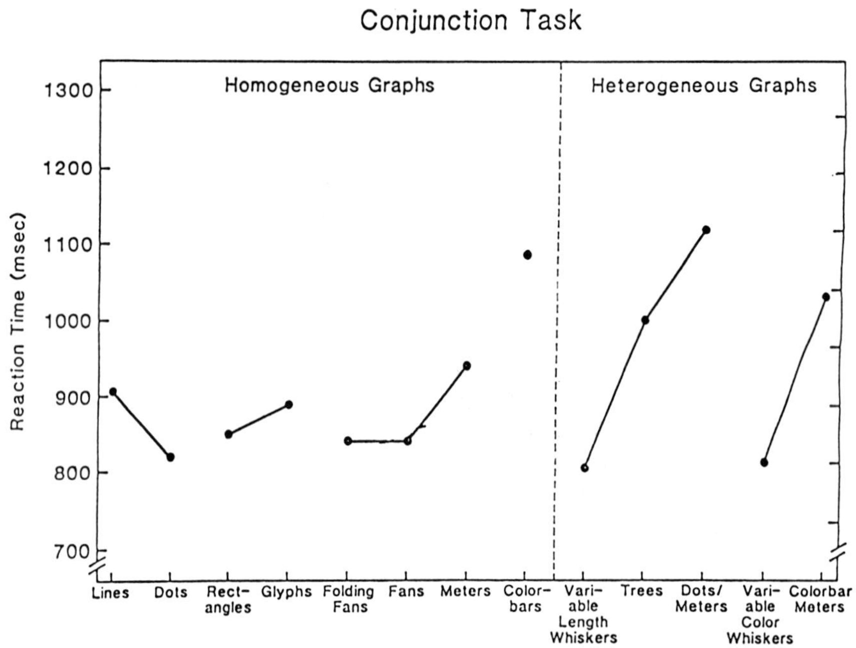

Figure 16.9. Mean reaction times for performing the conjunction task. Data are shown for each of the thirteen experimental graphs, collapsed across blocks.

description of display structure may have some merit for predicting graphical efficacy. However, performance in the second task, the conjunction task, provides evidence that immediately brings into question the generality of this assertion. The relative performance efficiency obtained with the thirteen graphs in the conjunction task seemed to reflect object-based rather than homogeneity-based (i.e., configurality) advantages. Further, object display benefits in the conjunction task occurred most strongly for heterogeneous object displays. This finding is critical for the dissociation of object-based and configurality-based effects of stimulus structure. Object display benefits for homogeneous dimensions might be attributed to the increased configurality of the component dimensions that results from increased spatial proximity. However, heterogeneous dimensions do not seem to configure regardless of their spatial proximity. Thus, performance in the conjunction task provides some support for an object-based definition of display proximity independent of configurality.

Taken together, the two experiments conducted by Carswell (1990) suggest that both object-based and configurality-based definitions of display structure must be included in any description of display proximity if the PCH is to increase in its predictive precision. It seems clear that for both integration tasks, homogeneous single-object displays appear to be among the most efficiently used graphs, while heterogeneous multi-object displays seem to consistently be among the worst. Homogeneous multi-object displays are generally intermediate in terms of their efficiency of use. However, the benefit of heterogeneous object displays depends very strongly on the nature of the task. It should be noted that both of the present tasks fit clearly the PCH's definition of high processing proximity; to perform either task, the subject had to utilize both dimensions, and was not required to focus on either. However, the comparison task required a computational or metric type of integration, while the conjunction task was essentially logical or nonmetric.

The present data are consistent with evidence from two additional studies comparing heterogeneous single object displays with other types of single- and multi-object formats. Zhang and Wickens (1990), for example, found that a heterogeneous object display which used color and height as its specifiers, was less efficiently used than either a homogeneous multi-object display (bar charts) or a homogeneous single-object display (a rectangle). However, in an additional experiment by Edgell and Morrissey (in press), heterogeneous object displays were found to support performance better than a heterogeneous multi-object format in some conditions. It should be noted, that the task of Zhang and Wickens (1990) was one in which subjects were required to multiply values of two variables, a metric integration task. The task used by Edgell and Morrissey (in press), on the other hand, involved multicue probability learning of nonmetric values.

To summarize the implications of the research on heterogeneous object displays for the validity of the PCH in general, it seems that they require a reconsideration of the processing distinctions made by the model. These results suggest that a processing taxonomy composed of characteristics other than the simple presence or absence of integration demands is necessary if the efficacy of both homogeneous and heterogeneous displays is to be adequately predicted. In addition, these results suggest that while configurality is certainly important in determining graphical efficacy, it is not a sufficient explanation for the performance enhancement that usually, but not always, occurs when graphical specifiers are combined into a single object.

IV. CONCLUSIONS

A review of the research in comparative graphics clearly reveals that the ease with which we can "read" any particular graph is multiply determined. In order to understand what constitutes a "good" graph, we must certainly look at the structural characteristics of the graphs themselves. We must consider the match formed between the specifiers chosen by the designer and the concepts that are to be represented by the display. We must also consider the resolution with which we can perceptually represent variation along the different physical dimensions we use as specifiers. Finally, we must consider the more global structures that result from the perceptual interaction of multiple specifiers--the unitariness, integrality, or configurality of graphical displays. However, dividing graphical formats into categories based on such structural distinctions is not sufficient to determine whether any particular graph will be easy to read. Instead, how good a particular graphical structure may prove to be depends largely on the nature of the task for which it will be used, and the match formed between these processing demands and the stimulus structure.

The Proximity Compatibility Hypothesis (PCH) was introduced as one framework for matching stimulus structure, as defined principally by the perceptual interactions that occur between combinations of specifiers, to the type of processing required of the graphical task. The PCH proposes that for "good" graphs, the level of processing proximity (multi-channel integration) required by a task should be matched by the level of structural proximity represented in the graphical display. The working definition of structural proximity used throughout this chapter, was the physical integration by the display designer of two or more specifiers into what appears to be a single perceptual object. When research in comparative graphics was reviewed from the standpoint of the PCH, the utility of the model was generally supported, although there were several notable exceptions to its predictions. Further research revealed the necessity of (1) expanding the definition of structural proximity to include the homogeneity of the specifiers in a graph and (2) further refining the processing taxonomy proposed by the PCH by specifying distinct types of information integration.

REFERENCES

Barnett, B.J., & Wickens, C.D. (1988). Display proximity in multicue information integration: The benefits of boxes. Human Factors, 30, 15 - 24.

Barnett, V. (1981). Interpreting multivariate data. Chichester: John Wiley and Sons.

Beniger, J.R., & Robyn, D.L. (1978). Quantitative graphics in statistics: A brief history. The American Statistician, 32, 1 - 11.

Bertin, J. (1983). Semiology of Graphics (William J. Berg, Trans.). Madison: The University of Wisconsin Press. (Original work published 1973).

Brinton, W.C. (1914). Graphic methods for presenting facts, New York: McGraw-Hill Book Co, Inc.

Brown, R.L. (1985). Methods for the graphic representation of systems simulated data. Ergonomics, 28, 1439 - 1454.

Carswell, C.M. (1990). Graphical information processing: the effects of proximity compatibility. Proceedings of the Human Factors Society 34th Annual Meeting (pp. 1494 - 1498), Santa Monica, CA: Human Factors Society.

Carswell, C.M., & Wickens, C.D. (1987a). Objections to objects: Limitations of human performance in the use of iconic graphics. In Mark, L.S., Warm, J.S., & Huston, R.L. (eds.), Ergonomics and human factors: Recent research. New York: Springer-Verlag, 253 - 260.

Carswell, C.M., & Wickens, C.D. (1987b). Information integration and the object display: An interaction of task demands and display superiority. Ergonomics, 30, 511 - 528.

Carswell, C.M., & Wickens, C.D. (1988). Comparative graphics: History and applications of perceptual integrality theory and the proximity compatibility hypothesis (Technical Report No. ARL-88-2/AHEL-88-1). Savoy, IL: University of Illinois Aviation Research Lab.

Carswell, C.M., & Wickens, C.D. (1990). The perceptual interaction of graphical attributes: Configurality, stimulus homogeneity, and object integration. Perception and Psychophysics, 47, 157 - 168.

Casali, J.G., & Gaylin, K.B. (1987). Selected graph design variables in four interpretation tasks: a microcomputer-based pilot study. Behaviour and Information Technology, 7, 31 - 4.

Casey, E.J., & Wickens, C.D. (1986). Visual display representation of multidimensional systems (Technical Report CPL-86-2/MDA903-83-K-0255). Champaign, IL: University of Illinois Cognitive Psychophysiology Laboratory.

Chambers, J.M., Cleveland, W.S., Kleiner, B., and Tukey, P.A. (1983). Graphical methods for data analysis, Belmont, CA: Wadsworth.

Chernoff, H. (1978). Graphical representation as a discipline. In F.C.C. Wang (Ed.), Graphical representation of multivariate data (pp. 1 - 12). New York: Academic Press.

Chernoff, H. (1973). The use of faces to represent points in K-dimensional space graphically. Journal of the American Statistical Association, 68, 361 - 368.

Cleveland, W.S. (1985). The elements of graphing data, Monterey, CA: Wadsworth.

Cleveland, W.S., Harris, C.S., & McGill, R. (1983). Experiments on quantitative judgments of graphs and maps. The Bell System Technical Journal, 62, 1659 - 1674.

Cleveland, W.S., & McGill, R. (1984). Graphical perception: Theory, experimentation, and application to the development of graphic methods. Journal of the American Statistical Association, 70, 531 - 554.

Cleveland, W.S., & McGill, R. (1985). Graphical perception and graphical methods for analyzing scientific data. Science, 229, 828 - 833.

Cole, W.G. (1986). Medical cognitive graphics. Proceedings of the ACM, SIGCHI '86.

Coury, B.G., Boulette, M.D., & Smith, R.A. (1989). Effect of uncertainty and diagnosticity on classification of multidimensional data with integral and separable displays of system status. Human Factors, 31, 351 - 369.

Coury, B.G., & Purcell, J. (1988). The bar graph as a configural and a separable display. In Proceedings of the Human Factors Society 32nd Annual Meeting (pp. 1361 - 1365). Santa Monica, CA: The Human Factors Society.

Croxton, F.E. (1927). Further studies in the graphic use of circles and bars: Some additional data. Journal of the American Statistical Association, 22, 36 - 39.

Croxton, F.E., & Stein, H. (1932). Graphic comparisons by bars, squares, circles, and cubes. Journal of the American Statistical Associates, 27, 54 - 60.

Croxton, F.E., & Stryker, R.E. (1927). Bar charts versus circle diagrams, Journal of the American Statistical Association, 22, 473 - 482.

Culbertson, H.M., & Powers, R.D. (1959). A study of graph comprehension difficulties. AV Communication Review, 7, 97 - 110.

DeSanctis, G. (1984). Computer graphics as decision aids: Directions for research. Decision Sciences, 15, 463 - 487.

Duncan, J. (1984). Selective attention and the organization of visual information. Journal of Experimental Psychology: General, 113, 501 - 517.

Edgell, S.E., & Morrissey, J.M. (in press). Separable and unitary stimuli in nonmetric multicue probability learning. Organizational Behavior and Human Decision Processes.

Eells, W.C. (1926). The relative merits of circles and bars for representing component parts. Journal of the American Statistical Association, 21, 119 - 132.

Everitt, B. (1978). Graphical techniques for multivariate data. New York: North-Holland.

Feinberg, S.E. (1979). Graphical methods in statistics. The American Statistician, 33, 165 - 178.

Funkhouser, H.G. (1937). Historical development of the graphical representation of statistical data. Osiris, 3, 269 - 404.

Garner, W.R. (1970). The stimulus in information processing. American Psychologist, 25, 350 - 358.

Garner, W.R. (1974). The processing of information and structure. Hillsdale, NJ: Erlbaum Associates.

Garner, W.R. (1976). Interaction of stimulus dimensions in concept and choice processes. Cognitive Psychology, 8, 98 - 123.

Garner, W.R. (1978). Selective attention to attributes and to stimuli. Journal of Experimental Psychology: General, 107, 287 - 308.

Garner, W.R. (1981). The analysis of unanalyzed perceptions. In M. Kubovy & J.R. Pomerantz (eds.), Perceptual Organization, Hillsdale, N.J.: Lawrence Erlbaum Associates, 119-139.

Garner, W.R., & Felfoldy, G.L. (1970). Integrality of stimulus dimensions in various types of information processing. Cognitive Psychology, 1, 225 - 241.

Goettl, B.P., Kramer, A.F., & Wickens, C.D. (1986). Display format and the perception of numerical data. Proceedings of The Human Factors Society 30th Annual Meeting (pp. 450 - 454). Santa Monica, CA: The Human Factors Society.

Goldsmith, T.E., & Schvaneveldt, R.W. (1984). Facilitating multicue judgments with integral information displays. In J. Thomas & M. Schneider (Eds.) Human factors in computer systems (pp. 243 - 270). Norwood, NJ: Ablex.

Hahn, G.J., Morgan, C.B, & Lorensen, W.E. (1983). Color face plots for displaying product performance. IEEE Computer Graphics and Applications, 3, 23 - 29.

Hutchingson, R.D. (1981). New horizons for human factors in design. New York: McGraw-Hill.

Jacob, R.J.K., Egeth, H.E., & Bevan, W. (1976). The face as a data display. Human Factors, 18, 189 - 200.

Jones, P.M., Wickens, C.D. & Deutsch, S.J. (1990). The display of multivariate information: An experimental study of an information integration task. Human Performance, 3, 1 - 17.

Kahneman, D., & Treisman, A.M. (1984). Changing views of attention and automaticity. In R. Parasuraman, R. Davies, and J Beatty (eds.), Varieties of attention. New York: Academic Press.

Kornblum, S., Hasbroucq, T., & Osman, A. (1990). Dimensional Overlap: Cognitive basis for stimulus-response compatibility -- A model and taxonomy. Psychological Review, 97, 253 - 270.

Kosslyn, S. (1989). Understanding charts and graphs. Applied Cognitive Psychology, 3, 185 - 226.

Kruskal, W. (1977). Visions of maps and graphs. Proceedings of the International Symposium on Computer-generated Cartography, Auto-Carlo II, 27 - 36.

MacDonald-Ross, M. (1977). How numbers are shown: A review of research on the presentation of quantitative data in texts. AV Communication Review, 25, 359 - 409.

MacGregor, D., & Slovic, P. (1984). Graphic representation of judgmental information. Human-Computer Interaction, 2, 179 - 200.

McCormick, E.J., & Sanders, M.S. (1982). Human factors in engineering and design. New York: McGraw-Hill.

Mezzich, J.E., & Worthington, D.R.L. (1978). A comparison of graphical representations of multidimensional psychiatric diagnosis data. In P.C.C. Wang (Ed.) Graphical representation of multivariate data (pp. 123 - 142) New York: Academic Press.

Moriarity, S. (1979). Communicating financial information through multidimensional graphics. Journal of Accounting Research, 17, 205 - 224.

Neurath, O. (1944). Visual aids and arguing. New Era. 25, 51 - 61.

Petersen, R.J., Banks, W.W., & Gertman, D.i. (1981). Performance-based evaluation of graphic displays for nuclear power plant control rooms. Proceedings of the conference on Human Factors in Computing Systems. Gaithersburg, MD, 182 - 189.

Playfair, W. (1801). Commercial and political atlas, Third Edition. London: Stockdale.

Pomerantz, J.R., & Garner, W.R. (1973). Stimulus configuration in selective attention tasks. Perception and Psychophysics, 14, 565 - 569.

Pomerantz, J.R., & Pristach, E.A. (1989). Emergent features, attention, and perceptual glue in visual form perception. Journal of Experimental Psychology: Human Perception and Performance, 15, 635 - 649.

Pomerantz, J.R., & Schwaitzberg, S.D. (1975). Grouping by proximity: Selective attention measures. Perception and Psychophysics, 18, 355 - 361.

Pomerantz, J.R., Sager, L.C., & Stoever, R.J. (1977). Perception of wholes and of their component parts: Some configural superiority effects. Journal of Experimental Psychology: Human Perception and Performance. 3, 422 - 435.

Pomerantz, J.R. (1981). Perceptual organization in information Processing. In M. Kubovy & J.R. Pomerantz (eds.) Perceptual Organization, Hillsdale, N.J.: Lawrence Erlbaum Associates, 141 - 180.

Sanderson, P.M., Flach, J.M., Buttigieg, M.A., & Casey, E.J. (1989). Object displays do not always support better integrated task performance. Human Factors, 31, 183 - 198.

Schmid, C.F., & Schmid, S.F. (1975). Handbook of graphic presentation, New York: Ronald Press.

Schutz, H.G. (1961a). An evaluation of formats for graphic trend displays. Human Factors, 3, 95 - 107.

Schutz, H.G. (1961b). An evaluation of methods for presentation of graphic multiple trends. Human Factors, 3, 108 - 119.

Siegel, J.H., Goldwyn, R.M., & Friedman, H.P. (1971). Pattern and process in the evolution of human septic shock. Surgery, 70, 232 - 243.

Sparrow, J.A. (1989). Graphical displays in information systems: some data properties influencing the effectiveness of alternative forms. Behaviour and Information Technology, 8, 43 - 56.

Strickland, R.G. (1948). A study of the possibilities of graphs as a means of instruction in the first four grades of the elementary school (Teachers College Contributions to Education No. 715). New York: Teachers College, Columbia University.

Tilling, L. (1975). Early experimental graphs. The British Journal for the History of Science, 8, 193 - 213.

Tufte, E.R. (1983). The visual display of quantitative information. Cheshire, Conn.: Graphics Press.

Tufte, E.R. (1990). Envisioning information. Cheshire, Conn.: Graphics Press.

Vernon, M.D. (1952). The use and value of graphical methods of presenting quantitative data. Occupational Psychology, 26, 22 - 34.

Wainer, H. (1980). A test of graphicacy in children. Applied Psychological Measurement, 4, 331 - 340.

Wainer, H., & Resier, M. (1978). Assessing the efficacy of visual displays. Graphic presentation of statistical information, papers presented to the 136th annual meeting of the American Statistical Association, U.S. Department of Commerce, Bureau of the Census, Technical Paper #43.

Wainer, H., & Thissen, D. (1981). Graphical data analysis. Annual Review of Psychology, 32, 191 - 241.

Washburne, J.N. (1927). An experimental study of various graphs, tabular and textual methods of presenting quantitative material. Journal of Educational Psychology, 18, 361 - 376 (part one), 465 - 476 (part two).

Wickens, C.D. (1986). The object display: Principles and a review of experimental findings (Tech. Report CPL-86-6). Champaign: Cognitive Psychophysiology Laboratory, University of Illinois.

Wickens, C.D., & Andre, A.D. (1990). Proximity compatibility and information display: Effects of color, space, and objectness on information integration. Human Factors, 32, 61 - 78.

Wickens, C.D., Kramer, A., Barnett, B., Carswell, M., Fracker, L., Goettl, B., & Harwood, K. (1985). Display/cognitive interface: The effect of information integration requirements on display formatting for C3 displays (Technical Report EPL-85-2/RADC-85-1). Urbana, IL: University of Illinois Engineering-Psychology Research Laboratory.

Wickens, C.D., & Scott, B.D. (1983). A comparison of verbal and graphical information presentation in a complex information integration decision task (Technical Report No. EPL-83-1/ONR-83-1). Urbana, IL: University of Illinois Engineering-Psychology Research Lab.

Wickens, C.D., & Todd, S. (1990). Three dimensional display technology for aerospace and visualization. Proceedings of the Human Factors Society 34th Annual Meeting (pp. 1479 - 1483). Santa Monica: The Human Factors Society.

Wilkinson, L. (1981). An experimental evaluation of multivariate graphical point representation. Proceedings of the Conference on Human Factors in Computing systems, Gaithersburg, MD.

Wood, D., Wise, J., & Hanes, L. (1981) An evaluation of nuclear power plant safety parameter display systems. Proceedings of the Human Factors Society 25th Annual Meeting Santa Monica, CA: The Human Factors Society.

Wrightstone, J.W. (1936). Conventional versus pictorial graphs. Progressive Education, 8, 460 - 462.

Zhang, K., & Wickens, C.D. (1990). Effects of noise and workload on performance with two object displays vs. a separated display. Proceedings of the Human Factors Society 34th Annual Meeting (pp. 1499 - 1503). Santa Monica: The Human Factors Society.

Commentary

Reading Graphs: Interactions of Processing Requirements and Stimulus Structure, C. M. Carswell

SUSAN E. BARRETT
Lehigh University

HERVE ABDI
University of Texas at Dallas
University of Bourgogne at Dijon

JILL M. SNIFFEN
Lehigh University

Carswell's paper focuses on a critical task facing human factors researchers, namely, specifying how processing requirements relate to the perceived structure of graphic displays. Carswell presents empirical support for the thesis that the effectiveness of a display is strongly task-dependent. The main theoretical notion behind her analysis is the Proximity Compatibility Hypothesis which predicts that tasks that require information integration will be best served by displays that are high in structural proximity whereas tasks that require independent decisions about each specifier will be best served by displays that are low in structural proximity.

There is considerable empirical support for this proposal and future work may make it possible to posit more precise links between the structural characteristics of the graphic display and the processing requirements of specific tasks. One major challenge involves specifying more clearly what is meant by display or structural proximity. As Carswell makes clear, a variety of graphic formats might be considered to be high in structural proximity. These include displays in which the various specifiers are contained in a single unitary object and displays composed of multiple heterogeneous specifiers that configure to create emergent features. One of the issues Carswell focuses on is whether the advantages afforded by single object displays in information integration tasks have been a function of locating the specifiers within a single object or whether this advantage has actually been a by-product of the emergent features that have arisen fortuitously in these displays.

To test the relative contribution of specific stimuli, Carswell asked subjects to perform two tasks involving high processing proximity. In the first task, subjects were required to make a decision about the relative value of the two specifiers. Here performance was best with the homogeneous displays and homogeneity was a more potent predictor of performance than the unitary nature of the display. In a second task, subjects were asked to check the exact value of each specifier. Under these conditions, the homogeneity of the display did not predict performance. Instead, performance seemed to depend on whether both specifiers were embodied in a single object and this effect was strongest with the heterogeneous displays.

As Carswell notes, the data from these two experiments suggest that a further refinement of the Proximity Compatibility Hypothesis might include both object-based and configurality-based definitions of display structure. In addition to refining the Proximity Compatibility Hypothesis to take into account more subtle variations in display proximity, it may also prove useful to include variables that take into account more complex differences in processing demands. One possibility might be to look more specifically at how the structure of the display maps onto the response requirements of various tasks. For example, certain tasks may allow the user to more readily capitalize on the configural properties of the stimuli. As an example, consider the advantages afforded by the configural stimuli in Pomerantz and Garner's (1973) condensation task. Here subjects were able to maximize the benefits of the stimuli's configural properties when emergent features were completely mapped onto the response categories. Perhaps a further refinement of the Proximity Compatibility Hypothesis which takes into account more subtle differences in display structure will also make it possible to predict the conditions under which observers are able to use emergent properties to shortcut the processing demands of the task.

Percepts, Concepts and Categories
B. Burns (Editor)

17

Search Process Versus Pattern Structure In Chess Skill

DENNIS H. HOLDING
University of Louisville

I. INTRODUCTION

As others in this volume have noted, the distinction between structure and process may be regarded as pervasive in cognitive psychology (Estes, 1975), shaping theoretical issues such as those concerning memory structures versus information processing. The distinction appears sufficiently fundamental that one expository text (Dodd and White, 1980) simply defines cognitive psychology as the study of mental structures and processes. Typical structural accounts deal with matters of organization such as representations in long-term memory or the nature of grammatical knowledge, while process descriptions tend to concern operations such as the conduct of retrieval in long-term memory or the function of the articulatory loop in working memory. The present chapter considers how theories of expertise, particularly in the domain of chess skill, may be distinguished in terms of their reliance on structure or process descriptions.

The study of expertise is a central topic in cognitive psychology, contributing both to the general theory and to specific applications of cognitive function. According to Anderson (1985), acquiring and practicing expertise requires the learning and deployment of both declarative and procedural knowledge. For present purposes, it should be noted that the distinction between declarative and procedural knowledge is loosely analogous to the structure-process dichotomy. However, although Anderson (1985) argues that the development of expertise proceeds similarly in different domains, much of the information regarding expertise has been derived from the domain of chess. Playing chess is a highly complex cognitive task which is nevertheless accessible to study, and it possesses the advantage that the skills of its proponents can be specified by a reliable rating scale.

The United States Chess Federation (USCF) scale devised by Elo (1978) weights the number of wins obtained in tournament play according to the caliber of the opponents, providing estimates of skill with a standard deviation of 200 relative to a mean at 1500 points. Classes E,D,C,B,A are defined at 200-point intervals from 1000-1800, and experts run from 2000-2200. Masters are rated 2200-2400, and senior masters from 2400 upward. Note that the grandmaster title is comparable but superior to the senior master qualification, and must be obtained through international play. The existence of this rather precise rating scale greatly facilitates comparisons of the structures or processes occurring at different levels of expertise.

II. THEORIES OF CHESS EXPERTISE

Although it may be admitted that chess skill involves the exercise of both declarative and procedural knowledge, the major theories tend to stress either structure or process. The earliest theoretical attempts were primarily structural, with the limited aim of classifying the faculties underlying chess skill. Binet (1984), despite his initial preconception that "blindfold" chess (played without seeing the board) would rely heavily on visualization, had to conclude that chess skill depended as much on experience ("l'erudition") and memory as on imagination.

The fertile but neglected work of Cleveland (1907) is more difficult to classify. The primary conclusion that chess depends on "position sense" accords well with the modern stress on the importance of evaluation and may be regarded as offering a structural account of chess expertise, although the work also draws attention to important processes such as the use of middlegame themes (such as pins, forks and skewers) and endgame routines (such as are required for queening a pawn). Later work by De Groot (1965) again emphasizes the part played by experience in developing chess skill, providing information concerning chess memory that contributed to a structural approach.

These early theories were couched in broad, general terms and failed to suggest detailed mechanisms for the selection of good chess moves. The theories were unsatisfactory in the sense that they offered no clear basis for experimental verification. Only two theories of chess expertise are sufficiently elaborate, and sufficiently specific, to permit the testing of critical predictions. The first of these, described below, is the recognition-association theory developed by Chase and Simon (1973), which may be classified as structural in emphasis. The second is the SEEK model (Holding, 1985), which relies primarily on the processes of forward search and evaluation. In forward search, a player considers what moves are available, what replies these moves would prompt from the opponent, what further moves would then be practicable, and so on. The resulting positions are evaluated at the end of each line of play in order to find the most favorable outcome.

The recognition-association theory assumes that differences in forward search can be discounted, for reasons to be discussed below. Instead, the theory suggests that chess skill depends on memorizing thousands of specific patterns of pieces. Recognizing one of these patterns during play is assumed to call forth an associated candidate move, also residing in memory, which can then be used or investigated by the player. Stronger players are assumed to be those with larger repertories of chess patterns. The part played by experience in chess skill is thus explained as the acquisition of a library of patterns, so that skill is said to depend primarily on memory. The recognition-association theory has received wide acceptance in cognitive psychology, perhaps because the reverse effect, the dependence of specialized memory on skill can be demonstrated in many fields. However, it will become apparent that there exists no evidence for the theory itself other than the indirect support provided by inferences and assumptions.

In contrast to the structural explanation offered by the recognition-association theory, the SEEK theory emphasizes process descriptions. The relevant processes (Search, EvaluatE, and Know) are those found important both in empirical research on chess expertise and in artificial intelligence work on computer chess. Essentially, the SEEK theory represents an extension of the commonsense view that stronger players excel at thinking ahead. Grandmasters, for example, are better than weaker players because they progress more deeply and efficiently through the search tree, and because their judgments produce better evaluations of the positions finally reached. Chess knowledge is viewed principally as a supply of general strategic principles and generally applicable move sequences, not simply as a repertory of specific patterns, and can be used either to supplement or to supplant the process of forward search.

III. RECOGNITION-ASSOCIATION

A. Antecedents

The idea that strong players see further ahead was originally dismissed by De Groot (1965), who measured various characteristics of the verbal protocols obtained from players during their analyses of chess positions. It was reported that grandmasters and experts did not differ in their search statistics, but that the grandmasters consistently made better moves. This conclusion has been widely cited, and it provided the basis for the development of the recognition-association theory. Nevertheless, there were a number of recorded differences, all favoring the grandmasters.

To explain the nature of such differences, it is necessary to briefly outline the ways in which search trees may be measured. More detailed descriptions can be found in Charness (1981b) or Holding (1985). A player speaking aloud while considering a chess position is likely to mention a number of moves. A "move" in a chess game consists of a pair, one for white and one for black, but a "move" in search usually refers to a "half-move," or *ply*. Each time the player reaches the end of a chain of replies and counterreplies (often called a "terminal node") and returns to an earlier stage in the search, an *episode* has been completed. It is possible for these episodes to overlap, or to constitute completely different lines of play. The moves that follow immediately from the game position are called *base* moves. Later moves may diverge, giving rise to *branches*, or may simply be reinvestigations of the same positions, or else may be left unspecified as *null* moves. All moves mentioned by the player, whether or not they involve reinvestigation, are counted toward *total* moves. The longest continuous sequence of moves (line of play) in the player's tree gives the *maximum depth* of search, while the *mean depth* is calculated as the average across episodes.

The titled players in De Groot's (1965) sample apparently searched more total moves (35.0 versus 30.8), and examined more different lines of play (4.2 versus 3.4) than did the lower-rated experts. There was little difference in the maximum depth of search (6.8 versus 6.6) but the difference in mean depth (5.3 versus 4.8, estimated as total/different base moves) is probably a more stable measure, less dependent on idiosyncracies of the presented positions. It should be noted that the reported gain of 0.5 ply at this depth, given the exponential growth of the search tree, potentially accounts for the consideration of many millions of additional positions and would represent a substantial achievement for any computer chess program. In addition to searching more, the grandmasters seemed to be searching faster. They took less total time (9.6 versus 12.8 minutes) to analyze the position, consequently mentioning potential moves at a faster rate (3.5 versus 2.5 moves per minute).

Although no tests were conducted, De Groot (1965) and subsequent authors assumed that these differences were nonsignificant. However, any failure to find differences in these circumstances should be taken simply to indicate lack of statistical power. There were only five subjects per group and the separation of values on the main variable was relatively small, since the estimated USCF ratings for the players might only vary from perhaps 2100-2500. In addition, the main position that was used required the calculation of series of exchanges, with little incentive to continue beyond that point. It is unlikely that search differences would be prominent in these circumstances, but it would be illegitimate to conclude that no differences existed. In any case, a reanalysis of the data by Holding (1992), making use of the Mann-Whitney test, showed that the difference in n* (different base moves/minute) was in fact significant ($p < .02$). Grandmasters explored new lines at the rate of .49 per minute, compared with only .29 per minute considered by the experts.

Having unfortunately dismissed search differences, skill effects were next sought in experiments on memory for briefly exposed positions. Djakow, Petrowski and Rudik (1927) had earlier examined titled players at the 1925 Moscow tournament, administering a number of tests with aim of characterizing chess talent. Apart from using instruments like the Rorschach, the experimenters devised several types of memory procedures, comparing the performance of the players with a sample of nonchess subjects. The players were far better at reconstructing a chess position from memory after seeing the board for one minute, although their scores were only average on other tests. The chess and nonchess subjects performed almost equally well at recalling two-digit numbers, at restoring the order of a set of geometric shapes, and even at remembering the positions of colored circles on a chessboard grid. It appeared that the benefits of lengthy chess experience were confined to chess-related tasks.

The memory experiment was modified by De Groot (1965), who presented 16 different chess positions to four players of varying skill levels. The exposure time was reduced to a few seconds (anywhere from 2-15 seconds), but the subjects were given half a minute to collect their thoughts. The two top players gave verbal recall, but the lower players were allowed to reconstruct the positions on a chessboard, and the scoring system included various bonus points. Despite the unstandardized conditions the results showed a fairly systematic trend across skill levels. A grandmaster correctly replaced 93 percent of the pieces, a master 91, an expert 70, and a class player 53. Similar work was subsequently performed by Jongman (1968), who put forward an explanation for the ability of skilled players to remember more pieces than can normally be held in short-term memory. If skilled players have learned to recognize and classify segments of a position, short-term memory has merely to hold references to the information already classified in long-term memory. This suggestion, modified and expanded, forms the basis of the recognition-association theory.

B. Chase and Simon

The brief memory experiment was repeated by Chase and Simon (1973) using shorter exposure times, with the additional manipulation of presenting random positions. It was confirmed that stronger players remembered the game positions better, although their advantage disappeared when random positions were used. It seems reasonable to expect that an expert in any field should remember relevant material, but not irrelevant material, better than an amateur, although the latter effect is not always obtained. Goldin (1979), for example, found skill differences in recognition memory for scrambled positions, and Reynolds (1982) showed that such differences depend on the degree to which the pieces converge on the central squares. Lories (1987) has shown that skill differences reappear when extended time is given for studying random positions, and Saariluoma (1989) has observed differences in memory for random positions that were verbally dictated. Nevertheless, it may be granted that the effects of skill are normally greater when real game positions are employed.

The most influential contribution by Chase and Simon (1973) came indirectly from their detailed analyses of the memory data. Their three players seemed to reconstruct the chess positions by replacing the pieces in groups of two or three at a time, separated by intervening pauses. These groups of pieces, which tended to share chess relationships such as color, proximity, and piece type, might be regarded as "chunks" for the purposes of memory storage. One problem with this assumption is that the apparent size of such chunks might be determined by the number of pieces that can be physically manipulated and placed. However, a master seemed to use more, and larger, such chunks than a class player or a beginner. Following Jongman (1968), the argument could therefore be made that memory for briefly exposed positions was achieved by rehearsing in short-term memory the labels for chunks already stored in long-term memory. Acquiring the requisite chunks would be a function of chess-specific experience.

If perhaps 50,000 chunks, or subpatterns, are required to simulate master performance, the further argument may be made that such a number is comparable to a natural language vocabulary, suggesting that as many years are needed to achieve mastery at chess as to learn a language. However, this estimate derives from a computer simulation by Simon and Gilmartin (1973), who actually found that using 447 or 572 chunks for both White and Black permitted the reproduction of 39 or 43 percent, respectively, of the game positions used in the memory task. The first problem was to perform an extrapolation from these closely adjacent figures. A linear projection would suggest that perhaps 2,500 chunks would be needed for perfect reproduction but, since many of these patterns would be rare, a linear equation could be considered an underestimate. Various nonlinear extrapolations gave figures ranging from 13,500 to as many as 363,000 chunks, and 50,000 was selected as a plausible guess. It can be argued (Holding, 1985) that this figure is inflated by counting White and Black patterns separately and by ignoring transpositions of identical chunks in different board locations. Given these considerations, a realistic figure might still be nearer 2,500 patterns.

Although exactly determining the required number of chunks will affect the argument that extensive experience is needed for mastery, it is not crucial to the next stage of the recognition-association theory. Chase and Simon (1973) proceeded from the observation that skill determines memory to the conclusion that memory determines skill. Although such an inference appears fallacious by illicit conversion, a mechanism was proposed to account for the control of move choice by memory processes. The suggestion was that the chunk organization in long-term memory also possesses the properties of a production system. Such a system is equivalent to a set of "if-then" statements, in the form of condition-action relationships. In the present application, each familiar pattern is viewed as the subject of a separate condition. Recognizing any given pattern satisfies the appropriate condition, thus evoking a plausible move already associated with the pattern and bringing it into short-term memory for consideration. Exercising chess skill is therefore seen as dependent on pattern recognition accompanied by learned association.

C. Theoretical Difficulties

There are a number of problems with this account of chess skill. One difficulty is that the chunks identified in the brief memory experiments may represent an inappropriate level of analysis. Reitman (1976) found many overlaps in the chunk boundaries indicated for a related board game, with poor prediction by the pause criterion, which also presented ambiguities in chess memory analyzed by Charness (1981a), perhaps because chunk organization has a hierarchical structure. More recent analyses of the manner in which players construe the board (Horgan, Millis and Neimeyer, 1989) rely on similarity judgments, finding wide qualitative differences in between players of different strengths. Masters use more abstract, superordinate categories than weaker players, whose limited knowledge seems to generate relatively few functionally independent clusters. The organization of expert memory appears quite complex. Hence, the relationships between short-term and long-term memory described by Chase and Simon (1973) no longer seem appropriate.

Furthermore, several types of evidence suggest that even briefly presented chess information receives rapid, permanent storage, and is unaffected by manipulations that normally interfere with short-term memory. Charness (1976) found no effect on chess memory from interpolated tasks such as repeating random digits or carrying out mental rotation, nor from chess-related activities like naming pieces or even solving problem positions. The implication that chess information is directly encoded by players into long-term memory can also be drawn from the work by Frey and Adesman (1976), who found that the first of two presented positions was as well remembered as the second. There are also demonstrations that chess memory depends on depth of processing. Goldin (1978) found that recognition memory for chess positions varied with the presentation of structural or semantic instructions, and Lane and Robertson (1979) observed the dependence of skill effects on orienting instructions. It therefore seems unlikely that chess information is passed from short-term to long-term storage in the manner suggested by the theory.

A severe problem for the recognition-association theory is that, even if storage proceeds by chunking, the postulated chunks would be incapable of generating the required moves. The chunks identified by Chase and Simon (1973) are not of the appropriate size, nor do they normally interrelate the White and

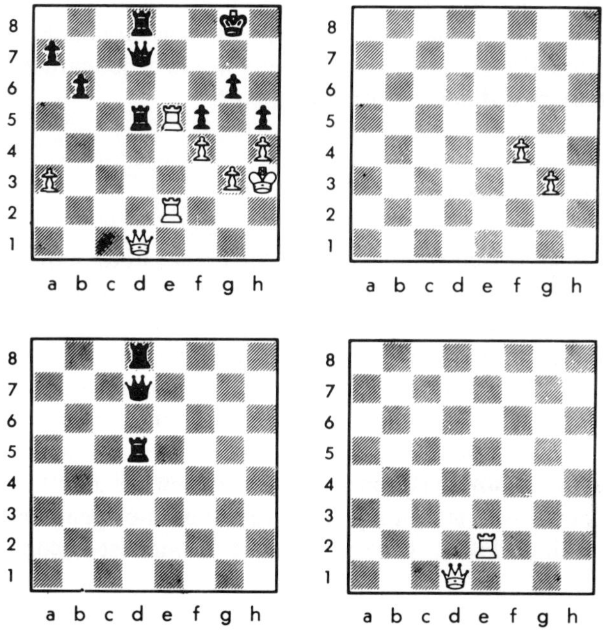

Figure 17.1. Encoding a chess position in chunks, following Chase and Simon (1973). The original position is shown on the top left, together with a representative sample of observed chunks.

Black pieces as would be required to generate useful moves. Typical chunks are composed of such groupings as a cluster of two or three pawns, same-color rooks and queens on a file, or perhaps a pair of opposing pieces, as illustrated in Figure 17.1. If these are regarded as the elements of a production system, they constitute conditions for which there are no appropriate actions. A disconnected chunk such as a triangle of pawns suggests no useful moves in a wider game setting, because such chunks do not embody the dynamic interrelationships that evoke potential moves.

Admittedly, an argument can be made that the chunks could be assembled into larger, more meaningful patterns by a series of transactions between long-term and working memory. These superchunks, which would correspond to the higher levels in a hierarchical storage mechanism, could then form workable conditions for the production system. However, a master would then have to learn only sufficient plausible moves for the number of resulting patterns. It would no longer be possible to derive quantitative estimates from the brief memory experiments, and the debate concerning the necessity for 50,000 patterns would become irrelevant. Consequently, the argument that extensive experience and practice are needed for mastery would be jeopardized. The entire theory would require a different form of support.

D. Visual and Verbal Factors

In any case it is far from clear that the emphasis on pattern recognition, or for that matter on the "mind's eye" (Chase and Simon, 1973), is justified by the facts. Indirect support for this approach rests on the assumption that chess skill is largely a product of visuospatial ability. This might be the case but, although authors such as De Groot (1965) have drawn the conclusion that spatial ability is primary, there is very little empirical evidence for the idea. The only recent support comes from Horgan and Morgan (1990), who found that children who did well at chess scored highly on Raven's progressive matrices and the Knight's Tour test. However, the Raven is typically used as a test of intelligence, which may also be a factor (Holding, 1985), while the Knight's Tour test is completely chess-specific.

On the other hand, there are several indications that chess requires verbal ability. It can be taken into consideration that players are accustomed to encoding and processing chess moves by means of symbolic descriptions, such as Rd7 (or R-Q7), and used such methods to demonstrate recall in some of the experiments reported above. Book knowledge of openings, middlegame themes and endgame techniques, which is of vital importance for mastery, is organized by means of verbal labels (Benoni defense; pawn lever; the Lucena position). Again, biographical surveys of chess titleholders (Elo, 1978) indicate superior verbal skills, as evidenced by the high incidence of professional writing occupations and widespread mastery of foreign languages. Among more formal studies, some of the best evidence comes from Pfau and Murphy (1988), who used a 75-item test of general chess knowledge on a range of chessplayers. Though purely verbal, the test correlated better with tournament ratings than did the brief visual memory test.

Chessplayers can be observed to use verbal mediation (muttering) while analyzing positions, so it appears natural to measure the effects of interfering with the process. Holding (1989b) persuaded players of different strengths to count backward by threes while attempting to analyze game continuations, comparing the results with a quiet condition. Also, in some cases the players were required to use an external representation, moving the pieces on the board, while in others all analysis was accomplished by internal representation. Moving the pieces tended to result in longer lines of search, with fewer side branches, but had little effect on the main manipulation. However, the effect of counting backward was devastating. At all skill levels, the interference abbreviated the search process, cutting down the total number of moves, the number of branches, and the depth of search. What is more the quality of the moves that were eventually chosen was halved, as measured by an objective index. Many players chose moves that would have lost material, and the stronger players were as badly affected as the weaker.

However, it may be that the counting manipulation is too powerful to implicate only verbal mediation. If working memory includes a central executive which directs the operations of an articulatory loop and a visuospatial scratchpad (Baddeley, 1983), the counting task might be considered sufficiently disruptive to affect the central executive. Later work by Saariluoma (1992) comparing visuospatial interference with an easier articulatory suppression task (simply repeating a Finnish word) found no effect of simple suppression on either counting minor pieces or finding short checkmate solutions. On the other hand, it seems clear that the counting task must interfere with processing during chess analysis by other than visuospatial means.

It should be noted that Chase and Simon (1973) themselves obtained information concerning verbal memory for chess. In addition to collecting data on brief visual memory for static game positions, separate measurements were reported concerning players' memory for the move sequences in actual games. One experiment investigated short-term memory, with immediate recall, and another long-term memory, with recall on a different day, but in both cases the games were dictated verbally. Again, players of different skill levels showed scores that differed in accordance with their degree of expertise. As Holding (1989a) showed, the different playing levels are at least as well discriminated by the verbal long-term memory scores as by the brief visual memory scores that formed the basis of the recognition-association theory. The theory makes no provision for verbal components in move choice, but it is obvious that a very different account of the cognitive processing underlying chess expertise could have been formulated on the basis of the information available at the time. Note that the theory prefers the structural information afforded by visual pattern data over the process explanations suggested by studying move sequences.

Apart from problems concerned with the visual chunking stage of the theory, it can be argued that there are difficulties over the status of the production system component. This aspect of the theory was never supported by direct evidence, but derived some credibility from the suggestion that a system of this type could facilitate rapid responding. Both De Groot (1965) and Chase and Simon (1973) propose that masters see the board differently from weaker players, immediately perceiving strong moves without having to calculate variations. If so, a possible underlying mechanism might be that players recognize familiar patterns that call forth immediate associations. There is no quantitative evidence for the rapid response hypothesis, although Charness (1981b) found that younger and more skilled players were more likely to mention the objectively best moves among the base moves in their search trees. However, the effect was not large, the time constraints were too relaxed to insure that the effect represented immediate perception, and the effect did not correlate with later move choice.

In any case, other explanations for immediate perception would be equally possible. For example, instead of having learned thousands of individual patterns, the player might have learned a variety of general principles underlying chess relationships. One way in which such knowledge might be implemented might be to incorporate experience of piece movement directly into board processing. Many of Binet's (1894) respondents denied that they visualized the board itself, but reported "seeing" lines of force, or the trajectories of the pieces. Holding (1992) has demonstrated that replacing a chess position with a diagram representing the piece trajectories draws attention to gaps in the pattern of forces, directly suggesting move possibilities in a manner that is independent of recognizing the position. Since other explanations are possible, support for the production system hypothesis is correspondingly weakened.

Finally, it should be noted that the recognition-association model itself has never been explicitly tested. A critical test (Holding, 1985) might consist of selecting a group of players varying in expertise, presenting a series of chess patterns, and asking for a choice of move. The patterns would be classified as either familiar or unfamiliar according to an acceptable criterion, and the chosen moves would be scored as good or poor. The recognition-association theory should predict that the familiar positions would rapidly elicit good moves, while the unfamiliar positions would elicit either bad moves or no moves at all. Furthermore, skill differences between players should be present in responding to the familiar positions, but not in responding to the unfamiliar positions.

An experiment that meets the most critical of these requirements was reported by Holding and Reynolds (1982). The technique was to present random positions which, by definition, should be unrecognizable, and to ask players to choose lines of best moves. The players were also given the brief memory test for the positions, with results that did not differ according to level of skill. Despite the absence of memory differences, the moves that were selected varied systematically with the strength of the players. Since move quality in these unrecognizable positions varied with playing skill, while short-term memory did not, the results appear to invalidate the recognition-association theory. It therefore seems appropriate to consider alternative models for chess skill.

IV. COMPUTER PLAY

It is instructive to consider how computer programs manage to play strong chess moves. Like human players, computers begin a game by consulting a "book" of opening moves that records the best lines known to theory. Once out of the opening book, most computers start by generating all of the moves and countermoves from the starting position, thus creating a tree of possible variations whose size depends on the time and memory available. The program will then make numerical evaluations of the positions reached at the endpoints, using criteria whose degree of sophistication and emphasis on material or positional considerations vary from program to program. The obtained values are then backed up through the tree, using a minimax procedure which insures that the move yielding the most favorable outcome for the playing side, given best play by the opponent, is selected for play. Many ways have been found to optimize the computer's path through the search tree, and to save time by pruning unnecessary branches. In general, the program that wins is the program that performs the deepest search while computing the most accurate evaluations.

There are many variations on this basic theme. As is well known, the exponential growth of the search tree is such that there are already 225 million possible combinations on Black's 3rd move (i.e., after W,B,W,B,W), and the entire tree potentially includes 10^{117} branches (Holding, 1985). Older programs attempted to reduce the computing load by performing selective searches, identifying perhaps half a dozen base moves for further exploration. Most of the successful mainframe programs, such as Cray Blitz or Deep Thought, now perform what are known as "full-width" searches, and can deal with millions of positions in a matter of seconds. However, modern pruning algorithms eliminate large sections of the tree, and one of the uses of updated hash tables (accessible records of prior computations) is to prioritize the order in which candidate moves are examined, so that the search is still effectively selective.

A different approach to reducing the volume of search is to make use of plans, goals, or themes, as explored by various experimental programs. For example, Church and Church (1977) described a program written for a small microcomputer which uses a technique of problem reduction. The program rapidly identifies one of several predetermined goals, perhaps finding an apparent weakness to be attacked or remedied, and organizes its search with reference to the goal. Two converging trees are constructed, one running forward from the current position and the other backward from the goal position, thus simplifying the search. The resulting program makes an extremely economical search, and was able to select reasonable moves in many positions after only a few seconds of calculation.

Although largely process-based, most computers also make use of knowledge to varying extents. Most programs enter a game by referring to an opening book, as mentioned above, many make use of devices such as transposition tables, and some employ middlegame and endgame routines known from earlier human play. These devices obviate the need for search to various extents. There is no need to perform a search when the path to a given outcome is already known. Berliner (1984) describes knowledge as possessing projection ability for search, and has attempted to quantify its effectiveness. For example, tactical knowledge can replace a 9-ply search procedure for the win of a particular pawn, positional knowledge can substitute for a 25-ply search, and strategic knowledge may be worth a 45-ply search. A well-known endgame principle, such as entering the square of the king, can be equivalent to between 2 and 10 plies of search. In addition, it should be remembered that applying chess knowledge at the terminal nodes of the search tree will enhance the effectiveness of the resulting evaluation function.

In outline, it can be concluded that a computer program selects a move on the basis of its anticipated consequences. The strength of any program will depend on several factors that determine how well those consequences are predicted and evaluated. The relevant factors include the depth of the search process, the manner in which search is conducted, the accuracy of the evaluation function, and the degree to which chess knowledge is incorporated into the search and evaluation processes. As it happens, the same factors seem to enter into human chess expertise. Although there are quantitative differences between human and computer play, it can be argued that the similarities are more important than the differences. The SEEK model of chess expertise acknowledges these similarities, emphasizing the dependence of human play on search, evaluation, and knowledge.

Table 17.1. Search statistics for weak, average and strong chessplayers, calculated from data by Charness (1981b).

	Tournament Rating		
Measure	1200	1569	2000
Total moves considered	4.1	25.0	49.5
Number of episodes	3.2	7.0	11.5
Terminal nodes reached	3.3	9.1	15.7
Repeated base moves	0.4	3.6	7.3
Repeated other moves	1.1	3.6	9.2
Mean depth of search	2.0	3.0	4.1

V. THE SEEK MODEL

According to the model, human chess skill relies on search and evaluation, modified or replaced by expert knowledge. The SEEK model is primarily a process model, basing its main emphasis on search and related processes, although the provision for knowledge use might be viewed as introducing structural considerations. The model derives its support from a broad range of empirical findings. There are many studies that exhibit skill differences in search processes, several experiments on evaluation and its use during search, and a variety of sources of information bearing on the use of knowledge.

A. Search Processes

As noted above, the expert and grandmaster players investigated by De Groot (1965) seemed to show meaningful differences in the dimensions of the searches that were made while choosing moves. Greater differences in the search statistics might be expected when wider differences between playing strengths are examined. In fact a more broadly based investigation by Charness (1981b), using a larger number of subjects under better controlled conditions, has found systematic differences in the verbal protocols of players rated from 1200 to 2000 on the Canadian scale. The expert players (at 2000) examined far more potential moves, taking the search to a greater mean depth and reaching more terminal nodes, than the weaker players. Although the results were presented as regression equations, the extreme values for some of the most important statistics have been calculated in Table 17.1 to provide a concrete illustration. The age of the player, which also played some part in determining the scores, has been neglected for this purpose. Notice that the relatively small numbers of separate search episodes, compared with a possible maximum of perhaps 38 move options, imply that human search is highly selective.

Another perspective on search efficiency is provided by the figures on the maximum depth of search attained. The regression of maximum depth on skill from Charness (1981b) is presented in Figure 17.2, extrapolated to master strength at 2200. For comparison, some of the data from the backward counting strategy

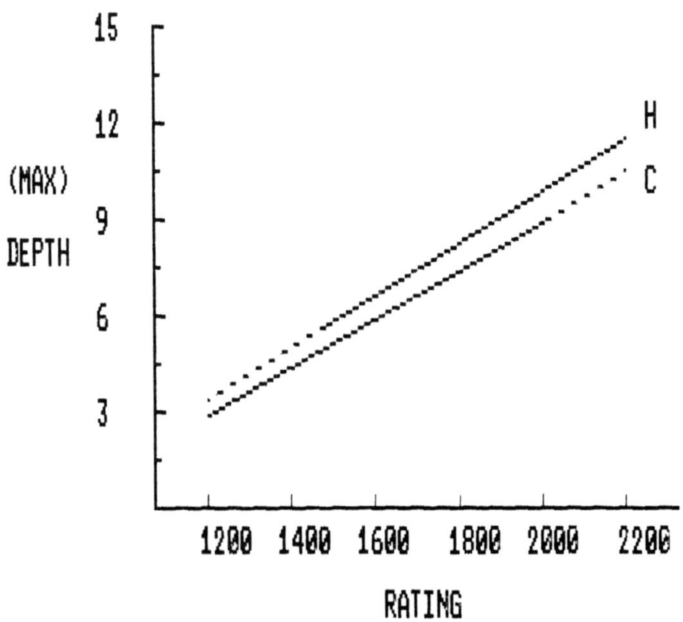

Figure 17.2. Maximum search depths associated with degrees of chess expertise, comparing regression lines from Charness (1981b) and Holding (1989b). Broken lines indicate extrapolations.

experiment by Holding (1989b) have also been plotted. For this purpose, only the data from the condition (quiet-still) in which subjects refrained from counting and from moving the pieces have been used. In this condition, the maximum depth of search correlated significantly with playing strength. The search trees in this experiment were generated by players rated at higher levels of skill, averaging 1830 points, and have been extrapolated downward. Despite the differences between the two experiments, it can be seen that the obtained slopes are surprisingly close. On this evidence it seems clear that, other things being equal, depth of search must play a major part in chess expertise.

It has been pointed out by Tikhomirov (1984) that verbal protocols, such as were used here, tend to underestimate the amount of search that is conducted. Using nonverbal techniques, including eye movement recording with sighted players and cyclographic measures that trace the hand movements made by blind players during tactile search, it can be shown that the number of pieces and squares actually explored are far in excess of the number mentioned in verbal statements. However, the reported data give no reason to counter the belief that the amount of exploration varies with playing strength. The same data also show that the amount and type of exploration conducted varies with the expectancies of the players. For example, if a given move is one of a highly predictable series of exchanges, the player will concentrate on the squares involved while neglecting peripheral squares. If a move by the opponent differs from what was anticipated, the search is enormously broadened, thus indirectly confirming that move choice is controlled by anticipated consequences.

On the other hand, Charness (1989) has questioned whether mere depth of search is sufficient to account for skill differences, on the basis of an opportunity that arose to reexamine one of the players first tested by Charness (1981b). Despite a considerable increase in rating since the original test, the player showed no increase in depth of search. To the extent that data from a single player are indicative of general trends, it might be necessary to conclude that advances in skill occur at different rates with respect to the subcomponents involved. For example, search depth might reach a plateau while "position sense," or evaluative judgments, are improved. In any case, what the SEEK model requires is that search efficiency, often represented by search depth, should improve.

Other factors in search efficiency have been discussed by Holding (1985). Apart from the possibilities afforded by plans and goal-directed search, there is evidence that players find their way through search trees in different ways. For example stronger players show a greater tendency to follow a win-stay/lose-shift strategy, or homing heuristic, better following the consequences of promising moves and eschewing the pursuit of unprofitable lines. This tendency can sometimes prove maladaptive, perhaps accounting for behavior in experiments by Saariluoma (1990) on smothered mate (where the king is hemmed in by its own pieces) and similar problems. The most highly skilled players followed familiar lines without exploring to find shorter solutions, possibly because the positive outcomes that were encountered provided no incentive to shift. The experiments were interpreted as supporting a recognition-association model, although it is not necessary to assume that specific chunk recognition mediated the effects of acquired knowledge for familiar move sequences.

Other work on search heuristics has shown, for example, that stronger players are more consistent from move to move in their judgments throughout the tree, and in choosing moves for play in accordance with their more remote judgments. Holding (1989c) has documented how weaker players often choose moves that are inconsistent with the evaluations of the anticipated endpoints of their search. Accurately foreseeing the values of anticipated moves appears to be an important component of chess expertise.

B. Evaluation Judgments

There is little point in looking ahead through a tree of variations without assessing the values of the endpoints. Understandably, the results of a computer search can become progressively worse at greater depths if the evaluation function is sufficiently poor, and one might expect similar results with human players. Hence, the accuracy with which players can discern whether chess positions are good or bad appears to be an essential component of chess expertise. The evaluation of a chess position includes assessing the material balance, which is relatively easy to accomplish, and making judgments of positional value, which requires more sophistication. Since determining material balance merely involves counting pieces, most research has been concerned with skill effects in positional judgments.

Holding (1979) used a rating scale varying from 10 (even game) to 20 (resignation overdue), asking players to rate the advantage for White or Black in a series of quiescent middlegames and endgame positions. Gross errors in determining which side was winning showed a clear relationship with playing class (Holding, 1989a), while the finer errors varied with the presence or absence of salient pieces or configurations, and with factors such as whether the playing side or the opponent had the advantage. The judgments of the stronger players, and to some extent of the weaker players, correlated highly with one of the more elaborate computer evaluation functions (Chess 4.5), although the weaker players agreed significantly with simpler assessments such as the piece mobility count.

The general relationship between evaluation error and playing skill can also be illustrated from data collected for other purposes. Subjects in the backward counting experiment (Holding, 1989b) performed evaluations at the conclusion of their analyses, without counting. The numerical error in positional rating, averaged over the four test positions, is shown as a regression line against level of skill in Figure 17.3. The same figure also depicts the results obtained by Holding (1989c) from an experiment in which moves from displayed starting positions were dictated through binary trees (branching twice at each of three full moves), and evaluations of the anticipated worths of the resulting positions were made at all eight terminal nodes. Evaluation error has been averaged over both the trees employed. This experiment produces a somewhat steeper slope, but it is evident in both cases that accuracy of evaluation is related to expertise.

During actual analysis, such evaluations must be clearly seen by the players before the final positions are reached. As a first step in investigating such anticipatory judgments, Holding and Pfau (1985) verbally dictated sequences of moves from seen positions, asking for evaluations at each unseen juncture. The judgments made by the players at each stage were compared with those eventually

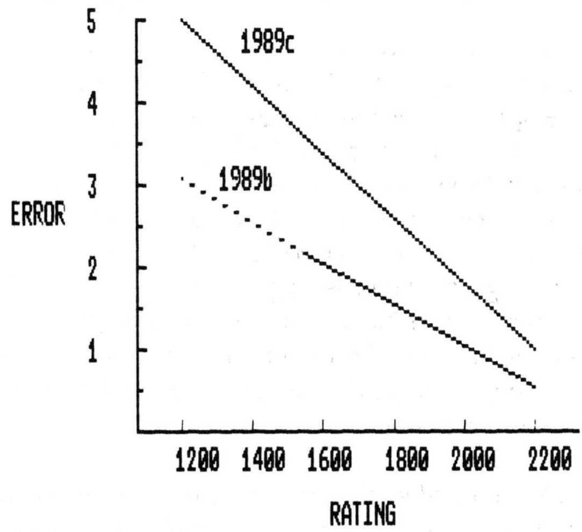

Figure 17.3. Errors in evaluation (scale values) related to level of chess skill, using regression lines from Holding (1989b,c). Broken lines indicate extrapolations.

made when the final position, 6 plies ahead, was exposed to view. The differences between current and final judgments diminished as each sequence progressed, for all players. However, the stronger players began with smaller discrepancies between the seen and foreseen positions, and maintained their advantage throughout the sequences, finally reaching zero error at least 2 plies ahead of the weaker players.

Instead of dictating a single line of moves, Holding (1989c) conducted players through entire binary trees as described above, each of which contained 29 nodes. In these circumstances, if stronger players can see more clearly ahead then one might expect their advantage to accumulate as each sequence progressed. As predicted, the discrepancy scores for the stronger players diminished across successive plies, just as in the previous experiment. However, the additional information load proved too much for anticipatory judgments of the weaker players, who showed little if any improvement as the tree developed. Not only were their judgments mutually inconsistent but, as stated above, their foreseen evaluations were neglected in making move choices. The data therefore confirm that the capacity to foresee later consequences is essential to chess expertise.

C. The Uses of Knowledge

Some of the consequences of series of moves do not require forward search, as they are known in advance by the chessplayer. In some cases players can rely on rote memory for previous games, whether played by themselves or reported in the chess literature, but much of the appropriate knowledge will come from generic rather than from episodic memory. Players are undoubtedly familiar with large numbers of chess patterns, but probably encode them in terms of prototypical themes with variations (Holding, 1985), rather than as originally encountered. Knowledge of patterns might serve a variety of uses, such as providing endpoints for transpositions of moves, or suggesting attacking themes. However, organizing chess knowledge on the basis of pattern information has the disadvantage of disrupting the natural move sequences, and the information stored in this way is less easy to generalize.

Much of the knowledge required for expertise in chess seems to be broadly applicable rather than episodic. For example, the player will know as a maxim that the Queen's Gambit (a standardized opening sequence that apparently loses a pawn) is not a true gambit, since the offered pawn can eventually be retaken. As stated here, this is a piece of verbalized, general knowledge possessed by any serious player. The player may well remember the visual pattern that constitutes the gambit, although he could in principle remember the sequence of opening moves that create the pattern; in any case the player will respond to recognizing the entire position rather than to separate chunks, and will recall a complete set of variations rather than a single candidate move. Again, the player may recognize that the necessary preconditions exist for a smothered mate, but the knowledge is transposable from the king's to the queen's side of the board, whether or not the situation has arisen during previous experience.

As Lesgold (1984) has argued, experts become proficient at seeing the essentials in a problem space, and such essentials must be abstracted from particular settings. The chess expert is known to have a fund of abstract knowledge, much of which can be verbalized. In the study by Pfau and Murphy (1988), playing skill was correlated with declarative knowledge of the openings (e.g., "Smyslov's variation of the Queen's Gambit Accepted is ...?"), the middlegame ("The best way to answer a wing attack is ...?"), and the endgame ("Triangulation is a technique used to ...?"). Such knowledge not only predicted tournament ratings better than a visual memory test, but the memory scores made no additional contribution over knowledge in the multiple-regression equation used to predict chess skill.

Declarative knowledge must also play a major part in forming evaluative judgments of chess positions. In order to evaluate seen or anticipated positions, players make use of a plethora of sometimes conflicting principles. The value of any position will depend, for example, on the presence of backward pawns, rooks on open files, centralization of knights, and factors concerned with king defense. Strong players also know when exceptions must be made to such principles, as in positions where two bishops are superior to a rook and bishop. Since gaining evaluative information is critical to the search process, the efficiency of move choice will depend heavily on such knowledge.

Note too that declarative knowledge of evaluative principles will help to formulate goals, which in turn may abbreviate and enhance the search process. Not a great deal is known about the formulation of goals, although the evidence by Horgan, Millis and Neimeyer (1989) suggests that there is some discontinuity across levels of skill. Estimates of the goals implied by move choices indicated an increase in the number of goals generated by players up to the expert level, followed by some decrease at the master level. Apparently masters cease to consider some of the goals attempted by players up to the expert level, perhaps as a consequence of formulating fewer but superordinate clusters in similarity judgments across positions.

Although there is scope for more research on knowledge functions, it seems evident that players must also possess wide repertories of procedural knowledge. It is important to know how to conduct a search, how to progress through the tree, and how to curtail the search where necessary. Probably knowing how to implement common routines, such as queening a pawn against a lone king, may be regarded as procedural skills. In addition the player needs procedural knowledge concerning ancillary requirements, such as keeping a score sheet and handling the time clock, and perhaps even for managing stress. Knowledge in various forms, both declarative and procedural, seems to pervade the entire process of move choice. Its primary function, however, is to insure that moves are chosen in accordance with their consequences.

VI. SPEED PLAY

Choosing moves according to the values of their outcomes is the central process of the SEEK model, which incorporates knowledge factors into search and evaluation. At this point, it seems clear that the structural assumptions of the recognition-association theory are unfounded, and that the theory cannot yield the necessary predictions. It might be added that the observed effects of verbal interference suggest the disruption of a process rather than a structural model. Furthermore, the recognition-association model makes no provision for the empirically demonstrated factors of search and evaluation, and limits the potential scope for the application of expert knowledge. However, before concluding that chess skill is fully understandable in terms of search and evaluation, one final objection must be met.

It is sometimes argued, for example by Frey (1977), that the conditions of speed chess do not allow sufficient time for the player to conduct a forward search. To the extent that the SEEK scheme makes provision for implicit search mediated by existing knowledge to replace explicit search, the objection loses some force. However, the model should predict that some degree of search is conducted in any meaningful chess transaction, so that if search appears impracticable in the conditions of speed chess, then doubt is thrown on the entire theory. Fortunately the question, whether search occurs in speed chess, can be answered empirically.

The most common form of speed chess is 5-minute chess, where each player has that time to complete all of the moves. The average tournament game lasts about 39 moves (Holding, 1985), although speed games might be somewhat

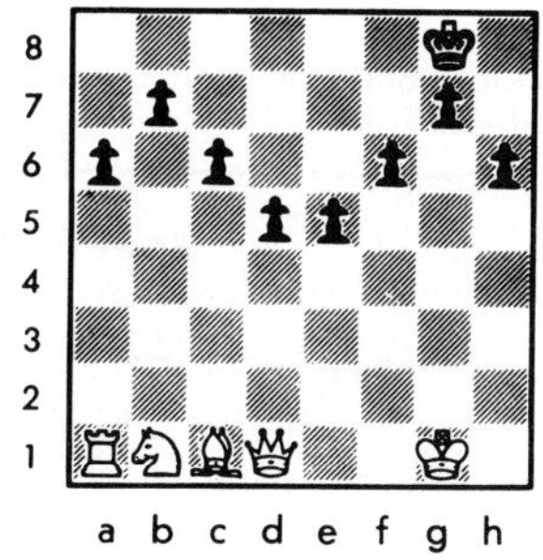

Figure 17.4. Test position for rapid search. The player counted the number of moves needed to transfer each White piece in turn to the opposite back rank.

shorter. Overall, the average time available is thus approximately 8 seconds per move, but far less will be required due to the use of rote memory in the openings and with many known algorithms in the endgame. Hence, it can be estimated that the player will have perhaps 10 seconds per move during the critical middlegame stages of play. Existing estimates of the time taken to mention a move, such as the 24 seconds reported by De Groot (1965) or the average of 15 seconds in the case of Charness (1981b), are based on leisurely search procedures that include extraneous comments. Consequently, it appeared necessary to measure the times taken by players under pressure.

A. An Experiment

Measurements were obtained from a group of 12 players rated from 1197 to 2005, with a mean USCF rating of 1637 and a mean age of 28, using two complementary techniques. One method was to set up speed games against these opponents, stopping each game without warning and asking for verbal protocols concerning the move that had just been executed. A stopwatch was used to find a break point, after castling (when the king had been transferred to safety) and while the material remained even, where the opponent had used approximately 10 seconds before moving. The method has more validity than the technique of presenting unknown positions, where time would be consumed in initial orientation, but carries some risk that players might conceive extra moves during the reporting period.

To control for this possibility a second, more artificial technique was employed. At the conclusion of the speed game, a test was administered using the position shown in Figure 17.4, preceded by a single practice trial with a different

Table 17.2. Search statistics for speed chess, in comparison with those during normal analysis (from Holding, 1989b).

	Speed Chess	Normal
Time available	11.3 sec	3 min
Mean rating	1637	1830
Tree statistics		
Total moves considered	7.3	14.5
Different base moves	2.6	2.7
Mean depth of search	2.3	5.4
Maximum depth of search	3.6	8.4

pawn chain. The subject's task was to call out, as fast as possible and without moving the pieces, how many moves would be needed to transfer the White R, N, B and Q to the eight rank (on the far side of the board), assuming that Black made no moves. The agreed total was 14 moves, which were imagined in a mean of 19 seconds. The average move time estimated by this technique was therefore less than 1.4 seconds per move, which would permit the consideration of over 7 moves in a 10-second period. As might be expected, the time taken varied with playing strength, $r(10) = -.58$, $p < .05$. The obtained regression line ($Y = 2.69 - .0008\ X$) suggests that any player rated above the expert level would require less than 1 second per move.

Despite the differences between the two techniques, the results from stopping the speed games appear very similar. The players' reports indicated that they had considered a mean of slightly over 7 moves, during an actual duration of 11.3 seconds. The average move time was therefore only 1.6 seconds per move in the context of an ongoing game. Again, the time required per move depended on the level of playing skill, $r(10) = -.62$, $p < .05$, and again the regression line ($Y = 3.1 - .0009\ X$) suggests that stronger players take less than 1 second per move. To interpret the resulting possibilities, it should be remembered that the amount of search that can be conducted at any point may be supplemented by analysis carried forward from prior moves. For any average move, the search statistics show that the speed players constructed small but coherent search trees.

Table 17.2 summarizes the most important characteristics of these trees, using as a comparison the trees obtained in the quiet condition of Holding (1989b).

Players in the earlier experiment had 3 minutes (180 seconds) instead of 11 seconds, and were rated slightly higher. Hence, the speed players had only 7.5 percent of the time, but nevertheless searched 50 percent of the total number of moves investigated by the slower players. It can be seen that the breadth of search, as shown by the number of different base moves, is extremely close in the two groups. The mean and maximum depths of search by the speed players are smaller, but not proportionately so. As a further comparison, referring back to Table 17.1 shows that these players as a group constructed larger trees at speed than the average 1200 player at leisure. The data do not show how this was achieved, but nevertheless suggest several possible explanations.

One consideration is that the speed players may have conducted rapid, goal-directed searches in the manner earlier described as the basis of the Church and Church (1977) speed chess program. The protocols do confirm that all players were able to identify goals when asked, although these varied from specific items such as preventing a discovered attack to more general aims such as strengthening the kingside. It is also possible that some goals were rapidly identified by the aid of lines of force, although such a mechanism is purely speculative. It is clear that some economies were made in the search process, since the data show none of the reinvestigations that commonly occur during search. However, the major factor might be that players severely curtail the time normally occupied in evaluating the endpoints of search, evincing a form of speed-accuracy tradeoff. There is no direct evidence for this hypothesis, although the fact that players often make blunders at speed suggests that the evaluation component has been treated as expendable.

VII. CONCLUSIONS

The finding that some degree of search is practicable during speed chess seems to remove the last barrier to accepting a theory based on search processes. A critical review of the structurally based recognition-association theory showed that many of its premises, such as the postulation of a production system, had no direct support. The underlying idea that memory directly determines chess skill has been shown to be incorrect, although skill may strongly influence memory, and a further difficulty is that the theory emphasizes visual to the exclusion of verbal mediation. The theory also neglects the processes of forward search and evaluation.

In contrast, process-oriented theories exemplified by the SEEK model emphasize the known differences in methods of move choice between players at different levels of expertise. In many respects, it appears that human players can be compared along the dimensions appropriate to computer programs. Skilled players have been shown to search more thoroughly, evaluate more accurately, and to have more extensive and flexible knowledge than weaker performers. Chess move choices, and hence chess skill attainments, are therefore described as depending on the accurate anticipation and comparison of move consequences with known values. Both declarative and procedural knowledge play some part in facilitating the process of looking ahead, and there are circumstances in which knowledge may substitute for search. Nevertheless, the reference activity is the process of forward search.

REFERENCES

Anderson, J.R. (1985). Cognitive psychology and its implications. New York: Freeman.

Baddeley, A.D. (1983). Working memory. Philosophical Transactions of the Royal Society of London, B302, 311-324.

Berliner, H.J. (1984). Search vs. knowledge: An analysis from the domain of games. In A. Elithorn and R. Banerji (Eds.), Artificial and human intelligence. New York: Elsevier.

Binet, A. (1894). Psychologie des grands calculateurs et joueurs d'echecs. Paris: Hachette.

Charness, N. (1976). Memory for chess positions: Resistance to interference. Journal of Experimental Psychology: Human Learning and Memory, 2, 641-653.

Charness, N. (1981a). Aging and skilled problem solving. Journal of Experimental Psychology: General, 110, 21-38.

Charness, N. (1981b). Search in chess: Age and skill differences. Journal of Experimental Psychology: Human Perception and Performance, 2, 467-476.

Charness, N. (1989). Expertise in chess and bridge. In D. Klahr and K. Kotovsky (Eds.), Complex information processing: The impact of Herbert A. Simon. New York: Academic Press.

Chase, W.G., and Simon, H.A. (1973). The mind's eye in chess. In W.G. Chase (Ed.), Visual information processing. New York: Academic Press.

Church, R.M., and Church, K.W. (1977). Plans, goals and search strategies for the selection of a move in chess. In P.W. Frey (Ed.), Chess skill in man and machine. New York: Springer-Verlag.

Cleveland, A.A. (1907). The psychology of chess and learning to play it. American Journal of Psychology, 18, 269-308.

De Groot, A.D. (1965). Thought and choice in chess. The Hague: Mouton.

Dodd, D.H., and White, R.M. (1980). Cognition: Mental structures and processes. Boston: Allyn & Bacon.

Djakow, I.N., Petrowski, N.W., and Rudik, P.A. (1927). Psychologie des Schachspiels. Berlin: de Gruyter.

Elo, A. (1978). The rating of chessplayers, past and present. New York: Arco.

Estes, W.K. (1975). Structural aspects of associative models for memory. In C.N. Cofer (Ed.), The structure of human memory. San Francisco: Freeman.

Frey, P.W. (1977). An introduction to computer chess. In P.W. Frey (Ed.), Chess skill in man and machine. New York: Springer-Verlag.

Frey, P.W., and Adesman, P. (1976). Recall memory for visually presented chess positions. Memory and Cognition, 4, 541-547.

Goldin, S.E. (1978). Effects of orienting tasks on recognition of chess positions. American Journal of Psychology, 91, 659-671.

Goldin, S.E. (1979). Recognition memory for chess positions: Some preliminary research. American Journal of Psychology, 92, 19-31.

Holding, D.H. (1979). The evaluation of chess positions. Simulation and Games, 10, 207-221.

Holding, D.H. (1985). The psychology of chess skill. Hillsdale, NJ: Erlbaum Associates.

Holding, D.H. (1989a). Adversary problem solving by humans. In K.J. Gilhooly (Ed.), Human and machine problem solving. New York: Plenum Press.

Holding, D.H. (1989b). Counting backward during chess move choice. Bulletin of the Psychonomic Society, 27, 421-424.

Holding, D.H. (1989c). Evaluation factors in human tree search. American Journal of Psychology, 102, 103-108.

Holding, D.H. (1992). Theories of chess skill. Psychological Research.

Holding, D.H., and Pfau, H.D. (1985). Thinking ahead in chess. American Journal of Psychology, 98, 271-282.

Holding, D.H., and Reynolds, R.I. (1982). Recall or evaluation of chess positions as determinants of chess skill. Memory and Cognition, 10, 237-242.

Horgan, D.D., Millis, K., and Neimeyer, R.A. (1989). Cognitive reorganization and the development of chess expertise. International Journal of Personal Construct Psychology, 2, 15-36.

Horgan, D.D., and Morgan, D. (1990). Chess expertise in children. Applied Cognitive Psychology, 4, 109-128.

Jongman, R.W. (1968). Het oog van de meester. Amsterdam: Van Gorcum.

Lane, D.M., and Robertson, L. (1979). The generality of the levels of processing hypothesis: An application to memory for chess positions. Memory and Cognition, 7, 253-256.

Lesgold, A.M. (1984). Acquiring expertise. In J.R. Anderson and S.M. Kosslyn (Eds.), Tutorials in learning and memory: Essays in honor of Gordon Bower. San Francisco: Freeman.

Lories, G. (1987). Recall of random and nonrandom chess positions in strong and weak chess players. Psychologica Belgica, 27, 153-159.

Pfau, H.D., and Murphy, M.D. (1988). Role of verbal knowledge in chess. American Journal of Psychology, 101, 73-86.

Reitman, J. (1976). Skilled perception in GO: Deducing memory structures from inter-response times. Cognitive Psychology, 8, 336-356.

Reynolds, R.I. (1982). Search heuristics of chessplayers of different calibers. American Journal of Psychology, 95, 383-392.

Saariluoma, P. (1989). Chess players' recall of auditorily presented chess positions. European Journal of Cognitive Psychology, 1, 309-320.

Saariluoma, P. (1990). Apperception and restructuring in chess players' problem solving. In K.J. Gilhooly, M.T.G. Keane, R.H. Logie and G. Erdos (Eds.), Lines of thinking, Vol. 2. Chichester: John Wiley.

Saariluoma, P. (1992). Visuospatial and articulatory interference in chess players' information intake. Applied Cognitive Psychology, 6, 77-89.

Simon, H.A., and Gilmartin, K. (1973). A simulation of memory for chess positions. Cognitive Psychology, 5, 29-46.

Tikhomirov, O.K. (1988). The psychology of thinking. Moscow: Progress Publishers.

Commentary

Search Process Versus Pattern Structure in Chess Skill, D. H. Holding

DORRIT BILLMAN
Georgia Institute of Technology

Holding considers whether chess skill is primarily based on declarative knowledge of board patterns in chess versus procedural knowledge about effective search. Contra the classical claims of Chase and Simon, Holding argues that the major component of skill is procedural. He provides two lines of evidence, empirical and analytic. First, he summarizes research by himself and others demonstrating that search skill improves with chess rating. Chess experts do search more moves, search deeper, and provide better evaluations of positions. Second, he argues that improved memory may be a byproduct rather than the cause of skill and that the "chunks" identified in memory research could not logically play the role demanded of them in the memory theory of skill. What role does declarative knowledge about specific prior experience play? Holding suggests that knowledge about familiar 'sequences' of moves--not static layout on a board--may be important in eliminating the need for search in particular cases. However, the more important knowledge may be about strategies, evaluation criteria, and goal setting.

This paper draws our attention to the interpenetration of declarative and procedural knowledge, structure and process, representation and use. It is much easier to talk about the content of information than whether it be structure or process. For example, a direct argument can be made against the "chunks" proposed by Chase and Simon. "Chunks" of several related pieces recalled from the board are supposed to index moves appropriate in that context.

Unfortunately, the patterns identified as chunks are very localized and do not contain the information necessary for successful move selection. This information won't do the job. When we turn to considering the form of knowledge rather than its content, clear distinctions are more difficult to come by. One structure/process contrast has to do with novelty: can the complete answer be looked up from memory, or must it be generated? Memory look-up sounds like use of structure, while generation sounds like process. In this sense processing means many steps carried out in working memory. A second contrast has to do with the form of information. Is it in the form of an asymmetrical, condition-action pairing accessed by one part of the information and executing another? Or is it in a symmetrical structure which can be accessed by any component but does not in itself execute anything? This contrast is clear from the point of view of constructing a computational model, but the implications for chess skill are not clear at all.

What, then, does Holding tell us about structure and process in chess? Because expert chess play is search intensive, skill cannot be based on any 'simple' match to specific structures stored in memory; the decisions encountered are novel. Rather it must draw on moderately abstract or general knowledge that specifies how to pose and evaluate possible chess moves.

SUBJECT INDEX

AUTHOR INDEX